Peter Jutzi, Ulrich Schubert (Eds.)

Silicon Chemistry

Further Titles of Interest

N. Auner, J. Weis (Eds.)
Organosilicon Chemistry V
From Molecules to Materials
2003, ISBN 3-527-30670-6

M. Driess, H. Nöth (Eds.)
Molecular Clusters of the Main Group Elements
2003, ISBN 3-527-30654-4

V. Lehmann
Electrochemistry of Silicon
Instrumentation, Science, Materials and Applications
2002, ISBN 3-527-29321-3

Peter Jutzi, Ulrich Schubert (Eds.)

Silicon Chemistry

From the Atom to Extended Systems

WILEY-VCH GmbH & Co. KGaA

Prof. Dr. Peter Jutzi
Faculty of Chemistry
University of Bielefeld
D-33501 Bielefeld
Germany
e-mail: peter.jutzi@uni-bielefeld.de

Prof. Dr. Ulrich Schubert
Institute of Materials Chemistry
Vienna University of Technology
Getreidemarkt 9
A-1060 Wien
Austria
e-mail: uschuber@mail.zserv.tuwien.ac.at

Cover picture:
Part of a polysilane $(SiH)_n$ nanotube

Library of Congress Card No.: applied for

A catalogue record for this book is available from the British Library.

Bibliographic information published by Die Deutsche Bibliothek
Die Deutsche Bibliothek lists this publication in the Deutsche Nationalbibliografie; detailed bibliographic data is available in the Internet at http://dnb.ddb.de

Printing: betz-druck gmbH, 64291Darmstadt
Bookbinding: Litges & Dopf Buchbinderei GmbH, Heppenheim

ISBN 3-527-30647-1

Foreword

In 1995 the *Deutsche Forschungsgemeinschaft* (DFG) started a focussed research program entitled "Specific Phenomena in Silicon Chemistry: New Experimental and Theoretical Approaches for the Controlled Formation and Better Understanding of Multidimensional Systems". The Austrian *Fonds zur Förderung der wissenschaftlichen Forschung* (FWF) established in 1996 a parallel funding program on silicon chemistry ("Novel Approaches to the Formation and Reactivity of Silicon Compounds") to improve collaboration between scientists from both countries. Both programs ended in 2002; 33 research groups in Germany and 6 research groups in Austria participated in the focussed programs during the years.

The intention of this book is twofold. First, an overview on the scientific results of the bi-national program is presented. However, the authors of the individual chapters were asked not to go into too much detail, but rather to embed their results in a broader perspective. For the latter reason, two "external" scientists, who had given invited talks at the final bi-national symposium on silicon chemistry in Werfenweng/Austria in 2002, were asked to contribute to this book. Thus, a book on topical developments in silicon chemistry came into being.

More so than for any other element, the development of two- or three-dimensional extended structures from molecular or oligomeric units can be studied ("bottom-up" syntheses) for silicon-based compounds. This aspect of silicon chemistry turned out to be a central topic of both focussed programs during the years. The book is thus organized in three sections. The first section deals with reactive molecular precursors and intermediates in silicon chemistry. Mastering their synthesis, understanding their molecular and electronic structures, and being able to influence their reactivity (including their kinetic stabilization) is essential for using them as building blocks for extended structures. In the second and third sections, the way from molecular building blocks via oligomeric compounds to extended networks is shown for several systems based on Si-O and Si-Si bonds.

The authors of the book thank the *Deutsche Forschungsgemeinschaft* and the *Fonds zur Förderung der wissenschaftlichen Forschung* for funding research in an exciting and topical area for many years.

Peter Jutzi, Ulrich Schubert
Editors

Acknowledgement

We gratefully acknowledge the generous financial support from the *Deutsche Forschungsgemeinschaft* (DFG) within the research program entitled "Specific Phenomena in Silicon Chemistry: New Experimental and Theoretical Approaches for the Controlled Formation and Better Understanding of Multidimensional Systems" and from the Austrian *Fonds zur Förderung der Wissenschaftlichen Forschung* (FWF) within the funding program entitled "Novel Approaches to the Formation and Reactivity of Silicon Compounds".
We are especially grateful to Dr. A. Mix, University of Bielefeld, for his engagement in formatting and editing the individual contributions.

Contents

I Reactive Intermediates in Silicon Chemistry – Synthesis, Characterization, and Kinetic Stabilization

II Si-Si-Systems: From Molecular Building Blocks to Extended Networks

III Si-O Systems: From Molecular Building Blocks to Extended Networks

Part I

Reactive Intermediates in Silicon Chemistry – Synthesis, Characterization, and Kinetic Stabilization

A detailed knowledge about important highly reactive intermediates is the key for a better understanding of fundamental mechanisms and for the optimization of established synthetic procedures; furthermore, it is regarded as a chance to develop synthetic strategies for novel molecules. Part 1 of the present book details recent results concerning the synthesis and characterization of the following short-lived silicon-containing species:

Si (atom)	Si_2	SiH_2	SiH_3
Si_2H_2	Si_2H_4	Si_4H_6	H_2SiCH_2
$(SiO)_{1,2,3}$	$(SiO_2)_{1,2}$	Si_2N	SiH_xO_y
RR'Si	$RR'SiO_2$	F_2SiS	

For the synthesis and characterization of transient species, rather sophisticated techniques have to be applied. These include high-temperature synthesis by element vaporization, vacuum thermolysis of precursor molecules, photolysis of matrix–entrapped precursor molecules, matrix isolation and spectroscopy (UV/Vis, IR, Raman), dilute gas-phase spectroscopy (including millimeter wave, microwave, high-resolution FTIR, IR spectroscopy), and gas-phase kinetics. The introduction of bulky substituents R instead of hydrogen atoms is the basis for the kinetic stabilization of highly reactive molecules; this strategy has been applied for the stabilization of the species SiH_2, Si_2H_2, Si_2H_4, and Si_4H_6.
Elemental silicon plays a very important role in solid-state physics (microelectronics, photovoltaic solar cells, etc.) as well as in inorganic and organic silicon chemistry (Müller–Rochow process, etc.). As expected, the reactivity of silicon depends drastically on the particle size (lump silicon < powder < nanoparticles (clusters) < atoms). In this context, fundamental silicon chemistry can be learned from the properties of silicon atoms. In Chapter 1, G. Maier et al. describe studies with thermally generated silicon atoms, which have been reacted in an argon matrix with the reactants SiH_4, CH_4, and O_2. Based on a combination of experimental and theoretical findings, the mechanisms of these reactions are discussed. In the reactions with SiH_4 and CH_4, the highly reactive double–bond species $H_2Si{=}SiH_2$ (disilene) and $H_2Si{=}CH_2$ (silaethene), respectively, are the final products. The reaction with O_2 mainly leads to SiO (the most abundant silicon oxide in the universe!) and to small amounts of SiO_2.

More about the matrix chemistry of SiO is reported by H. Schnöckel and R. Köppe in Chapter 2. Condensation of SiO prepared by high-temperature reaction of Si and O_2 gives rise to the formation of oligomers $(SiO)_n$ with n = 2,3,4. In the reaction of SiO with metal atoms, species of the type MSiO with M = Ag, Au, Pd, Al, Na are formed. Reactions with oxidizing agents yield monomeric species OSiX (X = O,S) and $OSiX_2$ (X = F, Cl). The synthesis of a dimeric SiO_2 molecule $(SiO_2)_2$ allows speculation about a possible formation of fibrous $(SiO_2)_n$. All presented structures have been deduced from spectroscopic data (IR, Raman) and from quantum chemical calculations.
Several highly important technical applications need elemental silicon in the form of thin films (microelectronics, photovoltaic solar cells, digital data storage and display devices, photocopy systems, X-ray mirrors). The preferred fabrication process for such thin films is deposition from the gas phase by physical vapor deposition (PVD) or chemical vapor deposition (CVD). Details of thin–film formation by PVD or CVD on the atomic or molecular scale are still scarce, due to the high reactivity, short lifetime, and low concentration of relevant gas–phase species, but would be very helpful to refine the processes and to regulate the film properties. In Chapter 3, H. Stafast et al. report on diagnostic methods, which allow the in situ characterization of gas–phase species such as Si, Si_2, SiH_2, and Si_2N during a-Si (amorphous silicon) thin–film deposition by PVD (thermal evaporation of Si) and by CVD (SiH_4 pyrolysis) and the measurement of mechanical stress in growing a-Si (during plasma CVD of SiH_4) thin films. Besides elemental silicon, silicon dioxide also finds many technical applications in the form of thin films due to the dielectric properties of this material. Thin SiO_2 layers are prepared by low–pressure oxidation of SiH_4 or Si_2H_6 with molecular oxygen in CVD processes. The most important step in the reaction manifold is the oxidation of SiH_3 radicals. This reaction has been investigated in detail by F. Temps et al. and this work is described in Chapter 4. In fast radical-radical reactions, several SiH_xO_y intermediates are formed, which show diverse consecutive reaction steps. The experimental results are supported by *ab initio* quantum chemical calculations.
Silylene SiH_2 and its derivatives SiR_2 with less bulky substituents R constitute another class of highly reactive silicon compounds. They play an important role as intermediates in several areas of silicon chemistry. In Chapter 5, W. Sander et al. report on the oxidation of silylenes SiRR' (RR' = F, Cl, CH_3 and R = CH_3,R' = Ph) with molecular oxygen, as studied by matrix–isolation spectroscopy. Dioxasiliranes were obtained as the first isolable products, whereas silanone *O*-oxides were most likely non-observable intermediates. In combination with DFT or *ab initio* calculations, IR spectroscopy once again proved to be a powerful tool to reliably identify reactive molecules.
Following the classical "double bond rule", compounds such as $H_2Si{=}SiH_2$, $H_2Si{=}CH_2$, and $HSi{\equiv}SiH$ (disilyne) with multiple bonding to silicon are too

reactive to be isolated under ordinary conditions and thus are predestined for investigations by more sophisticated methods (matrix experiments, dilute gas–phase studies). Interestingly, such studies have shown that the ground-state structures of these molecules differ to some extent from those found in the analogous carbon compounds. Explanations are given in the contributions of H. Beckers (Chapter 6), N. Wiberg (Chapter 7), and M. Weidenbruch (Chapter 8). The report of H. Beckers deals with the synthesis and characterization of the short-lived species $F_2Si{=}S$ and $H_2Si{=}CH_2$, which were obtained by coupling flash vacuum thermolysis (FVT) with matrix IR spectroscopy or with real-time high-resolution gas–phase spectroscopy.

The concept of "kinetic stabilization" has been applied very successfully to the class of compounds incorporating double or even triple bonds to silicon. The last two contributions deal with some recent highlights in this field. In Chapter 7, N. Wiberg reports on the introduction of very bulky silyl substituents R such as Si^tBu_3, $SiH(Si^tBu_3)_2$, and $SiMe(Si^tBu_3)_2$, which allow the synthesis of stable disilenes RR'Si=SiRR'. Elimination reactions possibly lead to the novel triply-bonded species RSi≡SiR, the final goal in the field of kinetic stabilization. In Chapter 8, M. Weidenbruch reports on the synthesis and characterization of a tetrasilabuta-1,3-diene containing two neighboring Si=Si double bonds (and also on the first tetragermabuta-1,3-dienes). Kinetic stabilization could be realized with the help of bulky 2,4,6-(triisopropyl)phenyl substituents. Several types of addition reactions are described, some of which lead to compounds with isolated Si=Si double bonds. Finally, novel types of conjugated compounds are presented, formed by the reaction of hexa-*tert*-butyl-cyclotrisilane with di- and polyynes.

Peter Jutzi, Ulrich Schubert

1 Investigations on the Reactivity of Atomic Silicon: A Playground for Matrix Isolation Spectroscopy

G. Maier, H. P. Reisenauer, H. Egenolf, and J. Glatthaar*

1.1 Introduction

During the past five years, we have studied the reactions of thermally generated silicon atoms with low molecular weight reactants in an argon matrix. The reaction products were identified by means of IR and UV/Vis spectroscopy, aided by comparison with calculated spectra. The method turned out to be very versatile and successful. The reactions that we have carried out to date cover a wide range of substrate molecules (Scheme 1.1).[1]

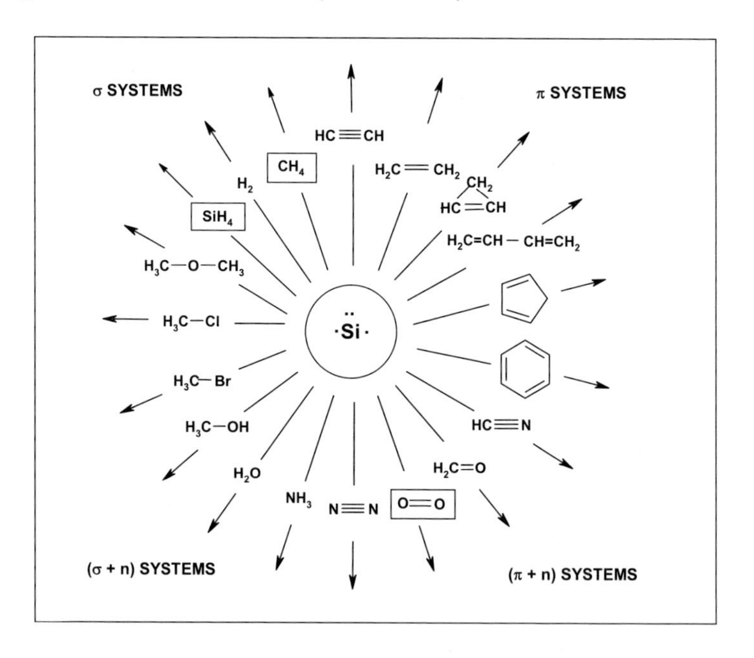

Scheme 1.1. Reactions of silicon atoms with different substrate molecules.

In order to get an idea about the potential of silicon atoms, we selected examples which belong to four different groups, namely (π) systems, (π + n) systems, (σ + n) systems, and pure (σ) systems. These reactions can be

* Address of the authors: Institut für Organische Chemie der Justus-Liebig-Universität, Heinrich-Buff-Ring 58, D-35392 Giessen, Germany

understood considering the basic features of atomic silicon. First, it has a triplet ground state. A diradical type of reaction can thus be anticipated. According to the law of spin conservation, the primary reaction product should be a triplet molecule. Second, the silicon atom has an empty 3p orbital. As a consequence, a strong electrophilic behavior can be expected.

The matrix study of silicon atoms is not merely an academic exercise, but also has practical relevance. This will be demonstrated by the selection of oxygen, silane, and methane as reaction partners (enframed in Scheme 1.1).

There are reports that porous silicon can be a dangerous material in the presence of oxygen[2] or even nitrogen,[3] depending on the grain size. Lump silicon is at one end of the scale, silicon atoms represent the other extreme. Silicon powder lies in between. Thus, the properties of silicon atoms can tell us something about the chemical behavior of "activated" silicon. Another example concerns simple silicon hydrides, which play an important role in silicon chemical deposition (CVD) processes, which are of significance to the semiconductor industry. Again, detailed study on the reaction of silicon atoms with silane will help to understand the mechanisms of these reactions. Last but not least, knowledge about the reaction pattern of atomic silicon in the presence of compounds such as chloromethane, methanol, or dimethyl ether can help us to understand the detailed features of the "direct process" (Müller–Rochow synthesis).

1.2 Matrix Isolation Spectroscopy

Matrix isolation is a very suitable technique for the synthesis and detection of highly reactive molecules. This method allows spectroscopic studies of the target species with routine spectroscopic instrumentation (IR, UV, ESR) without having to use fast, time-resolved methods. The reactive species are prevented from undergoing any chemical reaction by embedding them in a solid, provided that three conditions are fulfilled: (a) the solid has to be chemically inert, (b) isolation of the single molecules must be achieved by choosing concentrations which are sufficiently low, and (c) diffusion in the solid has to be suppressed by applying low temperatures during the experiments. In this way the kinetic instability inherent to the isolated molecules is counteracted.

In general, there are two possible means of creating a solid with the desired properties, the so-called *matrix*. The molecules of interest can be generated from suitable precursors by reactions in the gas phase. The routine method is the high-vacuum flash pyrolysis of thermally labile compounds followed by direct condensation of the reaction products and co-deposition with an excess of host material on the cold matrix holder. The second way is to produce the reactive species in situ by photolysis of the entrapped precursor molecules in the matrix material. Other procedures are also known. One such special case is presented in this article (see below).

For full information on the development of matrix isolation methods and their application, the reader is referred to several monographs.[4]

The properties of the matrix material determine the spectroscopic methods that can be applied. The use of solid rare gases like argon and xenon or solid nitrogen is well established, since they are optically transparent in the commonly observed spectral ranges. The fact that there is nearly no interaction between the host lattice and the enclosed guest species and that the rotational movements are frozen has an important effect on the IR spectra; under ideal conditions the recorded infrared spectra are reduced to spectra consisting of very narrow lines ($< 1\ cm^{-1}$). Each of them originates from the respective vibrational transition. UV/Vis spectra are also obtained easily and are likewise useful for the structural elucidation of unknown species. What is even more important is the fact that the UV absorptions allow the selection of the appropriate wavelengths for the induction of photochemical reactions, which under matrix conditions are often reversible.

1.3 Computational Methods

An important breakthrough in the development of matrix isolation was the construction of suitable cryostats, which goes back to the 1970s. A similar push, which, during the last few years, has opened a new dimension for the structure determination of matrix-isolated species, has come from theory. Quantum chemical computations of energies, molecular structure, and molecular spectra are nowadays no longer a task reserved to few specialists. The available programs, for instance the Gaussian package of programs,[5] have reached the degree of convenience and ease of application that nearly anyone can formulate the needed input data to obtain reliable information about the energy, electronic structure, geometry, and spectroscopic properties of the species of interest.

In our own experience, density functional calculations (B3LYP-DFT functional) are very well suited for a reliable prediction of vibrational spectra. TD (time dependent) calculations even give surprisingly good results for electronic transitions.

1.4 Identification of Matrix-Isolated Species

Comparison of the calculated and experimental vibrational spectra is in most cases (at least for molecules of moderate size) sufficient to identify an unknown molecule unequivocally. Examples are given below.

Special techniques can be applied if additional information is needed for the structural elucidation of the entrapped molecules (our study of the reactions of silicon atoms with nitrogen[1h] sets some shining examples in this respect). Since the matrix-isolated species are too reactive to be handled under standard conditions and therefore cannot be identified by routine methods, the structure determination has to rely exclusively on the matrix spectra. Sometimes, if a species can be reversibly photoisomerized upon matrix irradiation, the unchanged elemental composition provides valuable information. The advantages of IR compared with UV/Vis spectra are obvious: a) Calculated IR spectra are of high accuracy. b) FT-IR instruments allow the generation of difference spectra by subtraction of the measured spectra. In other words, if one of the matrix-isolated components is specifically isomerized to a new compound upon irradiation, one can eliminate all the bands of the photostable molecules. By these means it is possible to extract exclusively and separately the absorptions of the diminishing photolabile educt molecule and the newly formed photoproduct. c) If a compound has only a few (sometimes only one) observable IR bands, it may be dangerous to depend solely on the comparison of experimental and calculated spectra. In these cases isotopic labeling will help. Isotopic shifts of the IR absorptions are dependent on the structure, and on the other hand can be calculated with utmost precision.

1.5 Experimental Procedure

For matrix-isolation studies there is a minimum of necessary equipment. The essentials include: a) a refrigeration system (cryostat), b) a sample holder, c) a vacuum chamber (shroud) to enclose the sample, d) means of measuring and controlling the sample temperature, e) a vacuum-pumping system, f) a gas-handling system, g) devices for generating the species of interest, h) spectrometers for analysis of the matrices.

The generation of the species discussed in this article is different from the two classical methods mentioned above: Solid silicon is vaporized and the extremely reactive silicon atoms create the envisaged molecules in the moment of co-deposition by reaction with the selected partner (such as oxygen, silane, or methane) on the surface of the cold matrix holder. The products remain isolated in the argon matrix and can then be studied spectroscopically or transformed into other species upon irradiation.

Nowadays, the standard cryostats are closed-cycle helium refrigerators. They are commercially available. We use either the "Displex Closed-Cycle System CSA" from Air Products or the "Closed-Cycle Compressor Unit RW 2 with Coldhead Base Unit 210 and Extension Module ROK" from Leybold. These systems can run for thousands of hours with minimal maintenance. The sample holder can be cooled to temperatures from room temperature to about 10 K. A typical example of a closed-cycle helium matrix apparatus is shown in Figure 1.1.

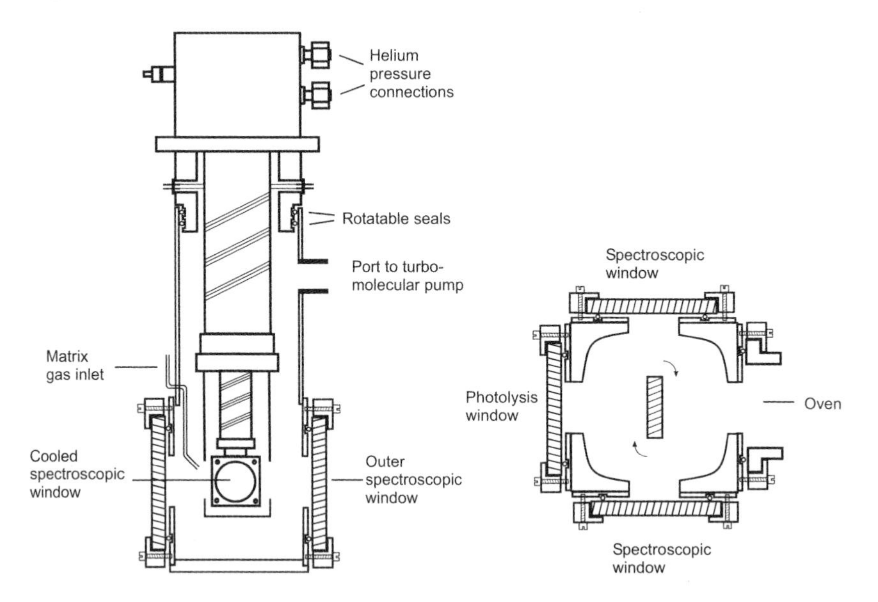

Figure 1.1. Closed-cycle helium cryostat.

1.6 Generation of Silicon Atoms

A critical point in the work presented in this communication was the generation of a steady stream of silicon atoms which have to be condensed together with the substrate molecule and an excess of matrix material onto the cooled window. In our early experiments, silicon was vaporized from a tantalum Knudsen cell (Figure 1.2, Type A) or a boron nitride crucible which was surrounded by an aluminum oxide tube. The oven was resistively heated to temperatures of 1490–1550 °C by means of a tungsten wire wound around the alumina tube (Figure 1.2, Type B). In later runs, a rod of dimensions 0.7·2·22 mm was cut out from a highly doped silicon wafer and heated resistively by using an electric current of 10 A at a potential of 10 V (Figure 1.2, Type C). Under these conditions, the surface temperature amounted to 1350–1380 °C.

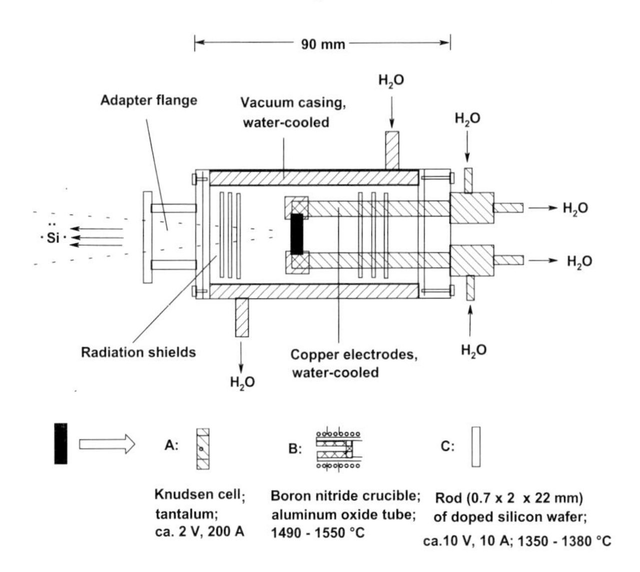

Figure 1.2. Oven for the evaporation of silicon.

The produced silicon atoms were quantified by applying a quartz crystal microbalance incorporated into the cryogenic sample holder. The amount of substrate molecules was determined by measuring the pressure decrease in the storage flask containing a gas mixture of argon and substrate. Annealing experiments, which allow a reaction between the isolated species by softening of the matrix, can be carried out by warm-up of the matrix to 27–38 K.

The construction of an oven for the evaporation of silicon on a larger scale turned out to be difficult, but quite recently we also found a solution for this technical problem.

1.7 Reactions of Silicon Atoms with Oxygen

The system Si/SiO_2 is very important for technical applications. Hence, it is no surprise that many studies have been carried out on low molecular weight silicon oxide and that much effort has been focussed on these intermediates on their way from molecule to solid.

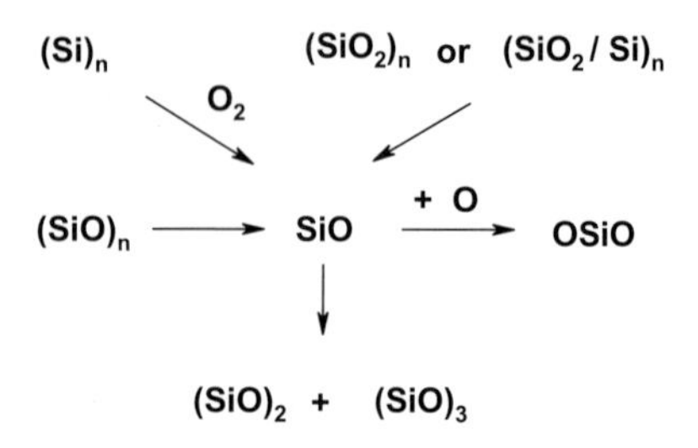

Schnöckel [6] has shown that co-condensation of gaseous SiO, which can be prepared by thermal depolymerization of solid $(SiO)_n$, and oxygen atoms, generated by a microwave discharge, yields molecular SiO_2. Not only molecular SiO, but also its oligomers $(SiO)_2$ and $(SiO)_3$ are well-known species.[7] They can be obtained upon heating solid quartz or a mixture of quartz and bulk silicon,[7] or when molecular oxygen is passed over heated silicon.[8]

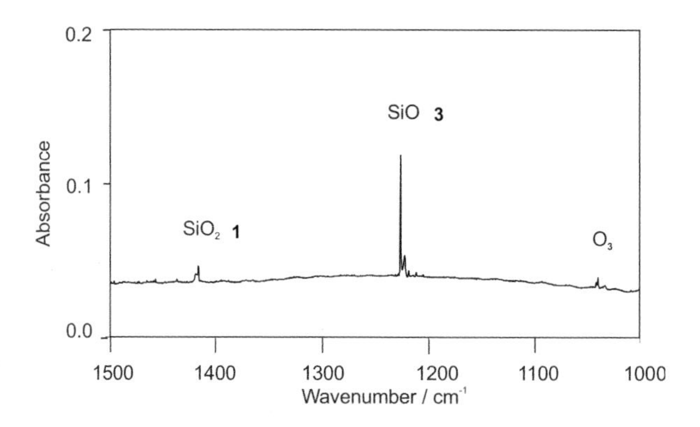

Figure 1.3. IR spectrum after co-condensation of silicon atoms and oxygen in argon (1:500).

Surprisingly, there was no known study of the reactions of silicon atoms with oxygen when we began an experimental and theoretical investigation of the SiO_2 energy hypersurface. In the meantime Roy et al.[9] carried out such experiments in connection with a search for silicon trioxide SiO_3.

As expected, the global minimum within the series of SiO_2 isomers is the linear OSiO molecule. The reaction of the components ^3Si and O_2 to give OSiO **1** is highly exothermic (ΔE = 153.2 kcal mol^{-1}). Astonishingly, even the splitting of molecular oxygen into two atoms and recombination of one of them with a silicon atom under formation of SiO **3** is exothermic (ΔE = 61.3 kcal mol^{-1}). So one can expect not only OSiO but also SiO upon reaction of silicon atoms with oxygen.

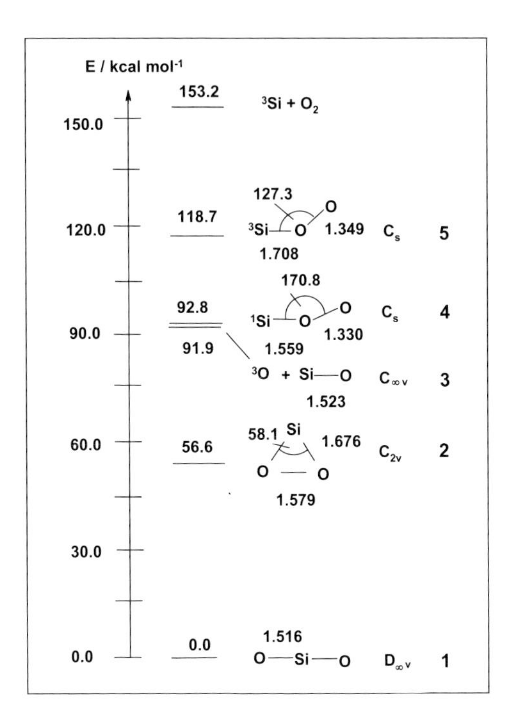

Scheme 1.2. Calculated energies and geometries of SiO_2 species (B3LYP/6-311+G**, zero-point energies included).

The calculations are in agreement with experiment (Figure 1.3). Co-deposition of silicon atoms and molecular oxygen in argon at 10 K mainly leads to SiO **3**. In addition, a small amount of SiO_2 **1** is formed. Traces of O_3 can be explained as the result of the capture of O atoms by O_2. The higher aggregates of SiO, namely Si_2O_2 and Si_3O_3, can also be detected (beyond the scale of Figure 3), especially after annealing of the matrix. The main products after warm-up to 30 K are OSiO **1** and O_3.

$$\underset{\mathbf{1}}{O{=}Si{=}O} \xleftarrow{O_2} \cdot\ddot{Si}\cdot \xrightarrow{2\,O_2} \underset{\mathbf{3}}{Si{=}O} + O_3$$

As far as the mechanism of the oxidation of silicon (yielding SiO **3** and OSiO **1**) is concerned it can be assumed that the first reaction product is triplet peroxide **5**, which either splits off an oxygen atom – even at 10 K – under formation of SiO **3** or forms singlet peroxide **4.** Silicon dioxide OSiO **1** can result from **4** via the cyclic peroxide **2**, in which the O,O bond should be broken very easily, but it is also possible that SiO **3** recaptures an oxygen atom.

For comparison, we also calculated the CO_2 energy hypersurface. The replacement of a silicon by a carbon atom leads to a very similar situation, although the energy differences are much greater in the case of carbon. The reaction of a carbon atom with molecular oxygen forming OCO **6**, the global minimum, is strongly exothermic (ΔE = 258.1 kcal mol^{-1}). Again, the formal splitting of molecular oxygen into two O atoms and addition of one of them to a C atom is also exothermic (ΔE = 131.8 kcal mol^{-1}).

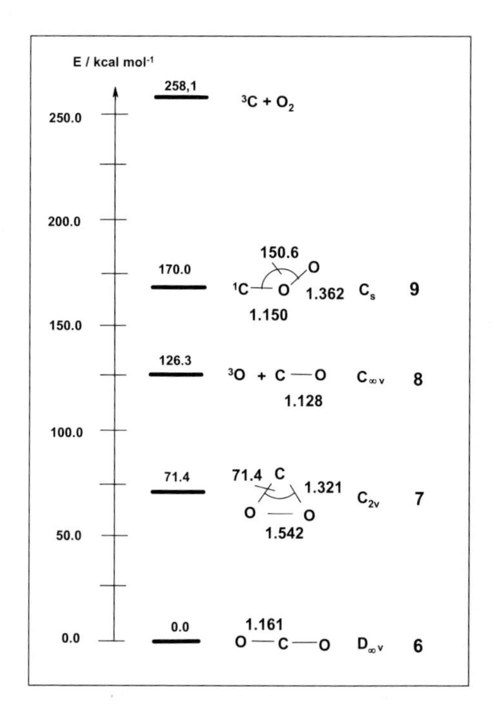

Scheme 1.3. Calculated energies and geometries of CO_2 species (B3LYP/6-311+G**, zero-point energies included).

It is trivial that CO_2 **6** and CO **8** are the combustion products of carbon. Nevertheless, there is a chance that other isomers of **6**, namely the cyclic form **7** and the singlet peroxide **9** might be detected in the reaction of atomic carbon with oxygen. Like the silicon analogues **2** and **4**, both are still unknown. Perhaps they are intermediates in the addition of oxygen to the carbon atom under formation of OCO **6**. The mechanistic implication would be similar to the silicon series. According to calculations the triplet peroxide ^{3}COO is not a minimum on the energy hypersurface.

1.8 Reactions of Silicon Atoms with Silane and Methane

A detailed study on the reaction of silicon atoms with silane will not only give us more insight into silicon CVD processes. Another appeal stems from the fact that starting with the first isolation of a disilene by West et al.[10] a new chapter in silicon chemistry was opened, yet the isolation and identification of the parent disilene was still missing. Last but not least, silicon hydrides are excellent target molecules to demonstrate the unique bonding characteristics of silicon compared to carbon, resulting very often in surprising "bridged" structures. These fascinating aspects explain the numerous experimental and theoretical studies covering silicon hydrides SiH_n and Si_2H_n.

On the other hand, a study of methane would also have scientific and practical relevance. It can be shown by calculation that methane **18** and silane **14** behave quite differently when attacked by a silicon atom (Figure 1.4). If a silicon atom in its triplet ground state approaches methane, the energy is continuously raised. There is no indication of any bonding interaction. On the contrary, the reaction coordinate for the approach between a ^3Si atom and silane descends steadily until the formation of a complex between the two partners is reached.

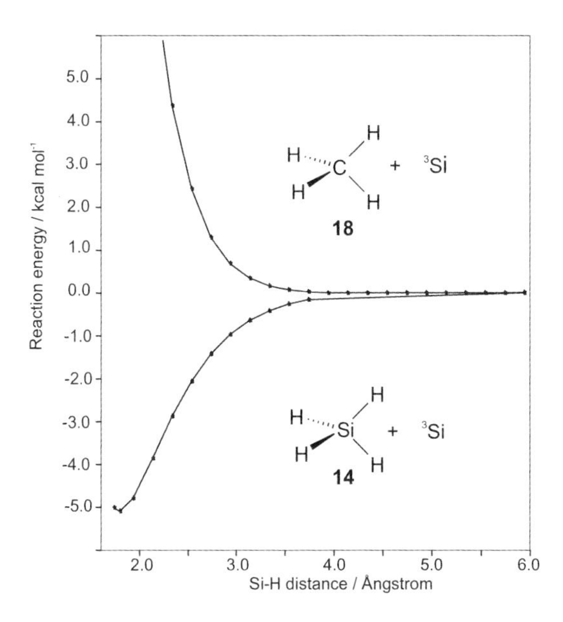

Figure 1.4. Calculated changes of the potential energy during the approach of a ^3Si atom to a methane (upper curve) or a silane molecule (lower curve); UB3LYP/6-311+G**; full optimization at each step.

For the structural identification of the expected species it was again necessary to obtain the calculated vibrational spectra. To get an overview of the Si_2H_4 potential energy surface, several stationary points, together with the corresponding vibrational spectra, were calculated. Scheme 1.4 shows the calculated relative energies of some relevant minima.

The global minimum is disilene **10**. The *trans*-bent geometry was suggested some twenty years ago.[11] The stabilization energy compared with the two components triplet silicon atom and silane **14** is 43.2 kcal mol^{-1}. Besides **10**, silylsilylene **11** should also be formed in an exothermic reaction from ^3Si atoms and **14**. On the triplet energy hypersurface, the starting components form, without an activation barrier, a loose complex **13**. Its stabilization energy amounts to 5 kcal mol^{-1}. In a subsequent step the primary complex **13** can be transformed into triplet silylsilylene **12**, followed by intersystem crossing to singlet silylsilylene **11** (T/S gap 14.1 kcal mol^{-1}). On the singlet energy hypersurface, isomerization of silylsilylene **11** to the thermodynamically more stable (ΔE = 5.7 kcal mol^{-1}) disilene **10** then takes place.

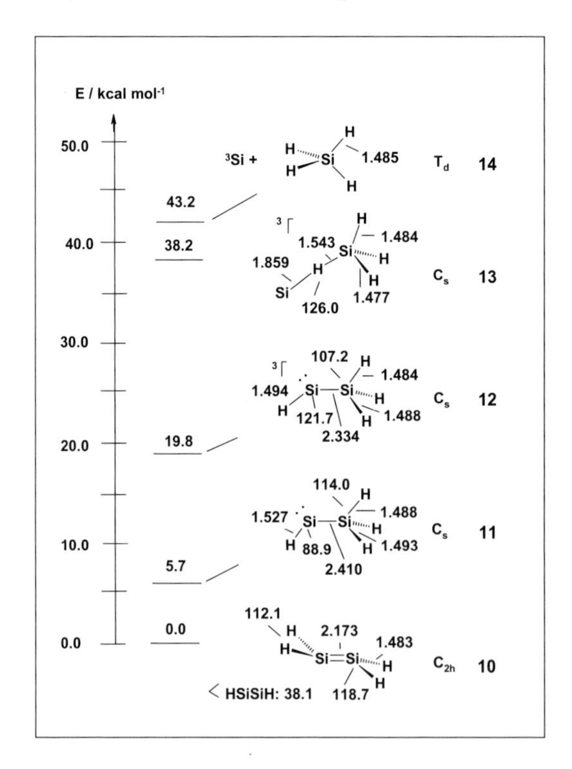

Scheme 1.4. Calculated energies and geometries of some Si_2H_4 species (B3LYP/6-311+G**, zero-point energies included).

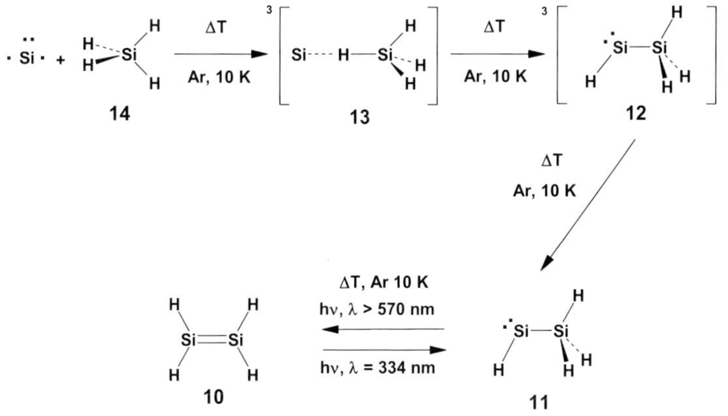

The analysis of the spectra shows that the reaction of silicon atoms with silane **14** leads to a mixture of silylsilylene **11** and disilene **10**. Upon irradiation of the matrix at long wavelengths (λ > 570 nm), **11** is isomerized to **10**. The backreaction can be induced by using shorter wavelengths. With λ = 334 nm, disilene **10** regenerates silylsilylene **11**.

Through a combination of the experimental and theoretical findings the mechanism of the reaction of ^{3}Si atoms with silane **14** can be summarized as follows. Via the triplet complex **13** triplet silylsilylene **12** is formed. Both are too short-lived to be detected. Intersystem crossing gives singlet silylsilylene **11**. The reaction leading to **11** releases enough energy to surpass the barriers on the pathway from **11** to **10**, even at 10 K.

Applying our standard procedure, we condensed methane **18** as a gaseous mixture with argon onto a spectroscopic window at 10 K. In all cases only the IR spectrum of the starting material could be registered. There was not even an indication for the existence of a complex **19** between methane **18** and a silicon atom. This was the first time that a substrate molecule had not reacted with silicon atoms in our experiments. This observation suggests that mineral oil may be a suitable medium to protect "activated" silicon from unforeseeable reactions with oxygen or nitrogen.

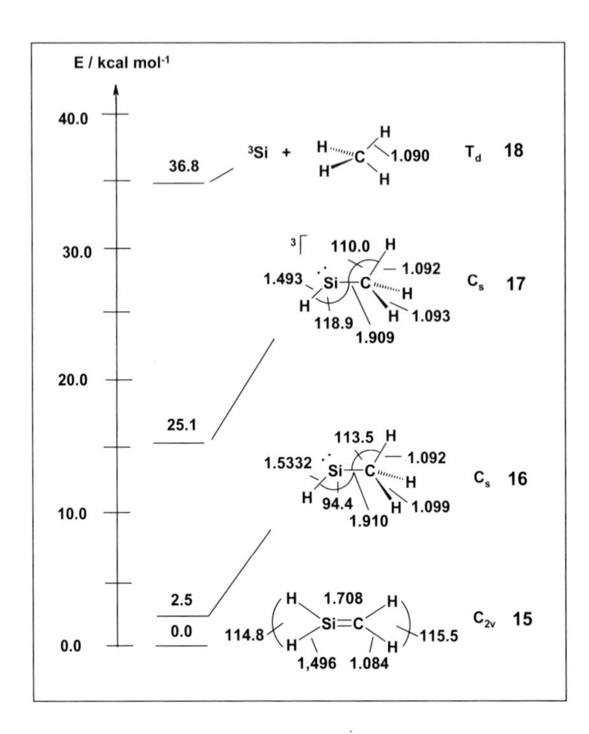

Scheme 1.5. Calculated energies and geometries of some $CSiH_4$ species (B3LYP/6-311+G**, zero-point energies included).

If one considers how a reaction can be enforced in such a case, one has to keep in mind that the silicon atom in the gas phase shows a weak UV transition at 220 nm and another, strong absorption at 251 nm. This suggests that irradiation with light of wavelength 254 nm might be a way to activate silicon atoms. Indeed, irradiation of the co-condensate of methane **18** and silicon atoms at short wavelengths (λ = 185 or 254 nm) leads to methylsilylene **16**. Obviously, photoexcitation leads to the insertion of a silicon atom into the C,H bond of methane **18**. The intensity of the IR bands of **16** arising during irradiation was relatively weak. Nevertheless, its structural elucidation has to be taken for granted, since it is possible to establish a photoequilibrium between **16** and the isomeric silaethene **15**. At $\lambda > 400$ nm the equilibrium is shifted to silaethene **15**; at $\lambda = 254$ nm it lies on the side of methylsilylene **16**.

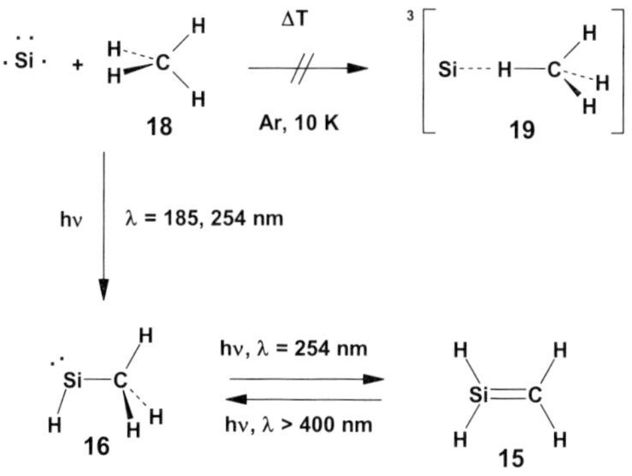

A thermal equilibrium between **16** and **15** should lie on the side of silaethene **15**. According to calculations, carried out in comparison to the silane system, methylsilylene **16** should be 2.5 kcal mol^{-1} higher in energy. The starting components ^3Si and methane **18** lie 36.8 kcal mol^{-1} higher than silaethene **15**. The calculated activation energy for the transformation of **15** into **16** amounts to 36.3 kcal mol^{-1}.

1.9 Conclusion

Studies of the reactions of atomic silicon with oxygen, silane, and methane illustrate that matrix-isolation spectroscopy can give valuable information on the basic properties of this element in both the microscopic and macroscopic dimensions.

1.10 References

[1] a) G. Maier, H. P. Reisenauer, A. Meudt, H. Egenolf, *Chem. Ber./Recueil* **1997**, *130*, 1043–1046; b) G. Maier, H. P. Reisenauer, H. Egenolf, in: *Organosilicon Chemistry III: From Molecules to Materials* (Eds.: N. Auner, J. Weis), VCH, Weinheim, **1998**, pp. 31–35; c) G. Maier, H. P. Reisenauer, H. Egenolf, J. Glatthaar, *Eur. J. Org. Chem.* **1998**, 1307–1311; d) G. Maier, H. P. Reisenauer, H. Egenolf, *Eur. J. Org. Chem.* **1998**, 1313–1317; e) G. Maier, H. P. Reisenauer, H. Egenolf, *Monatsh. Chem.* **1999**, *130*, 227–235; f) G. Maier, H. P. Reisenauer, H. Egenolf, *Organometallics* **1999**, *18*, 2155-2161; g) G. Maier, H. P. Reisenauer, H. Egenolf, in: *Organosilicon Chemistry IV: From Molecules to Materials* (Eds.: N. Auner, J. Weis), Wiley-VCH, Weinheim, **2000**, pp. 64–69; h) G. Maier, H. P. Reisenauer, J. Glatthaar, *Organometallics* **2000**, *19*,

4775–4783; i) Summary: G. Maier, A. Meudt, J. Jung, H. Pacl, *Matrix Isolation Studies of Silicon Compounds* in: *The Chemistry of Organic Silicon compounds*, Vol. 2 (Eds.: Z. Rappoport, Y. Apeloig), Wiley, New York, **1998**, Chapter 19, pp. 1143–1185; k) G. Maier, H. P. Reisenauer, H. Egenolf, J. Glatthaar in: *Organosilicon Chemistry V: From Molecules to Materials*, Wiley-VCH, Weinheim, in print; l) G. Maier, H. P. Reisenauer, J. Glatthaar, *Chem. Eur. J.* **2002**, in print.

[2] A. C. Crawford, *Silicon for the Chemical Industry V* (Eds.: H. A. Øye, H. M. Rong, L. Nygaard, G. Schüssler, J. Kr. Tuset), Trondheim, Norway, **2000**, p. 61–69.

[3] G. Tamme, L. Rösch, W. Kalchauer, *Silicon for the Chemical Industry V* (Eds.: H. A. Øye, H. M. Rong, L. Nygaard, G. Schüssler, J. Kr. Tuset), Trondheim, Norway, **2000**, p. 407–414.

[4] a) B. Meyer, *Low–Temperature Spectroscopy*, Elsevier, New York, 1971; b) H. E. Hallam (Ed.), *Vibrational Spectroscopy of Trapped Species*, Wiley, London, 1976; c) S. Cradock, A. J. Hinchcliffe, *Matrix Isolation*, Cambridge University Press, Cambridge 1975; d) M. Moskovits, G. A. Ozin (Eds.), *Cryochemistry*, Wiley, New York, 1976; e) L. Andrews, M. Moskowits (Eds.), *Chemistry and Physics of Matrix-Isolated Species*, Elsevier, Amsterdam, 1989; f) M. J. Almond, A. J. Downs, *Spectroscopy of Matrix-Isolated Species*, in: *Advances in Spectroscopy*, Eds.: R. J. H. Clark, R. E. Hester, Wiley, Chichester, 1989, Vol 17; g) I. R. Dunkin, *Matrix-Isolation Techniques*, Oxford Univ. Press, Oxford, 1998.

[5] Gaussian 98, Revision A.7, M. J. Frisch, G. W. Trucks, H. B. Schlegel, G. E. Scuseria, M. A. Robb, J. R. Cheeseman, V. G. Zakrzewski, J. A. Montgomery, Jr., R. E. Stratmann, J. C. Burant, S. Dapprich, J. M. Millam, A. D. Daniels, K. N. Kudin, M. C. Strain, O. Farkas, J. Tomasi, V. Barone, M. Cossi, R. Cammi, B. Mennucci, C. Pomelli, C. Adamo, S. Clifford, J. Ochterski, G. A. Petersson, P. Y. Ayala, Q. Cui, K. Morokuma, D. K. Malick, A. D. Rabuck, K. Raghavachari, J. B. Foresman, J. Cioslowski, J. V. Ortiz, A. G. Baboul, B. B. Stefanov, G. Liu, A. Liashenko, P. Piskorz, I. Komaromi, R. Gomperts, R. L. Martin, D. J. Fox, T. Keith, M. A. Al-Laham, C. Y. Peng, A. Nanayakkara, C. Gonzalez, M. Challacombe, P. M. W. Gill, B. Johnson, W. Chen, M. W. Wong, J. L. Andres, C. Gonzalez, M. Head-Gordon, E. S. Replogle, and J. A. Pople, Gaussian, Inc., Pittsburgh PA, 1998.

[6] H. Schnöckel, *Angew. Chem.* **1978**, *90*, 638–639; *Angew. Chem. Int. Ed. Engl.* **1978**, *17*, 616; see also: H. Schnöckel in: *Tailor-made Silicon-Oxygen Compounds*, Eds.: R. Corriu, P. Jutzi, Vieweg, Braunschweig, 1996, p. 131-140.

[7] J. S. Andersen, J. S. Ogden, *J. Chem. Phys.* **1969**, *51*, 4189–4196.

[8] H. Schnöckel, *Z. Anorg. Allg. Chem.* **1980**, *460*, 37–50.

[9] P. Roy, B. Tremblay, L. Manceron, M. E. Alikhani, D. Roy, *J. Chem. Phys.* **1996**, *104*, 2773–2781.

[10] a) R. West, J. Fink, J. Michl, *Science* **1981**, *214*, 1343; b) Summary: G. Raabe, J. Michl in: *The Chemistry of Organic Silicon Compounds* (Eds.: S. Patai, Z. Rappoport) Wiley, New York, **1989**, Part 2, pp. 1015–1114.

[11] a) H. Lischka, H.-J. Köhler, *Chem. Phys. Lett.* **1982**, *85*, 467–471; b) R. A. Poirier, J. D. Goddard, *Chem. Phys. Lett.* **1981**, *80*, 37–41.

2 Reactions with Matrix-Isolated SiO Molecules

Hansgeorg Schnöckel* **and Ralf Köppe**

2.1 Introduction

SiO and SiO_2 are the simplest binary compounds of silicon and oxygen, which are the most abundant elements in the Earth's crust. Whereas SiO is a well-characterized molecule at high temperatures and in interstellar space, there has hithero been much less research regarding the structure of solid SiO, even though this solid compound is frequently used in the technical sector, e.g. for the coating of optical construction elements. With SiO_2 the relationship is different: while crystalline and glass-like SiO_2 have been examined and characterized in detail, molecular SiO_2 has only become known since its synthesis in solid rare gas and through quantum chemical calculations.†

In this article we intend to address the following issues:

- What oligomeric SiO species are formed on the way from monomeric matrix isolated SiO to solid SiO? What deductions can be made from spectroscopic and quantum chemical examinations of the structures and the bonding arrangements of these intermediates?
- What reactions of matrix-isolated SiO species are possible, apart from oligomerization?
  - Reduction processes by reaction of metal atoms will lead to intermediates on the way to elemental Si. During these processes, differences to and analogies with corresponding conversions between metal atoms and CO become clear (e.g. we will clarify the matter as to whether, analogously to the metal carbonyls, there are comparable SiO compounds).
  - Oxidation processes of matrix-isolated SiO will lead to Si compounds in the oxidation state +4, as indicated above for SiO_2, which do not occur at all or only in barely detectable amounts in the gas phase. The reactions of SiO in solid rare gas therefore offer an ideal possibility to examine molecules such as, for instance, O=Si=O, S=Si=O, $O=SiX_2$ (X = F, Cl), and to find out the characteristics of the bonding arrangements on the basis of spectroscopic data and by using quantum chemical methods.

* Address of the authors: Universität Karlsruhe, Institut für Anorganische Chemie, Engesserstraße, D-76131 Karlsruhe, Germany

- The variation possibilities of the SiO multiple bonds in comparison to the situations in analogous CO bonds will be discussed on the basis of simple examples.
- Reflections about the synthesis of fibrous SiO_2, which can be deduced from the results of SiO_2 matrix examinations, show that a problem of matrix-isolation spectroscopy, initially of only academic interest, can lead to the solution of a central problem of solid-state chemistry and might also indicate possible applications of, e.g., nanostructured SiO_2 fibers.

2.2 The High-Temperature SiO Molecule

2.2.1 Synthesis, Matrix Isolation and Spectroscopy

On passing oxygen over elemental silicon at ca. 1200°C in a high temperature reactor made of Al_2O_3, gaseous SiO (Eq. 1) is produced.

$$Si_{(g)} + \frac{1}{2} O_{2\,(g)} \rightleftharpoons SiO_{(g)} \quad (1)$$

For this purpose, the O_2 flow has to be arranged in such a way that the resulting SiO partial pressure will be smaller than the partial pressure of SiO in the equilibrium according to the corresponding reaction between solid Si and SiO_2 (Eq. 2).

$$Si_{(s)} + SiO_{2\,(s)} \rightleftharpoons 2\ SiO_{(g)} \quad (2)$$

During the matrix experiments discussed here, the high temperature reactor is located in a high vacuum recipient, which also includes the cryostat (Figure 2.1). With the help of a cryostat a Cu block with a polished surface (13 cm^2) is cooled to ca. 10 K and positioned in such a way to enable the condensation of the SiO molecules together with a large surplus of argon or N_2 or CH_4 as a matrix (ca. 1:100 to 1:1000) on the polished surface.[1, 2] Thus, the SiO molecules are isolated in the solid inert gas like raisins in a cake and can be examined spectroscopically. For this purpose, we applied IR and Raman spectroscopy, because the observed vibrations of the atoms may permit conclusions to be made as regards the geometry and bonding arrangements of the isolated molecules.

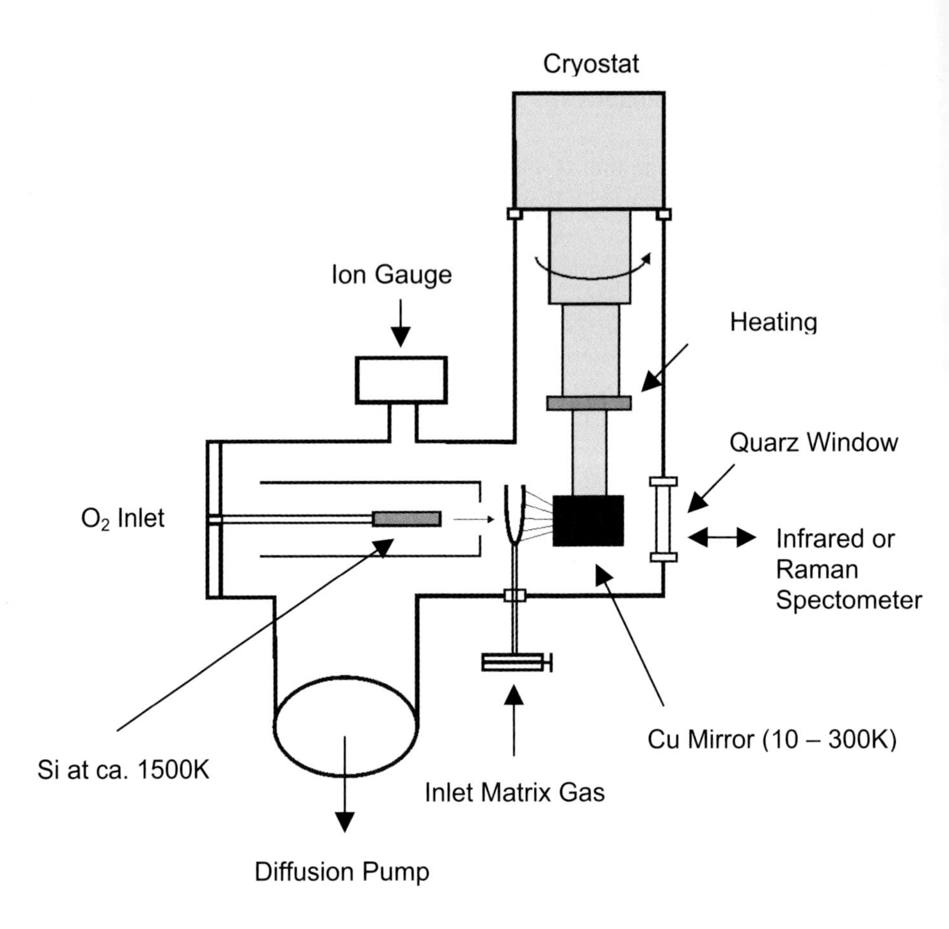

Figure 2.1. Schematic drawing of the matrix-isolation set-up.

It is especially helpful that these experiments can readily be conducted with ^{18}O-substituted SiO species, because the $^{16}O/^{18}O$ frequency shifts provide important additional information about structure and bonding arrangements. Since the absolute positions of the vibrational frequencies as well as their isotopic shifts (e.g. $^{16}O/^{18}O$, $^{28}Si/^{29}Si$) can be obtained by quantum chemical calculations, we are now able to assign vibrations to as yet unknown molecules with great reliability. Afterwards, the parameters of the individual vibrations can be quantified and the force constants can be determined as characteristic bonding constants. Thus, for the SiO bonds under discussion here, the corresponding SiO force constants are obtained, which constitute a measure of the restoring forces during the elongation of this bonding by 1 Å. The force constant $f_{(SiO)}$ for the bonding within the SiO molecule, derived simply from the observed frequency of 1226 cm^{-1} in solid argon and the masses of silicon (m_{Si})

and oxygen (m_O), amounts to 9.0 mdyn$Å^{-1}$, based on the following vibrational equation (3) for a diatomic SiO oscillator.[3]

$$5.891 * 10^{-7} \bullet 1226^2 = f\,(\mathrm{SiO}) \bullet \left(\frac{1}{m_{\mathrm{Si}}}\right) + \left(\frac{1}{m_{\mathrm{O}}}\right) \tag{3}$$

For polyatomic molecules, more complex vibrational equations result, which can only be solved by simplifying the interactions between the vibrations that reflect the particular electronic interactions during certain elongations. For many molecules discussed hereafter we quote the SiO force constants, as these are a very sensitive indication of the variation of the SiO bond strength.

2.2.2 Oligomeric SiO Species

During the condensation of the high-temperature SiO molecules with a rare gas surplus there will be interaction between the SiO molecules, and thus oligomers are observed. The simplest reaction leads to dimeric SiO with a D_{2h} ring structure (Figure 2.2).

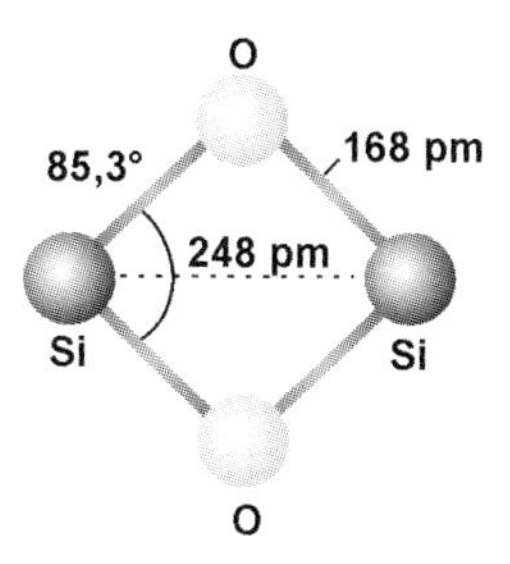

Figure 2.2. Structural parameters of dimeric SiO, calculated by quantum chemical methods.

The calculated dimerization energy of 249 kJ/mol for Si_2O_2 [4, 5] indicates that SiO single bonds are favored over SiO multiple bonds and clarifies the difference to CO, which, even at low temperatures, is still monomeric in the gas phase (boiling point 81.63 K [6]).

Figure 2.3 shows a Raman spectrum obtained after deposition of SiO in solid methane.[5] Further SiO oligomers can be deduced from the frequencies observed, the $^{16}O^{18}O$ shifts, and the band pattern, e.g. when using a $^{16}O_2/^{18}O_2$ mixture for the generation of SiO. The structures are shown in Figure 2.4. Ultimately, the geometrical data result from quantum chemical calculations, the reliability of which is verified by the correspondence of calculated and experimentally observed vibrational frequencies.

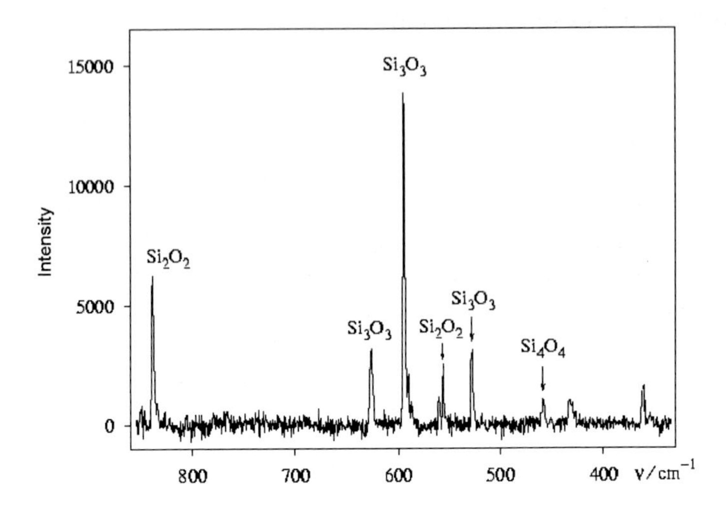

Figure 2.3. Raman spectrum of matrix-isolated SiO in solid CH_4 ($\lambda_{ecitation}$ = 514.5 nm).

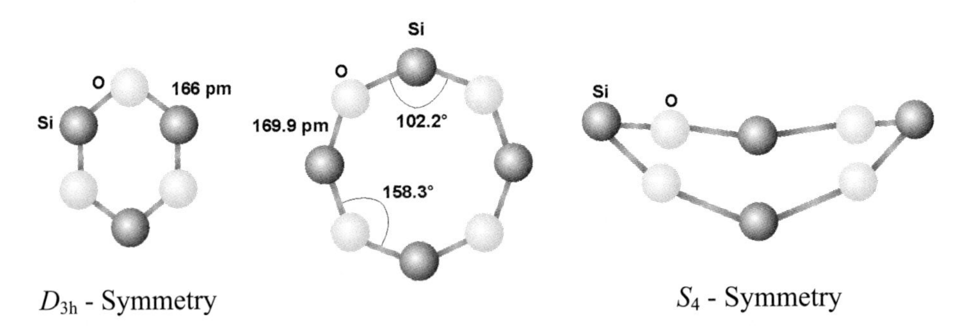

D_{3h} - Symmetry

S_4 - Symmetry

Figure 2.4. Structural parameters of trimeric and tetrameric SiO, calculated by quantum chemical methods.

There is as yet no experimental proof for the existence of higher oligomers formed by diffusion of SiO or $(SiO)_n$ species under matrix conditions. Since there are no SiSi bonds in Si_4O_4 (Figure 2.4) and, on the other hand, these bonds seem to be indispensable for the formation of a reasonable structure (see Chapter 32 in this book), a proof of further oligomerization steps would provide some essential information.

2.3 Reactions of SiO in Solid Noble Gases

2. 3.1 Reactions with Metal Atoms

The simultaneous deposition of gaseous SiO and metal atoms with a surplus of rare gas is difficult to realize experimentally as two high–temperature furnaces have to be located within a distance of a few centimeters from the cooling surface in order to achieve a ca. 1:1 ratio of SiO and metal atoms in the matrix by virtue of the metal's vapor pressure. In the last years, our group has performed reactions of SiO with the following metals: Ag,[7] Pd,[8] Na,[9] Al.[10] The structures shown in Figure 2.5 can be deduced from the spectra and by quantum chemical calculations in analogy to the facts stated above. The AgSiO molecule (Figure 2.5) shows an AgSiO angle of 109.50°, which was initially deduced from the spectra and has been the subject of many contradictory results obtained from quantum chemical calculations.[11,12]

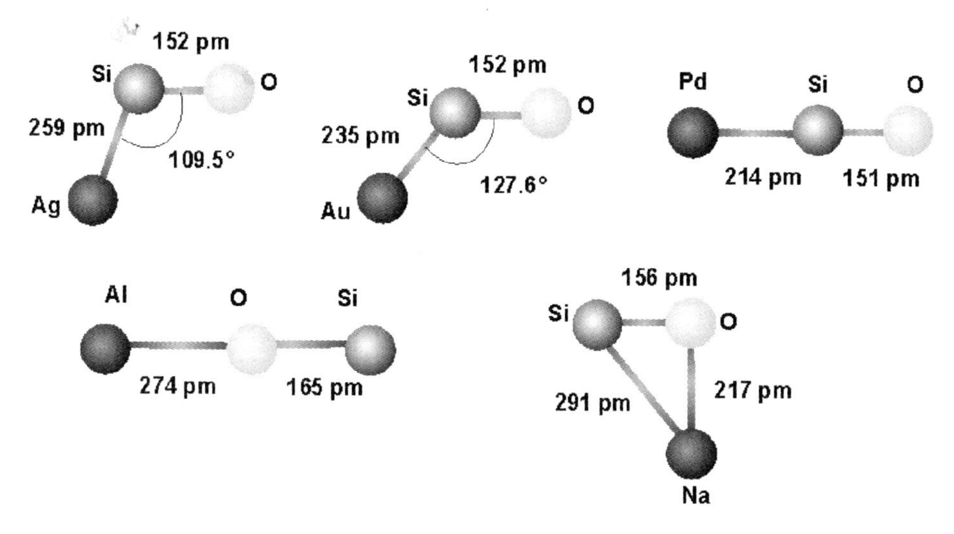

Figure 2.5. Calculated structural parameters of MSiO species (M = Ag, Au, Pd, Al, Na).

Only by using DFT methods could the originally suggested structure based on experiments be confirmed. In contrast to the reaction with precious Ag metal, the reaction of SiO with electropositive Na atoms leads to a structure that seems plausible in view of the expected ion-dipole interaction (Na^+SiO^-), in analogy to the bent NaCN molecule in the gas phase.[13] On the other hand, the linear structure of the Al-O-Si molecule is somewhat unexpected.[10] This linear arrangement of Al-O-Si can best be explained with reference to the also linear high temperature molecule Al-O-Al, in which the additional electron is located

in a π orbital. Thus, a molecule results, in which the Teller-Renner effect, as predicted by quantum chemical calculations, demands two slightly different deformation vibrations.
On the other hand, the also linear PdSiO molecule shows a metal-Si bond.[8] The bonding character is explained analogously to that of the carbonyls, whose existence has been proved by matrix isolation (e.g. PdCO). The calculated values of the corresponding dissociation energies [PdSiO $\rightarrow$ Pd + SiO: –182 kJ/mol; PdCO $\rightarrow$ Pd + CO: –162 kJ/mol] show that the bonding descriptions established for the carbonyls (σ-donor/π-acceptor bonding) can, in principal, also be applied for the Pd-Si bonding.[8] This bonding description is also proved by the force constants deduced from the vibrational spectra. According to our results, PdSiO is the first experimentally confirmed example of a CO-analogous bonding of an SiO molecule to a transition metal atom. The principal difficulty in trying to synthesize further $M(SiO)_n$ species (e.g. $Ni(SiO)_4$) is that SiO molecules – in contrast to CO molecules – tend to react with each other (see above) and, therefore, the required concentration of monomeric SiO units is difficult to attain.

2.3.2 Reactions of SiO with Oxidizing Species

2.3.2.1 SiO_2/OSiS: As long ago as the 1970s, it was shown that GeO_2 and SnO_2 molecules,[15, 16] as the heavy homologues of silicon, can be synthesized through insertion of the metal atoms into the O-O bond of di-oxygen under matrix conditions. The analogous reaction with silicon atoms failed. Therefore, we chose another method, in which SiO molecules were reacted with O atoms formed in a microwave discharge. By applying this method we were able to prove – simply with the help of IR spectra – the linearity of the SiO_2 molecule and derived a force constant of 9.2 mdyn/Å for the SiO bond, which is slightly higher than the force constant found for the SiO molecule (see below) [14] [Note: A few years later, other authors demonstrated that SiO and O_2 form the SiO_3 molecule [17]].

This strengthening of the SiO bonds in SiO_2 based on the force constants – which was unexpected – was also confirmed for the linear OSiS molecule which is synthesized in a matrix reaction from SiS and O atoms generated in a microwave discharge. For molecular S=Si=O, the three vibrational modes ($2\ \Sigma + 1\ \Pi$) and the corresponding $^{16}O/^{18}O$ shifts were observed, and therefore all force constants could be determined only on the basis of these experimental findings. In combination with quantum chemical calculations the following bonding parameters result for SiO_2 and SSiO (Table 2.1):[18]

Table 2.1. Comparison of force constant values and distances in SiO_2 and SSiO.

	d(SiO) (pm)	*d*(SiS) (pm)	*f*(SiO) mdyn/Å	*f*(SiS) mdyn/Å
SiO_2	148.3	–	9.2	–
SSiO	148.5	190.4	9.0	4.86

In both cases, we see molecules which, as isolated species (e.g. in interstellar space), are stable with respect to any dissociation (e.g. $SiO_2 \rightarrow SiO + O$ or $Si + 2\ O$). Only the exothermic dimerization of SiO_2 molecules and the final synthesis of three-dimensional cross-linked SiO_2 units as in quartz show that four SiO single bonds are significantly more stable than the two SiO double bonds. This means that the synthesis of compounds having an SiO multiple bond is only possible under certain conditions, and this will be demonstrated in the following examples.

2.3.2.2 $OSiX_2$/OSiX (X = F, Cl): Being polymers, the silicones $(OSiR_2)_n$ formally derived from the ketones, have no SiO double bonds but SiOSi bridges. The first example of such a monomeric species was the $OSiCl_2$ molecule synthesized in a matrix reaction:[1]

$$Si{=}O + Cl_2 \xrightarrow{h\nu} O{=}SiCl_2$$

From the IR spectra, the $^{16/18}O$ and the $^{35/37}Cl$ shifts, the C_{2v} structure and a strong SiO bond could be deduced: f(SiO) = 8,95 mdyn/Å.[1] This result was confirmed by quantum chemical calculations.[19]

These earlier matrix experiments recently helped us to prove the key role of the $OSiCl_2$ molecule during the technically important $SiCl_4$ combustion with O_2 forming SiO_2 and HCl.[19,20] A presumed possible participation of an OSiCl radical species could not be confirmed but, on the other hand, we were able to demonstrate by matrix experiments that OSiCl is only formed during a matrix reaction of SiO and Cl atoms. For the OSiCl molecule, the experimentally deduced bent structure with a strong SiO bond (*f*(SiO) = 8.0 mdyn/Å) could be confirmed by quantum chemical calculations as well:[19]

d(SiO) = 153.6 pm, *d*(SiCl) = 207.7 pm, <(OSiCl) = 125.2°

Analogously to $OSiCl_2$, an $OSiF_2$ molecule could be synthesized by a matrix reaction between SiO and F_2 and structurally characterized by observation of all permitted vibrations and their $^{16}O/^{18}O$ and $^{28}Si/^{29}Si$ shifts:[21] *f*(SiO) = 9.4, *f*(SiF) = 6.25 mdyn/Å. These results, and the subsequent quantum chemical calculations, which were recently reinvestigated with more sophisticated methods, essentially confirm the earlier results:[22]

d(SiO) = 147.8 pm; *d*(SiF) = 155.5pm); <(SiOF) = 127,2° §

Thus, in all $OSiX_2$ and SiOX species we encounter SiO bonds which essentially correspond to the bonds within the SiO molecule and, therefore, also exhibit bonding parameters completely different from the corresponding carbon compounds, e.g. CO and $OCCl_2$. In the following section, we examine this matter in more detail.

2.4 The SiO Multiple Bond

In Table 2.2, several bonding parameters of the SiO_2 and SiO molecules, which were obtained by experiments, by quantum chemical calculations, and by combining both methods, are compared to the analogous data of CO_2 and CO.

Table 2.2. Experimentally deduced force constants, distances, bond energies (BE), and calculated partial atomic atomic charges of MO_x (M = C, Si: x = 1, 2).

MO_x	*f*(MO) (mdyn/Å)	*d*(MO) (pm)	bond energy (kJ/mol)	*q*(M)
CO	18.6	110.6	1071.8	0.06
CO_2	15.6	113.6	799.0	0.68
SiO	9.02	148.9	794.1	0.34
SiO_2	9.2	148.3	621.7	1.22

For the carbon compounds, these data show that the bond within the CO molecule is stronger than those in the CO_2 molecule, thus justifying their common description as triple and double bond, respectively. On the other hand, the force constants and the distances in SiO_2 and SiO give an indication of similar bonding arrangements in both molecules with an apparently slightly stronger bond in SiO_2. The higher bond energy (BE) for SiO compared to SiO_2 on the other hand is suggestive of a drastic bonding increase in SiO, although this can be largely attributed to the different polarities of the molecules. Thus, during a separation into neutral fragments (atoms) (i.e. $SiO_{2(g)} \rightarrow Si_{(g)} + 2\ O_{(g)}$), for a polar bond the required BE will decrease according to the definition of BE due to the fact that the internal charge equalization is combined with a greater gain of Coulomb energy.

The parameters of distance and force constant permit a better evaluation of the bonding situation and demonstrate that in all SiO multiple bonds (including that in $OSiX_2$) the maximum bond strength is reached with force constants of about 9 mdyn/Å and short distances of ca. 149 pm, in contrast to the arrangements found in CO and CO_2. Recently, we interpreted the higher force constant of SiO_2 (compared to SiO) like this: by a charge equalization in the more polar SiO_2 the covalent bonding contributions will increase during the vibrational elongation, whereas during the elongation of SiO an increasing bonding polarity will weaken the covalent bonding contributions that have crucial influence on the distance as well as on the ascent of the potential function, and, therefore, on the force constants.[23]

2.5 Reflections on dimeric SiO_2 molecules and on the synthesis of solid, fibrous SiO_2

The reaction of SiO_2Si (Figure 2.2) with O_2 under matrix conditions leads to dimeric SiO_2 molecules with a D_{2h} structure (Figure 2.6):

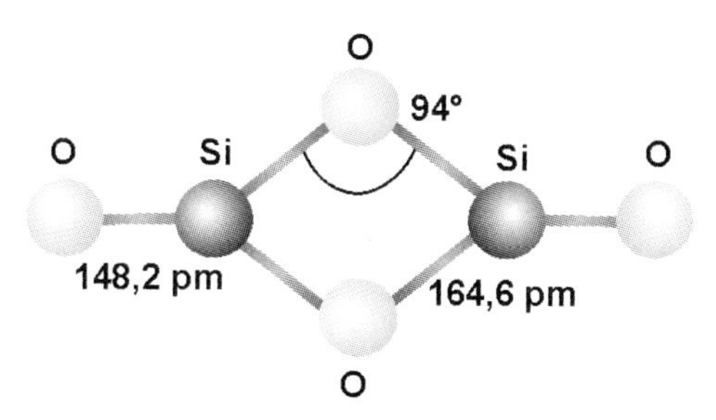

Figure 2.6. Calculated structural parameters of dimeric SiO_2.

From the vibrational spectrum and the $^{16}O/^{18}O$ and $^{28}Si/^{29}Si$ shifts, we could definitely infer this structure: f(SiO) = 9.2; f(SiO) = 3.65 mdyn/Å.[24] These results are confirmed by earlier and recent quantum chemical calculations (Figure 2.6). Apart from the structure, the mechanism of formation of Si_2O_4 seems to be of interest, since already existing SiO bonds are clearly broken during this photochemically promoted reaction. Thus, when using $^{18}O_2$, the ^{18}O atoms can be found in terminal as well as in bridging positions. A secondary reaction of the dimeric SiO_2 units with two further SiO molecules could not be observed due to the diminished concentration of this species after the primary matrix reaction. A hypothetically continued reaction with additional O_2 molecules should finally lead to a linear $(SiO_2)_n$ molecule constructed of edge-linked tetrahedral SiO_4 units which would correspond to a section of the SiS_2 structure (Figure 2.7).

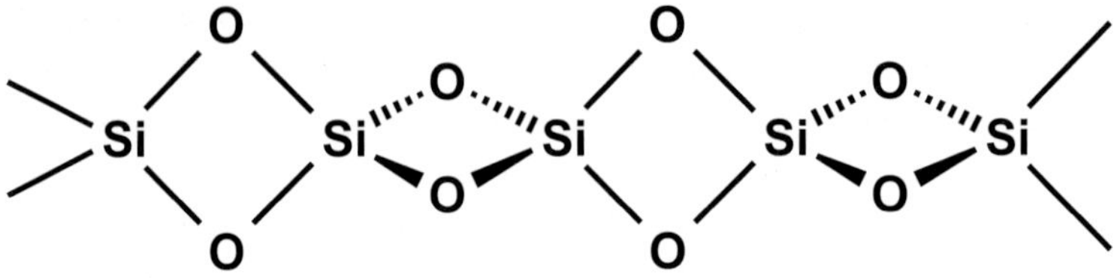

Figure 2.7. Solid-state structure of fibrous SiO_2.

These results demonstrate that, under certain conditions, during the reaction of SiO primary products of an SiS_2-analogous SiO_2 modification can be formed. As long ago as 1954, such a fibrous SiO_2 modification was described by Armin and Alarich Weiss.[25] Interestingly enough, in those early experiments, gaseous SiO molecules were also reacted with O_2 under special conditions. Unfortunately, the authors were unable to determine the bonding parameters in this crystalline modification with the required precision due to the inadequate methods available at that time. Until now, all experiments conducted in order to reproduce these results have been unsuccessful and so the fibrous SiO_2 seems to have steadily disappeared from textbooks. Since our recent matrix results can be interpreted as a confirmation of the results obtained by A. Weiss and A. Weiss, we will vary the conditions of the SiO oxidation by O_2 in order to confirm the pioneering examinations of 1954 and to obtain an exact structure of this SiO_2 modification (Figure 2.7).

2.6 Summary and Outlook

Examinations of the primary steps that occur during the formation of SiO bonds, i.e. the linking of two of the most abundant elements of the periodic system, with the help of matrix-isolation spectroscopy and of quantum chemical calculations have allowed us to

a) critically examine the possibilities for multiple bond formation between these elements for the first time,
b) make suggestions regarding the structure of solid SiO as a technically important but structurally yet unknown material by characterizing oligomeric SiO compounds,

c) make suggestions about the bonding changes in SiO upon varying the reduction conditions of SiO with different metals (these changes should also be important in the partial oxidation of the surface of Si wafers)
d) offer a plausible interpretation of earlier results concerning fibrous SiO_2 through the synthesis of a dimeric SiO_2 molecule, thereby also providing a stimulus for further optimized experiments.

In all these cases, problems of fundamental importance are discussed, concerning the very basics of chemistry, their potential technical applications, and, finally, their spectacular possibilities, e.g. the production of nanostructured SiO_2 fibers.

2.7 References

[1] H. Schnöckel, *Z. Anorg. Allg. Chem.* **1980**, *460*, 37–50.
[2] H. Schnöckel, S. Schunck, *Chem. in unserer Zeit* **1987**, *21*, 73–81.
[3] H. Siebert, *Anwendungen der Schwingungsspektroskopie in der Anorganischen Chemie*, Springer Verlag, Berlin, 1966.
[4] H. Schnöckel, T. Mehner, H. S. Plitt, S. Schunck, *J. Am. Chem. Soc.* **1989**, *111*, 4578–4852.
[5] M. Friesen, M. Junker, A. Zumbusch, H. Schnöckel, *J. Chem. Phys.* **1999**, *111*, 7887.
[6] M. W. Chase, C. A. Davies, J. R. Downey Jr., D. J. Frurip, R. A. McDonald, A. N. Syverud, *Janaf Thermochemical Tables, 3rd Edition,* New York, 1985.
[7] T. Mehner, H. Schnöckel, M. J. Almond, A. J. Downs, *J. Chem. Soc. Chem. Commun.* **1988**, 117–119.
[8] T. Mehner, R. Köppe, H. Schnöckel, *Angew. Chem.* **1992**, *104*, 653–655; *Angew. Chem. Int. Ed. Engl.* **1992**, *31*, 638–640.
[9] R. Köppe, H. Schnöckel, *Heteroatom Chem.* **1992**, *3*, 329–331.
[10] M. Junker, M. Friesen, H. Schnöckel, *J. Chem. Phys.* **2000**, *112*, 1444–1448.
[11] G. E. Quelch, R. S. Grev, H. F. Schaefer III, *J. Chem. Soc. Chem. Commun.* **1989,** 1498.
[12] J. S. Tse, *J. Chem. Soc. Chem. Commun.* **1990,** 1179.
[13] J. J. Van Veels, W. L. Meerts, A. Dymanns, *J. Chem. Phys.* **1982**, *77*, 5245.
[14] H. Schnöckel, *Angew. Chem.* **1978**, *90*, 638–639; *Angew. Chem. Int. Ed. Engl.* **1978**, *17*, 616.
[15] A. Bos, J. S. Ogden, L. Orgee, *J. Phys. Chem.* **1974**, *78*, 1763.
[16] A. Bos, J. S. Ogden, *J. Phys. Chem.* **1973**, *77*, 1513.

[17] B. Tremblay, P. Roy, L. Manceron, M. E. Alikhani, D. Roy, *J. Chem. Phys.* **1996**, *104*, 2773.
[18] H. Schnöckel, R. Köppe, *J. Am. Chem. Soc.* **1989**, *111*, 4583–4586.
[19] M. Junker, H. Schnöckel, *J. Chem. Phys.* **1999**, *110*, 3769–3772.
[20] M. Junker, A. Wilkening, M. Binnewies, H. Schnöckel, *Eur. J. Inorg. Chem.* **1999**, 1531–1535.
[21] H. Schnöckel, *J. Mol. Struct.* **1980**, *65*, 115–123.
[22] J. Breidung, W. Thiel, *Z. Anorg. Allg. Chem.* **2000,** *626*, 362.
[23] M. Friesen, M. Junker, H. Schnöckel, *Heteroatom. Chem.* **1999**, *10*, 658.
[24] T. Mehner, H. J. Göcke, S. Schunck, H. Schnöckel, *Z. Anorg. Allg. Chem.* **1990**, *580*, 121–130.
[25] A. Weiss, A. Weiss, *Z. Anorg. Allg. Chem.* **1954**, *276*, 95.

3 *In situ* Diagnostics of Amorphous Silicon Thin Film Deposition

H. Stafast*, G. Andrä, F. Falk, E. Witkowicz

3.1 Introduction

The deposition of thin solid films from the gas phase has achieved high technological importance. Its economically most prominent application is in microelectronics, i.e. in the manufacture of computers, digital data storage and display devices using pure and doped silicon (Si) either in its crystalline (c-Si) or amorphous state (a-Si). Additional applications of a-Si and c-Si thin films are in photocopy systems, X-ray detectors, and photovoltaic solar cells.[1] In each case, reliable and efficient device operation requires thin smooth films of high purity or well-defined doping, good adherence, and an absence of defects (e.g. particles, holes, segregation). In addition, the deposition process must be compatible with the fabrication requirements such as maximum device temperature, high throughput, and processing stability. Therefore, the preferred fabrication process is thin-film deposition from the gas phase by physical vapor deposition (PVD) [2] or chemical vapor deposition (CVD).[1, 3]

PVD processes typically use solid-state sources. The gas-phase species for thin-film deposition are generated from the source by thermal heating, electron beam evaporation, or sputtering. In CVD one or several gaseous precursors (gas mixtures) are activated to generate the reactive gas-phase species that forms the solid films. The precursor(s) are activated thermally ("standard" CVD), within a plasma (plasma CVD), or by optical excitation (photo or laser CVD).

The details of thin-film formation by PVD or CVD on the atomic and molecular scale are unknown in most cases, but such knowledge would be very helpful to design new processes and to tailor film properties. Information is lacking due to the high reactivities, short lifetimes, and low concentrations of the relevant transient gas-phase species. Furthermore, many of the thin-film properties *in statu nascendi* are unknown due to the experimental difficulties of thin-film characterization during deposition, particularly with non-crystalline films and if established methods such as RHEED and LEED cannot be applied. Among the film properties, mechanical stress in thin films can lead to unwanted (uncontrolled) instabilities and peel-off phenomena. Therefore, *in situ* diagnostic methods have been developed to

---

* Address of the authors: Institut fuer Physikalische Hochtechnologie (IPHT)
POB 100 239, D-07702 Jena, Germany

characterize transient (film-forming) gas-phase species and to determine the stress of thin films under their deposition conditions (temperature, atmosphere). The following text is focussed on the deposition of thermodynamically metastable a-Si. Its properties are more dependent and more sensitive to the precise conditions of thin film deposition than those of the thermodynamically stable c-Si state. Three different methods of a-Si deposition have been selected to demonstrate the potential of the *in situ* diagnostic methods: PVD by thermal evaporation of an Si wire and CVD by pyrolysis or plasma chemistry of SiH_4.

3.2 Experimental Section

The *in situ* diagnostic methods described here refer to the characterization of intermediate gas-phase species (Section 3.2.1) during a-Si thin-film deposition by PVD (thermal evaporation of Si) and by CVD (SiH_4 pyrolysis), as well as the determination of mechanical stress (Section 3.2.2) in growing a-Si during plasma CVD from SiH_4.

3.2.1 *In situ* Gas-Phase Diagnostics by TOF MS with Laser Ionization

The apparatus for *in situ* gas-phase diagnostics consists of three chambers, the processing chamber for thin-film deposition, a differential pumping stage, and an analysis chamber with a time-of-flight (TOF) mass spectrometer (MS) with pulsed laser ionization (Figure 3.1).

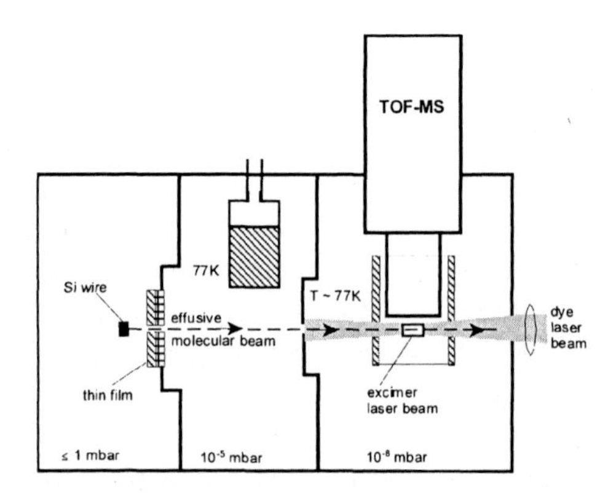

Figure 3.1. Scheme of experimental set-up comprising a deposition chamber, a differential pumping stage with LN_2 trap, and an analysis chamber equipped with a TOF-MS and laser ionization.[4–6]

This modular system is very flexible with respect to the type of thin-film deposition process (different methods of PVD and CVD) and the means of laser ionization (one-, two- or multiple-photon ionization with fixed or tunable laser wavelengths).

A special feature of this system is the gas-phase sampling by an effusive molecular beam leading to a collision-free transfer of the sample into the analysis chamber. Liquid nitrogen cooled traps in the differential pumping stage and around the ion source of the TOF-MS serve to ensure for a low background.[4–6]

In the deposition chamber, the film-forming gas-phase species are evidently hitting the film (substrate) surface and are therefore part of the gas sampled through the hole in the substrate surface (Figure 3.2). At first sight, a large hole appears attractive. However, with an orifice diameter *d* much larger than the mean free path length λ, many particles pass the orifice at the same time resulting in a continuous gas stream. This stream not only locally perturbs the deposition process but also enables collisions within the gas sample on its way to the analysis chamber (intermediate reactions). Therefore, the hole size is chosen to be small enough ($d < \lambda$) depending on the pressure in the deposition chamber, to sample by effusion only (independent single particles passing through the orifice), i.e. without perturbations and collisions. In this way, the analysed sample is representative of the processing gas in the region of film formation.

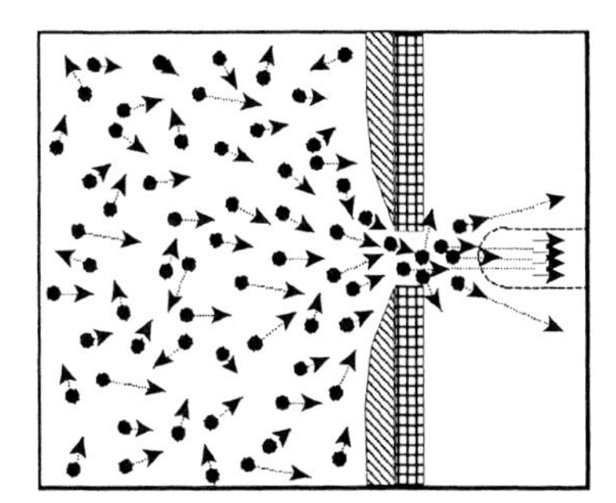

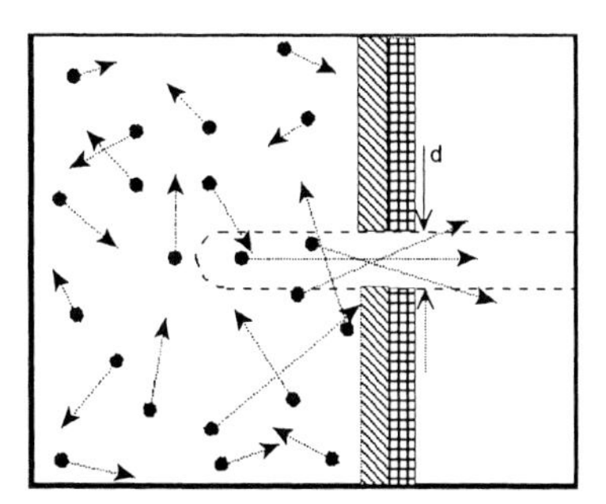

Figure 3.2. Different methods of gas sampling during thin-film deposition using a continuous gas stream (supersonic expansion, on the left) and effusion (on the right); the broken line indicates the different sampling volumes with collision-free transfer into the analysis chamber.[5]

PVD of a-Si is achieved by electrically heating an Si wire (0.4·2 mm^2 cross-Section, 25 mm length) at 10^{-5} mbar in front of the substrate surface at a distance of 4 mm. The substrate is moderately heated by radiation from the hot Si wire only. Typically, the Si wire can be used for 3 hours until thinning and breakage due to Si evaporation. CVD of a-Si can be achieved with the same set-up by pyrolyzing 0.5

mbar of an SiH_4/Ar mixture. In this case, the Si wire cross-Section increases due to c-Si deposition. The actual Si wire temperature is determined pyrometrically in both cases, PVD and CVD.

After passing the differential pumping stage, the gas sample enters the analysis chamber (10^{-8} mbar). In the ion source of the TOF-MS (reflectron type RFT10-TOF, Kaesdorf), it is ionized by an excimer laser alone (ArF (193 nm) or KrF (248 nm), EMG 201, Lambda Physik) or, in case of (1+1)-REMPI (resonance-enhanced multiple-photon ionization), by synchronizing a dye laser pulse (Nd:YAG pump laser, 503-D.NS 779/10, Soliton; Scan mate dye laser, Lambda Physik) with the KrF laser pulse. The ions are detected in the TOF-MS by a multichannel plate (MCP) and counted by a multichannel device (SR430, Stanford Research Systems) with time channels of 5 ns width. Typically, the signals generated by 1000 laser pulses are summed to obtain one mass spectrum.[4–6]

3.2.2 *In Situ* Measurement of Mechanical Stress in a-Si:H

A special set-up for glow discharge plasma CVD has been built to enable the *in situ* measurement of the mechanical stress in a-Si:H thin films (Figure 3.3).[5] The films are deposited from SiH_4 onto a thin fused silica substrate (1"·2"·0.15 mm) fixed at one end only to allow substrate bending. The direction and amount of substrate bending are detected by using two parallel He/Ne laser beams. These are reflected from the substrate surface and directed onto a screen, which is observed by a CCD camera. Substrate heating is performed via a contactless radiative heater and controlled pyrometrically. Temperature stabilization requires a feedback during deposition due to changing substrate properties with respect to optical absorption and heat flow. An inert gas flow from the heater towards the rear of the substrate is applied to achieve a-Si:H deposition selectively on the front face of the substrate.

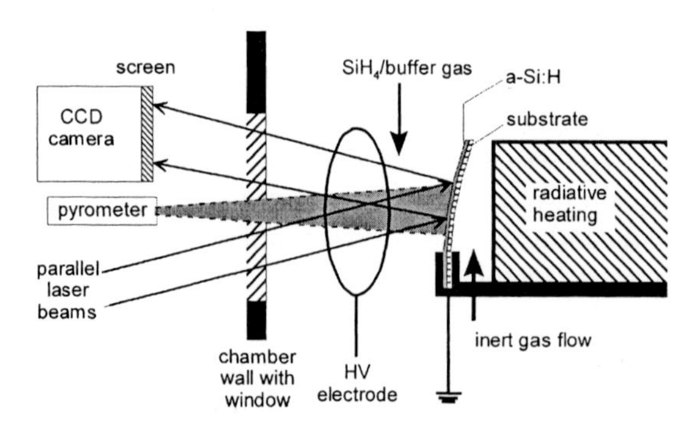

Figure 3.3. Scheme of optical set-up to measure substrate bending for the *in situ* determination of mechanical stress in thin films.[5]

3.3 Results and Discussion

3.3.1 Gas-Phase Species Detected During PVD of Amorphous Silicon (a-Si)

A simple and convenient method to prepare pure Si-containing gas-phase species for PVD of amorphous silicon uses an electrically heated Si wire under vacuum (Section 3.2.1). The vaporized Si species can be ionized by ArF (193 nm, 6.4 eV) or KrF (248 nm, 5.0 eV) laser irradiation and detected by TOF-MS (figure 6.4). As a common feature, "mass" peaks at flight times of 8.2–8.6 μs (mass:charge ratio $m/e = 28–30$) are detected in both spectra. In addition, ArF laser ionization leads to signals at 11.6–12.1 μs ($m/e = 56–58$) compared to those at 13.1–13.5 μs ($m/e = 70–74$) with KrF laser ionization. The masses 28–30 and 56–58 can be attributed to Si atoms and Si_2 molecules, respectively, whereas the parent molecule of mass 70 must contain an element different from Si (cf. Section 3.3.4).

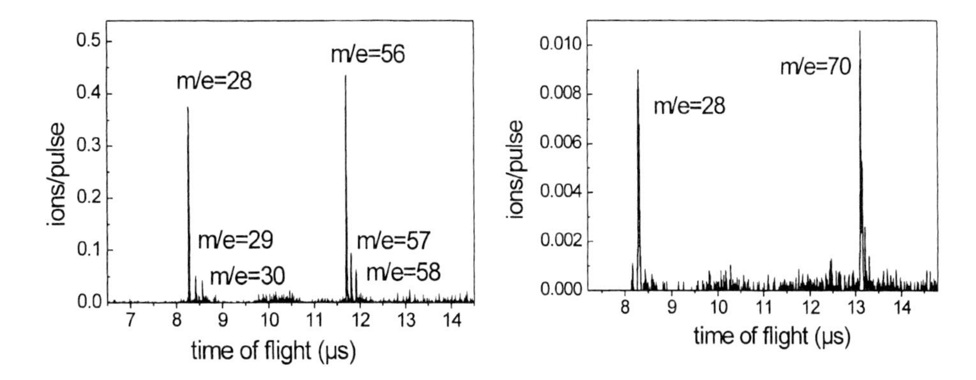

Figure 3.4. ArF (left) and KrF (right) laser ionization TOF-MS of the gas phase around the hot Si wire in vacuum (10^{-5} mbar).[5, 6]

For an unequivocal assignment of the masses 28–30, the (1+1)-REMPI method has been applied to identify the Si atoms in their electronic ground state (Figure 3.5) by their six well-known 4s $^3P \leftarrow$ 3p ^3P atomic resonances.[7] The mass scale (abscissa) shows three peaks (28–30) for each resonance, which are evidently related to the Si isotopic species ^{28}Si, ^{29}Si, and ^{30}Si.

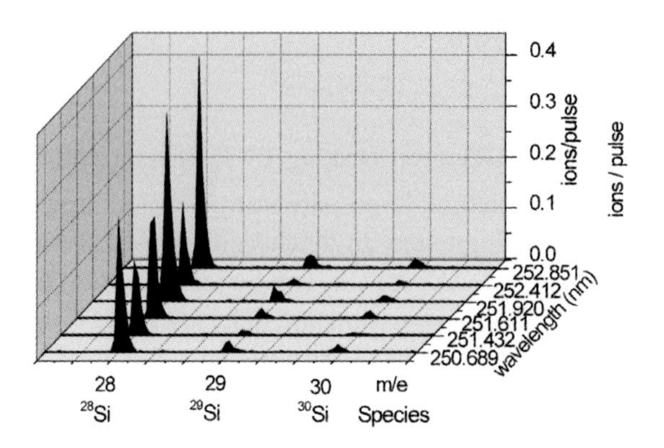

Figure 3.5. (1+1)-REMPI TOF-MS of Si atoms obtained by tuning a frequency-doubled dye laser through the 4s $^3P \leftarrow$ 3p ^3P atomic resonances (1st photon) and ionizing by synchronized KrF laser excitation (2nd photon).[4–6]

In a similar way, (1+1)-REMPI has been used to identify the Si_2 molecules desorbing from the Si wire surface (m/e = 56) by three rotational-vibrational-electronic (rovibronic) transition lines of the (v' = 5) $\leftarrow$ (v" = 0) vibrational transition in the H $^3\Sigma_u^-$ $\leftarrow$ X $^3\Sigma_g^-$ electronic band.[8] These signals are, however, very weak because the concentration of Si_2 is very low. Furthermore, the hot Si_2 molecules (Si wire at 1500 K) are distributed over several vibrational and many rotational states. Therefore, state-selective detection by REMPI only addresses a tiny part of the Si_2 entity.

3.3.2 Gas-Phase Species Detected During CVD of Amorphous Silicon (a-Si)

CVD of amorphous silicon has been performed by pyrolyzing SiH_4 into SiH_2 and H_2.[9] Pyrolysis is achieved by placing a hot Si wire in front of the substrate surface (Figure 3.1). Questions open to debate are whether: (i) pyrolysis occurs on the wire or in the gas phase and (ii) whether SiH_2 contributes to a-Si thin-film formation or is captured by the abundant SiH_4. The TOF-MS obtained with ArF laser ionization shows increasing signals at masses 2, 28, 29, and 56–58 when raising the wire temperature whereas only two peaks at masses 2 and 29 are obtained with KrF laser ionization (Figure 3.6).

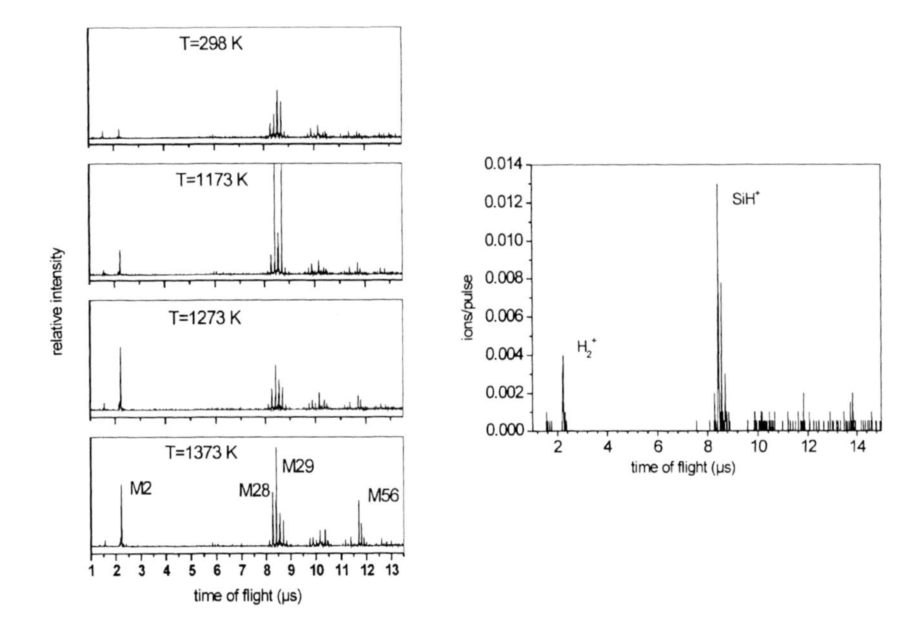

Figure 3.6. TOF-MS recorded during SiH_4 pyrolysis with a hot Si wire with ArF laser ionization at variable temperature (left) compared to KrF laser ionization (on the right, $T = 1423$ K).

The masses 56–58 are assigned to disilane Si_2H_6.[10] The masses 2 and 29 are assigned to stable H_2 molecules and SiH_2 intermediates, respectively.[4] SiH_2 is formed from SiH_4 on the Si wire surface and either reacts in the gas phase with abundant SiH_4 to yield Si_2H_6 or arrives at the substrate surface to form a-Si. This interpretation is based on the results in Figure 3.7. Figure 3.7 shows that masses 2 (H_2^+) and 29 (SiH^+) initially increase with rising SiH_4 pressure. Above 0.3 mbar SiH_4, the H_2 signal further increases whereas the SiH_2 signal decreases due to chemical reaction with SiH_4. The temperature dependences of the H_2 and SiH_2 signals are identical within experimental error showing an activation energy of ~ 75 kJ/mol, consistent with heterogeneous SiH_4 decomposition on Si. This low activation energy value excludes monomolecular gas phase decomposition (E_a ~ 250 kJ/mol).[4]

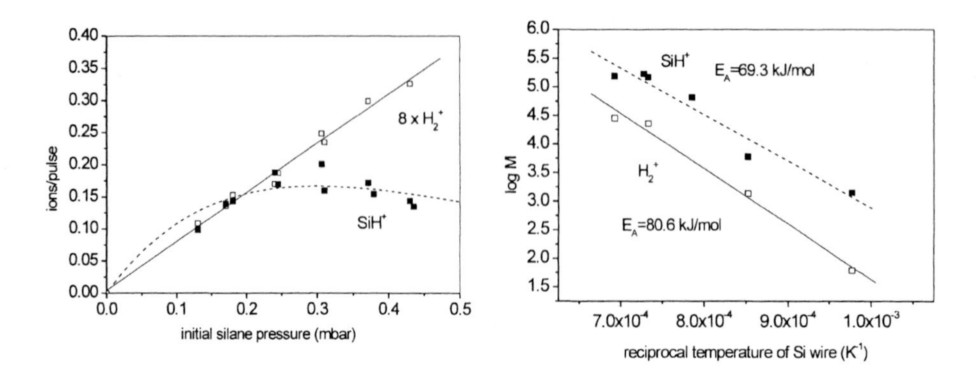

Figure 3.7. Mass signals 2 and 29 in the TOF-MS obtained with KrF laser ionization: their dependence on the initial SiH_4 pressure during pyrolysis (on the left, T_{wire} = 1450 K) and from the Si wire temperature (Arrhenius plots, on the right).[4]

On the basis of the increasing weight of the Si wire, the a-Si deposition rate, the SiH_4 consumption, and the mass spectra, it can be estimated that about 40% of the SiH_4 is decomposed on the Si wire to form c-Si. About 5% of that SiH_4 decomposes to SiH_2 and H_2. One half of SiH_2 reacts with the abundant SiH_4 to yield Si_2H_6, while the other half hits the substrate surface to form a-Si.[4]

3.3.3 Mechanical Stress in Growing a-Si:H Thin Films

Mechanical stress in hydrogenated amorphous silicon (a-Si:H) deposited by plasma CVD using He or Ar as buffer gas is compressive or tensile, respectively (Figure 3.8).[5] Cooling the substrate with the a-Si:H film after the end of CVD leads in both cases to a shift towards tensile stress, as is seen from the *ex situ* measurements. The opposite effects of He and Ar are confirmed by the effect of buffer gas changes in the plasma (Figure 3.8). The temporal curvature evolution of the substrate/a-Si:H film system with constant slope reveals constant internal stress for a given buffer gas. It appears attractive to optimize the buffer gas composition in plasma CVD systems to achieve a-Si:H thin films without, or with a minimum of, mechanical stress.

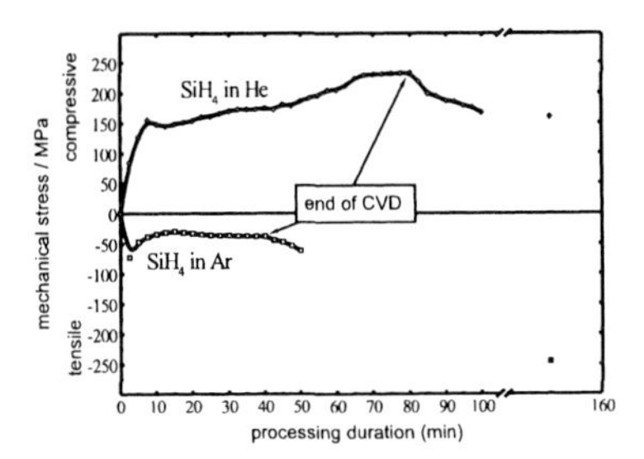

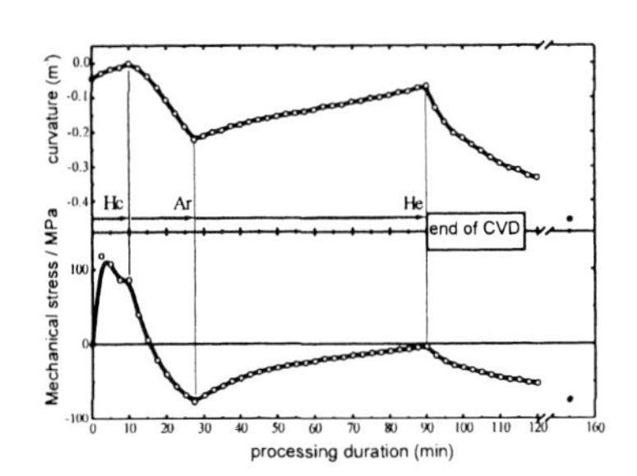

Figure 3.8. *In situ* measurements of mechanical stress in a-Si:H thin films during plasma CVD from SiH_4 with He or Ar as buffer gas: compressive stress with He and tensile stress with Ar (left part) and substrate curvature and stress with changing buffer gas (right part).[5]

3.3.4 Silicon-Nitrogen Reactions on Silicon Surface

The TOF-MS obtained upon hot Si wire evaporation at 10^{-5} mbar background pressure (Figure 3.4) shows mass 70.[5, 6] This cannot originate from a parent molecule consisting of Si atoms only. Furthermore, the wire temperature dependences of masses 28 and 70 in Figure 3.9 are inconsistent with a pure Si system. These phenomena can, however, be rationalized by considering the reactions between the Si wire and air (O_2 and N_2).

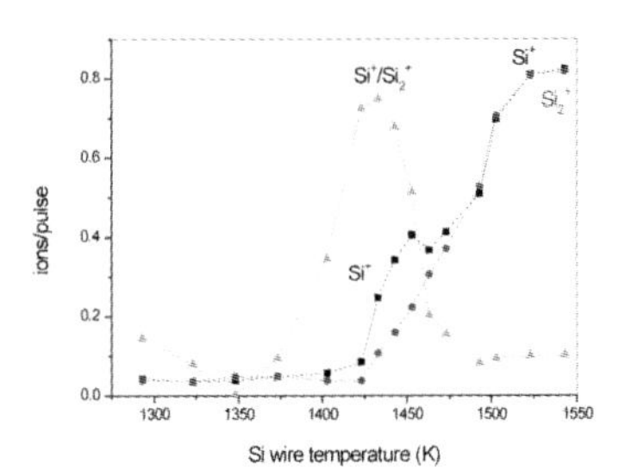

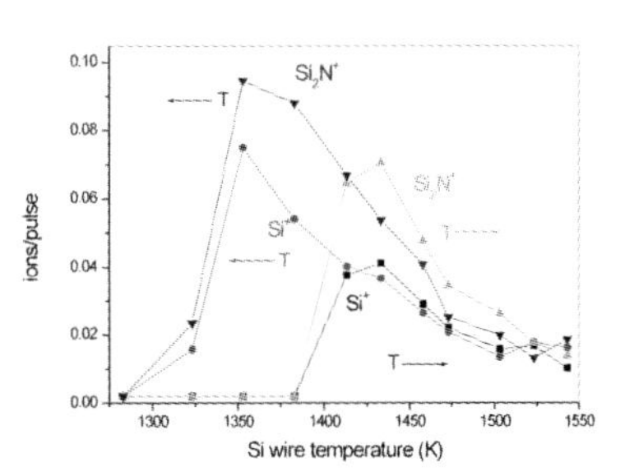

Figure 3.9. Yield and ratio of masses $28(Si^+)$ and $56(Si_2^+)$ obtained by ArF laser ionization (left) and of masses $28(Si^+)$ and $70(Si_2N^+)$ obtained by KrF laser ionization (right) as a function of the Si wire temperature.[5, 6]

According to the literature,[11–16] we suppose that the Si wire surface is initially covered with natural oxide. This oxide typically decomposes to SiO at 900 °C. The de-oxidized Si surface then reacts with N_2 to form Si-N compounds. Si_2N desorbs from the Si surface and is detected by its parent ion Si_2N^+ and Si^+ upon KrF laser ionization (two-photon ionization, 10.0 eV), whereas ArF laser ionization (two photons, 12.8 eV) leads to the "extra" Si^+ fragment ions (Si^+ peak at 1400–1450 K in Figure 3.9, left). Si_2N^+ and Si^+ ions are observed above 1380 K with increasing Si wire temperature, but down to 1320 K with decreasing temperature (Figure 3.9, right). This hysteresis is attributed to the presence and absence of the natural oxide with rising and falling temperature, respectively. The observed Si_2N gas-phase species must originate from the Si wire surface because gas-phase reactions between e.g. Si and N_2 cannot form SiN_2 under the prevailing conditions due to the lacking of a collision partner necessary to stabilize a collision complex. As a consequence, the gas-phase TOF-MS investigation provides a sensitive probe of the surface chemistry on Si.

The mass 56 signal obtained upon ArF laser ionization (Figure 3.9, left), on the other hand, shows Arrhenius behavior with $E_a = 575$ kJ/mol, in good agreement with literature values for Si_2 formation.[17, 18]

3.4 Summary and Conclusions

The importance, feasibility, and value of *in situ* investigations of thin-film deposition from the gas phase by PVD and CVD methods have been demonstrated using two selected techniques. In both cases, the measurement conditions ensure a direct relationship between the obtained data and thin-film deposition. Special efforts are made to avoid any interference of the processes from the measurements and *vice versa*. TOF-MS with laser ionization is applied to detect intermediate gas-phase species involved in thin-film formation. Deflection of two probe laser beams induced by reflection from a curved substrate is used to determine the direction and amount of mechanical stress in the growing layer.

PVD of a-Si is achieved by evaporating Si atoms and Si_2 molecules from an electrically heated Si wire. The Si atoms in their electronic ground state are easily identified by (1+1)-REMPI using the 4s $^3P \leftarrow$ 3p ^3P resonances (dye laser) and KrF laser ionization. The state-selective detection of the Si_2 molecules by (1+1)-REMPI has proved possible but difficult because the concentration of Si_2 molecules is relatively low. In addition, as a general phenomenon of molecules at high temperature, the Si_2 molecules are distributed over their large manifold of rotational and vibrational states. As a spin-off from the PVD experiments, information about the surface chemistry between Si and air (N_2 and O_2) has been obtained. After removal of the natural oxide on Si, N_2 reacts with the wire and finally forms gaseous Si_2N, which is detected as its parent ion upon KrF laser ionization. The evolution of Si_2N shows a temperature hysteresis that can be rationalized in terms of the presence or absence of oxide on the Si wire.

Looking at the CVD system with SiH_4 pyrolysis, the complementary information obtained with TOF-MS by selective ArF or KrF laser ionization together with their temperature and pressure dependences, as well as the deposition rates on the substrate and the Si wire have allowed us to establish a consistent reaction model. A large part of the SiH_4 decomposes on the wire according to $SiH_4 \rightarrow Si\downarrow + 2H_2\uparrow$ to form c-Si. A small part of the SiH_4 decomposes to SiH_2 and H_2, with half of the SiH_2 reacting with abundant SiH_4 to yield Si_2H_6 and the other half hitting the substrate surface to form a-Si. Under the applied conditions, the highly reactive SiH_2 species contributes to the a-Si film formation.

The *in situ* measurement of internal mechanical stress in the a-Si:H thin films obtained by plasma CVD from SiH_4 reveals that He as buffer gas induces compressive stress while Ar leads to tensile stress. Following this result, optimization of the buffer gas used in plasma CVD is recommended to obtain strongly adherent a-Si:H thin films without, or with a minimum of stress.

3.5 References

[1] W. G. J. H. M. van Sark in *Thin Films and Nanostructures, Vol. 30, Advances in Plasma-Grown Hydrogenated Films,* (Ed.: M.H. Francombe), Academic Press, San Diego, USA, **2002**, pp. 1–215

[2] J. E. Mahan, *Physical Vapor Deposition of Thin Films,* Wiley, New York, USA, **2000**.

[3] M. L. Hitchman, K. F. Jensen (Eds.), *Chemical Vapor Deposition*, Academic Press, London, England, **1993**.

[4] H. Stafast, F. Falk, E. Witkowicz, *Electrochem. Soc. Proc.* **2001**, 191–198.

[5] G. Andrä, F. Falk, H. Stafast, E. Witkowicz, *Final Report of DFG project STA 465/1-1,2*, Jena, Germany, **2001**

[6] E. Witkowicz, *PhD thesis*, University of Jena, Germany, **2001**.

[7] Landolt-Börnstein, Band 1, *Atom- und Molekularphysik*, Springer, Berlin, **1950**, 6. Edition, p. 117

[8] P. Ho, W. Breiland, *Appl. Phys. Lett.* **1984**, *44*, 51–53.

[9] H. Stafast, *Appl. Phys.* **1988**, *A45*, 93–102.

[10] G. Andrä, F. Falk, H. Stafast, E. Witkowicz, *Monatsh. Chem.* **1999**, *130*, 221–225.

[11] N. Waltenburg, J. T. Yates, *Chem. Rev.* **1995**, *95*, 1589–1673.

[12] Gmelin Handbook, *Silicon and Nitrogen*, B4 Si Suppl.: 1, **1989**

[13] A. G. Schrott, S. C. Fain, *Surf. Science* **1981**, *111*, 39–52.

[14] K. A. Gingerich, R. Viswanathan, R. W. Schmude, *J. Chem. Phys.* **1997**, *106*, 6016–6019.

[15] D. J. Brugh, M. D. Morse, *Chem. Phys. Lett.* **1997**, *267*, 370–376.

[16] M. D. Wiggins, R. J. Baird, P. Wynblatt, *J. Vac. Sci. Technol.* **1981**, *18*, 965–970.

[17] R. E. Honig, *J. Chem. Phys.* **1954**, *22*, 1610.

[18] H. Tanaka, T. Kanayama, *J. Vac. Sci. Technol.* **1997**, *B 15*, 1613–1617.

4 The Gas-Phase Oxidation of Silyl Radicals by Molecular Oxygen: Kinetics and Mechanisms

T. Köcher, C. Kerst, G. Friedrichs, and F. Temps*

4.1 Introduction

Silane, SiH_4, and disilane, Si_2H_6, are among the most common precursors for chemical vapor deposition (CVD) processes, which have found widespread application for the production of Si, Si:H, or SiO_2 thin films for microelectronic and photovoltaic devices, the synthesis of Si-containing nanoparticles, or the protection of metal surfaces by inert layers. The deposition processes can be initiated by electric discharges in plasma sources, pyrolysis, or photochemically by laser-induced reactions. The final products are usually formed through very rich reaction sequences. As illustrated schematically in Figure 4.1, many elementary reaction steps are involved in the gas phase (free radical reactions, ion-molecule reactions), at the gas-surface interface (adsorption, desorption), and on the surface (surface decomposition, hopping, lattice growth, annealing). There is great interest in the development of detailed reaction mechanisms and kinetic models which can be used to describe the overall processes based on quantitative experimental data for the important elementary reactions.[1, 2] Because of the widely differing environments and time scales, and in view of the different experimental techniques for laboratory diagnostics studies, it is useful to divide the mechanisms into subsets describing the gas-phase, gas-surface, and surface chemistries.

The gas-phase chemistry of Si-containing radicals forms one important sub-mechanism which is attracting considerable attention.[3] The two most important Si radicals generated in either discharge, pyrolysis, or photochemical CVD sources are silylene, SiH_2, and silyl, SiH_3.

SiH_2 is a highly reactive species which may undergo fast insertion reactions into σ bonds and addition reactions to π bonds. The available work on the elementary reactions of SiH_2 up to 1994 has been reviewed in detail by Jasinski et al.[3] Because of its intense and well-resolved $^1B_1 \leftarrow {}^1A_1$ electronic absorption in the visible spectrum,[4] SiH_2 is easily detected with high sensitivity by laser-induced fluorescence,[5] laser resonance absorption flash kinetic spectroscopy,[6, 7] or cavity ring-down spectroscopy.[8] These techniques

---

* Address of the authors: Institut für Physikalische Chemie, Christian-Albrechts-Universität zu Kiel, Olshausenstr. 40, D-24098 Kiel, Germany; Fax: +49-(0)431-880-1704; E-mail: temps@phc.uni-kiel.de

have been employed for detailed kinetic investigations.[3] Other detection methods for SiH_2 are infrared diode laser absorption [9] and mass spectrometry.[10–12]

SiH_3 is expected to be present in gas-phase systems in high concentrations because of its relatively long lifetime.[1, 2] However, despite significant efforts,[3] many aspects of its chemistry still provide considerable challenges. This is true in particular regarding the products and mechanisms of the ensuing elementary reactions.

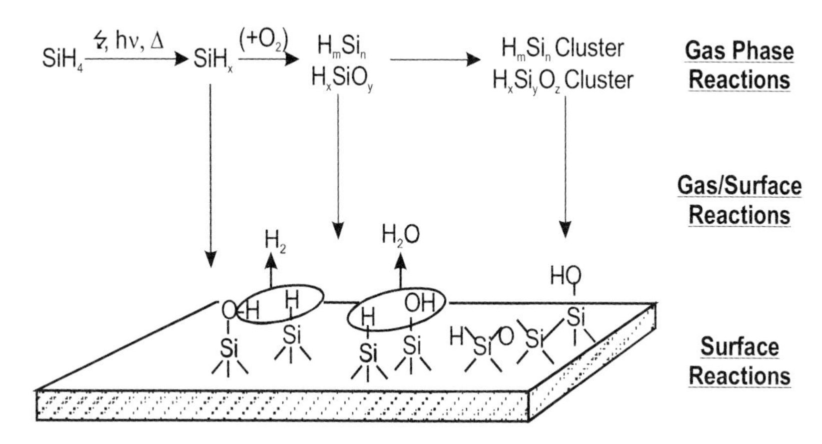

Figure 4.1. Schematic diagram of the different sub-mechanisms involved in the CVD of Si, Si:H, or SiO_2 layers.

This article is concerned with investigations of the kinetics, products, and mechanism of the oxidation of SiH_3 radicals by molecular oxygen,

$$SiH_3 + O_2 \rightarrow \text{products} \quad (1)$$

Reaction (1) is the most important step in the sequence of chain reactions in CVD processes which lead to the conversion of SiH_4 to SiO_2 layers.[3, 13] It is also critically important for the determination of the flammability and explosion limits and induction times of SiH_4-containing gas mixtures, issues of utmost concern for industrial applications.[14–17] The key question is whether the reaction products can be explained by the proposed channels [16, 17]

$$SiH_3 + O_2 \rightarrow SiH_2O + OH \quad (1a)$$
$$\rightarrow SiH_2O_2 + H \quad (1b)$$
$$\rightarrow SiH_3O + O \quad (1c)$$

or whether other products (e.g., SiO) arising from subsequent fragmentation steps have to be taken into account.

In view of the limited space, we focus on research on the title reaction conducted in our laboratory during the last few years. We describe measurements of the rate constant and Si-containing products of reaction (1) in the gas phase at pressures of a few mbar in the temperature range 298 K $\leq T \leq$ 890 K.[18] Results are presented in an exemplary fashion; the reader is referred to the original work[18] for details and to the review by Jasinski et al.[3] for work on other Si_xH_y reactions. After this introduction, we give a brief summary of the methods for generating SiH_3 radicals in laboratory kinetics studies and the available direct detection techniques for SiH_3. Next, we consider results for the mutual combination reaction

$$SiH_3 + SiH_3 \rightarrow \text{products} \tag{2}$$

which almost always occurs in parallel with reaction (1). The following main section concentrates on rate constant measurements and product observations for reaction (1), especially the detection of H_2SiO (silanone) and SiO, and the underlying reaction mechanism. We conclude with a brief outlook at further research directions.

4.2 Generation and Detection of SiH_3 for Kinetic Studies

The known direct sources of SiH_3 such as the photolysis of Si_2H_6, SiH_3Br, or SiH_3I at wavelengths of λ = 193 or 248 nm,[19] lead to serious complications in kinetic studies due to their low quantum yields for SiH_3 (ϕ = 0.05–0.2) compared with high backgrounds of other radicals (SiH_2, SiHBr, SiHI). Extreme care has to be taken to prevent the formation of such side products because of possible interference due to regeneration or fast consumption of SiH_3 by other products. Therefore, virtually all direct studies of SiH_3 reactions rely on H atom abstraction from SiH_4 by Cl atoms,

$$Cl + SiH_4 \rightarrow HCl + SiH_3 \tag{3}$$

Reaction (3) proceeds with a rate constant close to the gas kinetic collision frequency (k_3(298 K) = 2.0×10^{14} cm^3/mol s).[20, 21] The Cl atoms have frequently been obtained from the λ = 193 nm photodissociation of CCl_4 (σ = 8.7×10^{-19} cm^2, ϕ_{Cl} = 1.2).[19, 22] A more efficient Cl atom source is the photolysis of CCl_3F, which has a higher absorption cross-section (σ = 1.36×10^{-18} cm^2).[22] Although no evidence has been found for interfering reactions of the accompanying CCl_3 and CCl_2F radicals with SiH_3,[19, 23] their fate in the reaction systems is largely unknown.

A much cleaner and almost ideal source employed in our laboratory is the $\lambda = 193$ nm photodissociation of oxalyl chloride, $(COCl)_2$, which gives two Cl atoms ($\phi_{Cl} = 2.0$) with a very high absorption cross-section ($\sigma = 3.83 \times 10^{-18}$ cm^2) without other interfering radical products.[24] The reaction

$$SiH_3 + (COCl)_2 \rightarrow \text{products} \qquad (4)$$

has been found to be negligible ($k_4(298\text{ K}) \leq 10^{10}$ cm^3/mol s).[18] H atom abstraction from SiH_4 by F atoms from the $\lambda = 193$ nm photolysis of XeF_2 has also been used.[25] The reaction O (^{1}D) + SiH_4 with O (^{1}D) from the $\lambda = 193$ nm photolysis of N_2O has been found to be inferior due to the fast reaction O (^{1}D) + N_2O and too many side products.[25, 26]

The detection techniques for SiH_3 employed in kinetic studies are infrared laser magnetic resonance,[27] diode laser IR spectroscopy of the ν_2 and ν_3 bands,[28–30] mass spectrometry,[10, 31–32] and electronic $\tilde{A} \leftarrow \tilde{X}$ transient absorption spectroscopy in the 205–250 nm region.[33] Photoionization mass spectrometry, pioneered by Gutman and co-workers,[31] is extraordinarily sensitive, but the readily accessible photon energies are below the ionization limits of many interesting Si-containing reaction products. In view of the lack of spectroscopic information on these species, low energy electron impact ionization mass spectrometry remains the only universally applicable detection method.[10, 18, 32] The adiabatic ionization potentials of SiH_3 and SiH_4 have been reported by Berkowitz et al. (8.01 eV and 12.09 eV, respectively).[10] The detection of SiH_3 can thus be accomplished at a low electron energy (10.2 eV) to avoid fragmentation of the SiH_4.[18] Somewhat higher ionization energies (12–14 eV) are used for the detection of closed-shell reaction products.[34] More insight may be gained from future applications of resonantly enhanced multiphoton ionization.[35]

The experimental set-up for direct kinetic studies of SiH_3 reactions used in the author's laboratory is shown schematically in Figure 4.2. SiH_3 is produced by pulsed excimer laser photolysis of CCl_4, $CFCl_3$, or $(COCl)_2$ in the presence of a large excess of SiH_4 along the axis of a tubular slow-flow reactor and detected through a conical pinhole by time-resolved molecular beam sampling, near-threshold electron-impact ionization mass spectrometry. A detailed description can be found in [18]. Elementary reactions of Si containing radicals can be studied with this set-up using low radical concentrations (SiH_3 detection limit: 3×10^9 cm^{-3}) at pressures of a few mbar and temperatures up to $T = 1200$ K.

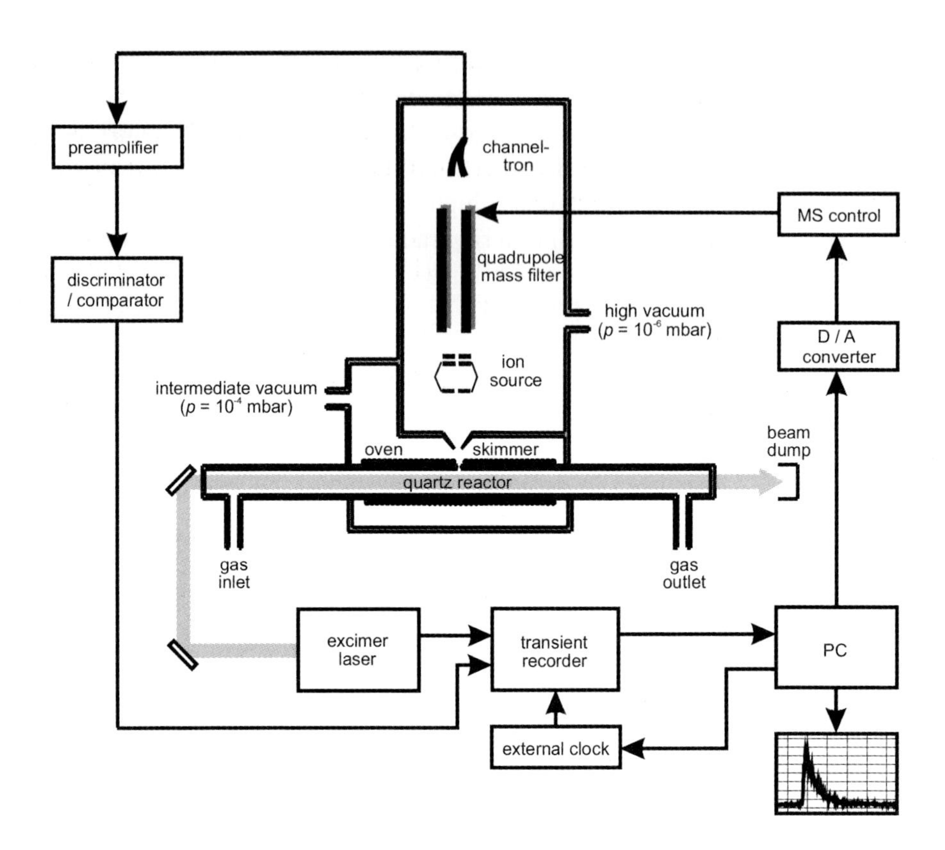

Figure 4.2. Experimental set-up for kinetic studies of SiH_3 reactions by excimer laser photolysis in a tubular slow-flow reactor with time-resolved detection by molecular beam sampling, near-threshold ionization electron impact mass spectrometry.[18]

4.3 The Reaction $SiH_3 + SiH_3 \rightarrow$ Products

The rate constant for the reaction $SiH_3 + SiH_3$ (2) has to be precisely known in order to model SiH_3 concentration profiles in CVD systems and in reaction (1). As a radical-radical reaction, reaction (2) is expected to proceed via formation of a short-lived vibrationally highly excited $Si_2H_6^*$ intermediate. Due to the high excess energy in this "complex", which is higher than the threshold for decomposition to $SiH_2 + SiH_4$ or $H_2 + Si_2H_4$, the primary $Si_2H_6^*$ complex can be collisionally stabilized only at very high pressures ($p >> 1$ bar). In the low-pressure regime, the only direct reaction products are $SiH_2 + SiH_4$ and H_2 + H_3SiSiH (silylsilylene). The alternative 1,2-elimination of H_2 with the slightly more stable H_2SiSiH_2 (disilene) as product requires a significant activation energy, but H_2SiSiH_2 can be produced in a subsequent isomerization of the H_3SiSiH.[36, 37] Formation of Si_2H_5 + H is endothermic and should not play a role.[37, 38] Si_2H_6 is formed as a final product by the insertion of SiH_2 into SiH_4. With the heats of formation of Sax and co-workers,[38] the reaction system may thus be summarized by the equations

$$SiH_3 + SiH_3 \rightarrow Si_2H_6^* \rightarrow Si_2H_5 + H \qquad \Delta_rH^o_{298} = +\ 53\ \text{kJ/mol} \qquad (2a)$$
$$\rightarrow SiH_2 + SiH_4 \qquad \Delta_rH^o_{298} = -\ 93\ \text{kJ/mol} \qquad (2b)$$
$$\rightarrow H_2 + H_3SiSiH \qquad \Delta_rH^o_{298} = -\ 81\ \text{kJ/mol} \qquad (2c)$$
$$\rightarrow H_2 + H_2SiSiH_2 \qquad \Delta_rH^o_{298} = -119\ \text{kJ/mol} \qquad (2d)$$
$$\xrightarrow{+M} Si_2H_6 \qquad \Delta_rH^o_{298} = -320\ \text{kJ/mol} \qquad (2e)$$

followed by

$$SiH_2 + SiH_4 + M \rightarrow Si_2H_6 + M \qquad \Delta_rH^o_{298} = -227\ \text{kJ/mol} \qquad (5)$$

with M as a third body collision partner.

Earlier measurements of the rate constant for reaction (2) gave rather divergent results (9.6×10^{12} cm^3/mol s $\leq k_2 \leq 9.0 \times 10^{13}$ cm^3/mol s).[19, 39–43] The discrepancies may be attributed to uncertainties as to the absolute SiH_3 concentrations, side reactions with other radicals, and influences from the heterogeneous SiH_3 wall loss reaction

$$SiH_3 + \text{wall} \rightarrow \text{products} \qquad (6)$$

which has sometimes been neglected. In flow reactors, the continuous chemical vapor deposition of an amorphous silicon hydride layer on the reactor surface observed after prolonged reaction times gives rise to increasing effective wall rate constants k_6 with increasing SiH_3 concentration, which have to be taken into account.[18]

A typical second-order SiH_3 decay plot due to reaction (2) measured in our laboratory [18] is shown in Figure 4.3a. The SiH_3 radicals were generated according to reaction (3) with Cl atoms from $(COCl)_2$. Their absolute concentration, which is needed for a rate constant determination, was calculated from the laser fluence and the $(COCl)_2$ absorption cross-section and, alternatively, from the calibrated HCl mass signals. The result from a standard second-order analysis (Figure 4.3b) is $k_2(298\ \text{K}) = (7.0 \pm 2.0) \times 10^{13}$ cm^3/mol s. However, due to reaction (6) this value constitutes an upper limit. A reanalysis with k_2 and k_6 as adjustable parameters gave a slightly lower rate constant value, $k_2(298\ \text{K}) = (4.4 \pm 1.5) \times 10^{13}$ cm^3/mol s. The excellent agreement with a measurement published during our investigations by the Krasnoperov group,[44] who monitored SiH_3 by UV absorption and found

$k_2(298\text{ K}) = 5.0 \times 10^{13}\text{ cm}^3/\text{mol s}$, leads us to recommend a value for future kinetic modeling work of $k_2(298\text{ K}) = 5.2 \times 10^{13}\text{ cm}^3/\text{mol s}$.

From the measured absolute Si_2H_6 yields, the branching ratios for the reaction channels to $SiH_2 + SiH_4$ (2b) and $H_2 + H_3SiSiH$ (2c) are $k_{2b}/k_2 = (0.8 \pm 0.1)$ and $k_{2c}/k_2 = (0.2 \pm 0.1)$.[18]

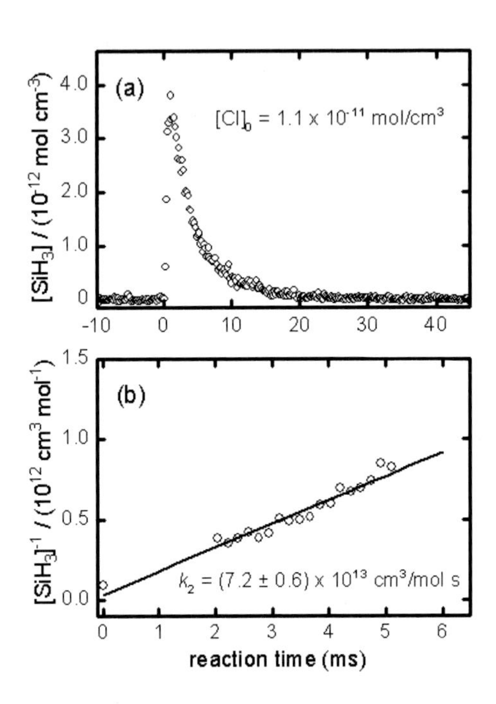

Figure 4.3. Measured second-order $[SiH_3]$ decay plot due to reaction (2). $T = 298$ K, $p = 4.4$ mbar, $[SiH_4] = 8.4 \times 10^{-11}\text{ mol/cm}^3$, $[(COCl)_2] = 8.58 \times 10^{-11}\text{ mol/cm}^3$, $[Cl]_0 = 1.11 \times 10^{-11}\text{ mol/cm}^3$.

4.4 The Reaction $SiH_3 + O_2 \rightarrow$ Products

4.4.1 Overall Reaction Rate Constant

The overall rate constant for reaction (1) can be determined under pseudo first-order conditions by monitoring the decay of the SiH_3 concentration in the presence of a large excess of O_2. The measurements in our laboratory are discussed in some detail in [18], preceding work is described in refs. [30–32, 45–46]. The rate constants are shown in an Arrhenius plot in Figure 4.4. The data points can be described by the rate expression $k_1(T) = (7.7 \pm 2.5) \times 10^{11} \exp[(650 \pm 130)\text{K}/T]\text{ cm}^3/\text{mol s}$ for $298\text{ K} \leq T \leq 690$ K.

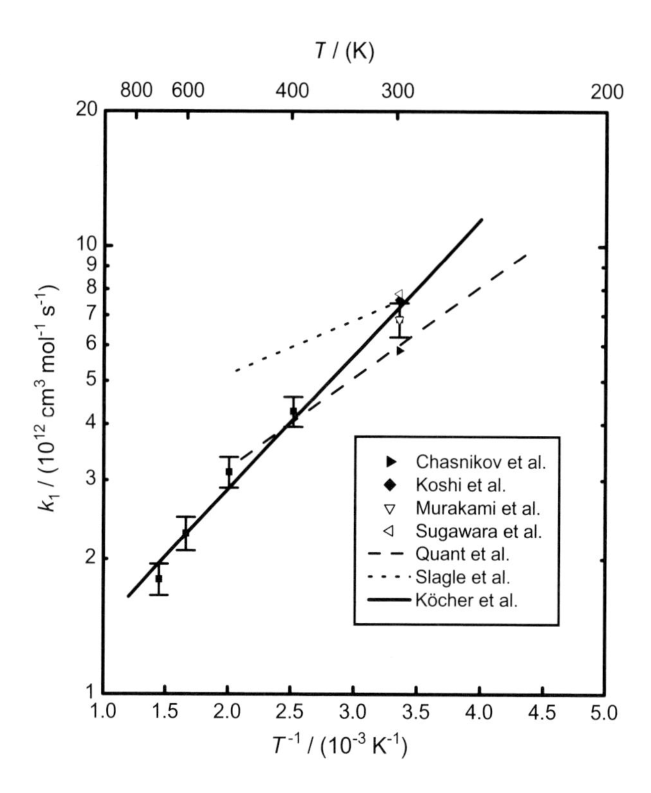

Figure 4.4. Arrhenius plot of the rate constant for reaction (1).

4.4.2 Reaction Products

The formation of Si containing reaction products has been investigated by mass spectrometry over a range of O_2 concentrations from room temperature up to $T = 890$ K.[18] Measured time profiles are displayed in Figure 4.5.

The interpretation of the results is complicated by the coincidences of the mass signals of Si_2H_6/SiH_2O_2, $Si_2H_5/SiHO_2$, and Si_2H_4/SiO_2 and by possible fragmentation steps in the ionization. The solid lines refer to kinetic simulations of the concentration profiles (SiH_3, HCl, Si_2H_6,) and multi-exponential fits to the mass signals (Si_2H_5, Si_2H_4, SiH_2O, SiHO, SiO), respectively. As can be seen from Figure 4.5, the Si_2H_6 is nicely described by the kinetic simulations. The ratios of the m/z = 62, 61, and 60 signals are consistent with the fragmentation pattern in the mass spectrum of Si_2H_6. Under slightly different conditions, evidence could also be observed for a production of SiH_2O_2 and SiO_2. However, these products could not be quantified.

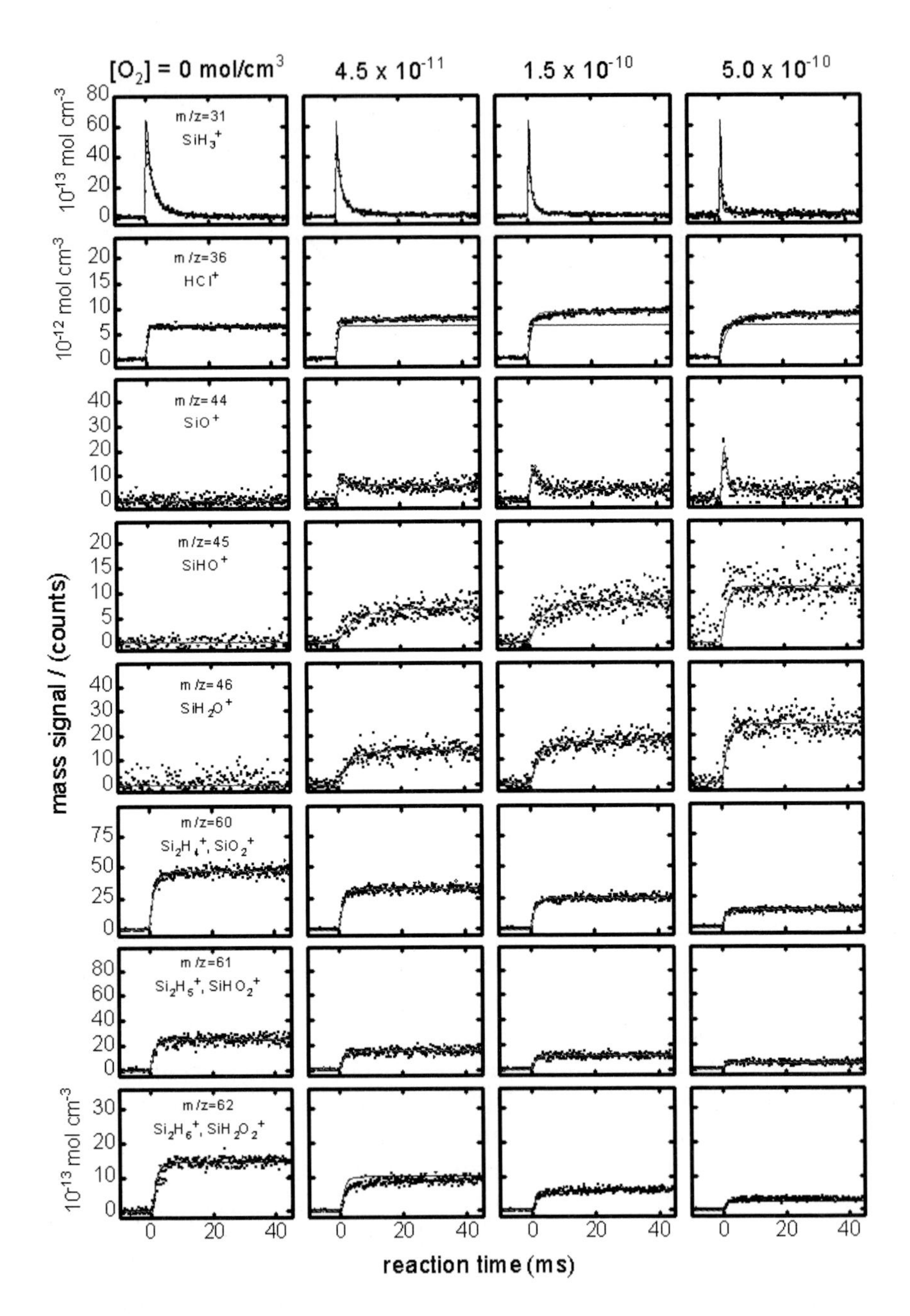

Figure 4.5. Measured time profiles of different Si-containing product species of reaction (1) at different $[O_2]$ concentrations; $T = 298$ K, $p = 3.9$ mbar, $[SiH_4] = 6 \times 10^{-11}$ mol/cm^3, $[CCl_3F] = 2.27 \times 10^{-10}$ mol/cm^3, $[Cl]_0 = 2.5 \times 10^{-12}$ mol/cm^3.

The mass signals at *m*/*z* = 46 (SiH_2O^+), 45 ($SiHO^+$), and 44 (SiO^+) are clearly due to products of reaction (1). At the low ionization energies used, fragmentation processes were minimal. The time profiles and the temperature dependencies of the *m*/*z* = 46 and 44 signals show that they belong to different species (SiH_2O and SiO, respectively), while the *m*/*z* = 45 signal could arise from fragmentation of SiH_2O. Based on unimolecular rate calculations (see below), the SiH_2O^+ signal can be attributed mainly to H_2SiO (silanone) rather than HSiOH (hydroxysilylene).

Temperature-dependent measurements showed a rapid decrease in the SiH_2O yield and a strong increase in the SiO yield with increasing temperature.[18] At $T > 650$ K, SiO was found to dominate. Further, the Si_2H_6, SiHO, SiH_2O, and SiO concentrations were observed to decay on time scales of several ms.

4.4.2 Reaction Mechanism

The radical-radical reaction (1) is expected to proceed via formation of a short-lived vibrationally highly energized SiH_3O_2* intermediate. The products are determined by the competition of the possible unimolecular isomerization and fragmentation steps and collisional stabilization.[47 – 49] The product yields may be rationalized in terms of unimolecular rate theory.[18, 50]

Figure 6 shows an energy diagram with the main minima and transition states on the potential energy hypersurface.[51–55] The reaction can follow two pathways, which proceed with comparable rates.[18] The first (Figure 4.6a) leads to an isomerization of the SiH_3O_2* to H_2SiOOH (hydroperoxysilyl), which then dissociates to give OH + H_2SiO (silanone). A subsequent isomerization of the H_2SiO to HSiOH (hydroxysilylene) or dissociation to SiO + H_2 can only occur surmounting high potential energy barriers.[55] Unimolecular rate calculations [18, 50] have shown that this is possible only if the OH is translationally and rovibrationally cold, i.e. the excess energy is concentrated in the H_2SiO fragment. In this case, the vibrationally excited H_2SiO isomerizes with a rate constant of $\approx 10^{11}$ s^{-1}, compared with an effective collisional stabilization rate of $\omega \approx 10^9$ s^{-1} (M = He, p = 1 bar). However, a sizable amount of excess energy in the dissociation of the H_2SiOOH should be partitioned into translation and rotation. Furthermore, the OH has been found to be vibrationally excited [$OH(v = 1)/OH(v = 0) = 0.49$], and some OH may be in states with $v > 1$.[56] In this case, the H_2SiO isomerization and decomposition become energetically forbidden. Thus, at room temperature, only a small fraction of the H_2SiO can isomerize and decompose. At higher temperatures, the internal energy distribution of the "chemically activated" SiH_3O_2* is much broader and it shifts to higher energies, so that the yield of H_2SiO is expected to drop, whereas the production of SiO should increase. These trends have been verified by experiments.[18]

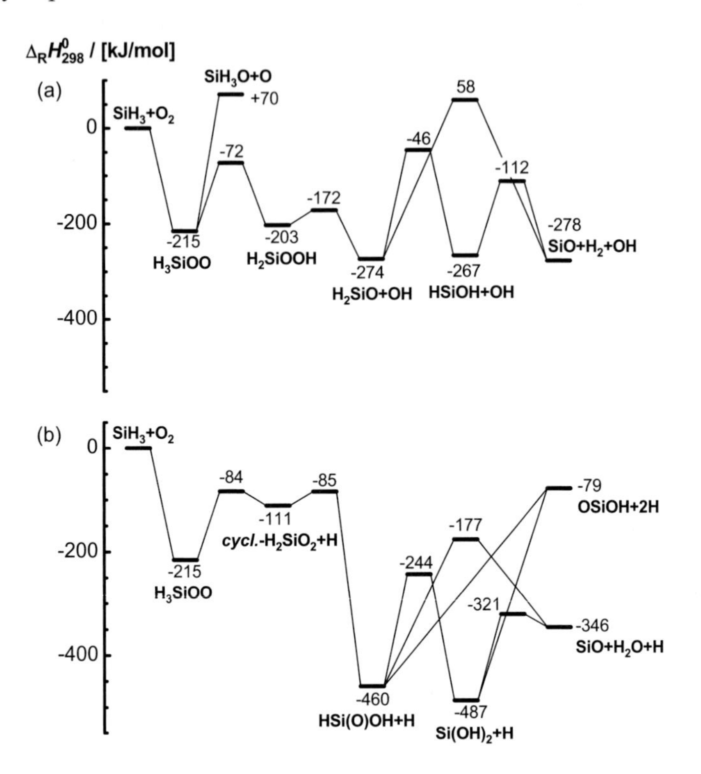

Figure 4.6. Energy diagram for the intermediates and transition states in reaction (1). See text for details.

The second pathway (Figure 4.6b) involves a cyclic H_2SiO_2 intermediate formed by H atom elimination. Subsequently formed intermediate structures are HSi(O)OH or $Si(OH)_2$, which may be collisionally stabilized if a large amount of the excess energy is partitioned to the H atoms. The energy barriers for decomposition to SiO + H_2O are, however, relatively low. Thus, this pathway is suggested to account for SiO in the room temperature experiments.

Direct bond fission in the SiH_3O_2* complex to give SiH_3O + O is endothermic. The reported mass spectrometric detection of SiH_3O [32] has to be due to some fragmentation of intermediate species. The chain branching role of the SiH_3O + O channel assumed to explain the SiH_4/O_2 explosion limits [16, 17] also appears questionable. H atoms and OH radicals from pathways (a) and (b) have been detected by laser induced-fluorescence in the VUV.[56] The reported relative ratio is OH/H = 0.38, but it is difficult to rule out contributions to the H and OH formation from secondary reactions.

4.4.4 Consecutive Reactions

Both H_2SiO and SiO are reactive species. H_2SiO decomposes to SiO + H_2 at high temperatures and it can react with H_2O.[18] The latter reaction was predicted by quantum chemical calculations.[55] The observed concentration and time dependences are consistent with the formation of an H_2SiO-H_2O complex that can reach an equilibrium value (the predicted stabilization energy is 88 kJ/mol [55]) or with a presence of two chemically different SiH_2O species in the system (H_2SiO as the main product and HSiOH as the minor product). SiO produced in reaction (1) was observed to decay with a rate that seems to depend on the O_2 concentration.[18] It is possible that this decay is due to a reaction of a precursor of the SiO with O_2. In addition, a fast dimerization of SiO has been predicted,[52] and it appears reasonable to assume a reaction with O_2 to form SiO_2 which eventually leads to the observed SiO_2 surface deposition. Attempts to detect larger $Si_xH_yO_z$ species have so far remained unsuccessful, but the formation of SiO_2 nanoparticles could be verified by Fourier transform infrared measurements and light-scattering experiments.[57]

4.5 Conclusions and Further Directions

The oxidation of SiH_3 by O_2 (1) proceeds by the way of fast radical-radical reactions with a rich variety of $Si_xH_yO_z$ intermediates and products. Many products can be stabilized only at high pressures, in the liquid phase, or in matrices. At low temperatures in the gas phase, H_2SiO (silanone) appears to be an important product. At high temperatures, the reaction leads mainly to SiO + H_2 + OH and SiO + H_2O + H. The consecutive reactions of the H atoms and OH radicals in the system still deserve further investigations. Interesting future research directions include the formation of $Si_xH_yO_z$ clusters and nanoparticles and the gas/surface deposition reactions of, e.g., SiO and SiO_2. That larger clusters can be detected has been shown by mass spectrometric observations of $Si_xH_yN_z$ species in the NH_2 + SiH_4 reaction system.[25]

4.6 References

[1] J. M. Jasinski, B. S. Meyerson, B. A. Scott, *Annu. Rev. Phys. Chem.* **1987**, *38*, 109.
[2] J. M. Jasinski, S. M. Gates, *Acc. Chem. Res.* **1991**, *24*, 9.
[3] J. M. Jasinski, R. Becerra, R. Walsh, *Chem. Rev.* **1995**, *95*, 1203.

[4] G. Herzberg, J. W. C. Johns, *Proc. Roy. Soc. London* **1966**, *295*, 107; I. Dubois, G. Herzberg, R. D. Verma, *J. Chem. Phys.* **1967**, *47*, 4262; I. Dubois, *Can. J. Phys.* **1968**, *46*, 2485.
[5] G. Inoue, M. Suzuki, *Chem. Phys. Lett.* **1985**, *122*, 361.
[6] J. M. Jasinski, *J. Phys. Chem.* **1986**, *90*, 555.
[7] J. E. Baggott, H. M. Frey, K. D. King, P. D. Lightfoot, R. Walsh, I. M. Watts, *J. Phys. Chem.* **1988**, *92*, 4025.
[8] M. Fikri, Y. Guo, G. Friedrichs, F. Temps, to be published.
[9] C. Yamada, H. Kanamori, E. Hirota, N. Nishiwaki, N. Itabushi, K. Kato, T. Goto, *J. Chem. Phys.* **1989**, *91* 5037.
[10] J. Berkowitz, J. P. Greene, H. Cho, B. Ruscic, *J. Chem. Phys.* **1987**, *86*, 1235.
[11] R. Robertson, D. Hils, H. Chatham, A. Gallagher, *Appl. Phys. Lett.* **1983**, *43*, 544.
[12] R. Robertson, M. Rossi, *J. Chem. Phys.* **1989**, *91*, 5037.
[13] S. Koda, *Progr. Energy Combust. Sci.* **1992**, *18*, 513.
[14] G. McLain, C. J. Jachimowski, R. C. Rogers, *NASA Report TP-2114*, 1983.
[15] S. Koda, O. Fujiwara, *Proc. Combust. Inst.* **1986**, *21*, 1861.
[16] J. R. Hartman, J. Famil-Ghiriha, M. A. Ring, H. E. O'Neal, *Comb. Flame* **1987**, *68*, 43.
[17] J. A. Britten, J. Tong, C. K. Westbrook, *Proc. Combust. Inst.* **1990**, *23*, 195.
[18] T. Köcher, C. Kerst, A. Benscura, G. Eshchenko, J. Roggenbuck, G. Friedrichs, F. Temps, *Z. Phys. Chem.*, to be published.
[19] S. K. Loh, J. M. Jasinski, *J. Chem. Phys.* **1991**, *95*, 4914.
[20] L. Ding, P. Marshall, *J. Phys. Chem.* **1992**, *96*, 2197.
[21] K. G. Kambanis, Y. G. Lazarou, P. Papagiannakopoulos, *J. Chem. Soc. Faraday Trans.* **1996**, *92*, 3299.
[22] W. B. DeMore et al., Chemical Kinetics and Photochemical Data for Use in Stratospheric Modeling, NASA-JPL Publication 97–4, Pasadena, 1997.
[23] H. Niki, P.D. Maker, C.M. Savage, L.P. Breitenbach, *J. Phys. Chem.* **1985**, *89*, 1752.
[24] V. Baklanov, L. N. Krasnoperov, *J. Phys. Chem. A* **2001**, *105*, 97.
[25] T. Köcher, *Dissertation*, Christian-Albrechts-Universität, Kiel, 2002.
[26] K. Okuda, K. Yunoki, T. Oguchi, Y. Murakami, A. Tezaki, M. Koshi, H. Matsui, *J. Phys. Chem. A* **1997**, *101*, 2366; A. Takahara, A. Tezaki, H. Matsui, *J. Phys. Chem. A* **1999**, *103*, 11315.
[27] L. N. Krasnoperov, V. R. Braun, V. V. Nosov, V. N. Panfilov, *Kinet. Kataliz.* **1981**, *22*, 1338.
[28] C. Yamada, E. Hirota, *Phys. Rev. Lett.* **1986**, *56*, 923.
[29] Y. Sumiyoshi, K. Tanaka, T. Tanaka, *Appl. Surf. Sci.* **1994**, *79/80*, 471.

[30] R. W. Quandt, J. F. Hershberger, *Chem. Phys. Lett.* **1993**, *206*, 355.
[31] R. Slagle, J. R. Bernhardt, D. Gutman, *Chem. Phys. Lett.* **1988**, *149*, 180.
[32] M. Koshi, A. Miyoshi, H. Matsui, *J. Phys. Chem.* **1991**, *95*, 9869.
[33] P. D. Lightfoot, R. Becerra, A. A. Jemi-Alade, R. Lesclaux, *Chem. Phys. Lett.* **1991**, *180*, 441.
[34] B. Ruscic, J. Berkowitz, *J. Chem. Phys.* **1991**, *95*, 2416.
[35] R. D. Johnson, B. P. Tsai, J. W. Hudgens, *J. Chem. Phys.* **1989**, *91*, 3340.
[36] R. Becerra, R. Walsh, *J. Phys. Chem.* **1987**, *91*, 5765.
[37] K. Tonokura, T. Murasaki, M. Koshi, *J. Phys. Chem. B* **2002**, *106*, 555.
[38] G. Katzer, M. C. Ernst, A. F. Sax, J. Kalcher, *J. Phys. Chem. A* **1997**, *101*, 3942.
[39] N. Itabashi, K. Kato, N. Nishiwaki, T. Goto, C. Yamada, E. Hirota, *Jpn. J. Appl. Phys.* **1989**, *28*, L325.
[40] S.K. Loh, D.B. Beach, J. M. Jasinski, *Chem. Phys. Lett*, **1990**, *169*, 55.
[41] M. Koshi, A. Miyoshi, H. Matsui, *Chem. Phys. Lett.* **1991**, *184*, 442.
[42] V. Baklanov, A. T. Chichinin, *Chem. Phys.* **1994**, *181*, 119.
[43] K. Matsumoto, M. Koshi, K. Okawa, H. Matsui, *J. Phys. Chem.* **1995**, *100*, 8796.
[44] V. Baklanov, L. N. Krasnoperov, *J. Phys. Chem. A* **2001**, *105*, 4917.
[45] S. A. Chasovnikov, L. N. Krasnoperov, *Chim. Fiz.* **1987**, *6*, 956.
[46] K. Sugawara, T. Nakanaga, H. Takeo, C. Matsumura, *Chem. Phys. Lett.* **1989**, *157*, 309.
[47] J. Troe, *J. Phys. Chem.* **1979**, *83*, 114.
[48] R. G. Gilbert, S. C. Smith, *Theory of Unimolecular and Recombination Reactions*, Blackwell Scientific, Oxford, 1990.
[49] K. A. Holbrook, M. J. Pilling, S. H. Robertson, *Unimolecular Reactions*, Wiley, Chichester, 1996.
[50] R. Roggenbuck, *Dissertation*, Christian-Albrechts-Universität, Kiel, 2000.
[51] L. Darling, H. B. Schlegel, *J. Phys. Chem.* **1994**, *98*, 8910.
[52] M. R. Zachariah, W. Tsang, *J. Phys. Chem.* **1995**, *99*, 5308.
[53] Y. Murakami, M. Koshi, H. Matsui, K. Kamiya, H. Umeyama, *J. Phys. Chem.* **1996**, *100*, 17501.
[54] S. Kondo, K. Tokuhashi, H. Nagai, A. Takahashi, M. Kaise, M. Sugie, M. Aioyagi, K. Mogi, S. Minamino, *J. Phys. Chem. A* **1997**, *101*, 6015.
[55] T. Kudo, S. Nagase, *J. Phys. Chem.* **1984**, *88*, 2833; S. Nagase, T. Kudo, T. Akasaka, W. Ando, *Chem. Phys. Lett.* **1989**, *163*, 23.
[56] M. Koshi, N. Nishida, Y. Murakami, H. Matsui, *J. Phys. Chem.* **1993**, *97*, 4473.
[57] S. Meyer, F. Temps, unpublished results.

5 Oxidation of Matrix-Isolated Silylenes

Wolfram Sander*, Holger F. Bettinger, Holger Bornemann, Martin Trommer, and Magdalene Zielinski

5.1 Introduction

Silylenes **1** and silenes **2** are generally highly reactive species that can only be isolated under normal laboratory conditions if protected by bulky substituents.[1–9] But even then these molecules rapidly react with molecular oxygen at room temperature and thus have to be stored under an inert gas atmosphere. The oxidation often leads to complex product mixtures or to complete combustion. At low temperatures, on the other hand, the reaction becomes slow and is difficult to follow. Thus, only a few studies on the mechanisms of the oxidation of stable silenes and silylenes have been published.

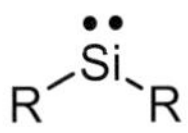

An alternative method of isolating **1** or **2** without sacrificing their reactivity is the matrix-isolation technique. In a low-temperature inert gas matrix, reactive molecules are immobilized, and thus bimolecular reactions are inhibited. In addition, the low temperatures prevent reactions with activation barriers larger than a few kcal/mol. In most of our experiments, argon matrices at 10 K have been used. Under these conditions, the diffusion of even small molecules like CO or O_2 is effectively suppressed. Warming the argon matrix from 10 K to temperatures above 30 K allows small molecules to slowly diffuse. Under these conditions, bimolecular reactions are observed, if the activation barrier is small enough. Thus, reactions of matrix-isolated reactive species such as silenes and silylenes can be effectively controlled by variation of the matrix temperature.

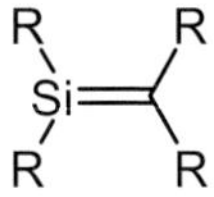

All the experiments described in this chapter consist of several steps. The first step is the matrix isolation of a precursor molecule in a large excess of argon (typical matrix ratios are >1000:1). The next step is the photochemical generation and spectroscopic characterization of the reactive intermediate. To

* Address of the authors: Lehrstuhl für Organische Chemie II der Ruhr-Universität, Universitätsstr. 150, D-44780 Bochum, Germany

prevent secondary photolyses of the intermediates, monochromatic light is usually used. This step is followed by a controlled warming of the matrix from 10 K to 32–40 K to allow the diffusion of molecular oxygen to induce bimolecular reactions. The oxidation can be continuously monitored by UV/vis or IR spectroscopy. In additional steps, the photochemistry of the oxidation products is investigated by selective irradiation. In many cases, this sequence of reaction steps allows one to cleanly synthesize and characterize a number of different intermediates in a single experiment and to gain a detailed insight into the reaction mechanisms. A prerequisite for a bimolecular reaction to occur rapidly at temperatures below 40 K is a low activation barrier, typically less than 2 kcal/mol. Thus, the oxidation of less reactive silicon species cannot be observed in cryogenic matrices.

By far the most important spectroscopic method for this purpose is IR spectroscopy. In combination with DFT or *ab initio* calculations matrix IR spectroscopy has become a very powerful tool for the reliable identification of reactive and unusual molecules. In addition, isotopic labeling with ^{18}O is frequently used to assign the IR spectra of oxidized species. However, a prerequisite for this technique is the availability of suitable photochemical or thermal precursor molecules of the reactive silicon species. During the last years, we have published details of the oxidation mechanism of alkyl-substituted silenes **2**.[10–15] In this chapter, our mechanistic studies on the oxidation of silylenes **1** using the matrix-isolation technique are summarized.

5.2 Oxidation of Silylenes

As primary products of the reaction of silylenes with molecular oxygen the formation of either silanone *O*-oxides **3** or the dioxasiliranes **4** can be expected. The homologous carbenes react exclusively to give carbonyl *O*-oxides **5** which rearrange photochemically to dioxiranes **6** (Scheme 5.1).[16, 17]

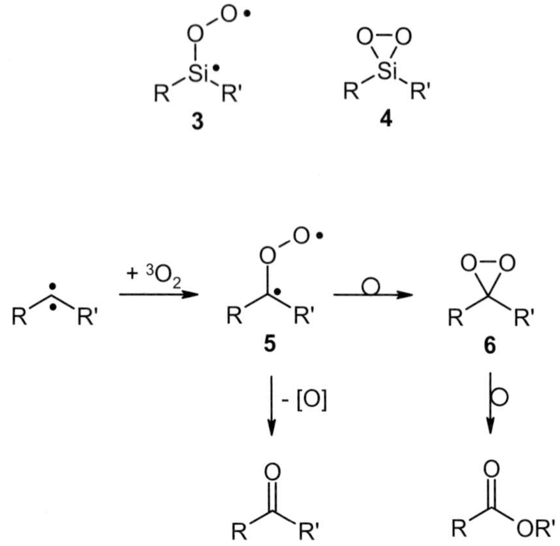

Scheme 5.1.

While many examples of carbene oxidations have been reported, only four papers on the reaction of silylenes **1** with molecular oxygen have been published. The limited number of experimental studies on the oxygenation of silylenes is mainly due to the lack of suitable precursors.[18] The photolysis of matrix-isolated trisilanes produces silylenes in close proximity to disilenes or other products of the precursor decomposition rather than matrix-isolated silylenes.[19–22] Gas-phase thermolysis of disilanes and other thermal precursors[23, 24] requires very high temperatures, while the photolysis of diazidosilanes[25, 26] requires short-wavelength UV irradiation. In all of these cases, the yields of silylenes are rather poor.

Ando et al. investigated the oxidation of dimesitylsilylene **1b** in solid oxygen at 16 K.[19, 27] During UV photolysis of the precursor of **1b** (the corresponding trisilane), a new IR absorption at 1084 cm^{-1} was observed, which was claimed to be due to the O-O stretching vibration of dimesitylsilanone *O*-oxide **3b** (Scheme 5.2). The assignment of the 1084 cm^{-1} vibration was based on comparison with RHF/6-31G(d) calculations for the triplet state of the parent silanone oxide **3a** (no minimum for singlet **3a** was located at this level of theory). The direct thermal reaction of **1b** with $^3O_2$ was not observed, and other IR bands of **3b** were not reported. In later communications a singlet ground state of **3a** and a small activation barrier for the thermal **3a** → **4a** rearrangement were predicted on the basis of MP2 calculations.[27, 28]

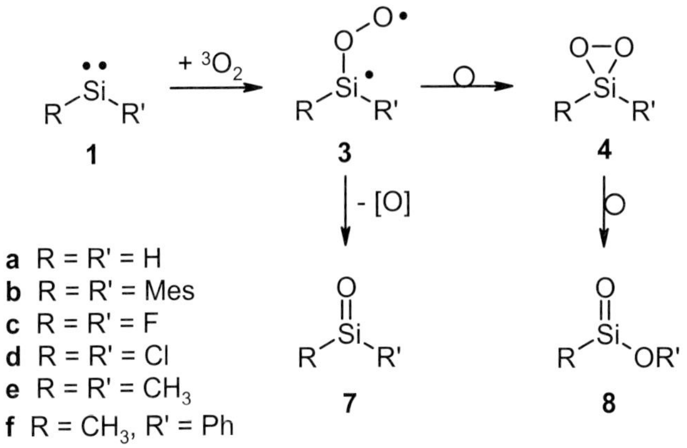

a R = R' = H
b R = R' = Mes
c R = R' = F
d R = R' = Cl
e R = R' = CH_3
f R = CH_3, R' = Ph

Scheme 5.2.

In our laboratory, the oxidation of difluoro-,[29] dichloro-,[29] dimethyl-,[28] and methyl(phenyl)silylene[30] **1c**–**f** was investigated in O_2-doped argon matrices (Scheme 5.2). The halogenated silylenes **1c** and **1d** were generated by FVP of the corresponding hexahalogenodisilanes with subsequent trapping of the products in argon or argon/oxygen at 10 K. Interestingly, despite the high affinity of divalent silicon species towards oxygen, no thermal reaction of **1c** or **1d** with O_2 was observed under these conditions. Thus, the silylenes **1c** and **1d** could be matrix-isolated in solid oxygen at 10 K, and even annealing of the oxygen matrix at 40 K did not result in the formation of products. This is in accordance with the observation that singlet carbenes with a large singlet-triplet splitting react exceedingly slowly with $^3O_2$ due to spin restrictions.[31] Irradiation with UV or visible light is required to induce the reaction of the halogenated silylenes with $^3O_2$ and the only products formed are the dioxasiliranes **4c** and **4d**, respectively (Scheme 5.3). Since neither the silylenes nor O_2 absorb in this spectral region, it is most likely that the excitation occurs into a charge-transfer absorption. The identification of **4c** and **d** was based on isotopic labeling studies using $^{16}O_2$, $^{16}O^{18}O$, and $^{18}O_2$. The IR spectra of the labeled compounds show the expected isotopic splittings of the Si-O and O-O stretching vibrations. From the number of isotopomers, it was clearly shown that the two oxygen atoms are equivalent, as required for **4**.[29] The carbonyl oxides **3c** and **3d**, respectively, were not observed. However, since the dioxasiliranes **4c** and **4d** are generated photochemically and oxides **3** are expected to be highly photolabile, this cannot be taken as evidence that silanone oxides **3** are not formed during the course of the reaction.

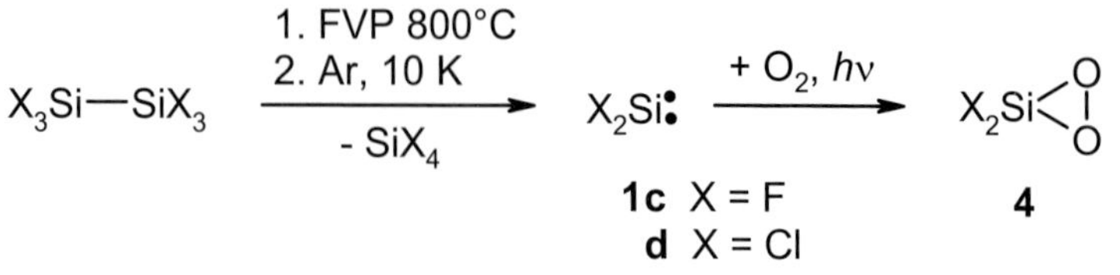

$$X_3Si{-}SiX_3 \xrightarrow[\text{- }SiX_4]{\text{1. FVP 800°C; 2. Ar, 10 K}} X_2Si\colon \xrightarrow{+\,O_2,\,h\nu} X_2Si{<}\begin{matrix}O\\|\\O\end{matrix}$$

1c X = F
d X = Cl

4

Scheme 5.3.

Dimethylsilylene **1e** was found to react thermally in 0.5% O_2-doped argon matrices at temperatures as low as 40 K. This indicates that the activation barrier for the oxidation of this silylene must be very close to zero. The only product is dimethyldioxasilirane **4e**, while dimethylsilanone *O*-oxide **3e**, the proposed precursor of **4e**, is not observed. This is in contrast to carbene oxidations, where carbonyl oxides **5** are the primary thermal products and dioxiranes **6** are formed on secondary photolysis of **5**. Under the assumption that the thermal reaction of silylenes **1** with molecular oxygen leads to silanone oxides **3** as the primary products, it follows that the barrier for the **3** → **4** rearrangement must be considerably smaller than that for the **5** → **6** rearrangement. This is in agreement with MP2 calculations, which predict a barrier of 22.8 kcal/mol for the carbonyl oxide, but only 6.5 kcal/mol for rearrangement of the silanone oxide **3a** (parent compounds, R = R' = H). The large excess energy released by the thermal reaction of **1e** with O_2 to give **3e** cannot be dissipated fast enough, and thus **3e** directly rearranges to **4e,** even under the conditions of matrix-isolation at cryogenic temperatures. In contrast to **4c** and **4d**, dioxasilirane **4e** is photolabile and 400 nm irradiation rapidly results in its rearrangement to methyl(methoxy)silanone **8e**. This is completely analogous to the photochemistry of dioxiranes, which rearrange to esters.[17] The dihalodioxasiliranes **4c** and **d**, on the other hand, are completely stable towards visible and UV irradiation, since a migration of a halogen atom from silicon to oxygen is energetically highly unfavorable.

Although dimethylsilylene **1e** reacts thermally with $^3O_2$, this molecule is not very suitable for the investigation of this reaction. The conditions for the matrix isolation of **1e** are quite harsh, and either short-wavelength UV irradiation or high pyrolysis temperatures (depending on the precursor used) are required for the synthesis. The yields of **1e** are therefore quite low. In addition, the methyl substituents in the proposed intermediate **3e** might not be efficient enough in dissipating the reaction energy to prevent the rearrangement to **4e**. This leads to the question as to whether large substituents, such as the mesityl groups used by Ando et al.,[19] can lead to a stabilization of the silanone *O*-oxides **3**.

We thus developed more efficient precursors for the synthesis and matrix isolation of silylenes **1** under mild conditions. Such a precursor is phenylsilyldiazomethane **9**, which can be easily synthesized by reaction of phenylsilyl trifluoromethanesulfonate with diazomethane.[30] Photolysis in the visible or near UV (400–350 nm) results in the loss of N_2 and formation of the highly labile carbene **10** (Scheme 5.4). Depending on the irradiation conditions, the diazirine **11** is also formed. Carbene **10** rapidly rearranges to silene **12** through a [1,2]-hydrogen migration and is thus not observed in the matrix. According to DFT calculations, only the triplet state of **10** is a minimum on the potential energy surface, while singlet **10** is a transition state directly leading to **12**. The formation of the isomeric silene **13** would require a [1,2]-phenyl migration and it is not observed. Prolonged irradiation under the same conditions ultimately leads to methyl(phenyl)silylene **1f** in excellent yields. This reaction sequence also allows the generation of **1f** in doped argon matrices and thus permits the investigation of the subsequent thermal reactions with small molecules such as O_2, CO, PH_3, etc.

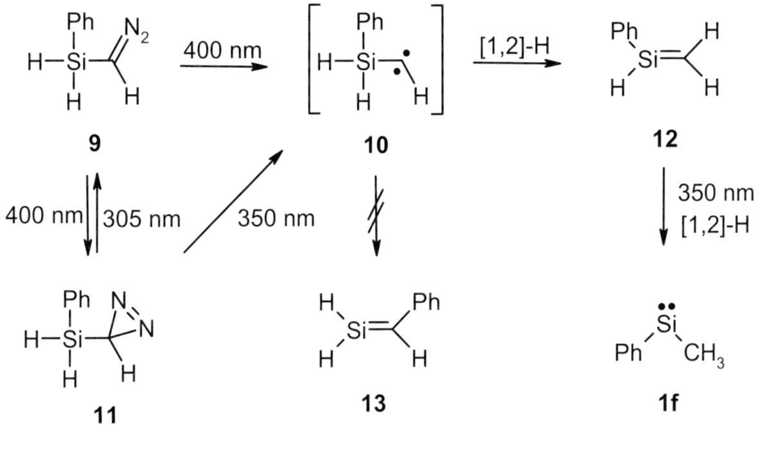

Scheme 5.4.

As with dimethylsilylene **1e**, the reaction of methyl(phenyl)silylene **1f** with O_2 results directly in the formation of the corresponding dioxasilirane **4f**, while silanone oxide **3f** could not be detected (Scheme 5.5). The dioxasilirane **4f** was identified by comparison of the experimental IR spectra with DFT calculations, by isotopic labeling with $^{18}O_2$, and by its subsequent photochemistry. The O-O stretching vibration of **4f** at 577 cm^{-1} shows the expected large isotopic shift on ^{18}O-labeling.[30] In dimethyldioxasilirane **4e,** this vibration is observed at 554 cm^{-1},[28] and in the fluorinated and chlorinated dioxasiliranes **4c** and **4d** at 615 and 576 cm^{-1},[29] respectively. Thus, with the exception of **4c,** the O-O stretching modes are found in a narrow range around 560 cm^{-1}. In dioxiranes **6,** the corresponding absorptions are weaker and less characteristic.[32–34] Another typical vibration of **4**, which exhibits a large isotopic shift, is the Si-O stretching vibration, in the case of **4f**, it is observed at 1002 cm^{-1}.

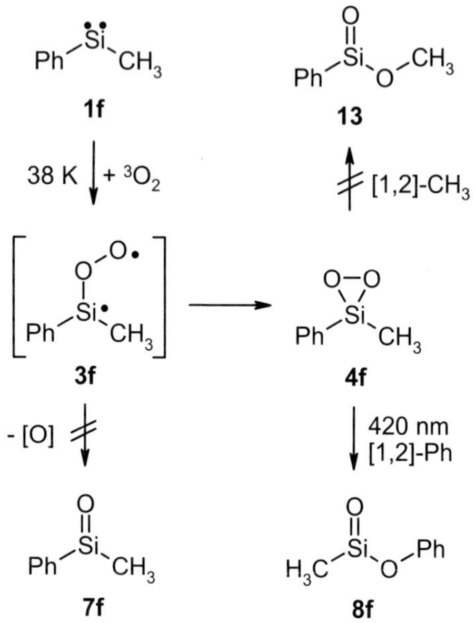

Scheme 5.5.

DFT calculations predict a very small barrier (ca. 1 kcal/mol) for the rearrangement of silanone oxide **3f** to dioxasilirane **4f**.[30] Although *ab initio* calculations at the MP2 level of theory predict a somewhat larger barrier (ca. 6 kcal/mol), it is safe to say that the barrier for the **3** → **4** rearrangement is much smaller than the barrier for the rearrangement of **5** to **6**, while the former rearrangement is considerably more exothermic. This explains why, even under the conditions of matrix isolation, only the siladioxiranes **4**, and not the proposed primary adducts of molecular oxygen and silylenes **1**, the silanone oxides **3**, are observed.

Dioxasilirane **4f** is photolabile, and visible light irradiation (420 nm) rapidly results in its rearrangement to methyl(phenoxy)silanone **8f**, the product of a [1,2]-phenyl migration. The isomeric silanone **13**, which would be produced by a methyl migration, could not be detected in the matrix. Presumably, the rearrangement of **4f** proceeds by rupture of the O-O bond to form a dioxy diradical, which subsequently rearranges. Although **8f** contains an unfavorable Si=O double bond, the rearrangement is estimated by DFT calculations to be exothermic by 62 kcal/mol, much more exothermic than the corresponding rearrangement of dioxiranes **6** to esters.

The question remains as to why, according to a previous report by Ando et al., the reaction of dimesitylsilylene **1b** with molecular oxygen results in the formation of silanone oxide **3b** and not dioxasilirane **4b**.[19] The reaction conditions in these experiments were quite different from ours. First of all, a trisilane was used, which on irradiation with short-wavelength UV light (254 nm) produces the silylene **1b**. An examination of the published IR spectra reveals that the products from this reaction were not matrix-isolated (very broad, overlapping UV absorptions). Moreover, the oxidation was carried out in neat oxygen during UV irradiation, while in our experiments the silylenes react in about 1% O_2-doped Ar matrices under controlled conditions. In our case, this means that in the first step the silylene is generated and spectroscopically characterized and in a second step the oxidation is induced by either warming the matrix from 10 K to 35 K to allow the diffusion of $^3O_2$ (**1e** and **f**) or by visible or long-wavelength UV irradiation into a charge-transfer absorption of the silylene-oxygen complex (**1c** and **d**). Since the silanone oxide **3b** is expected to be not only thermo- but also photolabile, neither **3b** nor the isomeric dioxasilirane **4b** should survive 254 nm irradiation. We thus suggest a re-evaluation of the oxidation of **1b**.

5.3 Computational Studies on Silanone oxides 3 and Dioxasiliranes 4

The parent silanone oxide **3a** has been well investigated by Nagase et al. using perturbation theory up to order four (MP2 to MP4SDTQ), generalized valence bond (GVB), and complete-active space self consistent field (CASSCF) methods.[27] In contrast to the structure of carbonyl oxides (**5**), the structure of **3a** is not planar, but rather it is significantly pyramidalized. The sum of bond angles at the Si atom is 318.7° (close to three times 109.1°) rather than the 360° of a trigonal-planar coordination.[27] The O-O and O-Si distances are 1.306 Å and 1.748 Å, respectively, while the O-O-Si angle is 122.5° (all data are MP2(full)/6-31G(d), see Figure 5.1). As silyl radicals are pyramidal, in contrast to alkyl radicals, the nonplanar structure of **3a** has been ascribed to the silyl radical character of the SiH_2 moiety. Nagase et al. analyzed the CASSCF and GVB wavefunctions and concluded from the spin polarizations and orbital overlaps that **3a** has a markedly greater biradical character than **5a**. This significant biradical character of **3a** makes a description of its electronic structure rather difficult, as noted by Nagase et al., who obtained significantly different results for **3a** from single configuration (HF, MP*n*, and CI) and multiconfiguration treatments.

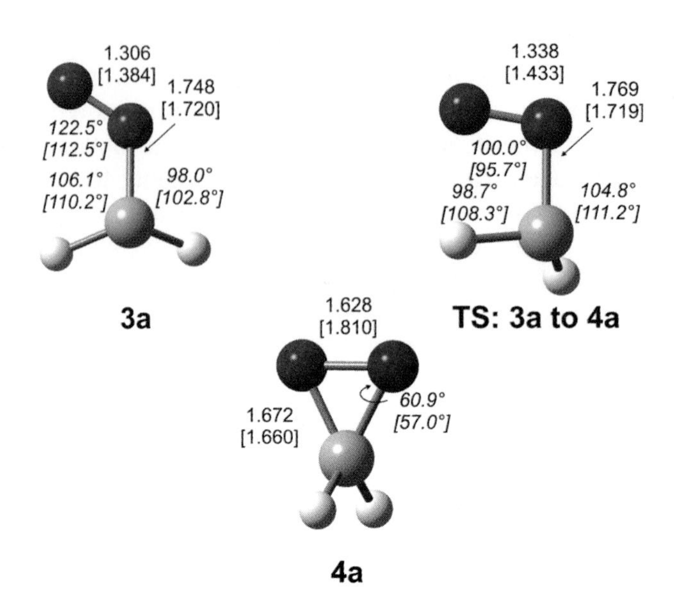

Figure 5.1. Structures of silanone oxide (**3a**), siladioxirane (**4a**) and the transition state for ring closure as computed at the MP2(full)/6-31G(d) and GVB/6-31G(d) (in brackets) levels of theory. All data are taken from reference [27].

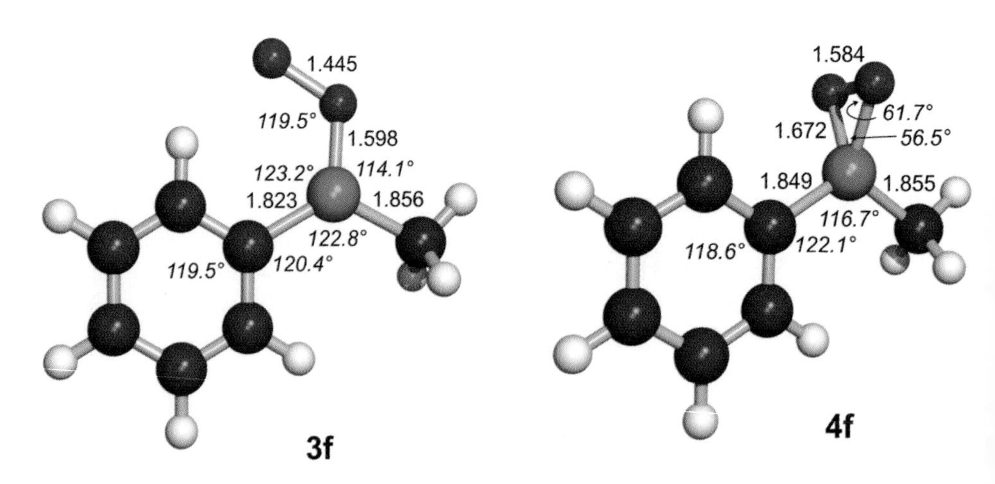

Figure 5.2. Structures of phenylmethylsilanone oxide (**3f**) and the corresponding siladioxirane (**4f**) as computed at the RB3LYP/6-311++G(d,p) level of theory. All data are taken from reference [30].

The ring-closure reaction from **3a** to **4a** was studied independently by Nagase et al. and Patyk et al. in 1989.[27, 28] Both groups reported an

MP2(full)/6-31G(d) barrier of 6.5 kcal/mol. A somewhat lower barrier was obtained by Nagase et al. at MP4/6-31G(d)//MP2/6-31G(d) (5.0 kcal/mol) and at GVB/6-31G(2d,p) and CASSCF/6-31G(d) (2.5 and 2.4 kcal/mol). As mentioned above, a barrier ranging in the order of 2–6 kcal/mol is significantly smaller than that for the **5a** → **6a** rearrangement, and in agreement with the experiments, in which **3** cannot be detect as an intermediate in the oxidation of **1**. The syn isomer of phenylmethylsilanone *O*-oxide **3f** is 0.7 kcal/mol more stable than the *anti* form. The silicon atom in phenylmethylsilanone *O*-oxide **3f** is not pyramidalized at the RB3LYP/6-311++G(d,p) level of theory.[30] Nonetheless, it was found that dipolar resonance structures are not in agreement with charges derived from the natural population analysis at the B3LYP/6-31G(d,p) level and thus that **3f** also has a strong biradical character. Although it is difficult to compare directly structures obtained at different levels of theory, it is obvious from the computed O-O and Si-O distances that the RB3LYP description of **3f** is much more delocalized (1.445 Å and 1.598 Å, respectively) than that of **3a** at MP2 (1.306 Å and 1.748 Å, respectively). As Nagase et al. concluded from their investigation of **3a** that Hartree-Fock and MP2 overestimate the delocalized planar geometry, it remains to be probed whether the structural differences of **3f** and **3a** are due to substituent effects.

As mentioned above, ring closure of **3f** to **4f** has an extremely small barrier of less than 1 kcal/mol at RB3LYP/6-311++G(d,p), in agreement with its elusive nature in the matrix-isolation study. The reactions **3a** → **4a** and **3f** → **4f** are strongly exothermic, by around 60 kcal/mol [MP4SDTQ/6-31G(2d,p)] and 50 kcal/mol, respectively. An interesting feature of **4a** is its rather long O-O bond (1.63 Å at MP2 and 1.81 Å at GVB),[27] which is significantly longer than in dioxiranes (microwave 1.516 Å;[32] computation 1.52–1.55 Å [33]). In contrast, the O-O distance in **4f** is only 1.58 Å, and is thus much closer to the value in the carbon system. Again, it is unclear as to whether the difference in geometry parameters of **3f** and **4f** reflects substituent effects or whether it arises from the differing computational models. In order to better understand the silanone oxide and siladioxirane species, a computational investigation at a uniform and adequate level of theory appears highly desirable.

5.4 Summary

The oxidation of silylenes **1c**–**f** with molecular oxygen has been investigated by means of matrix-isolation spectroscopy and *ab initio* and DFT methods. With all four silylenes, the dioxasiliranes are obtained as the first isolable products. The halogenated silylenes **1c** and **d** require photochemical activation to be oxidized, while the silylenes **1e** and **f** react thermally with $^3O_2$ even in the solid state (O_2-doped argon matrix) at temperatures below 40 K. Thus, halogenation increases the activation barrier for the oxygenation substantially. The thermal reaction of the singlet molecules **1e** and **f** with triplet O_2 to give singlet products requires an intersystem crossing (ISC) during the reaction. However, our experiments give no clue as to at which step the ISC occurs. Although there is no direct evidence, silanone *O*-oxides **3** are most likely the primary products of the addition of $^3O_2$. In this case, oxides **3** might be formed in their triplet states, which subsequently decay to the lower-lying singlet states. Alternatively, the ISC might occur along the reaction path and not at a stationary point.

The silanone oxides **3** formed in these reactions are not observed, even under the conditions of matrix isolation. This is rationalized in terms of a very small activation barrier for the cyclization and the high excess energy released during the primary addition of oxygen to the silylene molecule, which cannot be dissipated to the matrix sufficiently rapidly. The IR spectra of **3** are nicely reproduced by DFT calculations. This suggests that this computational method also reliably reproduces the geometric and electronic properties of **3**. These results suggest that **3** is highly polar, in accordance with the difference in electronegativity between silicon and oxygen, but not as polar as would be expected for a zwitterion. On the other hand, the wavefunction of **3** has a significant diradical-type character. The dioxasiliranes **4** exhibit properties similar to those of the dioxiranes **6**. Photolysis of **4** and **6** results in their rearrangement to "sila-esters" and esters, respectively. So far, the reactivity of **4** in solution has not been explored due to the lack of suitable precursors. The properties of **4** as reagents for the transfer of oxygen atoms, in analogy to the most important reaction of **6**, thus remain to be investigated.

5.5 References

[1] A. G. Brook, S. C. Nyburg, F. Abdesaken, B. Gutekunst, G. Gutekunst, R. K. M. R. Kallury, Y. C. Poon, Y. M. Chang, W. Wong-Ng, *J. Am. Chem. Soc.* **1982**, *104*, 5667.

[2] A. G. Brook, K. M. Baines, *Silenes*, in Advances in Organometallic Chemistry, Academic Press, **1985**.

[3] A. G. Brook, M. A. Brook, *The Chemistry of Silenes*, Academic Press, San Diego, **1996**.

[4] N. Wiberg, G. Preiner, O. Schieda, *Chem. Ber.* **1981**, *114*, 3518.

[5] N. Wiberg, G. Wagner, G. Mueller, *Angew. Chem.* **1985**, *97*, 220.

[6] N. Wiberg, T. Passler, K. Polborn, *J. Organomet. Chem.* **1997**, *531*, 47.

[7] N. Tokitoh, H. Suzuki, R. Okazaki, K. Ogawa, *J. Am. Chem. Soc.* **1993**, *115*, 10428.

[8] M. Denk, R. Lennon, R. Hayashi, R. West, A. V. Belyakov, H. P. Verne, A. Haaland, M. Wagner, N. Metzler, *J. Am. Chem. Soc.* **1994**, *116*, 2691.
[9] R. West, M. Denk, *Pure Appl. Chem.* **1996**, *68*, 785.
[10] W. Sander, M. Trommer, *Chem. Ber.* **1992**, *125*, 2813.
[11] M. Trommer, W. Sander, A. Patyk, *J. Am. Chem. Soc.* **1993**, *115*, 11775.
[12] M. Trommer, W. Sander, C. H. Ottosson, D. Cremer, *Angew. Chem.* **1995**, *107*, 999.
[13] W. Sander, M. Trommer, A. Patyk, *Oxidation of Silenes and Silylenes : Matrix Isolation of Unusual Silicon Species*, Organosilicon Chem. III, N. Auner, J. Weis, Wiley-VCH **1998**, 86–94.
[14] M. Trommer, W. Sander, *Organometallics* **1996**, *15*, 189.
[15] M. Trommer, W. Sander, *Organometallics* **1996**, *15*, 736.
[16] W. Sander, *Angew. Chem.* **1986**, *98*, 255; *Angew. Chem. Int. Ed. Engl.* **1986**, *25*, 254.
[17] W. Sander, *Angew. Chem.* **1990**, *102*, 362; *Angew. Chem. Int. Ed. Engl.* **1990**, *29*, 344.
[18] O. M. Nefedov, M. P. Egorov, A. I. Ioffe, L. G. Menchikov, P. S. Zuev, V. I. Minkin, B. Y. Simkin, M. N. Glukhovtsev, *Pure Appl. Chem.* **1992**, *64*, 266.
[19] T. Akasaka, S. Nagase, A. Yabe, W. Ando, *J. Am. Chem. Soc.* **1988**, *110*, 6270.
[20] H. Vancik, G. Raabe, M. J. Michalczyk, R. West, J. Michl, *J. Am. Chem. Soc.* **1985**, *107*, 4097.
[21] S. G. Bott, P. Marshall, P. E. Wagenseller, Y. Wang, R. T. Conlin, *J. Organomet. Chem.* **1995**, *499*, 11.
[22] T. Miyazawa, S.-y. Koshihara, C. Liu, H. Sakurai, M. Kira, *J. Am. Chem. Soc.* **1999**, *121*, 3651.
[23] H. P. Reisenauer, G. Mihm, G. Maier, *Angew. Chem.* **1982**, *94*, 864; *Angew. Chem. Int. Ed. Engl.* **1982**, *21*, 854.
[24] G. Maier, H. P. Reisenauer, A. Meudt, *Eur. J. Org. Chem.* **1998**, 1285.
[25] G. Raabe, H. Vancik, R. West, J. Michl, *J. Am. Chem. Soc.* **1986**, *108*, 671.
[26] K. M. Welsh, J. Michl, R. West, *J. Am. Chem. Soc.* **1988**, *110*, 6689.
[27] S. Nagase, T. Kudo, T. Akasaka, W. Ando, *Chem. Phys. Lett.* **1989**, *163*, 23.
[28] A. Patyk, W. Sander, J. Gauss, D. Cremer, *Angew. Chem.* **1989**, *101*, 920; *Angew. Chem. Int. Ed. Engl.* **1989**, *28*, 898.
[29] A. Patyk, W. Sander, J. Gauss, D. Cremer, *Chem. Ber.* **1990**, *123*, 89.
[30] H. Bornemann, W. Sander, *J. Am. Chem. Soc.* **2000**, *122*, 6727.
[31] G. A. Ganzer, R. S. Sheridan, M. T. H. Liu, *J. Am. Chem. Soc.* **1986**, *108*, 1517.

[32] R. D. Suenram, F. J. Lovas, *J. Am. Chem. Soc.* **1978**, *100*, 5117.
[33] D. Cremer, T. Schmidt, J. Gauss, T. P. Radhakrishnan, *Angew. Chem.* **1988**, *100*, 431; *Angew. Chem. Int. Ed. Engl.* **1988**, *27*, 427.

6 Short-Lived Intermediates with Double Bonds to Silicon: Synthesis by Flash Vacuum Thermolysis, and Spectroscopic Characterization

H. Beckers*

6.1 Introduction

The synthesis and characterization of compounds with multiple bonds to silicon and its heavier neighbours in the periodic table has attracted enduring interest in recent decades.[1, 2] Above all, remarkable progress has been achieved in the synthesis of isolable compounds whose reactive multiple bonds are protected by bulky groups. On the other hand, the price to be paid when resorting to tailored highly crowded substituents is that the properties of such kinetically stabilized compounds are governed primarily by what the bulky ligands permit rather than the nature of the reactive centers linked by the multiple bond in question.

Recent gas-phase spectroscopic studies, particularly on transient silicon compounds, have established that their ground-state structures differ to some extent from those of highly substituted derivatives.[3] Moreover, in sharp contrast to molecules with stabilized multiple bonds that can be obtained at ambient temperatures, unprotected derivatives may have non-rigid structures owing to a low barrier to rearrangement of the small substituents.[3a, 4a] Above all, according to timely *ab initio* calculations,[4] silicon with a low coordination number tends to avoid multiple bonds as are commonly found for carbon, favoring instead structures which permit an increased population of its valence s-orbital. One striking example is Si_2H_2, the disilicon analogue of acetylene. The ground-state structure of the most stable isomer of Si_2H_2 (**1**) contains twofold H-bridged Si atoms in a butterfly-type non-planar arrangement rather than triply-bonded silicon.[3a] More recently, the rotational spectrum of CH_2Si, formed in an SiH_4/CO plasma, has been observed by Izuha *et al.*[3b] For the CH_2Si molecule the non-classical singlet silylidene structure (**2**) predicted by *ab initio* calculations [4b] has been confirmed.

* Address of the author: Bergische Universität-GH, FB 9, Anorganische Chemie, Gaußstr. 20, D-42097 Wuppertal, Germany; *E-mail: beckers@uni-wuppertal.de

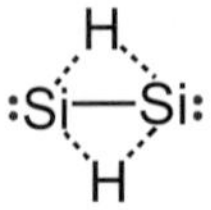

:Si(μ-H)₂Si: **1**   :Si=CH₂ **2**

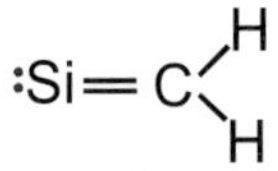

Widely used techniques to produce reactive species such as **1** or **2** are dc glow discharges coupled to sensitive high-resolution spectroscopic techniques such as millimeter wave (mmw), microwave (mw), or tunable diode laser (TDL) IR spectroscopy.[5] However, even in cooled electric discharges, not only the precursor, but also the target species thus formed, are destroyed rather unselectively, and their concentrations in discharges are often very low.[5]

Short-lived species may be obtained selectively in the gas phase by either UV photolysis or flash vacuum thermolysis (FVT). However, both photolysis and FVT have their limitations. In order to generate transients by photolysis *in situ* in the absorption cell, the precursor should have absorption bands in the UV or visible region. By FVT, transients are produced outside the absorption cell. This method is thus limited to transients that are both thermally sufficiently stable, and whose lifetimes are not shorter than a few ms. Being generated in the gas phase, such transients may be either stabilized in a matrix at low temperature and characterized by IR spectroscopy, or observed in real time by high-resolution spectroscopic measurements. The latter techniques provide information about the lifetime, the gas-phase structure, and the rotational and vibrational energy patterns of the target species.

High-resolution spectroscopic methods cannot, however, be successfully applied to molecules composed of more than a few atoms. Our work was therefore aimed at synthesizing novel small transients in the gas phase and characterizing them by their rovibrational and rotational spectra. For this goal, it was deemed highly desirable to develop productive and selective routes directed towards the transient targets. This report is restricted to some short-lived intermediates with unprotected double bonds formed by silicon, while a few results of our work on related transient phosphines are mentioned in Sections 6.4 and 6.5 for comparison. The respective species have been obtained by coupling FVT with matrix-IR, mmw, and high-resolution FTIR spectroscopies.

6.2 Rotational Spectrum and Equilibrium Structure of Silaethene, $H_2Si{=}CH_2$ (3a)

The first structural study of a molecule with a silicon-carbon double bond was the electron-diffraction investigation of 1,1-dimethylsilaethene, $Me_2Si{=}CH_2$ (**3b**), reported in 1980.[6a] In that first work, an extremely long Si=C bond length was erroneously determined for **3b**. This was even longer than those determined by X-ray crystallography [7] of highly substituted stable derivatives. This prompted Gutowski *et al.* [6b] to reinvestigate the structure of **3b** by mw spectroscopy. A complete structure determination of **3b** was, however, impossible in that work. Thus, only structural parameters of the planar $C_2Si{=}C$ heavy atom skeleton were determined experimentally, and an effective Si=C bond length of 169.2(3) pm (r_0) was obtained.

In order to obtain more reliable structural data on silaethenes with unstabilized Si=C double bonds, we tried to obtain the mmw spectrum of the parent species **3a**.[8] Contrary to the synthesis of substituted silaethenes such as **3b**, the thermolysis of the corresponding silacyclobutane (**4**) proved to be less useful for obtaining the parent species **3a** (route A in Scheme 6.1 [8a, 9]).

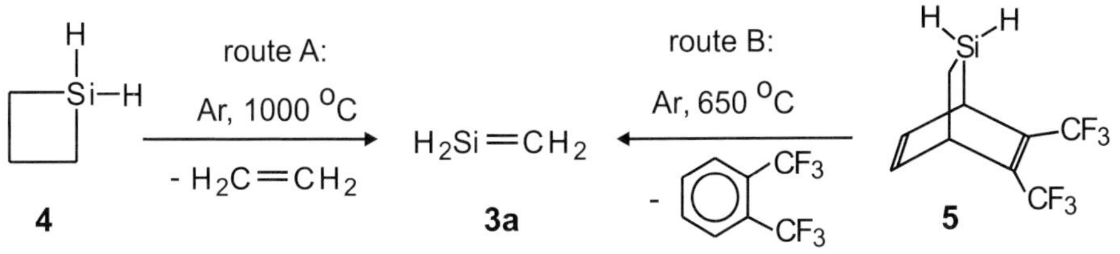

Scheme 6.1. Synthesis of silaethene (**3a**) by FVT of **4** (route A) and **5** (route B) [8].

We obtained the mmw rotational spectrum of **3a**,[8a] produced by low-temperature FVT of precursor **5** (route B in Scheme 6.1). Both the lower temperature required for the pyrolysis of **5** compared to that of **4**, and rigorous cooling of the absorption cell in order to trap condensable products, proved to be crucial for the first detection of rotational lines belonging to gaseous **3a**. Concomitantly with the spectrum of **3a**, rather strong lines arising from methylsilane, CH_3SiH_3 were observed. The latter might be formed from **3a** in two steps by the well-known silaethene-to-methylsilylene rearrangement [9] followed by hydrogenation of the reactive silylene thus formed.

Altogether seven isotopomers of **3a** were generated from isotopically labeled species **5**, and examined by mmw spectroscopy. Based on the different sets of experimental molecular parameters obtained and on predicted corrections for vibrational effects provided by *ab initio* methods, a very accurate equilibrium structure of **3a** has been determined.[8b] The length of the Si=C double bond in **3a** (170.39(18) pm, r_e) was found to be similar to that in the silylidene **2** (169.5(9) pm, r_0),[3b] as estimated from the rotational spectrum of the main isotopomer, but shorter than that found for the $X^3\Pi$ ground state of diatomic SiC (171.9(1) pm, r_e [10]).

6.3 Gas-Phase Studies of Intermediates with Unstabilized Silicon-Chalcogen Double Bonds

Both monomeric oxo- and thioxosilanes, $X_2Si{=}O$ and $X_2Si{=}S$, respectively, contain silicon in a trigonal-planar coordination. Their strongly polarized silicon-chalcogen double bonds, $Si^{\delta+}{=}O^{\delta-}$ and $Si^{\delta+}{=}S^{\delta-}$,[11, 12] render them kinetically unstable unless efficient protecting substituents X are present at silicon. The first kinetically stabilized thioxosilane, stable at ambient temperature, has recently been described.[12a] Stable oxosilanes have, however, not been isolated to date, and their synthesis continues to challenge experimentalists.[12b, 12c]

Although a few transient oxosilanes in low-temperature matrices have been observed by IR spectroscopy,[13] successful gas-phase studies of such intermediates have hitherto been limited to the examination of the unsubstituted parent species H_2SiO (**6**).[14] Oxosilane **6** was detected by means of its sub-mmw rotational spectrum in a silane/oxygen/argon low-power plasma cooled to liquid-nitrogen temperature.[14c] The experimental structure of **6** has been reported.[14d] The Si=O bond length in **6** (151.5(2) pm, r_m [14d]) was found to be similar to that in diatomic SiO (150.97 pm, r_e [15a]).

Subsequent efforts to obtain **6** by FVT coupled with mmw spectroscopy were, however, unsuccessful. A search for **6** in the products of the thermolyses of both the retro-ene precursor **7** (Scheme 6.2) [16a] and disiloxane **8** [16b] at temperatures between 600 and 1000°C was fruitless. In the course of both experiments, strong rotational lines attributable to SiO [15b] were observed, and their intensity increased with temperature. It thus appears that **6**, if formed at all, decomposes at the temperatures required for FVT. At least partial dehydrogenation occurs to yield the observed SiO and hydrogen. However, unimolecular destruction of **6** by H-migration to give slightly more stable HSiOH isomers is predicted by *ab initio* calculations [11c] to be more facile than dehydrogenation. Poor thermal stability has also been observed for the substituted derivative $Me_2Si{=}O$.[13f] This transient, obtained by low-temperature thermolysis in the gas phase, has been characterized by matrix IR spectroscopy. It decomposes under FVT conditions at temperatures > 850°C.

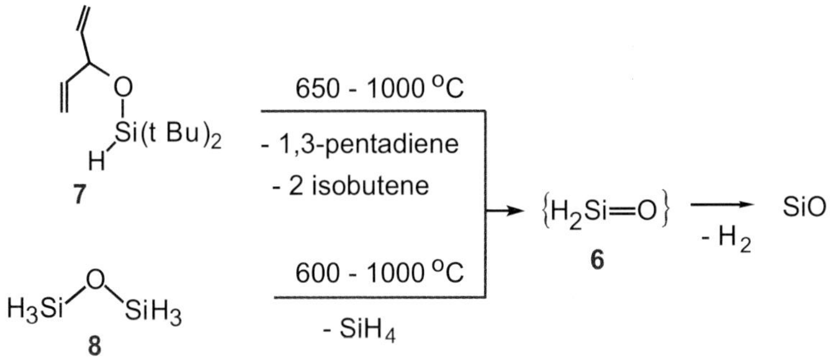

Scheme 6.2. FVT of **7** [16a] and **8**.[16b]

Compared to spectroscopic studies of oxosilanes, direct observations of transient thioxosilanes have remained particularly scarce.[17, 18] Photolysis of matrix-isolated SiS in the presence of Cl_2 and HCl provided the first thioxosilanes, $Cl_2Si{=}S$ [17a] and Cl(H)Si=S,[17b] respectively. Previous attempts to detect $Me_2Si{=}S$ [17c] by matrix IR spectroscopy during FVT of either its cyclodimer or the cyclotrimer failed. Recently the rotational spectrum of the HSiS radical has been reported,[19] while that of the parent thioxosilane $H_2Si{=}S$ is still unknown.[18b, 19] The HSiS radical was produced on striking an electrical discharge in a mixture of SiH_4 with either OCS or H_2S.[19]

More recently, substituted transient thioxosilanes have been obtained by a retro-ene reaction according to Scheme 6.3.[18a] In this study, the thermolysis of several thiosilanes such as **9a** and **9b** (Scheme 6.3) was examined using both high-resolution mass spectrometry (HRMS) and photoelectron spectroscopy (PES). While monomeric $(i\text{-Pr})_2Si{=}S$ (**10b**) was obtained from **9b** and identified independently by both methods employed, **10a** was not observed in the mass spectrum of the thermolysis products of **9a**. Moreover, the detection of monomeric **10a** by its PE spectrum was hampered by strong bands due to allene and undecomposed precursor obscuring bands assumed to stem from **10a**.[18a]

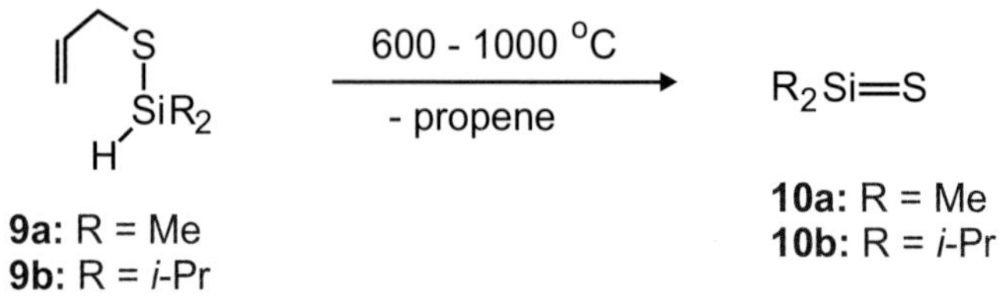

Scheme 6.3. FVT of allylthiosilanes **9a, b**.[18a]

In order to overcome the limited thermal stability of unprotected monomeric, four-atom thioxosilanes we synthesized the fluorine-substituted derivative $F_2Si{=}S$ (**11**), incorporating two particularly strong silicon-fluorine bonds.[20] First, we intended to synthesize **11** by thermal decomposition of its cyclodimer $(F_2SiS)_2$ (**12**), which had been reported in the literature.[21] However, exploring several routes directed towards this promising precursor we were unable to obtain any **12**.[22] Monomeric **11** was finally obtained according to Eq. 1 by FVT of $(F_3Si)_2S$ (**13**) [22] at ≥ 500°C. Being formed in the gas phase, **11** was trapped at cryogenic temperatures and characterized by its FTIR spectrum.[20]

$$\underset{\textbf{13}}{(F_3Si)_2S} \xrightarrow{\Delta T,\ Ar} \underset{\textbf{11}}{F_2Si{=}S} + SiF_4 \qquad (1)$$

The infrared spectrum of the deposit obtained by FVT is displayed in Figure 6.1.[20] Evidently, the thermolysis of **13** according to Eq. 1 was rather selective and almost quantitative. Nevertheless, the most intense absorptions associated with **11**, and assigned to the SiF_2-stretching vibrations at 995.7 and 969.1 cm^{-1}, are rather weak compared to the intensity of the ν_3 band of $^{28}SiF_4$ at 1023.3 cm^{-1}. This peak is accompanied by two satellites of $^{29}SiF_4$ at 1014.4 cm^{-1} (4.7%) and of $^{30}SiF_4$ at 1006.2 cm^{-1} (3.1%; marked by an asterisk in Figure 6.1). It thus appears that the majority of **11**, once generated, was destroyed under the FVT conditions before being trapped in the low-temperature matrix.

In Table 6.1 theoretical results reported for the fundamental vibrational wavenumbers and infrared band intensities of **11** are compared with their experimental counterparts determined in this work. The assignment of the measured bands is fully consistent with these calculations. Further support for this assignment comes from the calculated intensity pattern of the infrared absorptions, which is in very good agreement with the experimental relative intensities (see Figure 6.1 and Table 6.1).

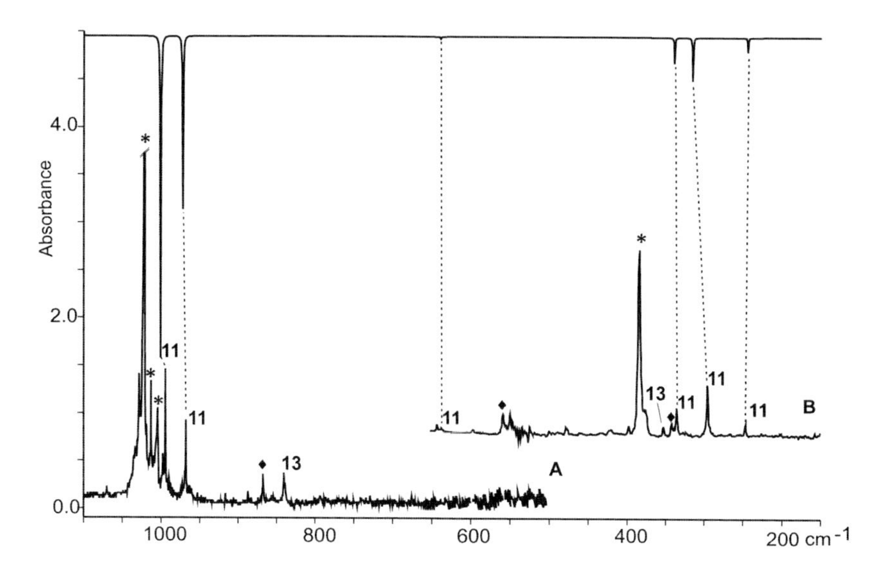

Figure 6.1. Matrix IR spectra of a deposit obtained by FVT of $(F_3Si)_2S$ (**13**) [20]; A: 500–1100 cm^{-1}; B: 150–650 cm^{-1}; on top: predicted *ab initio* spectrum of **11**; SiF_4 peaks are marked by an asterisk and peaks due to contamination by F_3SiSH with a filled rhomboid.

Table 6.1. Computed and experimental fundamental vibrational wavenumbers (cm^{-1}) of $F_2Si^{32}S$ [a].

Mode	Experimental [b]	*Ab initio* [c]
a_1; $\nu_s(SiF_2)$	996 (2.60)	1003 (262)
$\nu(SiS)$	638 (0.02)	647 (1)
$\delta(SiF_2)$	337 (0.30)	339 (25)
b_1; $\nu_{as}(SiF_2)$	969 (1.30)	967 (142)
$\rho(SiF_2)$	247 (0.12)	247 (10)
b_2; $\gamma(SiF_2)$	296 (0.54)	316 (33)

[a] Ref. 20. [b] Extinction in parentheses. [c] MP2/VTZ+1 values; infrared band intensities (km/mol) in parentheses.

The hitherto unknown cyclodimer **12** [22] might be a secondary product formed by dimerization of **11**. The search for absorptions associated with **12** in the spectrum of the deposit (Figure 6.1) was, however, unsuccessful. No peak could be definitely assigned to this elusive species, even after annealing the matrix and the disappearance of lines due to **11**. Hence, if formed at all, only traces of **12** might have been co-condensed in the low-temperature matrix. Thus, the fate of **11** differs strongly from that of the methyl-substituted derivative **10a**. The crowded spectra obtained in related experiments directed towards the synthesis of **10a** [17c, 18a] are dominated by bands attributable to the corresponding cyclodimer. The latter was assumed to be formed by dimerization of kinetically unstable, monomeric **10a** on the walls.[17c]

Due to the lack of experimental data on the structure of sterically unhindered thioxosilanes, we have to resort to a high-level ab initio structure of **11**.[20] The SiS distance in **11** (191.1(1) pm) is predicted to be shorter than the experimental distance in diatomic SiS (192.93 pm, r_e [23]). These bond lengths differ considerably from that recently estimated for the HSiS radical. The reported effective Si=S bond length of this radical, 195.37(19) pm (r_0), is similar to those determined by X-ray crystallography of a stable, sterically overcrowded derivative (194.8(4) pm and 195.2(4) pm [12a]).

6.4 Exploration of the Time Domains of Transients

Gas-phase studies have provided compelling evidence of the transient nature of the reactive species investigated. Their high reactivity may certainly be attributed to their polar and unprotected double bonds. However, apart from a few kinetic studies on substituted silaethenes,[24] any detailed knowledge on their chemical stability is lacking. Nevertheless, the temporal fate of a transient is certainly of high chemical significance and interest since it strongly influences the choice of methods for its production and spectroscopic characterization.

In spite of the successful detection of **3a** by its mmw spectrum, we have not as yet been able to record its FTIR spectrum.[25] While the FTIR technique is certainly less sensitive than both TDL and pure rotational experiments, the wide and continuous scan range makes FTIR spectra extremely valuable. FTIR spectroscopy has, however, only been applied in very few cases to study transient molecules.[26] Recently, we have been able to observe both the high-resolution FTIR and mmw rotational spectra of short-lived monomeric chalcogenophosphines FP=O (**14a**) [27] and FP=S (**14b**).[28] It is instructive to compare effective lifetimes of alkylidene- and chalcogenosilanes with those of corresponding transient phosphines (Scheme 6.4).

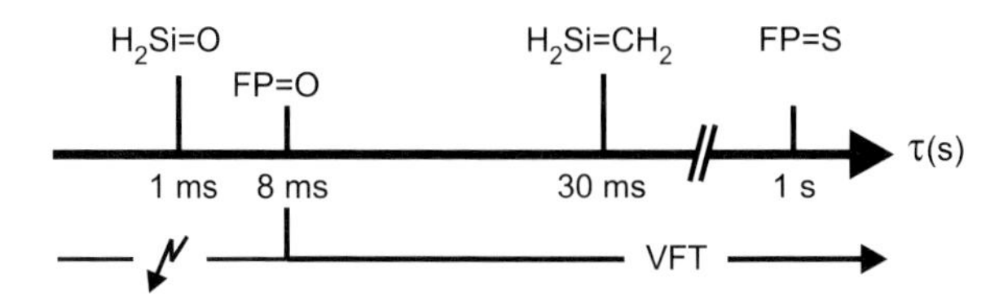

Scheme 6.4. Effective 1/e lifetimes τ(s) (see text) observed for some transient species.

The time-resolved decay of ground-state mmw lines of the relevant transients, produced either by modulated discharges or by pulsed UV photolysis, has been measured under analogous experimental conditions. A typical decay plot is shown in Figure 6.2. The intensities of rotational lines due to the fluorophosphines **14a** [27b] and **14b**,[28b] observed in modulated discharges in (20:1) mixtures of a noble gas with **15a** and **15b**, respectively (Eq. 2), at 8–10 Pa and at room temperature, decrease exponentially, with time constants of about 8 ms for **14a** (Figure 6.2), and 2 s for **14b**.

$$F_2PYPF_2 \longrightarrow FP{=}Y + PF_3 \qquad (2)$$

15a: Y = O — **14a**: Y = O

15b: Y = S — **14b**: Y = S

The 1/e lifetime of silanone **6**, produced in a pulsed discharge cooled to liquid nitrogen temperature, was reported to be 1.2 ms.[14c] Silaethene **3a**, generated by pulsed UV photolysis of pure **5** (Scheme 6.1) at 193 nm with an ArF excimer laser at a pressure of 0.8 Pa and at ambient temperature, revealed a 1/e lifetime of 30 ± 2 ms.[8b]

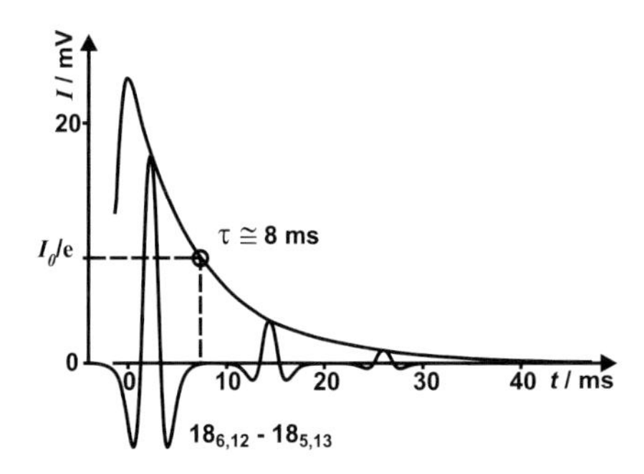

Figure 6.2. Decay characteristics of the $18_{6,12}$–$18_{5,13}$ mmw line of FPO (**14a**).[27b]

The absorbance versus time profiles thus observed comprise all reactions of the transient with any of the species present in the mixture, and these reactions may be different for the different transients investigated. We found **14b** to be less reactive than the other transients investigated in this work (Scheme 6.4), since its double bond is expected to have the lowest polarity. However, **14b** is nevertheless a highly reactive intermediate compared to compounds with unprotected carbon-phosphorus multiple bonds such as RCP, with R = H or CH_3. The latter were extensively investigated during the 1980s,[2] and are known to be stable in the gaseous state at room temperature.[29] A higher polarity of the double bond of the transient may facilitate both dimerization and insertion of these intermediates into polar bonds, and hence shorten their lifetime in the presence of polar reagents.

Although the decay of the obtained transients is believed to show trends in relative total reactivity at ambient temperature, other reaction channels of the relevant transients, unavailable at ambient temperature, may well be favored at high FVT temperatures. Although the FTIR technique demands more rigorous FVT conditions (higher partial pressures and higher FVT temperatures) than those employed for mmw measurements, it appears that robustly bound molecules such as **14a** survive even under these more rigorous conditions at partial pressures high enough permit their studies by FTIR spectroscopy, while the thermally less stable **3a** may not be observable. Preliminary gas-phase mmw and matrix IR spectroscopic studies on the difluorinated intermediates **11** and $F_2Si{=}CH_2$, respectively, have established that even these more robust molecules are thermally unstable with respect to elimination of the kinetically more stable F_2Si under the chosen FVT conditions.

6.5 Conclusion

Productive and highly selective FVT routes directed towards the short-lived intermediates $H_2Si{=}CH_2$ (**3a**),[8] $F_2Si{=}S$ (**11**),[20] FP=O (**14a**),[27] and FP=S (**14b**) [28] with unprotected double bonds involving silicon or phosphorus have been explored, and new gas-phase syntheses of **11**, **14a**, and **14b** have been developed. The transients **3a**, **14a** and **14b** have been obtained directly in the gas phase, and their rotational spectra, as well as high-resolution FTIR spectra of **14a** and **14b**, have been recorded. The silane **11** has been trapped from the gas phase in a low-temperature matrix and unambiguously characterized by its matrix FTIR spectrum.

Highly accurate experimental equilibrium lengths of unstabilized Si=C bonds, as well as of P=O and P=S double bonds of compounds with Si(IV) and P(III), respectively, have been determined. These will serve as references for forthcoming studies. This work includes the first structural investigation of a thioxophosphine in the gas phase. The double-bond lengths are compared to those of known corresponding diatomic molecules. Thus, the double-bond length in silaethene **3a** is shorter than that in the diatomic species SiC, while that in $H_2Si=O$ (**6**) is similar to that in SiO. In particular, the behavior of these silicon-oxygen bond lengths differs significantly from that of the corresponding carbon-oxygen homologues. This observation accounts for the known reluctance of silicon to form multiple bonds at the expense of the population of its valence s-orbital.[4] Owing to the electron withdrawing effect of fluorine, the double-bond lengths in the phosphines **14a** (145.28(2) pm, r_e [27b]) and **14b** (188.86(4) pm, r_e ([28b]) are also found to be even shorter than those in the corresponding diatomic species PO (147.64 pm, r_e [30a]) and PS (189.77 pm, r_e [30b]), respectively.

The investigated transients with double bonds to silicon exhibit short effective 1/e lifetimes ranging from a few ms (for $H_2Si=O$) up to ca. 30 ms for **3a**. In contrast to these silanes, the kinetic instability of multiply-bonded transient phosphines depends more strongly on the electronegativity of the element involved in the multiple bond to phosphorus. The observed 1/e lifetimes span a substantial range from 8 ms for highly reactive **14a** up to stable derivatives with unprotected phosphorus-carbon multiple bonds. FVT coupled with FTIR spectroscopy has proved to be a competitive technique to study robustly bound transients with 1/e lifetimes as short as 8 ms.

6.6 Acknowledgment

Contributors to this work are S. Bailleux, M. Bogey, J. Demaison, P. Dréan, and A. Walters (Laboratoire de Physique des Lasers, Atomes et Molécules, CERLA, Université Lille), R. Fajgar and J. Pola (Institute of Chemical Process Fundamentals, Academy of Science, Prague), J. Breidung and W. Thiel (Max-Planck-Institut für Kohlenforschung, Mülheim), C. Kötting and W. Sander (Lehrstuhl für Organische Chemie II der Ruhr-Universität, Bochum), R. Köppe, and H. Schnöckel (Institut für Anorganische Chemie der Universität, Karlsruhe), H. Bürger, M. Senzlober, and P. Paplewski (Anorganische Chemie, FB 9, Bergische Universität-GH, Wuppertal). Professor P. Jutzi, Bielefeld, is thanked for his valuable support.

6.7 References

[1] For a recent review on multiple bonds between heavier main group elements see: P. P. Power, *Chem. Rev.* **1999**, *99*, 3463–3503.

[2] For a comprehensive review on multiply bonded phosphorus, see: M. Regitz, O. J. Scherer, *Multiple Bonds and Low Coodination in Phosphorus Chemistry,* G. Thieme, Stuttgart, New York **1990**.

[3] a) M. Bogey, H. Bolvin, C. Demuynck, J. L. Destombes, *Phys. Rev. Lett.* **1991**, *66*, 413–416; b) M. Izuha, S. Yamamoto, S. Saito, *J. Chem. Phys.* **1996**, *105*, 4923–4926.

[4] a) H. Jacobsen, T. Ziegler, *J. Am. Chem. Soc.* **1994**, *116*, 3667–3679; b) R. Stegmann, G. Frenking, *J. Comput. Chem.* **1996**, *17*, 781–789; c) K. Kobayashi, S. Nagase, *Organometallics* **1997**, *16*, 2489–2491; d) Y. Apeloig, M. Karni, *Organometallics* **1997**, *16*, 310–312; e) D. Danovich, F. Ogliaro, M. Karni, Y. Apeloig, D. L. Cooper, S. Shaik, *Angew. Chem.* **2001**, *113*, 4146–4150; *Angew. Chem. Int. Ed. Engl.* **2001**, *40*, 4023–4026; f) K. Kobayashi, N. Takagi, S. Nagase, *Organometallics* **2001**, *20*, 234–236.

[5] P. B. Davies, *Chem. Soc. Rev.* **1995**, 151–157.

[6] a) P. G. Mahaffy, R. Gutowsky, L. K. Montgomery, *J. Am. Chem. Soc.* **1980**, *102*, 2854–2856; b) H. S. Gutowsky, J. Chen, P. J. Hajduk, J. D. Keen, C. Chuang, T. Emilsson, *J. Am. Chem. Soc.* **1991**, *113*, 4747–4751.

[7] a) S. C. Nyburg, A. G. Brook, F. Abdesaken, G. Gutekunst, W. Wong-Ng, *Acta Crystallogr., Sect. C: Cryst. Struct. Commun.* **1985**, *C41*,1632–1635; b) N. Wiberg, G. Wagner, J. Riede, G. Müller, *Organometallics* **1987**, *6*, 32–35.

[8] a) S. Bailleux, M. Bogey, J. Breidung, H. Bürger, R. Fajgar, Y. Liu, J. Pola, M. Senzlober, W. Thiel, *Angew. Chem.* **1996**, *108*, 2683–2685; *Angew. Chem. Int. Ed. Engl.* **1996**, *35*, 2513–2515; b) S. Bailleux, M. Bogey, J. Breidung, H. Bürger, J. Demaison, R. Fajgar, J. Pola, M. Senzlober, W. Thiel, *J. Chem. Phys.* **1997**, *106*, 10016–10026.

[9] a) G. Maier, G. Mihm, H. P. Reisenauer, *Angew. Chem.* **1981**, *93*, 615–616; *Angew. Chem. Int. Ed. Engl.* **1981**, *20*, 597–598; b) G. Maier, G. Mihm, H. P. Reisenauer, D. Littmann, *Chem. Ber.* **1984**, *117*, 2369–3381.

[10] P. F. Bernath, S. A. Rogers, L. C. O'Brien, C. R. Brazier, A. D. McLean, *Phys. Rev. Lett.* **1988**, *60*, 197–199.

[11] a) M. A. Alikhani, B. Silvi, *J. Comput. Chem.* **1998**, *19*, 1205–1214; b) J. Kapp, M. Remko, P. v. Ragué Schleyer, *J. Am. Chem. Soc.* **1996**, *118*, 5745–5751; c) T. Kudo, S. Nagase, *Organometallics* **1986**, *5*, 1207–1215.

[12] a) H. Suzuki, N. Tokitoh, R. Okazaki, S. Nagase, M. Goto, *J. Am. Chem. Soc.* **1998**, *120*, 11096–11105; b) N. Takeda, N. Tokitoh, R. Okazaki, *Chem. Lett.* **2000**, 244–245; c) M. Kimura, S. Nagase, *Chem. Lett.* **2001**, 1098–1099.

[13] a) H. Schnöckel, *Angew. Chem.* **1978**, *90*, 638–639; *Angew. Chem. Int. Ed. Engl.* **1978**, *17*, 616–617; b) H. Schnöckel, *J. Mol. Struct.* **1980**, *65*, 115–123; c) R. Whithnall, L. Andrews, *J. Am. Chem. Soc.* **1985**, *107*, 2567–2568; d) R. Whithnall, L. Andrews, *J. Phys. Chem.* **1985**, *89*, 3261–3268; e) R. Withnall, L Andrews, *J. Am. Chem. Soc.* **1986**, *108*, 8118–8119; f) V. N. Khabashesku, Z. A. Kerzina, E. G. Baskir, A. K. Maltsev, O. M. Nefedov, *J. Organomet. Chem.* **1988**, *347*, 277–293; g) M. Junker, A. Wilkening, M. Binnewies, H. Schnöckel, *Eur. J. Inorg. Chem.* **1999**, 1531–1535.

[14] a) R. J. Glinski, J. L. Gole, D. A. Dixon, *J. Am. Chem. Soc.* **1985**, *107*, 5891–5894; b) R. Srinivas, D. K. Böhme, D. Sülzle, H. Schwarz, *J. Phys. Chem.* **1991**, *95*, 9836–9841; c) S. Bailleux, M. Bogey, C. Demuynck, J.-L. Destombes, A. Walters, *J. Chem. Phys.* **1994**, *101*, 2729–2733; d) M. Bogey, B. Delcroix, A. Walters, J. C. Guillemin, *J. Mol. Spectrosc.* **1996**, *175*, 421–428.

[15] a) E. L. Manson, Jr., W. W. Clark, F. C. De Lucia, W. Gordy, *Phys. Rev. A* **1977**, *15*, 223–226; b) F. J. Lovas, P. H. Krupenie, *J. Phys. Chem. Ref. Data,* **1974**, *3*, 245–257.

[16] a) A. Chive, V. Lefevre, A. Systermans, J.-L. Ripoll, M. Bogey, A. Walters, *Phosphorus, Sulfur, and Silicon*, **1994**, *91*, 281–284, b) M. Bogey, A. Walters, H. Beckers, unpublished.

[17] a) H. Schnöckel, H.-J. Göcke, R. Köppe, *Z. Anorg. Allg. Chem.* **1989**, *578*, 159–165; b) R. Köppe, H. Schnöckel, *Z. Anorg. Allg. Chem.* **1992**, 607, 41–44; c) L. G. Gusel'nikov, V. V. Volkova, V. G. Avakyan, N. S. Nametkin, M. G. Voronkov, S. V. Kirpichenko, E. N. Suslova, J. *Organomet. Chem.* **1983**, *254*, 173–187.

[18] a) A. Chrostowska, S. Joantéguy, G. Pfister-Guillouzo, V. Lefèvre, J. L. Ripoll, *Organometallics* **1999**, *18*, 4759–4799; b) M. Letulle, A. Systermanns, V. Lefèvre, J. L. Ripoll, V. Métail, S. Joantéguy, A. Chrostowska-Senio, G. Pfister-Guillouzo, A. Walters, M. Bogey, *COST Action D6; Office for Official Publications of the European Communities,* Luxembourg, **1998**, 161–168.

[19] F. X. Brown, S. Yamamoto, S. Saito, *J. Mol. Struct.* **1997**, 413–414, 537–544.

[20] H. Beckers, J. Breidung, H. Bürger, R. Köppe, C. Kötting, W. Sander, H. Schnöckel, W. Thiel, *Eur. J. Inorg. Chem.* **1999**, 2013–2019.

[21] a) V. Gutmann, P. Heilmeyer, K. Utvary, *Monatsh.* **1961**, *92*, 942–943; b) B. J. Aylett, I. A. Ellis, J. R. Richmond, *J. C. S. Dalton Trans.* **1973**, 981–987.

[22] H. Beckers, H. Bürger, *Z. Anorg. Allg. Chem.* **2001**, *627*, 1217–1224.

[23] E. Tiemann, H. Arnst, W. U. Stieda, T. Törring, J. Hoeft, *Chem. Phys.* **1982**, *67*, 133–138.

[24] For a recent review, see: T. L. Morkin, T. R. Owens, W. J. Leigh, in *The Chemistry of Organic Silicon Compounds*, (Eds.: Z. Rappoport, Y. Apeloig), Vol. 3, Ch. 17, John Wiley, Chichester **2001**.

[25] S. Bailleux, H. Beckers, M. Bogey, H. Bürger, J. Pola, M. Senzlober, unpublished.

[26] a) K. Kawaguchi, *J. Chem. Phys.* **1992**, *96*, 3411–3415; b) I. K. Ahmad, P. A. Hamilton, *J. Mol. Spectrosc.* **1995**, *169*, 286–291; c) P. F. Bernath, *Chem. Soc. Rev.* **1996**, 111–115.

[27] a) P. Paplewski, H. Bürger, H. Beckers, *Z. Naturforsch. A* **1999**, *54*, 507–512; b) H. Beckers, H. Bürger, P. Paplewski, M. Bogey, J. Demaison, P. Dréan, A. Walters, J. Breidung, W. Thiel, *Phys. Chem. Chem. Phys.* **2001**, *3*, 4247–4257; For previous work on FPO, see references therein.

[28] a) H. Beckers, M. Bogey, J. Breidung, H. Bürger, P. Dréan, P. Paplewski, W. Thiel, A. Walters, *Phys. Chem. Chem. Phys.* **2000**, *2*, 2467–2469; b) H. Beckers, M. Bogey, J. Breidung, H. Bürger, J. Demaison, P. Dréan, P. Paplewski, W. Thiel, A. Walters, *J. Mol. Spectrosc.* **2001**, *210*, 213–223; for previous work on FPS, see references therein.

[29] a) K. Ohno, H. Matsuura, D. McNaughton, H. W. Kroto, *J. Mol. Spectrosc.* **1985**, *111*, 415–424; b) K. K. Lehmann, S. C. Ross, L. L. Lohr, *J. Chem. Phys.* **1985**, *82*, 4460–4469.

[30] a) J. E. Butler, K. Kawaguchi, E. Hirota, *J. Mol. Spectrosc.* **1983**, *101*, 161–166; b) K. Kawaguchi, E. Hirota, M. Ohishi, H. Suzuki, S. Takano, S. Yamamoto, S. Saito, *J. Mol. Spectrosc.* **1988**, *130*, 81–85.

7 Kinetic Stabilization of Disilenes >Si=Si< and Disilynes –Si≡Si–

Nils Wiberg*

7.1 Introduction

As silicon represents the next homologue to carbon, which is widely admired as the "king of elements", the question of the carbon analogy of silicon arose early, and chemists, bearing the group connections in mind, have tried to find chemical similarities between them. In fact, carbon and silicon form compounds with analogous compositions which are more comparable in their properties than analogous compounds of element pairs to the left and right in the periodic table.[1]

Whereas saturated silicon compounds Si_nH_{2n+2} (e.g. disilane $H_3Si–SiH_3$) have been known for a long time, the same is not true for unsaturated silicon compounds Si_nH_{2n} (e.g. disilene $H_2Si=SiH_2$) or Si_nH_{2n-2} (e.g. disilyne HSi≡SiH). Their existence has been regarded as the central touchstone not only for the aforementioned analogy between carbon and silicon, but – in a broader sense – also for the double bond rule.[1] In fact, disilenes first became accessible in 1981 after many unsuccessful experiments with the isolation of $Mes_2Si=SiMes_2$ by R. West et al., when it was realized that substitution of the hydrogens in $H_2Si=SiH_2$ with bulky groups reduces drastically the tendency of the disilene to di-, oligo- or polymerize.[2] Since this discovery, the chemistry of disilenes has made rapid progress.[2] At the same time, it strongly stimulated the search for an isolable, dimerization-stable disilyne RSi≡SiR.

Clearly, the substituents R used for a kinetic stabilization of disilenes or disilynes must not only hinder the intermolecular dimerization of the reactive species through steric crowding, but, in addition, must also be chemically inert with respect to intramolecular isomerization processes such as insertions or migrations, by which the disilenes are transformed into disilanes, and the disilynes into disilenes as well as disilanes. According to ab initio calculations, electrophilic groups R (silyl less electronegative than organyl) are favorable to the aforementioned carbon analogy of silicon. In place of a bulky substituent, we selected the chemically very inert tri-*tert*-butylsilyl group Si^tBu_3, called supersilyl and symbolized by R*. Indeed, this moiety has a lot of merits, as we described in a review in 1997,[3] which summarized our knowledge regarding

---

* Address of the author: Ludwig-Maximilians-Universität München, Department Chemie, Butenandtstr. 5–13 (Haus D), 81377 München, Germany

supersilyl compounds at that time. In addition, the even more sterically crowded disupersilylsilyl group R′ = $SiHR*_2$ and the extremely bulky methyldisupersilyl group R** = $SiMeR*_2$, which we jokingly call “megasilyl”, have been used. We also considered using the sterically very crowded trimesitylsilyl group, but the extreme insolubility of $SiMes_3$-containing starting materials for the preparation of disilenes and disilynes, stopped us from persuing this further.

In the following text, results concerning the syntheses and reactions of disilenes of types **1**–**4** are presented. In addition, dehalogenations of the disilenes **2** and **4**, and dehydrobromination of the disilane R′HBrSi–SiBrHR′, which possibly lead to disilynes of types **5**–**7**, will be described (in some cases, these reactions obviously proceed via cyclotrisilenes as well as cyclotetrasilenes and even tetrasilabutadienes). The disilenes **1** (R = Ph), **3**, and **4** and, (obviously) the disilyne **7**, could be obtained under normal conditions due to an adequate steric shielding of the reactive >Si=Si< and -Si≡Si– entities with bulky silicon-bound groups.

R*(R)Si=Si(R)R*	R*(X)Si=Si(X)R*	R′(R)Si=Si(R)R′	R**(X)Si=Si(X)R**
1	**2**	**3**	**4**
(R = H, Me, Ph)	(X = Cl, Br, I)	(R = H)	(X = Cl)

R*—Si≡Si—R*	R′—Si≡Si—R′	R**—Si≡Si—R**
5	**6**	**7**
(R* = Si^tBu_3)	(R′ = $SiHR*_2$)	(R** = $SiMeR*_2$)

Incidentally, the silylene $SiR*_2$, generated by dehalogenation of $R*_2SiX_2$ (X = halogen), does not dimerize due to steric reasons (R = R* in **1** is not possible). Instead, it decomposes by an intramolecular insertion of the silylene Si atom into a CH bond of supersilyl.[4]

7.2 Syntheses and Reactions of Disilenes R*RSi=SiRR* (1)

7.2.1 Formation of 1 from R*RXSi–SiXRR*

The disilanes R*RBrSi–SiBrRR* (R = H, Me, Ph; also R*PhClSi–SiBrPhR*), in the presence of equimolar amounts of supersilyl sodium NaR* in THF at –78°C, form – according to Scheme 7.1 – the disilanides R*RBrSi–SiNaRR* (**1**•NaBr), which have been identified by protonation, bromination, and silylation [5, 6] (for the preparation of R*RBrSi–SiBrRR* cf. [7]). The disilanides are transformed by elimination of NaBr (at about –70 to –50°C) into the disilenes R*RSi=SiRR* (**1** with R = H, Me, Ph). Thereby, $\mathbf{1}_H$ and $\mathbf{1}_{Me}$ are obtained as reactive species, the intermediate existences of which have been proved by trapping the disilenes with diphenylethyne or anthracene (Scheme 7.1).[5, 6] According to the structures of the obtained [2+2] and [2+4] cycloadducts **8** and **9**, with both the R* and R groups in *trans* positions, there is every reason to believe that the disilenes $\mathbf{1}_H$ and $\mathbf{1}_{Me}$ are *trans* configured. On the other hand, $\mathbf{1}_{Ph}$ is metastable at room temperature and – as a disilene which is more overcrowded than $\mathbf{1}_H$ and $\mathbf{1}_{Me}$ – reacts neither with C_2Ph_2 nor $C_{10}H_8$, even at 80 °C.

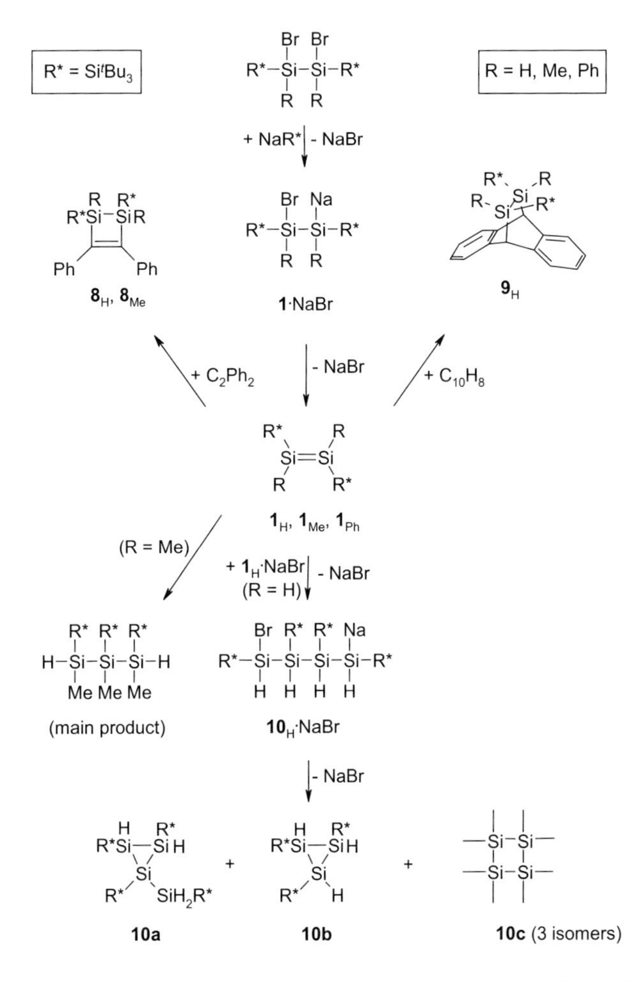

Scheme 7.1. Dehalogenation of R*RBrSi–SiBrRR*; formation of the disilenes R*RSi=SiRR* (R = H, Me, Ph); reactions of the disilenes $\mathbf{1}_H$ and $\mathbf{1}_{Me}$.

Heating the silanide R*HBrSi–SiNaHR* (1_H•NaBr) in THF from –78 °C to room temperature leads – according to Scheme 7.1 – obviously via the tetrasilanide 10_H•NaBr (which is an adduct of 1_H•NaBr and the disilene 1_H, generated from 1_H•NaBr) to a mixture of the *cis,trans*-cyclotrisilanes **10a** and **10b** along with the *cis,cis,trans-*, *cis,trans,cis-* and *trans,trans,trans*-cyclotetrasilanes **10c** (Scheme 7.1).[5] At –78 °C, 1_H•NaBr transforms only into *cis,trans*-**10a** and *cis,trans,cis*-**10c** (mole ratio 3:1), whereas at 66 °C *cis,trans*-**10b** and *cis,trans,cis*-**10c** (mole ratio 3:1) are formed.[5] On the other hand, on heating a solution of the silanide R*MeBrSi–SiNaMeR* (1_{Me}•NaBr), prepared in THF at –78 °C, a mixture of compounds is obtained via unknown reaction pathways, with the trisilane R*MeHSi–SiMeR*–SiHMeR* being the main product (Scheme 7.1).[6] Finally, heating the silanide R*PhBrSi–SiNaPhR* (1_{Ph}•NaBr) in THF from –78 °C to room temperature leads – according to Scheme 7.1 – to the disilene R*PhSi=SiPhR* (1_{Ph}). It could be isolated as light-yellow, hydrolysis- and air-sensitive crystals and characterized (i) spectroscopically (Raman: $\tilde{\nu}$ (Si=Si) = 591 cm^{-1}; UV/Vis: λ_{max} = 398 nm (ε = 1560); ^{29}Si NMR: δ (>Si=) = 128 ppm), (ii) by X-ray structure analysis ($d_{Si=Si}$ = 2.182 Å; planar and *trans* configured >Si=Si< framework), and (iii) by its thermolysis and by further reactions.[8]

7.3 Syntheses and Reactions of Disilenes R*XSi=SiXR* (2)

The action of supersilyl sodium NaR* in THF on the silanes R^*SiX_3 or disilanes $R^*X_2Si–SiX_2R^*$ (X = Cl, Br, I) leads to the *tetrahedro*-tetrasupersilyltetrasilane $Si_4R^*_4$ (**13**) shown in Scheme 7.2 [5] (for the preparation of R^*SiX_3 or $R^*X_2Si–SiX_2R^*$, see [7, 9]). As the tetrahedrane is obtained in quantitative yields from the bromides and iodides, the routes for the formation of **13** must be straightforward. Indeed, the question arose as to whether the starting materials are first transformed by NaR* into disilenes R*XSi=SiXR* and then into the disilyne R*Si≡SiR*, which finally dimerizes or reacts with its precursors with formation of **13**. These ideas stimulated detailed studies, the results of which are partly represented in Scheme 7.2.

Incidentally, it is worth mentioning that the action of NaR* in THF on the tribromodisilane $R^*HBrSi–SiBr_2R^*$ rather than the tetrabromodisilane $R^*Br_2Si–SiBr_2R^*$, leads cleanly to the *bicyclo*-tetrasilane **15**, shown in Scheme 7.2[5].

7.3.1 Formation of R*XSi=SiXR* from R^*SiX_3 or $R^*X_2Si–SiX_2R^*$

Trapping experiments with acids, dienes, triethylsilane etc.[4, 5] indicate that the silanes R^*SiX_3 (X = Br, I) react with equimolar amounts of NaR* in THF at –78 °C according to $R^*SiX_3 + NaR^* \rightarrow R^*SiX_2Na + R^*X$ to form initially the silanides R^*SiX_2Na, which then add silylenes R*SiX (generated according to $R^*SiX_2Na \rightarrow R^*SiX + NaX$ at about –50 °C) with formation of disilanides $R^*X_2Si–SiXNaR^*$ (**2**•NaX; Scheme 7.2). The disilanides transform by elimination of NaX (at about –30 to –10 °C) into the reactive disilene intermediates R*XSi=SiXR* (**2**).

The disilanides **2**•NaX, which are also obtained from $R^*X_2Si–SiX_2R^*$ (X = Br, I) and NaR* in THF at –78 °C (the chlorides $R^*Cl_2Si–SiCl_2R^*$ and R^*SiCl_3 react only slowly with NaR*) stabilize in different ways.[5] The chloride $\mathbf{2}_{Cl}$•NaCl (formed from $R^*Cl_2Si–SiBrClR^*$/NaR*) adds to the disilene $\mathbf{2}_{Cl}$, generated from **2**•NaCl, with formation of the tetrasilanide **11**•NaCl, which then transforms into the cyclotetrasilane **11** (Scheme 7.2). The decomposition pathways of the bromide $\mathbf{2}_{Br}$•NaBr have hitherto not been studied in detail, but the iodide **2**•NaI cleanly gives product $(\mathbf{2}_I)_2$. Due to steric reasons, insertion of the disilenes **2** into the NaSi bond of their precursors **2**•NaX is possible only for $\mathbf{2}_{Cl}$•NaCl. The same is obviously true for a reaction of less bulky R*HBrSi–SiNaBrR* with R*HSi=SiBrR* (both generated from $R^*HBrSi–SiBr_2R^*$ and NaR*; see above), leading via the tetrasilanide R*HBrSi–SiBrR*–SiHR*–SiNaBrR* to a cyclotrisilane of type **10a** (cf. Scheme 7.1: exchange of H/SiH_2R^* for Br/SiBrHR*), which, in the presence of NaR*, ultimately affords the *bicyclo*-tetrasilane **15** (Scheme 7.2).[5]

The disilenes **2** are formed as reactive (not yet isolated) intermediates. X-ray structure analyses of products **2**•DMB and **2**•C_2Ph_2 obtained by trapping **2** with 2,3-dimethylbutadiene (DMB) or with diphenylethyne, respectively, show both R* and X in *trans* positions,[5] indicating that the disilenes are *trans* configured. Obviously, the disilenes R*XSi=SiXR* may isomerize to some extent to form silylenes $R^*X_2Si–SiR^*$, as follows from a trapping experiment with Et_3SiH leading to the product $R^*X_2Si–SiHR^*(SiEt_3)$.[5]

The decomposition pathways of the disilenes **2** themselves are, in fact, not clear, while the unsaturated compounds form in the presence of their precursors **2**•NaX, which may act as traps for **2**. It seems most probable that $\mathbf{2}_{Cl}$ dimerizes with formation of the cyclotetrasilane $(\mathbf{2}_{Cl})_2$ (Scheme 7.2), whereas $\mathbf{2}_I$, due to steric reasons, dimerizes only with formation of the cyclotrisilane $(\mathbf{2}_I)_2$ (Scheme 7.2), which can be rationalized as the [2+1] cycloadduct of the disilene R*ISi=SiIR* and its isomer $R^*I_2Si–SiR^*$.[5]

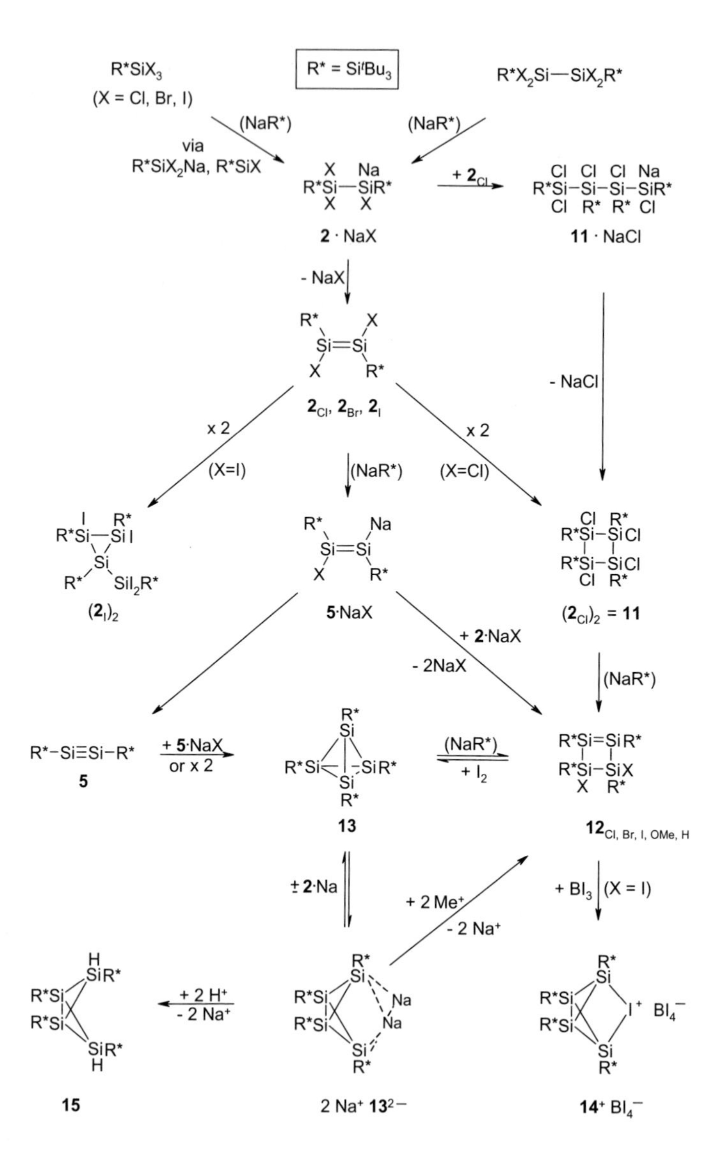

Scheme 7.2. Syntheses and reactions of disilenes R*XSi=SiXR* (**2**).

The metastabilities of **2** must clearly increase in the order $\mathbf{2}_{Cl} < \mathbf{2}_{Br} < \mathbf{2}_{I}$ due to an increasing shielding of the Si=Si double bond in the same direction, a consequence of which is that the isolated $(\mathbf{2}_I)_2$ is indeed not a product of $\mathbf{2}_I$ with $\mathbf{2}_I$•NaI, but with itself. Naturally, it should be favorable to catalyze the transformation of the precursors **2**•NaX into **2** (with Me_3SiX?) at temperatures at which the disilenes are still metastable.

7.3.2 Formation of $Si_4R^*_4$ (13) from R*XSi=SiXR* (2)

The intermediately formed disilenes R*XSi=SiXR* (**2**) react with NaR* – according to Scheme 7.2 – via disilenides R*XSi=SiNaR* (**5**•NaX) and probably via cyclotetrasilenes $R^*_4Si_4X_2$ (**12**) with formation of the orange tetrahedrane $Si_4R^*_4$ (**13**) , the structure of which has been determined by X-ray structure analysis.[5] In fact, the cyclotetrasilane **11** in contact with NaR* cleanly affords the tetrahedrane **13** (Scheme 7.2), evidently through the intermediacy of the unsaturated ring compound $\mathbf{12}_{Cl}$.[5] In addition, the red cyclotetrasilene $\mathbf{12}_{I}$ is not only formed from **13** with equimolar amounts of I_2 (Scheme 7.2), but has also been proven to be a product of the reaction of $R^*I_2Si–SiI_2R^*$ with a sub-stoichiomctric amount of NaR*, i.e insufficient for the formation of **13** in quantitative yield.[5, 10] The cyclotetrasilene $\mathbf{12}_{I}$ is dehalogenated with re-formation of **13**, and takes up H_2O or MeOH via $R^*_4Si_4I^+$ ($\mathbf{14}^+$; isolable as $\mathbf{14}^+BI_4^-$) with formation of the light-yellow tetrahedrane oxide $R^*_4Si_4O$ ($\mathbf{14}^+$ with O for I^+) or the orange cyclotetrasilene $\mathbf{12}_{OMe}$.[10] The mentioned unsaturated ring compounds $\mathbf{12}_{I}$ and $\mathbf{12}_{OMe}$ show Si=Si distances of 2.26 Å,[10] whereby the >Si=Si< groups with approximately planar Si atoms are evidently twisted about the double bonds, with the R* and X groups at the saturated Si atoms in a *trans* arrangement.

The disilenides **5**•NaX may be transformed into $\mathbf{12}_{Cl,\ Br,\ I}$ by reaction with any of the other compounds present in solution, namely the disilanides **2**•NaX or the disilenes **2** as precursors of **5**•NaX, and the disilyne **5** may be a possible product of **5**•NaX (cf. Scheme 7.2). In fact, insertions of **2** into the NaSi bond of **5**•NaX or of **5** into the NaSi bond of **2**•NaX or **5**•NaX are quite probable. The formed tetrasilenides Na–*R*XSi–SiXR**–R*Si=SiXR* or Na–*R*Si=SiR**–R*XSi–SiX₂R* may then transform via R*XSi=SiR*–R*Si=SiXR* or directly into $\mathbf{12}_{Cl,\ Br,\ I}$. So far, we have been unable to trap the disilyne **5**, a result that neither proves or disproves the intermediacy of such a disilyne.[5]

It is interesting to note that the tetrahedrane **13** is also formed by the dehalogenation of R^*SiX_3 and $R^*X_2Si–SiX_2R^*$ (X = Cl, Br, I) with alkali metal naphthalenides in THF at low temperatures, and may be reduced further with $NaC_{10}H_8$.[5] Thus, **13** is transformed at –100 °C – according to Scheme 7.2 – into the tetrasilanediide $\mathbf{13}^{2-}$ which is easily oxidized by I_2 with re-formation of the tetrahedrane **13**. The existence of the thermolabile sodium salt 2 $Na^+ \cdot \mathbf{13}^{2-}$ in THF (not yet isolated) has been established by its reaction with MeOH, leading to the *endo,endo-bicyclo*-tetrasilane **15**, as would be expected if $\mathbf{13}^{2-}$ has the structure shown in Scheme 7.2 (the *endo,endo* species slowly isomerizes at room temperature to some extent into the *exo,endo* species).[5] On the other hand, the formation of the cyclotetrasilene $\mathbf{12}_{Me}$ from $\mathbf{13}^{2-}$ and Me_2SO_4 is better realized if the central structured element of 2 Na^+ $\mathbf{13}^{2-}$ is a (puckered) Si_4^{2-} ring with the Na^+ ions each coordinated to Si atoms on opposite sides of the Si_4^{2-} ring, which then constitutes a 6π aromatic system.

7.4 Syntheses and Reactions of the Disilenes R′HSi=SiHR′ (3) and R**ClSi=SiClR** (4) as well as the Disilynes R′Si≡SiR′ (6) and R**Si≡SiR** (7)

Based on the formation of the *tetrahedro*-tetrasilane $Si_4R^*_4$ (**13**) from R^*SiX_3 or $R^*X_2Si–SiX_2R^*$ and NaR* via disilenes R*XSi=SiXR* (**2**) (see above), it is quite certain – regardless of whether or not the the disilyne R*Si≡SiR* (**5**) represents the direct precursor of **13** – that the steric crowding of supersilyl groups R* = Si^tBu_3 would not suffice to stabilize a halogen-containing disilene RXSi=SiXR against further reactions or a disilyne RSi≡SiR against dimerization. Therefore, we planned to synthesize disilynes R′Si≡SiR′ (**6**) and R**Si≡SiR** (**7**) with the very bulky groups R′ = $SiHR^*_2$ and the extremely bulky "megasilyl" groups R** = $SiMeR^*_2$. On the way to these disilynes, two hitherto unknown disilenes, R′HSi=SiHR′ (**3**) and R**ClSi=SiClR** (**4**), were obtained.

In fact, another indication of the formation of a bulky substituted disilyne has recently been published.[11] Dehalogenation of the silane $RSiF_3$ (R = 2,6-$Mes_2C_6H_3$) with Na in boiling THF was found to afford the product **18** in Scheme 7.3. One may envisage formation of this product either via the disilyne **16** or via the silylene **17** as reaction intermediates.

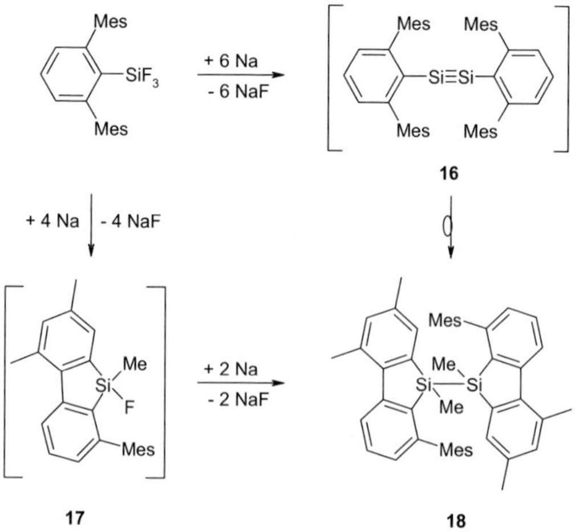

Scheme 7.3. Dehalogenation of 2,6-$Mes_2C_6H_3$-SiF_3 with Na.

7.4.1 Formation of R′HSi=SiHR′ (3) from R′HBrSi–SiBrHR′ – On the Route to an Unstable Disilyne R′Si≡SiR′ (6)

We have observed [13] that the disilane R′HBrSi–SiBrHR′ (R′ = $SiHR^*_2$), generated according to Scheme 7.4 from the silane $R'SiH_2Cl$ (obtained by the action of SiH_2Cl_2 on NaR′ [12]) and Na via the disilane $R'H_2Si–SiH_2R'$, reacts with supersilyl sodium NaR* in THF at –78 °C to form mainly the disilene R′HSi=SiHR′ (**3**) besides the cyclotetrasilene $R^*_4Si_4H_2$ (**20**). The hydrolysis- and air-sensitive disilene **3**, which is in fact the first silicon compound with hydrogen atoms bound to unsaturated silicon atoms that has been isolated as a pure substance (^{29}Si NMR for >Si=: d of d at δ = 141.32 ppm with $^1J_{SiH}$ = 149.9 and $^2J_{SiH}$ = 0.9 Hz), decomposes slowly at room temperature ($\tau_{½}$ • 3 h) with formation of the colorless isomer **19** (Scheme 7.4), and adds MeOH with formation of the colorless compound R′HSi–SiH(OMe)R′. Unfortunately, crystals of **3** suitable for an X-ray structure analysis have not yet been obtained.

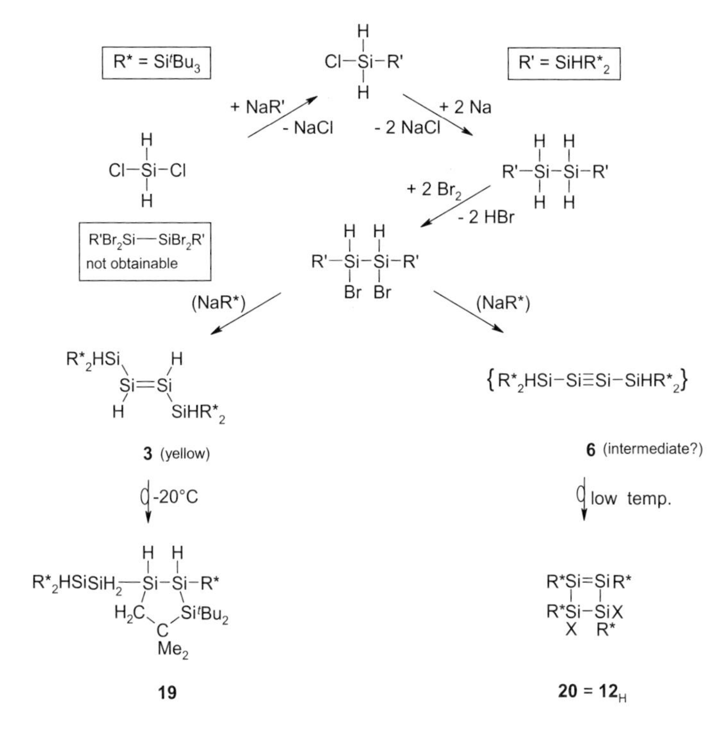

Scheme 7.4. Dehalogenation of R'HBrSi-SiHBrR'; formation and reactions of the disilene R'HSi=SiHR' (**3**) and the cyclotetrasilene $R^*_4Si_4H_2$ (**20**). (NaR*) = + NaR*/–R*Br, NaBr or –R*H, NaBr.

The red cyclotetrasilene **20** (puckered Si_4 ring; $d_{Si=Si}$ = 2.36 Å [11]) may arise via an intermediate disilyne R′Si≡SiR′ (**6**), which may then transform via the tetrasilabutadiene R*HSi=SiR*–SiR*=SiHR* ($\mathbf{6}_{iso}$) into **20**. Thus, the intermediate $\mathbf{6}_{iso}$ must be formed by a migration of two R* groups associated with a relief of the steric crowding of **6**, and must then react further by an electrocyclic process in a conrotatory sense with formation of **20**.

7.4.2 Formation of R**ClSi=SiClR** (4) from R**SiX_3 – On the Route to a Stable Disilyne R**Si≡SiR** (7)

To obtain the silanes R**SiX_3 (X = Cl, Br), which could function as precursors for R**XSi=SiXR** and R**Si≡SiR** (cf. R*SiX_3 as precursors for R*XSi=SiXR* and Si_4R*$_4$, Scheme 7.2), we – according to Scheme 7.5 – first brominated the compound R**SiH_2Cl (prepared by the action of SiH_2Cl_2 on NaR** [12]) with Br_2 in CCl_4 at 0 °C.[14] This led to an unseparated mixture of R**$SiBrCl_2$ (**21a**) and R**$SiBr_2$Cl (**21b**) in a molar ratio of ca. 2 : 1 (obviously, CCl_4 also acts as a chlorinating agent). On subsequent warming of mixture **21** with NaR* in THF from –78 °C to room temperature, the initially dark-red solution of the formed silanides **22**•NaCl was transformed into an orange solution, which may have contained the silylene R**SiCl (**22a**) and at best traces of the silylene R**SiBr (**22b**), and from which the disilene R**ClSi=SiClR** (**4**) slowly (over hours) precipitated.[14]

According to X-ray structure analysis, the disilene **4**[14] shows, like the disilene R*PhSi=SiPhR* ($\mathbf{1}_{Ph}$), a planar >Si=Si< framework ($d_{Si=Si}$ = 2.162 Å) with both the R* groups and Cl atoms in *trans* arrangements (Raman: $\tilde{\nu}$ (Si=Si) = 589 cm^{-1}). It represents the first silicon compound with halogen atoms bound to unsaturated silicon atoms that has been isolated as a pure substance. It forms very thermostable orange-red crystals (m.p. 228°C with dec.), which are practically insoluble in organic solvents. In contrast to the disilene $\mathbf{1}_{Ph}$, the disilene **4** is unreactive towards water, methanol, hydrogen fluoride, or supersilyl sodium, and even oxygen decolorizes the disilene only over a period of days.[14] Obviously, there is only a small step from R′ = SiHR*$_2$ to R** = SiMeR*$_2$, but the additional effect of the latter group in stabilizing disilenes is evident. Therefore, megasilyl R** should also be a suitable substituent for stabilizing disilyne Si_2H_2.

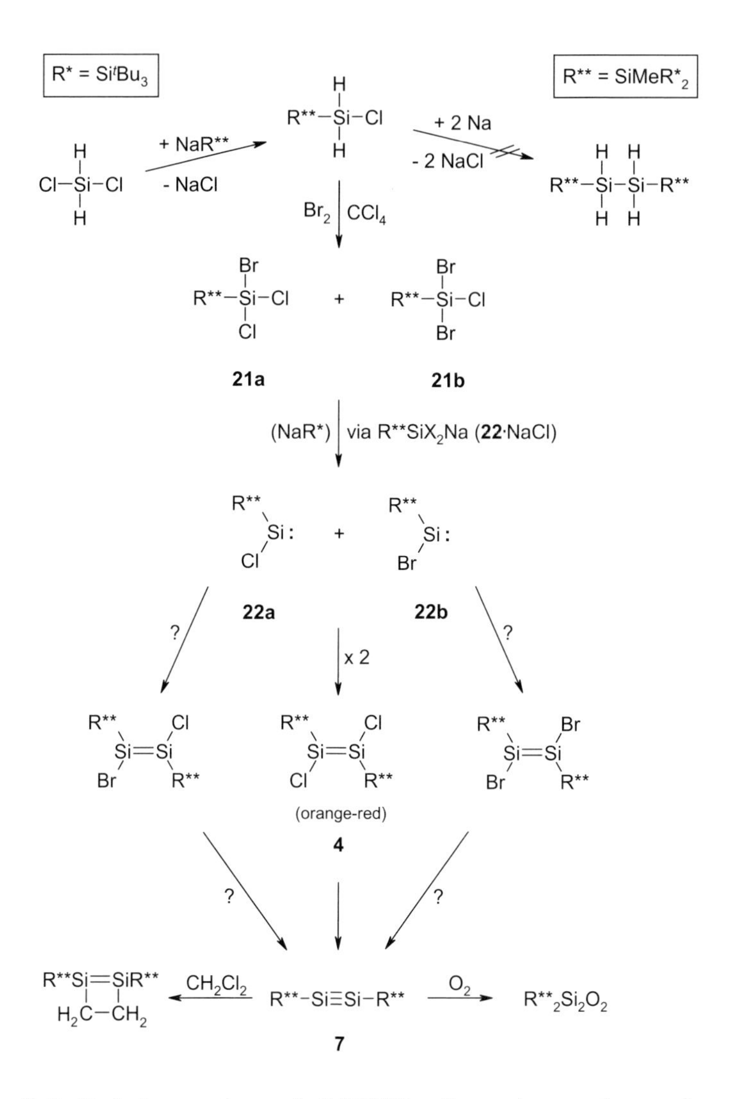

Scheme 7.5. Dehalogenation of $R^{**}SiX_3$; formation and reactions of the disilene R**ClSi=SiClR** (**4**) and the disilyne R**Si≡SiR**(**7**). (NaR) = +NaR*/–R*Br, NaCl

As a consequence of increasing bulkiness in the direction R′ → R**, only the disilane $R'H_2Si–SiH_2R'$, and not the disilane $R^{**}H_2Si–SiH_2R^{**}$, may be prepared from $RSiH_2Cl$ (R = R′, R**) and Na (cf. Scheme 7.4 and 7.5). Obviously, disilanes >R**Si–SiR**< with four-coordinated central Si atoms are not favored for steric reasons, and this also may explain the chemical inertness of **4**. Steric reasons are undoubtedly also responsible for the fact that **22a** is favoritely formed from **21** and NaR* (Scheme 7.5).

To dehalogenate **4**, we combined a suspension of the disilene in THF at –78 °C with $LiC_{10}H_8$ in THF.[14] After evaporation of all volatile products at room temperature and treating the residue with benzene, we obtained an orange-red solution. Chemical ionization (NH_3) mass spectra indicated the presence of chlorine-free molecules with the mass of the desired disilyne R**Si≡SiR** (**7**) and in addition **7** plus two oxygen atoms. This was confirmed by EI high resolution measurements. Indeed 7 is extremely oxygen-sensitive and also reacts cleanly with ethene (cf. Scheme 7.5); d_{SiSi} in the very insoluble ethene adduct 2.18 Å; $\delta(^{29}Si)$ in the dioxygen adduct +81 ppm.

7.5 Concluding remarks

Syntheses, geometric as well as electronic structures and reactions of disilenes have been well studied in the last 20 years, as is reflected in many published review articles.[2] To date, ca. 40 acyclic and several cyclic disilenes have been isolated, and many disilenes have been proved as intermediates.[2] Among the more thoroughly investigated stable acyclic disilenes, the following are worth mentioning: $^tBu_2Si{=}Si^tBu_2$, $Mes_2Si{=}SiMes_2$, and $(^tBuMe_2Si)_2Si{=}Si(SiMe_2{}^tBu)_2$, besides the isolable disilenes R*PhSi=SiPhR* (**1**), R′HSi=SiHR′ (**3**) and R**ClSi=SiClR** (**4**), which have been highlighted above. Only a few disilenes $R_2Si{=}SiR_2$ have a planar structure like **1** and **4**; most of these compounds show a *trans* bent and/or a twisted >Si=Si< central element.[2]

As shown herein and elsewhere,[1, 2] carbon and silicon exhibit many analogies with regard to their bonding and chemical properties. Indeed, they form not only saturated, but also unsaturated compounds of analogous compositions (cf. acyclic as well as cyclic alkenes and disilenes,[2] butadienes and tetrasilabutadienes,[2, 15] alkynes and disilynes), which, in addition, show analogous reaction pathways (e.g. additions, ene reactions, cycloadditions). Silicon compounds are generally much more reactive than analogous carbon compounds. This, as in other cases, is exemplified by ethene C_2H_4 and disilene Si_2H_4, both of which are thermodynamically unstable with respect to polymerization, but only the former may be isolated under normal conditions due to its metastability.

Moreover, in many cases, unsaturated carbon and silicon compounds have different conformations, the former possessing planar >C=C< or linear –C≡C– units, the latter having *trans*-bent >Si=Si< or non-linear –Si≡Si– entities, based on structure analyses [2] and ab initio calculations.[16] Furthermore, nonlinear H–Si≡Si–H does not represent the global minimum on the potential energy surface: facile isomerizations to more stable H-shifted or H-bridged isomers have been calculated.[16] Upon substitution of the H atoms by R, the isomerized species disappear from the potential energy surface of R–Si≡Si–R as R becomes bulkier.

The differences in the conformations of the >C=C< and –C≡C– groups on the one hand, and the >E=E< and –E≡E– groups (E = Si, but also the heavier tetrels Ge, Sn, Pb) on the other, can be traced back to fundametally or gradually different electronic ground states of tetrelylenes or tetrelylynes, these being a triplet or singlet for CR_2 or ER_2 and a doublet energetically near to or far away from a quartet for CR or ER.[1] As shown in Scheme 7.6, the dimerizations of tetrelylenes or tetrelylynes in the triplet/singlet state or the quartet/doublet state lead to ditetrelenes >E=E< with planar/*trans*-bent double bonds or to ditetrelynes with linear/non linear triple bonds (classical or non-classical multiple bonds).[16] Bulky substituents R may, in addition, force a twist of classical or non-classical double as well as triple bonds due to van der Waals repulsions or attractions.

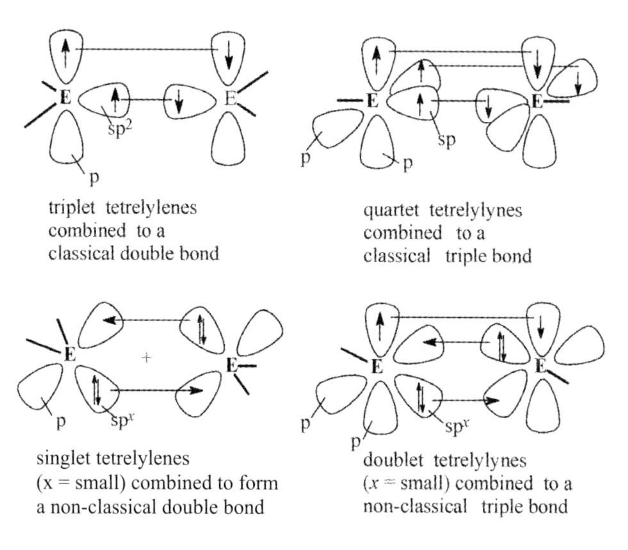

Scheme 7.6. Illustration of the formation of ditetrelenes $R_2E{=}ER_2$/ditetrelynes RE≡ER with planar/linear and non-planar/non-linear multiple bonds from tetrelylenes ER_2/tetrelylynes ER.

The non-classical double bonds $R_2E{=}ER_2$ become more and more pronounced if the singlet→triplet promotion energy of two ER_2 species increases, that is with increasing mass of E and electronegativity of R.[1] In fact, while alkenes show as a rule classical structures, only a few disilenes (see above) and no digermenes, distannenes, or diplumbenes that have been studied so far, possess a classical structure.[2, 17] Indeed, the bond shortening on going from ditetrelanes to comparable ditetrelenes is much less pronounced for disilenes and digermenes than for alkenes, and even changes to a bond lengthening in the case of distannenes and diplumbenes. In addition, the *trans* bent angles increase in direction >Si=Si<, >Ge=Ge<, >Sn=Sn<, >Pb=Pb<. The same dependences of bond lengths and *trans*-bent angles on the nature of the element and the substituent have been calculated for non-classical triple bonds RE≡ER. Interestingly, besides the triply-bonded ditetrelynes, singly-bonded isomers RE–ER were calculated to be energetically less/comparable/more stable in the cases of Si_2R_2/Ge_2R_2/Sn_2R_2,Pb_2R_2. In the latter isomers, each tetrel atom uses two of its valence p orbitals for bonding with the group R and the other tetrel atom E; the remaining p orbital is unoccupied, and the lone electron pair is thus accommodated with the s orbital. As a consequence, the central EE distance and the non-linearity are subject to an additional increase in the direction RE≡ER → RE–ER.

7.6 Acknowledgement

The work on disilenes and disilynes, described above, was undertaken together with my coworkers Dr. H. Auer, Dr. G. Fischer, Dr. W. Niedermayer, and Prof. Dr. S.-K. Vasisht. In addition, it has been achieved with the cooperation of specialists in X-ray structure analysis (Prof. H. Nöth, Dr. K. Polborn and their groups). All of them are gratefully acknowledged.

7.7 References

[1] Cf. textbooks of inorganic chemistry, e.g. N. Wiberg, *Holleman-Wiberg/Inorganic Chemistry,* 1st English Edition, Academic Press, London, **2001** (for 101st German Edition, cf. de Gruyter, Berlin **1995**).

[2] R. Okazaki, R. West, *Adv. Organomet. Chem.* **1996**, *39*, 231–273 and refs. cited therein; M. Weidenbruch in *"The Chemistry of Organic Silicon Compounds"* (Eds.: Z. Rappoport, Y. Apeloig), Vol. 3, Wiley, Chichester, **2001**, 391–428 and refs. cited therein.

[3] N. Wiberg, *Coord. Chem. Rev.* **1997**, *163*, 217–252 and refs. cited therein.

[4] N. Wiberg, W. Niedermayer, *J. Organomet. Chem.* **2001**, *628*, 57–64 and refs. cited therein.
[5] N. Wiberg, H. Auer, S. Wagner, K. Polborn, G. Kramer, *J. Organomet. Chem.* **2001,** *619*, 110–131.
[6] N. Wiberg, W. Niedermayer, K. Polborn, *Z. Anorg. Allg. Chem.* **2002**, *628*, 1045–1052.
[7] N. Wiberg, H. Auer, W. Niedermayer, H. Nöth, H. Schwenk-Kircher, K. Polborn, *J. Organomet. Chem.* **2000**, *612*, 141–159 and refs. cited therein.
[8] N. Wiberg, W. Niedermayer, K. Polborn, P. Mayer, *Chem. Eur. J.* **2002**, *8*, 2730–2739 and refs. cited therein.
[9] N. Wiberg, W. Niedermayer, H. Nöth, J. Knizek, W. Ponikwar, K. Polborn, *Z. Naturforsch., Part B* **2000**, *55*, 389–405.
[10] N. Wiberg, H. Auer, H. Nöth, J. Knizek, K. Polborn, *Angew. Chem., Int. Ed.* **1998**, *37*, 2869–2872.
[11] R. Pietschnig, R. West, D. R. Powell, *Organometallics* **2000**, *19*, 2724–2729.
[12] N. Wiberg, W. Niedermayer, H. Nöth, M. Warchhold, *J. Organomet. Chem.* **2001**, *628*, 46–46 and refs. cited therein.
[13] N. Wiberg, W. Niedermayer, H. Nöth, M. Warchhold, *Z. Anorg. Allg. Chem.* **2001**, *627*, 1717–1722.
[14] N. Wiberg, W. Niedermayer, G. Fischer, H. Nöth, M. Suter, *Eur. J. Inorg. Chem.* **2002**, 1066–1070 and refs. cited therein.
[15] M. Weidenbruch, chapter 8 herein.
[16] M. Karni, Y. Apeloig in *"The Organic Chemistry of Silicon Compounds"* (Eds.: Z. Rappoport, Y. Apeloig), Vol. 3, Wiley, Chichester, **2001**, 1–163 and refs. cited therein; N. Tagaki and S. Nagase, *Organometallics* **2001**, *20*, 5498–5500 and refs. cited therein.
[17] K. W. Klinkhammer in *The Organic Chemistry of Germanium, Tin and Lead"* (Eds.: Z. Rappoport, Y. Apeloig), Vol. 2, Wiley, Chichester, **2002**.

8 A Tetrasilabuta-1,3-diene and Related Compounds with Conjugated Multiple Bonds

Manfred Weidenbruch*

8.1 Introduction

The first molecule with an Si=Si double bond, the classical disilene **2**, was prepared more than 20 years ago by dimerization of dimesitylsilylene generated photolytically from the trisilane **1**.[1]

$$\mathrm{Mes_2Si(SiMe_3)_2}\ (\mathbf{1}) \xrightarrow[-(\mathrm{Me_3Si})_2]{h\nu} [\mathrm{Mes_2Si:}] \xrightarrow{\times 2} \mathrm{Mes_2Si{=}SiMes_2}\ (\mathbf{2}) \qquad (1)$$

Mes = 2,4,6-trimethylphenyl

Since then, over 40 more acyclic and cyclic disilenes have been prepared and more than half of them have been structurally characterized. Most of these compounds are tetraaryldisilenes, although some examples with alkyl, silyl, or amino substituents are also known.[2,3]

$$\mathrm{Si} \overset{d}{=} \mathrm{Si} \qquad \tau \qquad \theta \qquad (2)$$

Characteristic features of the structures of the disilenes are the length of the Si=Si double bond d, the twist angle τ, and the *trans*-bent angle θ. In contrast to the double bonds of sterically crowded alkenes, in which variations of the bond lengths are small, the Si=Si bond lengths of disilenes vary between 214 and 229 pm. A twisting of the two SiR_2 planes can also occur, as reflected by the twist angle τ, which varies between 0° and 25°. As a result of this type of distortion, which has also been observed in alkenes with bulky substituents, close contacts between the usually voluminous groups are avoided.

A further peculiarity of the disilenes, not observed in alkenes, is the possibility of *trans*-bending of the substituents, which is described by the *trans*-bent angle Θ between the R_2Si planes and the Si=Si vector. This effect, which

* Address of the author: Fachbereich Chemie, Universität Oldenburg, Carl-von-Ossietzky-Straße 9-11, D-26111 Oldenburg, Germany

can lead to *trans*-bent angles of up to 34°, can be rationalized as follows. Carbenes have either a triplet ground state (T) or a singlet ground state (S) with relatively low $S \rightarrow T$ transition energies. The familiar picture of a C=C double bond results from the approach of two triplet carbenes, as shown in Figure 8.1a.

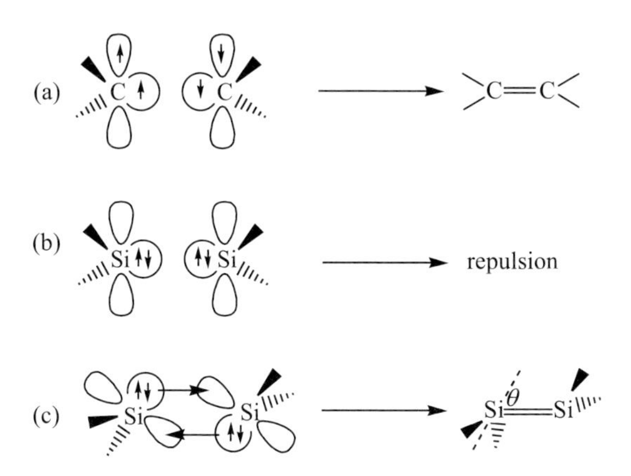

Figure 8.1. Double bond formation from two triplet carbenes and singlet silylenes.

In contrast to carbenes, all silylenes have singlet ground states with a relatively large $S \rightarrow T$ excitation energy.[3] Very recently, it was found that the sterically encumbered silylene $i\mathrm{Pr_3SiSiSi}t\mathrm{Bu_3}$ is an exception and probably has a triplet ground state.[4] Approach of two singlet silylenes should usually result in repulsion rather than in bond formation (Figure 8.1b). However, if the two electron-sextet molecules are rotated with respect to each other, interactions between the doubly-occupied s-type orbitals and the vacant p orbitals can also lead to the creation of a double bond, which, in contrast to that in alkenes, arises by way of a double donor-acceptor adduct formation. This is accompanied by a *trans*-bending of the substituents about the Si=Si vector (Figure 8.1c).

The authors group addressed the question as to whether molecules with two or more neighboring Si=Si double bonds can be synthesized. This chapter details the isolation of a tetrasilabutadiene and a tetragermabutadiene, as well as the formation of stable or intermediate compounds with conjugated Si=C and Ge=C double bonds.

8.2 Hexaaryltetrasilabuta-1,3-diene

8.2.1 Synthesis and Properties

Recently, we obtained the first and, as yet, only tetrasilabuta-1,3-diene **6** from the tetraaryldisilene **3**[5,6] as follows. The disilene was treated with excess lithium to give the putative disilenyllithium compound **4**. In the second step of the reaction sequence, mesityl bromide was added in the expectation that the bulk of this aryl group and the poor solubility of mesityllithium would favor halogenation over the competing transarylation. In fact, the bromodisilene **5** does appear to be formed smoothly but, like **4**, has not yet been unambiguously identified. Intermolecular cleavage of lithium bromide from the two intermediates **4** and **5** then furnished the tetrasilabutadiene **6** in up to 60% yield.[7]

$$R_2Si{=}SiR_2 + 2\,Li \xrightarrow{-LiR} R_2Si{=}Si(R)Li$$

3 → **4**

$$\mathbf{4} + MesBr \xrightarrow{-LiMes} R_2Si{=}Si(R)Br \xrightarrow[-LiBr]{+\mathbf{4}} R_2Si{=}Si(R){-}Si(R){=}SiR_2 \qquad (3)$$

5 → **6**

R = 2,4,6-triisopropylphenyl

The reddish-brown crystals of **6** are thermally very stable but extremely sensitive to air. In the solid state, and presumably also in solution, **6** exists in the s-*gauche* form with a dihedral angle for the Si_4 framework of 51°. Similar to the situation in other molecules with conjugated double bonds, the Si=Si double bonds are slightly elongated [217.5(2) pm] while the central Si–Si single bond length of 232.1(2) pm is remarkably short in view of its bulky substituents. The preference of **6** for the s-*gauche* form is in sharp contrast to the analogous carbon compounds, which mostly prefer the s-*trans* configuration. Conjugation between the two Si=Si double bonds has been confirmed by the electronic spectrum of **6**. The band at longest wavelength experiences a bathochromic shift of more than 100 nm in comparison of those of disilenes with analogous substitution patterns.[7]

8.2.2 Reactions

8.2.2.1 Cycloadditions. Although **6** exists in the solid state in the s-*gauche* form, which should favor [4+2] cycloadditions of the Diels-Alder type, to date all attempted reactions with alkenes, acetylenes, and the C=O bonds of ketones have remained unsuccessful. The reasons for this are probably the steric overcrowding in **6** as well as the large 1,4-separation of 540 pm between the terminal silicon atoms. Thus, it was at first surprising to find that compound **6** reacted with maleic anhydride through a formal [4+4] cycloaddition to furnish the tetracyclic compound **7**.[8]

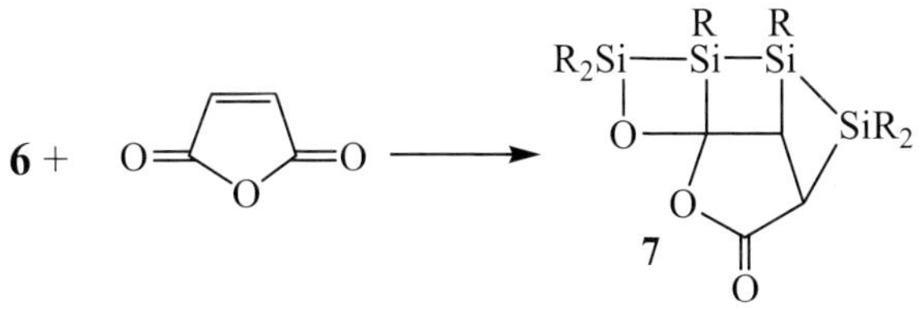

(4)

The mechanism of formation of **7** is not yet known. A feasible sequence involves a [2+2] cycloaddition of one of the Si=Si double bonds of **6** to the highly reactive C=O group, followed by a second cycloaddition of the remaining Si=Si double bond across the C=C double bond to complete the formation of the final product **7**.[8] To obtain a better insight into the course of this reaction, we have also treated a large number of structurally related compounds with **6** under otherwise identical conditions. However, in no case did we find any sign of a reaction.[8]

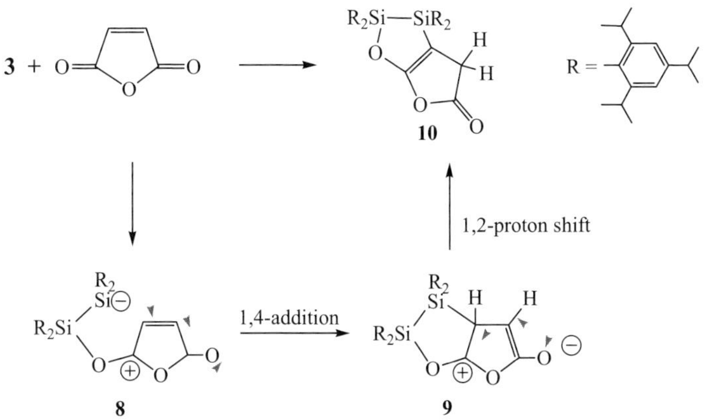

(5)

Since maleic anhydride apparently exhibits a high reactivity towards Si–Si multiple bonds, we have also examined its reaction with the disilene **3**, from which the bicyclic compound **10** was isolated. The driving force for the formation of **10** is assumed to be the oxophilicity of silicon. This leads to the dipolar addition product **8** that, in turn, affords the new intermediate **9** through 1,4-addition. The last step of the sequence would then be a 1,2-proton shift to produce **10**.[8]

$$6 + 1/8\ S_8 \longrightarrow \text{cyclo-}[R_2Si\text{–}Si(R){=}Si(R)\text{–}SiR_2\text{–}S]\quad \mathbf{11}$$

$$6 + Et_3P\text{—}X \xrightarrow{-Et_3P} \text{cyclo-}[R_2Si\text{–}Si(R){=}Si(R)\text{–}SiR_2\text{–}X]\quad \mathbf{12}\ X = Se;\ \mathbf{13}\ X = Te \tag{6}$$

In view of the low tendency of **6** to undergo cycloaddition reactions, it was surprising that treatment of **6** with sulfur afforded the five-membered ring **11** in a formal [4+1] cycloaddition. Upon addition of catalytic amounts of triethylphosphane, selenium and tellurium reacted with **6** to give the analogous rings **12** and **13** in high yield. The five-membered rings **11**–**13** are nearly planar (sum of angles 539.5°) and reveal Si=Si double bond lengths ranges from 217.0(1) pm to 219.8(2) pm.[9]

Although the mechanism of formation of **11**–**13** cannot be proven experimentally, the following proposal seems to be reasonable. In analogy to the reaction of disilenes,[10] the reaction sequence could be initiated by a [2+1] cycloaddition of a chalcogen atom to one of the Si=Si double bonds, followed by a rearrangement of these intermediates into the less strained five-membered rings **11**–**13**.[9]

$$\mathbf{14} \xleftarrow{O_2} 6 \xrightarrow{m\text{CPBA}} \mathbf{15} \tag{7}$$

Passage of dry air through a solution of **6** leads to the formation of colorless crystals of **14**. This molecule consists of two 2,4-disiladioxetane rings linked by an oxygen atom. In analogy to the reactions of disilenes,[2,10] the reaction sequence involves a two-fold cycloaddition of atmospheric oxygen to the Si=Si double bonds, followed by a rearrangement to furnish the 2,4-disiladioxetane rings. In this context the two double bonds of **6** behave like isolated disilenes. However, the insertion of oxygen into the highly shielded Si–Si single bond with formation of a disiloxane unit at room temperature is very unusual and can be considered as additional evidence for the high reactivity of the tetrasilabutadiene.[12]

On the other hand, treatment of **6** with *meta*-chloroperbenzoic acid (*m*CPBA) afforded the product **15** consisting of two disiladioxetane rings with retention of the Si–Si single bond.[12]

8.2.2.2 1,2-Addition Reactions. While cycloadditions to **6** are still exceedingly rare, reactions of **6** with small molecules, as summarized in Scheme 8.1, were more successful. Treatment of the tetrasilabutadiene with small amounts of water led, via the 1,2-addition product **16** to the rearranged oxatetrasilacyclopentane **17**, an analogue of tetrahydrofuran. With an excess of water the tetrasilane-1,4-diol **18** was obtained, which showed no tendency to eliminate water with the formation of **17**.[12]

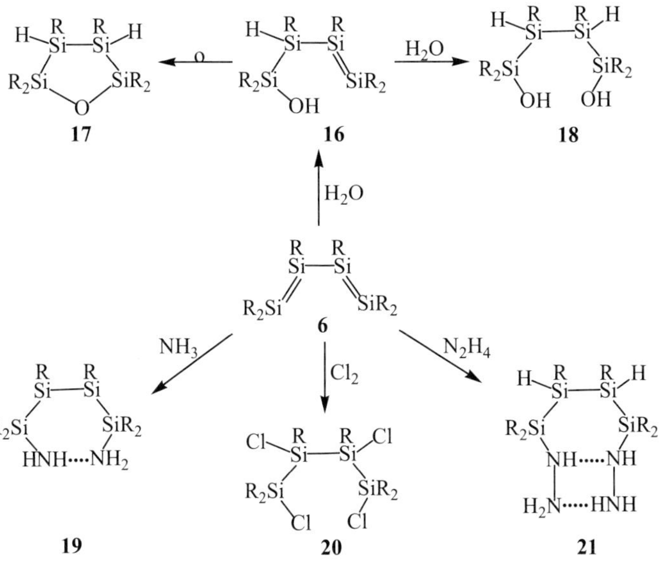

Scheme 8.1. 1,2-Addition reactions to the tetrasilabutadiene.

When dry ammonia is passed over the dark red solution of **6**, yellow crystals of the twofold 1,2-addition product **19** are formed in high yield. This result is somewhat surprising because attempted additions of ammonia to the disilene **3** were unsuccessful. This observation again reflects the increased reactivity of **6** in comparison to the isolated double bond in **3** and other disilenes.

On account of the presence of two stereogenic centers in **19**, the existence of a diastereomeric *meso*-form in addition to the enantiomeric *R*,*R*- and *S*,*S*-forms is possible. An X-ray crystallographic analysis of the crystals of **19** revealed the existence of a conglomerate of enantiomerically pure substances. Such conglomerates are still rather rare in comparison to the large number of similar molecules that crystallize as racemates.[13] A further special feature in the structure of **19** is the small separation between the nitrogen atoms of 300 pm, which is indicative of a dynamic hydrogen bridging system NH···N.[14]

The 1,4-dihydrazinotetrasilane **21**, obtained from **6** and anhydrous hydrazine, also exists in the solid state as a conglomerate of enantiomerically pure crystals. The separations between the two pairs of nitrogen atoms are very short (303 pm and 315 pm) and thus suggest the presence of two dynamic NH···N bridges.[15]

In order to determine whether other 1,2-addition products at the two double bonds of **6** also crystallize as conglomerates, we treated **6** with chlorine. Although the individual molecules of the tetrachlorotetrasilane **20** are also chiral, this compound crystallizes as a racemate.[14]

*8.2.2.3 1,4-Additions***.** In contrast to the numerous 1,2-additions, 1,4-additions to the terminal silicon atoms of **6** were unknown until quite recently. Since X–H and X–Cl bonds participate in smooth addition processes with the double bonds of **6**, we attempted to address the question as to which of these bonds reacts preferentially with **6**. Thus, trichlorosilane, which contains not only Si–H but also Si–Cl bonds, was added to a solution of **6**. However, the data obtained for the isolated bright-orange crystals revealed that the expected addition of trichlorosilane to **6** had not occurred. Instead, a 1,4-addition of hydrogen chloride with formation of **22** had taken place. Analogously, the reaction of **6** with hydrogen bromide, slowly generated by the action of trifluoroacetic acid on lithium bromide, furnished intensely orange-colored crystals of the 1,4-adduct **23**.[16]

R R
Si=Si
R₂Si SiR₂
H Cl
22

$\xleftarrow{HSiCl_3}$ **6** $\xrightarrow{LiBr/CF_3COOH}$

R R
Si=Si
R₂Si SiR₂
H Br
23

(8)

The X-ray structure analyses of the unsymmetrically substituted disilenes **22** and **23** revealed the presence of somewhat stretched Si=Si and Si–Si bonds. It is not clear as to why no 1,2-additions of the hydrogen halides to the double bonds of **6** occur in these reactions or why instead the previously unobserved 1,4-additions with formation of new double bonds take place. The closest precedent involves the action of small amounts of water on **6**, which is presumably initiated by a 1,2-addition and subsequent 1,3-hydrogen shift to form compound **17**.

8.3 Hexaaryltetragermabuta-1,3-diene

The successful synthesis of the tetrasilabutadiene **6** raised the question as to whether an analogous compound with conjugated Ge=Ge double bonds would be accessible by following a similar synthetic approach. As a starting compound we chose the tetraaryldigermene **25**, which was previously prepared in a multistage procedure starting from a germanium tetrahalide.[17,18] We have found that the reaction of the germanium(II) amide **24**[19] with the appropriate aryllithium compound affords the digermene **25** in one step and in an acceptable yield. The structural integrity of **25** in solution was confirmed by its reactions with diazomethane and trimethylsilyldiazomethane, from which the three-membered rings **26a,b** were obtained in high yield.[20]

$:Ge[N(SiMe_3)_2]_2 + 2\ LiR \longrightarrow R_2Ge: \xrightarrow{x\ 2} R_2Ge{=}GeR_2$ (**24** → **25**)

25 + $CHR'N_2$ ⟶ **26**

26a R' = H
26b R' = $SiMe_3$ (9)

R =

Treatment of **25** with an excess of lithium did not afford the digermenyllithium compound **28**, but instead we obtained dark-red crystals of the ionic compound **27**, which contains an allyl-like Ge_3 anion as part of a four-membered ring. Within the planar ring, the Ge–Ge multiple bond lengths (236.79 pm) differ substantially from the Ge–Ge single bond lengths (251.16 pm).[21]

$R_2Ge{=}GeR_2$ (**25**) $\xrightarrow[-\ 3\ LiR]{+\ 4\ Li/DME}$ $[Li(dme)_3]^+$ [RGe(Ge R)(Ge R_2)GeR]$^-$ (**27**)

25 $\xrightarrow[-\ LiR]{+\ 2\ Li}$ $R_2Ge{=}Ge(R)Li$ (**28**) $\xrightarrow[-\ LiMes]{MesBr}$ $R_2Ge{=}Ge(R)Br$ (**29**) $\xrightarrow[-\ LiBr]{+\ \mathbf{28}}$ **30**

Scheme 8.2. R = 2,4,6-iPr$_3$C$_6$H$_2$, Mes = 2,4,6-Me$_3$C$_6$H$_2$.

Shortening the reaction time to such an extent that most of the digermene had reacted before the formation of **27** could become the main reaction did indeed lead to compound **28**, which, by subsequent reaction with mesityl bromide, presumably resulted in the bromine derivative **29**, from which the tetragermabutadiene **30** was formed by intermolecular coupling with elimination of LiBr.

Similar to **6**, compound **30** adopts the s-*gauche* form with a dihedral angle of the Ge_4 framework of merely 22°. The Ge=Ge double bonds (average value 235 pm) and the Ge–Ge single bond (245.81 pm) are in the typical ranges for such bond lengths. Both double bonds display a considerable *trans*-bending of the substituents from the respective Ge=Ge vector (35.4° and 31.1°) and torsion angles of 22.4° and 21.3°. The conjugation between the two double bonds is confirmed by the electronic spectrum of the dark blue solution of **30** which shows a longest-wavelength absorption of 560 nm. Compared with the yellow or orange digermenes, this absorption is shifted bathochromically by about 140 nm.[21]

4 GeCl2 · dioxane + 6 RMgBr + Mg $\xrightarrow{-\ MgX2}$ $R_2Ge{=}Ge(R){-}Ge(R){=}GeR_2$ (10)

31 **30**

A disadvantage of the method depicted in Scheme 8.2 is the low yield of the tetragermabutadiene of merely 10%. Very recently, we have found that the reaction of the germanium(II) chloride dioxane complex with the Grignard compound **31** affords compound **30** in one step and in a yield of more than 30%.[22] We will next investigate whether **30** can participate in similar reactions to those of its silicon analogue **6**.

8.4 Acetylene-Linked Bis(silaethenes) and Bis(germaethenes)

Photolysis of hexa-*tert*-butylcyclotrisilane **32**[23] leads to the silylene **33** and the disilene **34** through simultaneous cleavage of two Si–Si bonds. In contrast to the aryl-substituted silylenes, which undergo dimerization to give further disilene, the short-lived silylene **33** and the marginally stable disilene **34** remain together in the form of a cage and only separate when appropriate trapping reagents are available.[24]

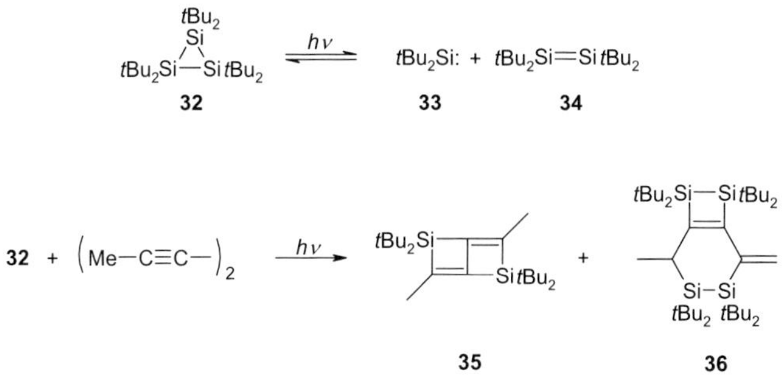

(11)

The high reactivity of both intermediates is demonstrated by their reactions with di- and polyynes from which different products can be isolated depending on the irradiation time. For example, photolysis of **32** in the presence of hexa-1,3-diyne furnishes the bicyclic compounds **35** and **36**, which are presumably formed through [2+1] and [2+2] cycloadditions of **33** and **34** to the triple bonds of the diyne.[25]

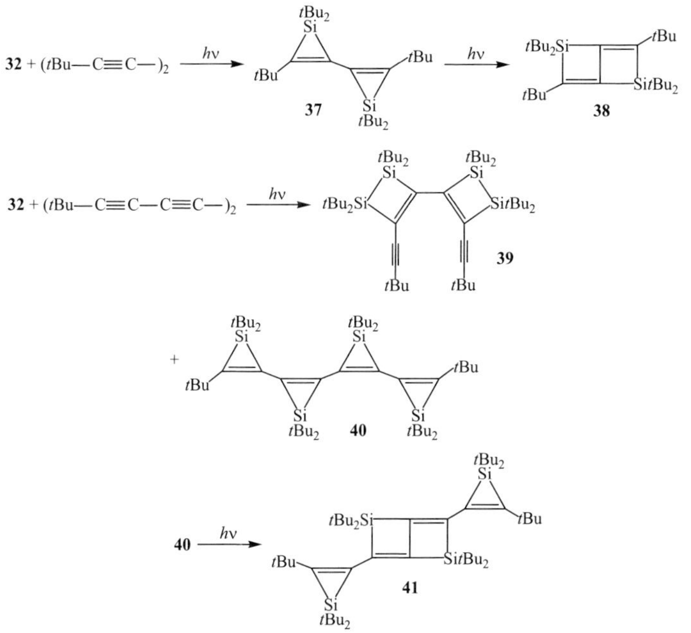

(12)

Insights into the reactivity of **33** are provided by its reaction with 2,2,7,7-tetramethylocta-3,5-diyne, from which the C–C-linked bis(silacyclopropene) **37** can be isolated. Renewed photolysis of **37** furnishes the rearranged bicyclic compound **38**.[26] The reaction behavior of **33** and **34**, formed by photolysis of **32**, is even more transparent when the photolysis of **32** in the presence of an octa-1,3,5,7-tetrayne is considered. This reaction affords three products. [2+2] Cycloadditions of two molecules of the disilene **34** to the two inner triple bonds of the tetrayne lead to product **39** having an s-*cis*-butadiene framework bridged by two disilene moieties. Addition of the silylene **33** to each of the four C≡C triple bonds affords the C–C linked quatersilirene **40**, whose carbon skeleton has the s-*cis-trans-cis* configuration, presumably in order to minimize steric interactions between the bulky *tert*-butyl groups. Upon prolonged photolysis, **40** rearranges to the bicyclic compound **41**.[27]

$$R_2Ge{=}GeR_2 \;\rightleftharpoons\; 2\,R_2Ge: \qquad \mathbf{42} \qquad \mathbf{43} \qquad R = \text{2-}tert\text{-butyl-4,5,6-trimethylphenyl}$$

$$(R'{-}C{\equiv}C{-})_2 + 2\,\mathbf{43} \longrightarrow R_2Ge{=}C(R'){-}C{\equiv}C{-}C(R'){=}GeR_2 \qquad (13)$$

44 R' = *n*Bu; **45** R' = Ph

The reaction of the germylene **43**, formed from the digermene **42** in solution, with 1,3-diynes proceeds differently and gives the products **44** and **45** as the first molecules with conjugated Ge=C double bonds.[28] The conjugation of the two Ge=C double bonds across the linking acetylene bridge is reflected in the UV/vis spectra rather than in the bond lengths. The absorptions at longest wavelength at 518 nm for **44** and 595 nm for **45**, in comparison to the yellow or orange germaethenes,[29] are bathochromically shifted by 100 nm or more.[28] In analogy to the formation of C–C-linked bis(silacyclopropenes), it can be assumed that the reaction is again initiated by a twofold [1+2] cycloaddition of the germylene **43** to the two triple bonds of the 1,3-diyne. Unlike with the bis(silacyclopropenes), however, the next step does not involve a rearrangement to the bicyclic molecules, but instead affords the acyclic molecules **44** and **45** through cleavage and the formation of new bonds. However, the conjugation between the two double bonds does not result in a higher reactivity. Thus, for example, these products do not react with phosphaalkynes or dienes, whereas simple germaethenes form the corresponding addition products.[30] Only with compounds containing electron-poor multiple bonds do addition reactions occur.[31]

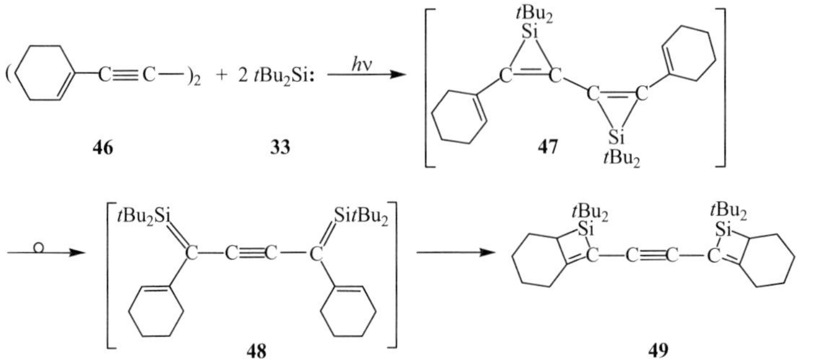

(14)

The successful synthesis of the acetylene-linked bis(germaethenes) posed the question as to whether analogous compounds with conjugated Si=C double bonds could at least be detected through the use of suitable substituents on the 1,3-diynes. The reaction of the silylene **33** with the 1,4-bis(cyclohex-1-enyl)-substituted 1,3-diyne **46** furnished colorless crystals of compound **49** in high yield. Like all other additions of silylenes to di- and oligoynes, the reaction sequence probably begins with the formation of the cycloadduct **47**. Subsequent opening of the Si–C bonds would then lead to intermediate **48**, which, in this case, does not react further to afford a bicyclic product of type **35**, but rather undergoes twofold [2+2] cycloadditions with the double bonds of the cyclohexene rings to furnish the isolated product **49**.[32]

In an attempt to make this intramolecular [2+2] cycloaddition less easy and thus, perhaps, to effect stabilization of an acetylene-linked bis(silaethene) of type **48**, we increased the steric bulk of the substituents at the double bonds. However, in this case, only a rearranged bicyclic compound of type **35** was isolated.[32]

8.4 Acknowledgement

Financial support of our work by the Fonds der Chemischen Industrie is gratefully acknowledged.

5 References

[1] R. West, M. J. Fink, J. Michl, *Science* **1981**, *214*, 1343.

[2] For recent reviews, see: a) R. Okazaki, R. West, *Adv. Organomet. Chem.* **1996**, *39*, 231; b) M. Weidenbruch in *The Chemistry of Organic Silicon Compounds, Vol. 3* (Eds.: Z. Rappoport, Y. Apeloig), Wiley, Chichester, **2001**, p. 391.

[3] For recent reviews, see: a) P. P. Gaspar, R. West, in *The Chemistry of Organic Silicon Compounds, Vol. 2* (Eds.: Z. Rappoport, Y. Apeloig), Wiley, Chichester, **1998**, p. 2463; b) M. Weidenbruch, *Coord. Chem. Rev.* **1994**, 275.

[4] P. Jiany, P. P. Gaspar, *J. Am. Chem. Soc.* **2001**, *123*, 8622. However, see: M. Yoshida, N. Tamaoki, *Organometallics* **2002**, *21*, 2587.

[5] H. Watanabe, K. Takeuchi, N. Fukawa, M. Kato, M. Goto, Y. Nagai, *Chem. Lett.* **1987**, 1341.

[6] A. J. Millevolte, D. R. Powell, S. G. Johnson, R. West, *Organometallics* **1992**, *11*, 1091.

[7] M. Weidenbruch, S. Willms, W. Saak, G. Henkel, *Angew. Chem.* **1997**, *109*, 2612; *Angew. Chem. Int. Ed. Engl.* **1997**, *36*, 2503.

[8] S. Boomgaarden, W. Saak, M. Weidenbruch, H. Marsmann, *Organometallics* **2001**, *20*, 2451.

[9] A. Grybat, S. Boomgaarden, W. Saak, H. Marsmann, M. Weidenbruch, *Angew. Chem.* **1999**, *111*, 2161; *Angew. Chem. Int. Ed.* **1999**, *38*, 2010.

[10] a) R. West, D. J. De Young, K. J. Haller, *J. Am. Chem. Soc.* **1985**, *107*, 4942; b) J. E. Mangette, D. R. Powell, R. West, *Organometallics* **1991**, *10*, 546.

[11] K. L. McKillop, G. R. Gillette, D. R. Powell, R. West, *J. Am. Chem. Soc.* **1992**, *114*, 5203.

[12] S. Willms, A. Grybat, W. Saak, M. Weidenbruch, H. Marsmann, *Z. Anorg. Allg. Chem.* **2000**, *626*, 1148.

[13] E. L. Eliel, S. H. Wilen, L. N. Mander, *Stereochemistry of Organic Compounds*, Wiley, New York, **1994**.

[14] S. Boomgaarden, W. Saak, M. Weidenbruch, H. Marsmann, *Z. Anorg. Allg. Chem.* **2001**, *627*, 349.

[15] S. Boomgaarden, W. Saak, H. Marsmann, M. Weidenbruch, *Z. Anorg. Allg. Chem.* **2001**, *627*, 805.

[16] S. Boomgaarden, W. Saak, H. Marsmann, M. Weidenbruch, *Z. Anorg. Allg. Chem.* **2002**, *628*, 1745.

[17] J. Park, S. A. Batcheller, S. Masamune, *J. Organomet. Chem.* **1989**, *367*, 39.

[18] W. Ando, H. Itoh, T. Tsumuraya, *Organometallics* **1989**, *8*, 2759.

[19] M. J. S. Gyane, D. H. Harris, M. F. Lappert, P. P. Power, P. Riviére, M. Riviére-Baudet, *J. Chem. Soc., Dalton Trans.* **1977**, 2004.

[20] H. Schäfer, W. Saak, M. Weidenbruch, *Organometallics* **1999**, *18*, 3159.

[21] H. Schäfer, W. Saak, M. Weidenbruch, *Angew. Chem.* **2000**, *112*, 3847; *Angew. Chem. Int. Ed.* **2000**, *39*, 3703.

[22] G. Ramaker, A. Schäfer, W. Saak, M. Weidenbruch, *Organometallics* **2003**, *22*, in press.

[23] A. Schäfer, M. Weidenbruch, K. Peters, H. G. von Schnering, *Angew. Chem.* **1984**, *96*, 311; *Angew. Chem. Int. Ed. Engl.* **1984**, *23*, 302.
[24] M. Weidenbruch, *Chem. Rev.* **1995**, *95*, 1479.
[25] L. Kirmaier, M. Weidenbruch, H. Marsmann, K. Peters, H. G. von Schnering, *Organometallics* **1998**, *17*, 1237.
[26] D. Ostendorf, L. Kirmaier, W. Saak, H. Marsmann, M. Weidenbruch, *Eur. J. Inorg. Chem.* **1999**, 2301.
[27] D. Ostendorf, W. Saak, M. Weidenbruch, H. Marsmann, *Organometallics* **2000**, *19*, 4938.
[28] F. Meiners, W. Saak, M. Weidenbruch, *Organometallics* **2000**, *19*, 2835.
[29] For germaethenes, see: a) H. Meyer, G. Baum, W. Massa, A. Berndt, *Angew. Chem.* **1987**, *99*, 790; *Angew. Chem. Int. Ed. Engl.* **1987**, *26*, 798; b) M. Lazraq, J. Escudié, C. Couret, J. Satgé, M. Dräger, R. Dammel, *Angew. Chem.* **1988**, *100*, 885; *Angew. Chem. Int. Ed. Engl.* **1988**, *27*, 826; c) N. Tokitoh, K. Kishikawa, R. Okazaki, *J. Chem. Soc., Chem. Commun.* **1985**, 1425; d) M. Stürmann, W. Saak, M. Weidenbruch, A. Berndt, D. Scheschkewitz, *Heteroat. Chem.* **1999**, *10*, 554; e) F. Meiners, W. Saak, M. Weidenbruch, *Chem. Commun.* **2001**, 215.
[30] Review: K. M. Baines, W. G. Stibbs, *Adv. Organomet. Chem.* **1996**, *99*, 275.
[31] F. Meiners, D. Haase, R. Koch, W. Saak, M. Weidenbruch, *Organometallics* **2002**, 3990.
[32] D. Ostendorf, W. Saak, D. Haase, M. Weidenbruch, *J. Organomet. Chem.* **2001**, *636*, 7.

Part II

Si-Si Systems: From Molecular Building Blocks to Extended Networks

Networks with Si-Si linkages do not occur in Nature due to the sensitivity of the Si-Si bond towards air and moisture. Nevertheless, there are several man-made materials, which have already gained technological importance or look set to become important in the near future. In the following 14 chapters, Si-Si bond containing species of different size and dimensionality are presented, from low–molecular weight units to three-dimensional extended networks.

For the formation of Si-Si bonds, silyl- and oligosilylanions are useful reagents in nucleophilic substitution or electron-transfer reactions. In Chapter 9, C. Marschner et al. describe a novel access to oligosilyl potassium compounds by splitting of Si-Si bonds with potassium alkoxides. Further metal-exchange reactions allow a tuning of the nucleophilicity of the respective oligosilyl anion. In Chapter 10, K. Hassler et al. report on the competition between Si-Si and Si-P bond splitting in the reaction with potassium *tert*-butoxide; the steric demand of the silyl groups determines the regiospecificity.

Elongation of the Si-Si chain in oligosilanes leads to the important class of polysilanes. Conjugation within the all-silicon backbone is the reason for the interesting physical and chemical properties of these polymers. They are photoluminescent, photoconductive, conductive on chemical doping, and photosensitive to UV light. In Chapter 11, R. G. Jones gives a basic introduction to the synthesis, structure and properties of polysilanes of the type $(R_2Si)_n$. Several properties of these polymers depend on the conformation they adopt. The mobility of alkyl side chains and the resulting phase behavior of n-alkyl–substituted polysilanes is described in the contribution of H. Frey, C. Schmidt et al. in Chapter 12.

Quite different numbers and arrangements of Si atoms are present in the anionic components of metal silicides belonging to the class of Zintl phases. In Chapter 13, R. Nesper presents electron counting rules and bonding concepts to describe the different kinds of silicide structures and to explain the electronic behavior of the respective compounds (isolator, semiconductor, conductor). In Chapter 14, J. Evers and G. Oehlinger describe very high pressure effects on the Zintl phases MSi_2 with M = Ca, Sr, Ba, Eu. The high–pressure phases of these compounds, prepared in a belt apparatus at 40 Kbar, crystallize in structures which are related to that of the superconducting compound MgB_2.

The Zintl phase $CaSi_2$, with its puckered $(Si^-)_n$ polyanion layers, has been the starting point for an interesting development, which began as long ago as in 1863 with a contribution from F. Wöhler: in a topochemical reaction with aqueous HCl, the sheet polymer $(Si_2HOH)_n$, called “Wöhler-siloxene”, is formed, the photoluminescence of which resembles that recently observed for “porous silicon”. The search for electroluminescence in such materials is the motivation for current research activities. More information is given in the contributions of M. S. Brandt et al. (Chapter 15) and H. Stüger (Chapter 16). The group of M. S. Brandt could verify the epitaxial growth of $CaSi_2$ and other Zintl phases on crystalline silicon or germanium substrates and the topotactic transformation of these materials into novel puckered sheet polymers. Partial substitution of the backbone silicon by germanium is used to tune the luminescence properties. Modified conditions in the reaction of $CaSi_2$ with aqueous HCl lead to the so-called “Kautsky-siloxene”. H. Stüger has prepared polymeric materials with similar properties, including intense photoluminescence, by selective hydrolysis and subsequent condensation of cyclic oligosilanes bearing hydrolytically labile substituents.

The next step from Silicon-based sheet polymers to tubular structures has been performed so far only on the computer. In Chapter 17, Th. Frauenheim et al. describe the structure and the electronic properties of silicide, silane and siloxane nanotubes. The structures can be understood in terms of conventional graphitic carbon nanotubes by replacing the flat hexagons by puckered rings. The electronic properties depend on the tube diameter. Potential applications are discussed.

The next two chapters deal with investigations concerning solid silicon monoxide. The application of thin films of this material is based on its unique mechanical, chemical, and dielectric properties. It is related to Si-Si systems in so far as solid SiO consists of small particles of Si and SiO_2. Depending on the conditions for synthesis, the material has different local structures. In the contribution of U. Schubert and T. Wieder (Chapter 18), the structure and reactivity of a special SiO modification (Patinal®) is described. This material consists of Si and SiO_2 regions of 0.25 – 0.5 nm in diameter, which are connected by a thin interface. Most of the SiO reactions are also observed for elemental silicon. H. Hofmeister and U. Kahler (Chapter 19) show that thermal processing of solid SiO (from BALZERS) up to 1300°C leads to phase separation into Si nanocrystallites embedded in an SiO_x matrix. Their internal structure is determined by solid–phase crystallization processes.

Clusters of small and intermediate size play an important role in the explanation of the chemical and physical properties of matter on the way from molecules to solids. The great interest in silicon clusters stems from the importance of silicon in solid-state physics and from the photoluminescence properties of porous silicon. In Chapter 20, A. F. Sax presents theoretical models to describe bare and hydrogenated small silicon clusters and discusses two quite different approaches to describe solid silicon and localized defects such as vacancies with the help of clusters. Sophisticated molecular beam experiments for the size-selective preparation of neutral silicon clusters are presented by J. A. Becker in Chapter 21. Calorimetric studies indicate the presence of spherically and non-spherically shaped cluster isomers. The existence of several isomers might play an important role with regard to the growth kinetics of solid silicon.

One of the great issues in the field of silicon clusters is to understand their photoluminescence (PL) and finally to tune the PL emission by controlling the synthetic parameters. The last two chapters deal with this problem. In experiments described by F. Huisken et al. in Chapter 22, thin films of size-separated Si nanoparticles were produced by SiH_4 pyrolysis in a gas–flow reactor and molecular beam apparatus. The PL varies with the size of the crystalline core, in perfect agreement with the quantum confinement model. In order to observe an intense PL, the nanocrystals must be perfectly passivated. In experiments described by S. Veprek and D. Azinovic in Chapter 23, nanocrystalline silicon was prepared by CVD of SiH_4 diluted by H_2 and post-oxidized for surface passivation. The mechanism of the PL of such samples includes energy transfer to hole centers within the passivated surface. Impurities within the nanocrystalline material are often responsible for erroneous interpretation of PL phenomena.

Peter Jutzi, Ulrich Schubert

9 Chemistry of Metalated Oligosilanes

Roland Fischer, Dieter Frank, Christian Kayser, Christian Mechtler, Judith Baumgartner, Christoph Marschner*

9.1 Introduction

Analogous to carbanions there are also compounds with silicon atoms bearing negative charges known as silylanions.[1] However, methods for the synthesis of carbanions can rarely be applied for the synthesis of silylanions. This is mainly due to the different electronegativities of silicon and carbon, as well as due to the different chemical reactivity of carbon-carbon bonds compared to silicon-silicon bonds. Therefore, proton abstraction, which is an important method for the generation of carbanions is generally not a useful method for the synthesis of silylanions. Similarly, silicon-silicon bond cleavage with alkali metals, which is the most important method for the generation of silylanions,[2] is not a very practical procedure for obtaining carbanions.
The chemistry of silylanions has been extensively reviewed.[1] The goal of this article is to present some fundamental basics, a few recent highlights, and a more exhaustive coverage of the developments in our laboratory since the publication of the latest comprehensive accounts.

9.2 Silylanions

The single most important reaction for the generation of silylanions is the cleavage of disilanes with alkali metals. Lithium is most often used for this purpose (Eq. 1),[1,2] but examples that make use of sodium/potassium alloy,[3] potassium graphite (C_8K),[4] or other sources of alkali metals are also known. Frequently, disilanes are first generated *in situ* by treatment of chlorosilanes with an excess of alkali metal. The major drawback of this reaction is the requirement of a charge stabilizing group (e.g. R' = phenyl) on the silicon atom, if the reaction is to be conducted in ether solvents such as THF or DME (dimethoxyethane). Hexamethyldisilane, for example, cannot be cleaved in this way.

Besides alkali metals, certain very strong anionic compounds, such as alkyllithiums[5] and alkali metal alkoxides[6] can also be used to cleave disilanes (Eq. 2). Hexamethyldisilane can only be cleaved in strongly coordinating

* Address of the authors: Institut für Anorganische Chemie, Technische Universität Graz, Stremayrgasse 16, A-8010 Graz, Austria

solvents such as HMPA,[6a] DMI (1,3-dimethyl-2-imidazolidone),[6b] or DMPU (*N*,*N*-dimethylpropyleneurea).[6c]

$$R'R_2Si{-}SiR_2R' \xrightarrow{2\,Li} 2\ R'R_2Si{-}Li \qquad R' = \text{charge stabilizing group} \tag{1}$$

$$R_3Si{-}SiR_3 \xrightarrow[-\,R'SiR_3\ \text{or}\ R'OSiR_3]{R'M\ \text{or}\ R'OM} R_3Si{-}M \qquad (M = Li, Na, K) \tag{2}$$

A disadvantage of both methods is that the obtained silylanions are usually adducts of the donor solvents used for their synthesis. These donor molecules can sometimes cause problems in subsequent reactions. Base-free silylanions can be obtained by means of transmetalation from silylmercury compounds (eq. 3).[7,14] In this case, hydrocarbon solvents can be used to generate the silylanion. While this is a very flexible and reliable reaction, the use of the toxic and frequently very sensitive bis(silyl)mercury compounds diminishes the value of the method.

$$R_3Si{-}Hg{-}SiR_3 \xrightarrow[-\,Hg]{2\,Li} 2\ R_3Si{-}Li \tag{3}$$

9.3 Oligosilylanions

Compounds containing silicon-silicon bonds, the so-called oligosilylanions, are of special interest among the silylanions. The prototypical compound, $Li[Si(SiMe_3)_3]$, was first prepared by Gilman et al. by treatment of $Si(SiMe_3)_4$ with methyllithium (Eq. 4).[8]

$$Me_3Si{-}Si(SiMe_3)_2{-}SiMe_3 \xrightarrow[-\,SiMe_4]{MeLi} Me_3Si{-}Si(SiMe_3)_2{-}Li \tag{4}$$

This reagent has become increasingly popular over the last 20 years for the nucleophilic introduction of the $Si(SiMe_3)_3$ group. A number of interesting, kinetically labile structural units have been stabilized with this substituent. However, it was found that, in some cases, even the bulk of the $Si(SiMe_3)_3$ group is not sufficient to provide the necessary steric protection.[9] An example is Wiberg's ingenious synthesis of a tetrasilatetrahedrane, where the $Si(Si^tBu_3)_3$ group was used.[10] Another example is Apeloig's reaction of $Li[Si(SiMe_3)_3]$ with adamantanone, which leads to dimerization of the resulting silene.[11] Repeating the reaction with $Li[Si(SiMe_3)_2(SiMe_2{}^tBu)]$ resulted in the isolation of a stable silene.[12]

These and other examples called for a systematic study, varying three parameters of the tris(trimethylsilyl)silylanion:

- *size* – through the incorporation of substituents of different steric demand,
- *electronic properties* – through the incorporation of substituents with different electronegativities and resonance capabilities,
- *reactivity* – through the choice of counterions and donor molecules.

To generate the silylanions, a potassium alkoxide was used (Eq. 5) instead of the alkyllithium compounds used by Gilman et al.[8] The reaction with $Si(SiMe_3)_4$, which is very clean and almost quantitative, was not described in the literature prior to our studies.[13] THF or DME are the solvents of choice.

$$Si(SiMe_3)_4 \xrightarrow[-\ tert\text{-}BuOSiMe_3]{tert\text{-}BuOK\ /\ THF\ or\ DME} (Me_3Si)_3Si{-}K \quad (5)$$

The resulting $K[Si(SiMe_3)_3]$ was previously prepared base-free by Klinkhammer et al. by transmetalation from $M[Si(SiMe_3)_3]_2$ (M = Hg or Zn) with potassium.[14] The THF or DME adducts of the compound are slightly more reactive than the lithium compound and, rather unexpectedly, show improved solubility in apolar organic solvents. The marked nucleophilic behavior of this compound makes it easy to obtain derivatives.

9.3.1 Sterically More Demanding Oligosilylanions

Other neopentasilanes can be easily obtained by the reaction of different silyl halides or triflates with $K[Si(SiMe_3)_3]$ (Eq. 6).

$$(Me_3Si)_3Si{-}K \xrightarrow{R'R''_2SiX} (Me_3Si)_3Si{-}SiR'R''_2 \quad (6)$$

R'	R''
tert-Bu	Me
thex	Me
iso-Pr	*iso*-Pr
Ph	Me
Me	Ph
Ph	Ph

When these compounds were subjected to further treatment with potassium *tert*-butoxide, it was found that they do not all behave in the same way.

Compounds without phenyl groups reacted, as expected, with very selective removal of $SiMe_3$ groups to give products, that differ from the $K[Si(SiMe_3)_3]$ in such a way as befits the introduction of a sterically more demanding substituent. On the other hand, $(Me_3Si)_3Si{-}SiMe_2Ph$ reacted with potassium *tert*-butoxide to give a mixture of $K[Si(SiMe_3)_2SiMe_2Ph]$ and $K[Si(SiMe_3)_3]$ in an approximate ratio of 1:1. The reactions with the higher phenylated compounds showed a similar picture, complicated by the additional formation of $K[SiMePh_2]$ or $K[SiPh_3]$, respectively.

9.3.2 Secondary Oligosilylanions

The approach employed in the previous section was also used for the generation of isotetrasilanes and their subsequent transformation into silylpotassium compounds. Reactions of $K[Si(SiMe_3)_3]$ or the above mentioned bulkier analogues with various alkylating agents usually proceed smoothly to form the respective alkylated isotetrasilanes (Eq. 7). An exception are reactions with alkyl iodides. In these cases, nucleophilic substitution is not observed. Instead, metal/halogen exchange takes place, leading to the formation of silyl iodides in substantial amounts.

$$R''_2R'Si{-}Si(SiMe_3)_2{-}K \xrightarrow[R''' = \text{Me, Et, } iso\text{-Pr}]{\text{alkylating agent}} R''_2R'Si{-}Si(SiMe_3)_2{-}R''' \tag{7}$$

Reactions of methyl, ethyl, and isopropyl isotetrasilanes (obtained according to Eq. 7), as well as the phenylated compound, with potassium *tert*-butoxide proceed analogously to those with neopentasilanes. Again, the expected silylpotassium compounds are formed in almost quantitative yields (Eq. 8).

$$R{-}Si(SiMe_3)_2{-}SiMe_3 \xrightarrow[- \textit{tert}\text{-BuOSiMe}_3]{\textit{tert}\text{-BuOK / THF or DME}} R{-}Si(SiMe_3)_2{-}K \qquad R = \text{Me, Et, } iso\text{-Pr, Ph, H} \tag{8}$$

However, the corresponding hydrosilane, which is easily available from $K[Si(SiMe_3)_3]$ by protonation, does not react with potassium *tert*-butoxide to give a uniform product. While the expected compound $K[SiH(SiMe_3)_2]$ is formed in ca. 65% yield, the remainder of the product consists of $K[Si(SiMe_3)_3]$, apparently formed by deprotonation.

9.3.3 Alkynylated Oligosilylanions

One objective for the synthesis of new silylanions is to incorporate certain functionalities into the oligosilyl moiety that would allow further functionalization at a later stage. This can be achieved by a variety of strategies. One is to incorporate an Si-H unit as exemplified above. This unit allows for further radical chemistry, hydrosilylation, or the introduction of a leaving group. The latter can, of course, also be incorporated through a phenyl group and a subsequent protodesilylation reaction with a strong acid.[15] The same is probably true for aminosilanes, which will be treated in the next section.

However, if it is not advisable to react the central silicon atom, some other group needs to be introduced that can be further functionalized. This can be an alkynyl group, for example. To test the compatibility of these groups with the formation of silylanions, a number of model compounds were prepared and subjected to metalation with potassium *tert*-butoxide (Eq. 9). The results were surprising in that the reactions with the alkoxide proceeded substantially faster than in the previously described examples, with the phenylethynyl derivative being completely consumed almost instantaneously after mixing of reagents and addition of the solvent.[16] Reaction of potassium *tert*-butoxide with $(Me_3Si)_3Si–C{\equiv}CH$ showed that deprotonation of the alkynyl substituent by the alkoxide is not a competitive reaction.

$$(Me_3Si)_3Si{-}Br \xrightarrow{Li-C\equiv C-R} (Me_3Si)_3Si{-}C{\equiv}C{-}R \xrightarrow{tert\text{-}BuOK} K{-}Si(SiMe_3)_2{-}C{\equiv}C{-}R \qquad (9)$$

R= H, Me, Ph, C_8H_{17}

9.3.4 Heteroatom-Substituted Oligosilylanions[17]

Amino-substituted silylanions were introduced by Tamao and have proved to be valuable synthetic building blocks.[18] In our case, the objective to prepare aminooligosilylanions was driven by the intention to incorporate a site suitable for further functionalization, as well as by the prospect of obtaining transition metal silyl ligands with additional coordination sites.

This was accomplished by coupling either an amine or a phosphide with a silyl halide, followed by metalation. In the case of the obtained aminosilane, the subsequent metalation proceeded smoothly resulting in a stable aminosilylanion. In the case of phosphorus, we observed the formation of the phosphinosilyl anion by NMR, but the compound was not stable under the reaction conditions and rearranged to give the potassium phosphide.

$$(Me_3Si)_3Si{-}Cl \xrightarrow{Et_2NH \text{ or } Ph_2PK} (Me_3Si)_3Si{-}ER_2 \xrightarrow[-\,tert\text{-}BuOSiMe_3]{tert\text{-}BuOK\,/\,THF} K{-}Si(SiMe_3)_2{-}ER_2 \quad (10)$$

9.3.5 Higher Oligosilylanions

Most of the chemistry outlined so far would probably also work using Gilman's method[8] employing methyllithium instead of potassium alkoxides. However, the main restriction in this case is that it is not suitable for the generation of higher oligosilylanions. For example, the reaction of $Si_2(SiMe_3)_6$ with methyllithium leads to cleavage of the central silicon bond and generates $Li[Si(SiMe_3)_3]$ and $MeSi(SiMe_3)_3$.[19]

In contrast, reaction of the same starting material with potassium *tert*-butoxide only causes cleavage of one trimethylsilyl group, thus leading to the clean formation of $K[Si_2(SiMe_3)_5]$ (Eq. 11).[13] In this reaction, there is no indication of cleavage of the central Si-Si bond. It is assumed that the two reactions follow different mechanisms. While the alkoxide attack is probably a nucleophilic attack and therefore rather selective with regard to the steric shielding of the attacked silyl center, the reaction with methyllithium appears to be an electron-transfer reaction and is therefore governed by the position of the LUMO.

MeLi *tert*-BuOK

$$R{-}Si(SiMe_3)_2{-}Si(SiMe_3)_2{-}SiMe_3 \xrightarrow[-\,tert\text{-}BuOSiMe_3]{tert\text{-}BuOK\,/\,THF} R{-}Si(SiMe_3)_2{-}Si(SiMe_3)_2{-}K \quad (11)$$

$R = SiMe_3$, Ph

It was surprising to detect additional selectivity in the reaction of $Si_2Ph(SiMe_3)_5$ with with potassium *tert*-butoxide: only a trimethylsilyl group in the ß-position to the phenyl group was attacked.[20]

9.3.6 Multiply-Charged Oligosilylanions

The feature of the reaction with potassium alkoxide not to cleave inner Si-Si bonds made us optimistic that with larger silanes multiply metalated products could be obtained. Such compounds have been prepared before by cleavage of strained cyclosilanes with alkali metals (Eq. 12).[21]

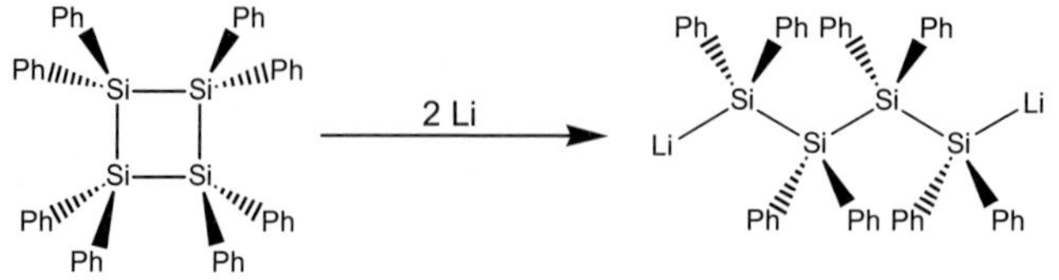

(12)

More recently, the synthesis of a geminal dianion by Sekiguchi et al.[22] has attracted some attention (Eq. 13).

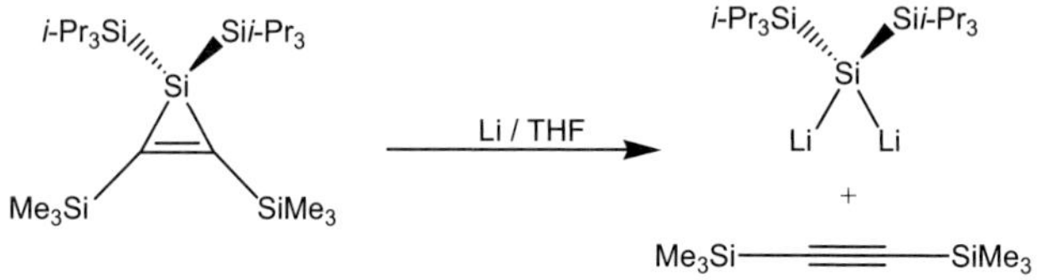

(13)

Kira et al. have since reported the formation of a 1,2-dianion by reaction of a disilene with alkali metal (Eq. 14).[23]

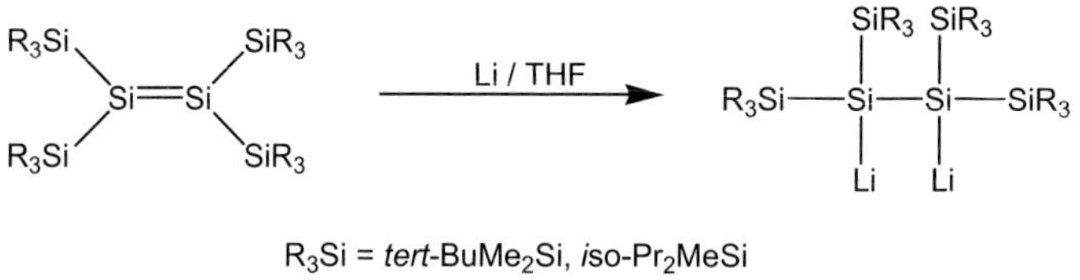

R_3Si = *tert*-$BuMe_2Si$, *iso*-Pr_2MeSi

(14)

In another interesting approach, Apeloig et al. showed that double metalation with methyllithium becomes feasible without cleavage of any inner Si-Si bonds once the silyl bridge between two tris(trimethylsilyl)silyl units consists of two dimethylsilyl groups (Eq. 15).[24]

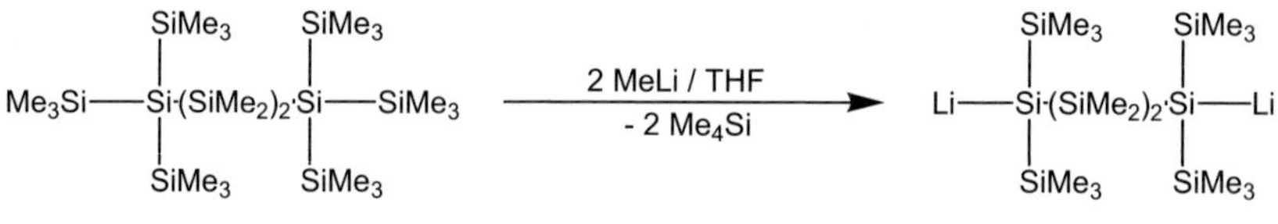

(15)

In order to test our system with respect to multiple metalation, we tried to react compounds like $Si_2(SiMe_3)_6$ and some higher homologues with spacer units between the $Si(SiMe_3)_3$ units with two equivalents of potassium *tert*-butoxide.

9.3.6.1 Silylene-bridged dianions. In the compounds having two, one, or no dimethylsilylene units (n = 0–2) between the $Si(SiMe_3)_3$ groups, it was found that a second metalation step can take place at 60 °C in THF.[25] While the compounds with n = 1 or 2 yielded rather clean products (Eq. 16) with some additional formation of $K[Si(SiMe_3)_3]$, the reaction with $Si_2(SiMe_3)_6$ (n = 0) was rather messy. Reaction of the compound with n = 3 also did not result in the formation of the expected dianion.

$$\mathrm{Me_3Si{-}Si(SiMe_3)_2{-}(SiMe_2)_n{-}Si(SiMe_3)_2{-}SiMe_3} \xrightarrow[\substack{-\,2\ \textit{tert}\text{-BuOSiMe}_3 \\ n\,=\,1,2}]{2\ \textit{tert}\text{-BuOK / THF, 60°C}} \mathrm{K{-}Si(SiMe_3)_2{-}(SiMe_2)_n{-}Si(SiMe_3)_2{-}K} \quad (16)$$

The use of an alternative protocol eventually made it possible to carry out bismetalations with all the mentioned substrates (n = 0–3) under very mild conditions. This required the use of crown ether to coordinate the potassium ion and allowed switching from the previously required ether solvents to aromatic solvents such as benzene or toluene. The reaction proceeds even at room temperature and is much faster than previously observed. An additional advantage is that in almost all cases the products precipitate and can be subjected to X-ray crystallographic analysis.[26]

$$\mathrm{Me_3Si{-}Si(SiMe_3)_2{-}(SiMe_2)_n{-}Si(SiMe_3)_2{-}SiMe_3} \xrightarrow[\substack{-\,2\ \textit{tert}\text{-BuOSiMe}_3 \\ n\,=\,0,1,2,3}]{2\ \textit{tert}\text{-BuOK /18-Cr-6/ C}_6\text{H}_6\text{, rt}} \text{18-Cr-6}\rightarrow\mathrm{K{-}Si(SiMe_3)_2{-}(SiMe_2)_n{-}Si(SiMe_3)_2{-}K}\leftarrow\text{18-Cr-6} \quad (17)$$

9.3.6.2 Alkylidene-bridged dianions. Alternatively, alkyl chains can also be used as spacers between the $Si(SiMe_3)_3$ groups.[27] Again, the conditions outlined above result in the formation of the expected dianions. The compounds exhibit interesting chemistry as they cyclize upon cautious hydrolysis and are transformed into cyclosilylanions.[27a]

$$\mathrm{Me_3Si{-}Si(SiMe_3)_2{-}(CH_2)_n{-}Si(SiMe_3)_2{-}SiMe_3} \xrightarrow[\substack{-\,2\ \textit{tert}\text{-BuOSiMe}_3 \\ n\,=\,2,3,4}]{2\ \textit{tert}\text{-BuOK / THF, 60°C}} \mathrm{K{-}Si(SiMe_3)_2{-}(CH_2)_n{-}Si(SiMe_3)_2{-}K} \quad (18)$$

9.3.6.3 Alkynylidene and phenylenealkynylidene-bridged di- and trianions. The alkynylidene unit is a special kind of spacer. With its sp-hybridized carbon atoms it leads to the formation of a linear conjugated dianion. The 1,3- and 1,4-phenyldialkynylidene spacers behave similarly. The enhanced reactivity of the alkynylsilanes facilitates silylanion formation. All dianions in this series are formed extremely smoothly in ether solvents (THF, DME) at room temperature. This fortunate situation sets the stage for the synthesis of a triply-metalated silane starting from 1,3,5-$C_6H_3[C{\equiv}CSi(SiMe_3)_3]_3$. The metalation also proceeds very readily in this case.

9.3.7 Differently Metalated Oligosilylanions

The variations in the tris(trimethylsilyl)silylanion reactivity so far addressed have mainly been due to different substituents. However, a real regulation of reactivity is not easily feasible this way. As mentioned above, the potassium compounds are more reactive than the corresponding lithium compounds. While this increased reactivity is sometimes beneficial, it also can impose some difficulties. For example, in the reactions with some metal halides, the high reactivity, coupled with a substantial reduction potential can lead to unintentional electron-transfer reactions. Therefore, it is desirable to somehow tune the reactivity. This is in fact easily possible by employing metathesis reactions of the silyl potassium compounds with the halides of less electropositive alkali or alkaline earth metals. The most convenient reactions include those with lithium or magnesium bromide. In the latter case, a sila-Grignard reagent can be obtained.[28] Use of only half an equivalent of magnesium bromide facilitates the formation of the $Mg[Si(SiMe_3)_3]_2{\cdot}THF$ adducts. Similarly, Grignard reagents can also be used in the transmetalation process (Eq. 19). This then allows the isolation of mixed silyl-alkyl(aryl)magnesium compounds.

$$(Me_3Si)_3Si{-}K \xrightarrow[\text{- KBr}]{\text{MgBrR}} (Me_3Si)_3Si{-}MgR \qquad (19)$$

R = Br, Me, Ph, $Si(SiMe_3)_3$

9.4 Conclusions

It has been demonstrated that by the use of potassium alkoxides as metalating agents an impressive variety of silylanions can be prepared. Starting from simple $K[Si(SiMe_3)_3]$, the substituents, the reactivity, and also the charge of the compounds can be manipulated in a rather straightforward fashion.

9.5 References

[1] P. D. Lickiss, C. M. Smith, *Coord. Chem. Rev.* **1995**, *145*, 75. K. Tamao, A. Kawachi, *Adv. Organomet. Chem.* **1995**, *39*, 1. J. Belzner, U. Dehnert in Z. Rappoport, Y. Apeloig (Eds.) *The Chemistry of Organic Silicon Compounds*, J. Wiley & Sons, New York, 1998, Vol. 2, p. 779. A. Sekiguchi, V. Y. Lee, M. Nanjo, *Coord. Chem. Rev.* **2000**, *210*, 11.

[2] D. D. Davis, C. Gray, *Organometal. Chem. Rev. A* **1970**, *6*, 283.

[3] A. G. Brook, A. Baumegger, A. J. Lough, *Organometallics* **1992**, *11,* 310.

[4] A. Fürstner, H. Weidmann, *J. Organomet. Chem.* **1988**, *354,* 15.

[5] W. C. Still, *J. Org. Chem.* **1976**, *41*, 3063. K. Krohn, K. Khanbabaee, *Angew. Chem.* **1994**, *106*, 100; *Angew. Chem. Int. Ed. Engl.* **1994**, *33*, 99.

[6] a) H. Sakurai, A. Okada, M. Kira, K. Yonezawa, *Tetrahedron Lett.* **1971**, *19*, 1511. b) H. Sakurai, F. Kondo, *J. Organomet. Chem.* **1975**, *92*, C46. c) E. Buncel, T. K. Venkatchalam, E. Edlund, *J. Organomet. Chem.* **1992**, *437*, 85.

[7] A. Sekiguchi, M. Nanjo, C. Kabuto. H. Sakurai, *Organometallics* **1995**, *14*, 2630. M. Nanjo, A. Sekiguchi, H. Sakurai, *Bull. Chem. Soc. Jpn.* **1998**, *71*, 741.

[8] H. Gilman, J. M. Holmes, C. L. Smith, *Chem. Ind. (London)* **1965**, 848. G. Gutekunst, A. G. Brook, *J. Organomet. Chem.* **1982**, *225,* 1.

[9] J. Frey, E. Schottland, Z. Rappoport, D. Bravo-Zhivotovskii, M. Nakash, M. Botoshansky, M. Kaftory, Y. Apeloig, *J. Chem. Soc., Perkin Trans. 2* **1994**, 2555.

[10] N. Wiberg, Ch. M. M. Finger, K. Polborn, *Angew. Chem.* **1993**, *105*, 1140; *Angew. Chem. Int. Ed. Engl.* **1993**, *32*, 1054.

[11] D. Bravo-Zhivotovskii, V. Braude, A. Stanger, M. Kapon, Y. Apeloig, *Organometallics* **1992**, *11*, 2326.

[12] Y. Apeloig, M. Bendikov, M. Yuzefovich, M. Nakash, D. Bravo-Zhivotovskii, D. Bläser, R. Boese, *J. Am. Chem. Soc.* **1996**, *118*, 12228.

[13] Ch. Marschner, *Eur. J. Inorg. Chem.* **1998**, 221.

[14] K. W. Klinkhammer, W. Z. Schwarz, Z. *Anorg. Allg. Chem.* **1993**, *619*, 1777; K. W. Klinkhammer, *Chem. Eur. J.* **1997**, *3*, 1418.

[15] W. Uhlig, *Chem. Ber.* **1992**, *125*, 47.

[16] Ch. Mechtler, Ch. Marschner, unpublished results.

[17] P. Buchgraber, D. Frank, Ch. Marschner, unpublished results.

[18] A. Kawachi, K. Tamao, *J. Am. Chem. Soc.* **2000**, *122*, 1919, and references therein.

[19] H. Gilman, R. L. Harrel, *J. Organomet. Chem.* **1967**, *9*, 67. Y. Apeloig, M. Yuzefovich, M. Bendikov, D. Bravo-Zhivotovskii, K. Klinkhammer, *Organometallics* **1997**, *16*, 1265.

[20] Ch. Kayser, R. Fischer, J. Baumgartner, Ch. Marschner, *Organometallics* **2002**, *21,* 1023.

[21] A. W. Jarvie, H. J. S. Winkler, D. J. Peterson, H. Gilman, *J. Am. Chem. Soc.* **1961**, *83*, 1921. H. Gilman, R. A. Tomasi, *Chem. Ind. (London)* **1963**, 954. E. Hengge, D. Wolfer, *J. Organomet. Chem.* **1974**, *66*, 413. G. Becker, H. M. Hartmann, E. Hengge, F. Schrank, *Z. Anorg. Allg. Chem.* **1989**, *572*, 63.

[22] A. Sekiguchi, M. Ichinohe, S. Yamaguchi, *J. Am. Chem. Soc.* **1999**, *121*, 10231.

[23] M. Kira, T. Iwamoto, D. Yin, T. Maruyama, H. Sakurai, *Chem. Lett.* **2001**, 910.

[24] Y. Apeloig, G. Korogodsky, D. Bravo-Zhivotovskii, D. Bläser, R. Boese, *Eur. J. Inorg. Chem.* **2000**, 1091.

[25] Ch. Kayser, G. Kickelbick, Ch. Marschner, *Angew. Chem.* **2002**, *114*, 1031; *Angew. Chem. Int. Ed.* **2002**, *41*, 989.

[26] R. Fischer, D. Frank, J. Baumgartner, Ch. Marschner, unpublished results.

[27] a) Ch. Mechtler, Ch. Marschner, *Tetrahedron Lett.* **1999**, *40*, 7777. b) For an almost identical approach, see: J. R. Blanton, J. B. Diminnie, T. Chen, A. M. Wiltz, Z. Xue, *Organometallics* **2001**, *20*, 5542.

[28] C. Krempner, H. Reinke, H. Oehme, *Chem. Ber.* **1995**, *128*, 143. J. D. Farwell, M. F. Lappert, Ch. Marschner, Ch. Strissel, T. D. Tilley, *J. Organomet. Chem.* **2000**, *603*, 185.

10 Oligosilyl Substituted Heptaphosphanes – Syntheses, Reactions and Structures

Judith Baumgartner, Vittorio Cappello, Alk Dransfeld, Karl Hassler[*†]

10.1 Introduction

The occurrence of homoatomic anions $[E_n]^{x-}$ of the elements of groups IV (C, Si, Ge, Sn, Pb) and V (P, As, Sb, Bi) as constituents of Zintl phases is well known.[1] Despite numerous attempts, just a few of the E_n clusters have been isolated as uncharged molecules E_nR_x from the Zintl phases by reaction with suitable agents such as alkyl chlorides.

With the heavier elements of group V, the $[E_7]^{3-}$ Zintl ion with a nortricyclene structure (Figure 10.1) is exceptionally stable, and numerous derivatives P_7R_3 and As_7R_3 with R = alkyl have been synthesized over the years from these Zintl ions.[2]

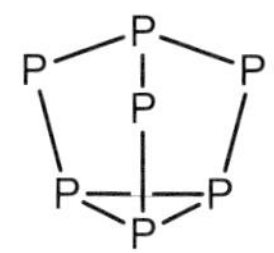

Figure 10.1. Arrangement of the phosphorus atoms in the $[P_7]^{3-}$ Zintl ion.

A few P_7 derivatives with silyl substituents, such as $SiMe_3$,[3] $SiPh_3$,[4] and Si^tBu_3 [5] have also been reported. Attempts to functionalize the P_7 cluster to use it as a building block for larger systems were, however, largely unsuccessful. For example, when $P_7(SiMe_3)_3$ is treated with BuLi, MeLi, ^{t}BuLi or $LiP(SiMe_3)_2$, a mixture of degradation products including $P(SiMe_3)_3$, $LiP(SiMe_3)_2$ and Li_3P_7 is obtained.[6] $LiP_7(SiMe_3)_2$ is initially present in the reaction mixture but decomposes quickly. Mixtures of heteroleptic cages such as $P_7(SiMe_3)_n(CMe_3)_{n-3}$ have also been prepared, but no pure compounds could be isolated.[7] It should be mentioned that by reaction of crystalline K_3P_7 with R_4N^+ (R = Me, Et, n-Bu), the $[R_2P_7]^-$ ions and, subsequently, the heteroleptic cages P_7RR_2 (R = alkyl) were obtained,[8] but X-ray structure analyses were not possible.

---

* Corresponding author

† Address of the authors: Institute of Inorganic Chemistry, University of Technology, Stremayrgasse 16, A-8010 Graz, Austria.

The anticipation that larger oligosilyl substituents at the P_7 system would not only exert stabilizing effects, but also slow down reaction rates and improve the tendency to crystallize was the starting point for this work. We furthermore expected that the Si-Si bonds surrounding the cage would allow reactions to take place without complete destruction of the inner core, leading to functionalized cages which then could be used for the preparation of larger systems.

10.2 Preparation of Oligosilyl Substituted Heptaphosphanes P_7R_3

When sodium potassium phosphide, prepared from red phosphorus and sodium potassium alloy in DME, reacts with Me_3SiCl, the heptaphosphane $P_7(SiMe_3)_3$ is obtained in good yields.[9] By replacing Me_3SiCl with $Me(Me_3Si)_2SiCl$, $(Me_3Si)_3SiCl$ (hypersilyl chloride) and $Ph(Me_3Si)_2SiCl$, the heptaphosphanes **1**–**3** can be synthesized. The large oligosilyl groups effectively shield and protect the P_7 cage (Figure 10.2). For example, **2** (R = $Si(SiMe_3)_3$, which was prepared previously [10]) is inert towards oxygen and moisture and can be stored in air for several weeks without detectable decomposition. It does not react with water or PCl_5, even when refluxed in benzene.

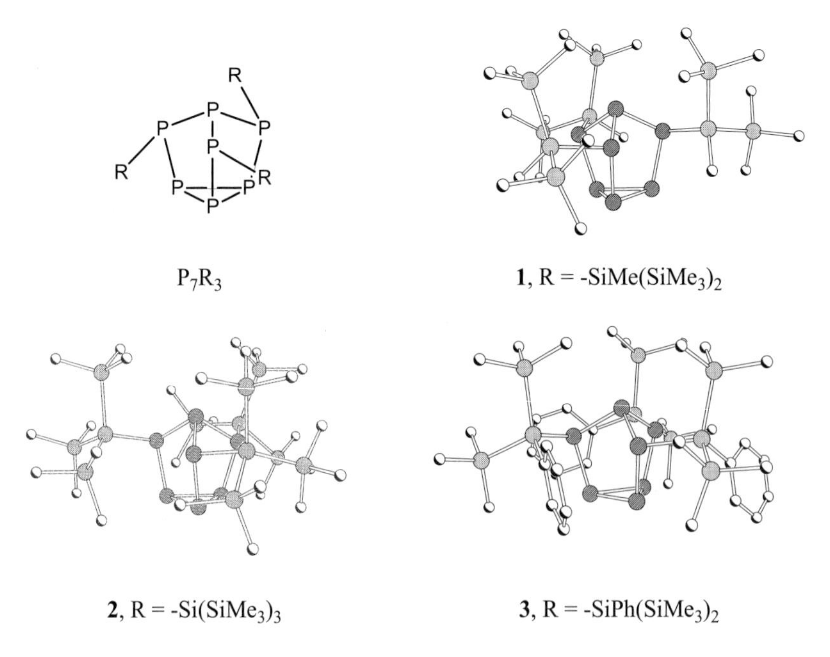

Figure 10.2. Molecular structures of the heptaphosphanes **1**–**3**.

Due to the large size of the substituents, only the symmetric isomers P_7R_3-sym (Figure 10.3) are formed during the syntheses. The steric demand of the substituents is expected to decrease in the order $Si(SiMe_3)_3 > SiPh(SiMe_3)_2 > SiMe(SiMe_3)_2$.

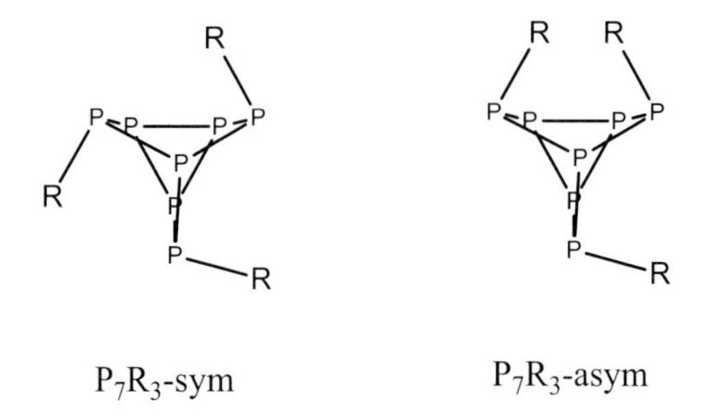

Figure 10.3. Symmetric and asymmetric isomer of P_7R_3 viewed down the C_3-axis.

10.3 Reaction of $P_7[SiMe(SiMe_3)_2]_3$ with *tert*-BuOK

In toluene and in the presence of 18-crown-6, $P_7[SiMe(SiMe_3)_2]_3$ (**1**) reacts with one or two equivalents of *tert*-BuOK by cleavage of one or two Si-P bonds to quantitatively afford the phosphanide ions $[P_7R_2]^-$ and $[P_7R]^{2-}$, respectively. The by-product *tert*-BuOSiMe$(SiMe_3)_2$ was identified by ^{29}Si NMR spectroscopy. The $[K(18\text{-crown-6})]^+$ salts of **4** and **5** crystallize readily from toluene.

The molecular structure of **5** was determined by X-ray crystallography (Figure 10.4) and is the first example of a structurally characterized $[P_7R]^{2-}$ dianion.

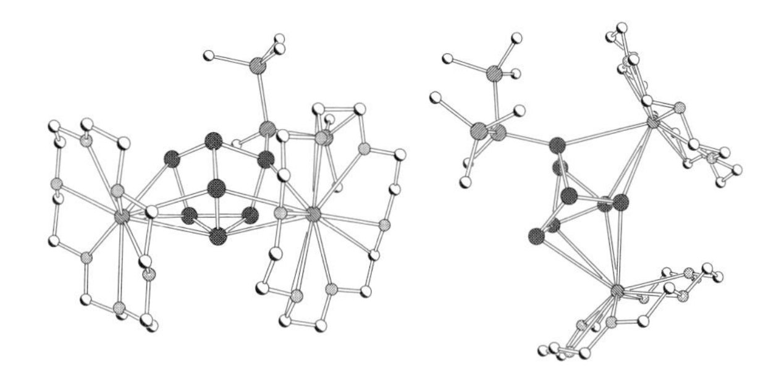

Figure 10.4. Two views of the molecular structure of $[K(18\text{-crown-6})]^+_2[P_7R]^{2-}$.

As expected, the P_7-cage is somewhat distorted. The longest P-P bonds are those of the three-membered ring (221.5–226.2(5) pm) and between the apical and the silicon-substituted phosphorus atom (221.1(3) pm). The bonds which involve the negatively charged phosphorus atoms are considerably shorter (213–210(5) pm). This shortening is also observed in the $[P_7]^{3-}$ ion. The complex cations serve as 'face-capping' units that effectively surround and protect each anionic cage.

The existence of the $[P_7R_2]^-$ ion **4** is deduced from the ^{31}P NMR spectrum (Figure 10.5). The spectrum of **1**, shown for comparison, is typical for all P_7R_3 systems and displays three second-order resonances with relative intensities 3:1:3 at –160 ppm (P_3 ring), –80 ppm (apical P atom), and 0 ppm (Si-substituted P atoms). For **4**, five resonances with relative intensities 1:1:2:1:2 at –18, –32, –56, –95 and –188 ppm, respectively, are observed.

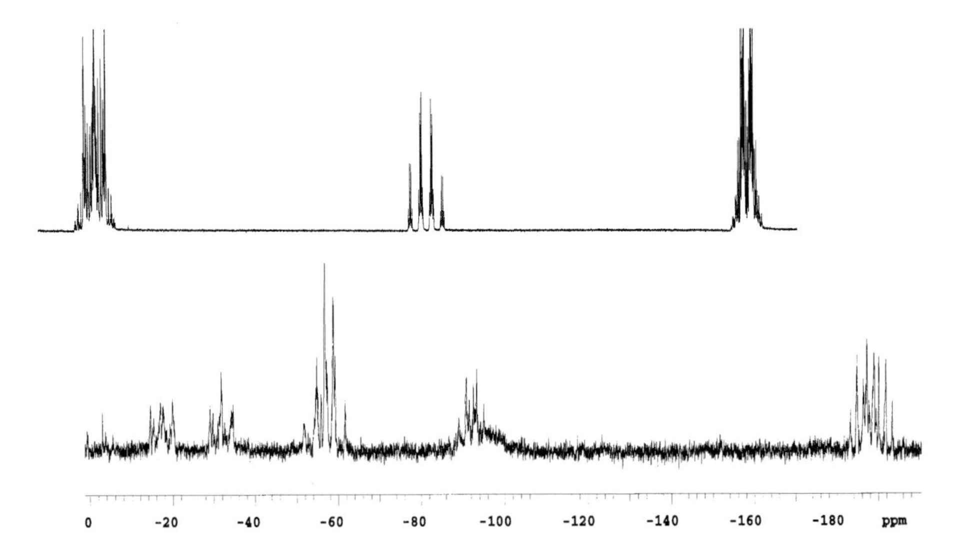

Figure 10.5. $^{31}P\{^1H\}$ spectra of **1** (top) and **4** (bottom).

Figure 10.6 shows three possible isomers for **4**. Isomer II (C_s symmetry) can be ruled out on the basis of steric arguments as the size of the oligosilyl substituents is too large. In isomer III, all seven phosphorus atoms are non-equivalent and hence seven resonances would be expected. For isomer I (C_s symmetry) with two pairs of equivalent phosphorus atoms, five resonances with an intensity ratio 1:1:1:2:2 would be expected, in accordance with observation. There is no evidence for the formation of isomer III. The formation of the anion upon reaction of **1** with *tert*-BuOK proceeds with a simultaneous inversion of one of the phosphorus atoms.

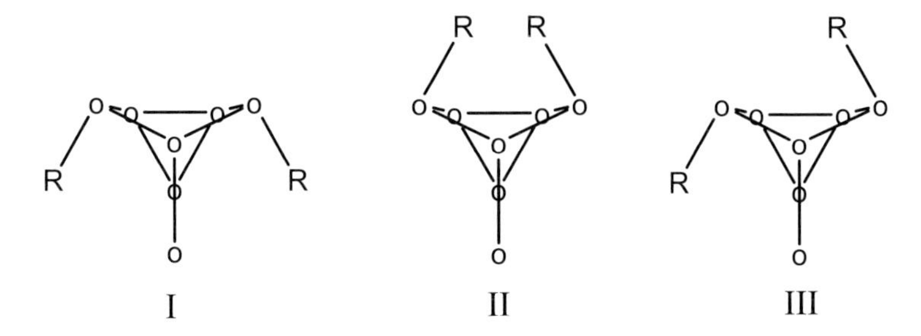

Figure 10.6. The three possible isomers of **4**, neglecting rotational isomerism about the SiP bonds.

The spectrum of **4** gives no indication of fluxional processes in solution; its general features are quite similar to those of the $[P_7H_2]^-$ anion.[2] In contrast, **5** undergoes rapid bond fluctuation in analogy to dihydrobullvalene at room temperature and can be described by two valence tautomeric forms (Figure 10.7). The $^{31}P\{^1H\}$ spectrum at room temperature shows broad signals without detectable $^{31}P^{31}P$ couplings. At –60 °C, the usual second-order resonances are observed. The general features of the spectrum at low temperature are again similar to those of the $[P_7H]^{2-}$ dianion.[11]

Figure 10.7. Cope rearrangement of **5.**

10.4 Reaction of $P_7[Si(SiMe_3)_3]_3$ with *tert*-BuOK

The reaction of $P_7[Si(SiMe_3)_3]_3$ **(2)** with *tert*-BuOK proceeds in a completely different way compared to the reaction of **1**. Due to the bulkier hypersilyl groups, a Si-Si bond instead of a Si-P bond is cleaved.[12] *tert*-$BuOSiMe_3$ and a silyl anion are initially formed. The latter quickly rearranges into a phosphanide anion and bis(trimethylsilyl)silylene. The silylene then inserts into a P-P bond of the three-membered ring of the phosphanide anion forming compound **6**. This is the only process which takes place at low temperatures. Figure 10.8 shows the proton $^{31}P\{^1H\}$ NMR spectrum of **6** together with the spectrum of **2**. The signals of the phosphorus atoms of the three-membered ring around –180 ppm are clearly missing in **6**. The second-order splitting of all signals and their relative intensity ratio of 1:1:1:2:2 prove that seven phosphorus atoms are still present in the anion **6** and that it possesses C_s symmetry.

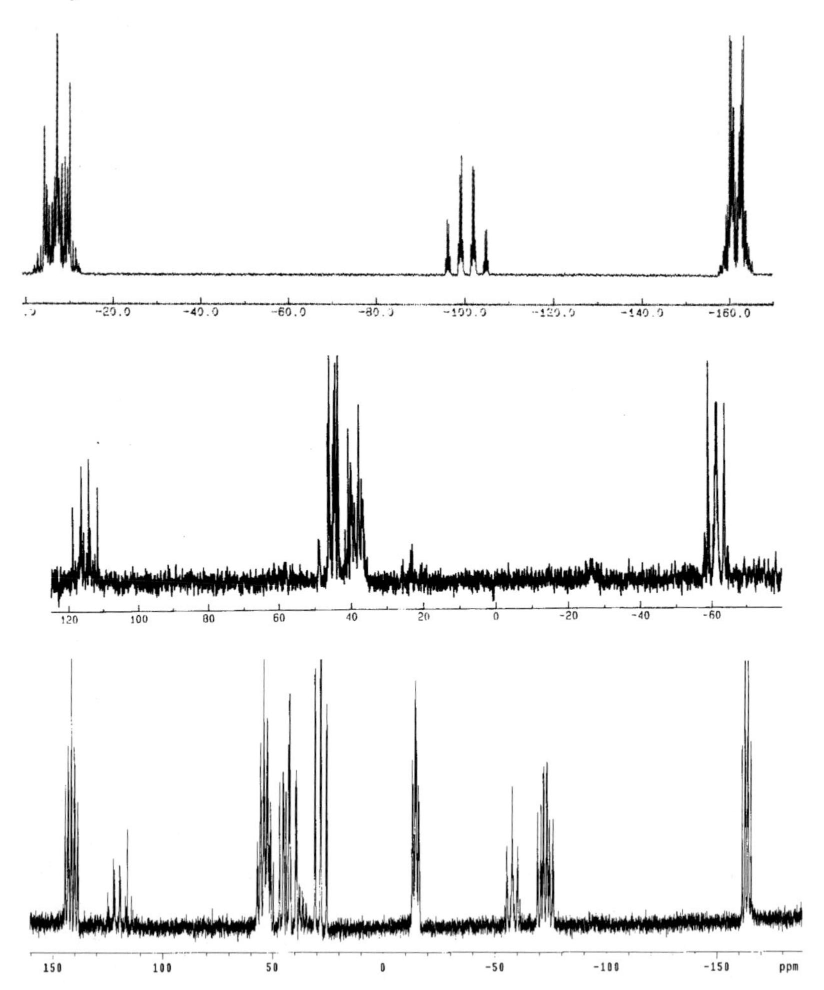

Figure 10.8. $^{31}P\{^{1}H\}$ NMR spectra of **2** (top), **6** (center), and a mixture of **6** and **7** (bottom).

The second-order splitting of the NMR signals can be reproduced accurately by a simulation of the spectrum based upon the structure shown on the left side of Figure 10.9. As in the formation of **4**, reaction of **2** with *tert*-BuOK proceeds with a simultaneous inversion of one of the phosphorus atoms, and therefore **6** has C_s symmetry.

When the reaction is carried out at ambient temperatures, a second product **7** is formed together with **6**. The spectrum consists of seven second-order resonances at 140, 55, 40, 30, –15, –70 and –160 ppm with equal intensity ratios. The molecular structure of **7** (Figure 10.9) shows that this compound also contains the SiR_2 unit (R = $SiMe_3$) which is inserted into a P-P bond of the three membered ring, but differs from the structure of **6** by a shift of the negative charge and a shift of a P-P bond.

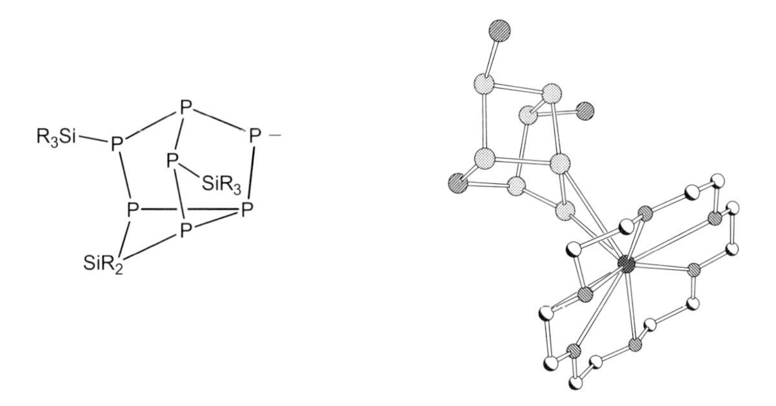

Figure 10.9. Proposed structures of **6** (left) and molecular structure of **7** (right; the $SiMe_3$ groups have been omitted for clarity).

The proposed rearrangement of a PSiK fragment into a silylene and a phosphanide anion is supported by results of *ab initio* calculations and also by unpublished results of C. Marschner.[13] Figure 10.10 shows the minimized structure of $(PH_2)_2PSiH_2Li$ (the triphosphane unit was chosen as an approximation for the P_7 cage) together with bond lengths and Mulliken charges. The lithium atom bridges the silicon atom and the two phosphorus atoms not directly bonded to it. These two phosphorus atoms also bear large negative charges. Furthermore, the P-Si bond is stretched. The results make the formation of a silylene and a phosphanide anion at least plausible and can also explain the observation that the negative charge is not necessarily located on the phosphorus atom from which the silylene is expelled.

10.5 Synthesis and Reactions of Hypersilylmonophosphanes

To further gain some insight into the reactivity of the P-$Si(SiMe_3)_3$ bond, the $Si(SiMe_3)_3$-substituted monophosphanes **8** and **9** were prepared.

$(Me_3Si)_2P–Si(SiMe_3)_3$ (**8**, $\delta(^{31}P) = -267.0$ ppm) was obtained in excellent yield by the reaction of $LiP(SiMe_3)_2$ with $ClSi(SiMe_3)_3$. Its reaction with MeOH afforded $(Me_3Si)HP–Si(SiMe_3)_3$ (**11**). Treatment of $(CF_3O_2SO)Si(SiMe_3)_3$ with PH_3 gives $H_2P–Si(SiMe_3)_3$ (**9**, $\delta(^{31}P) = -264.9$ ppm, $^1J(P,H)= 176.8$ Hz) as the only product (Scheme 10.1).

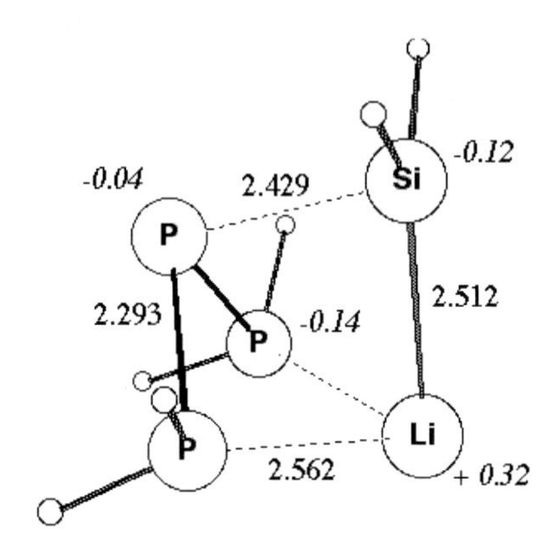

Figure 10.10. Calculated bond lengths [Å] and Mulliken charges for $(PH_2)_2PSiH_2Li$.

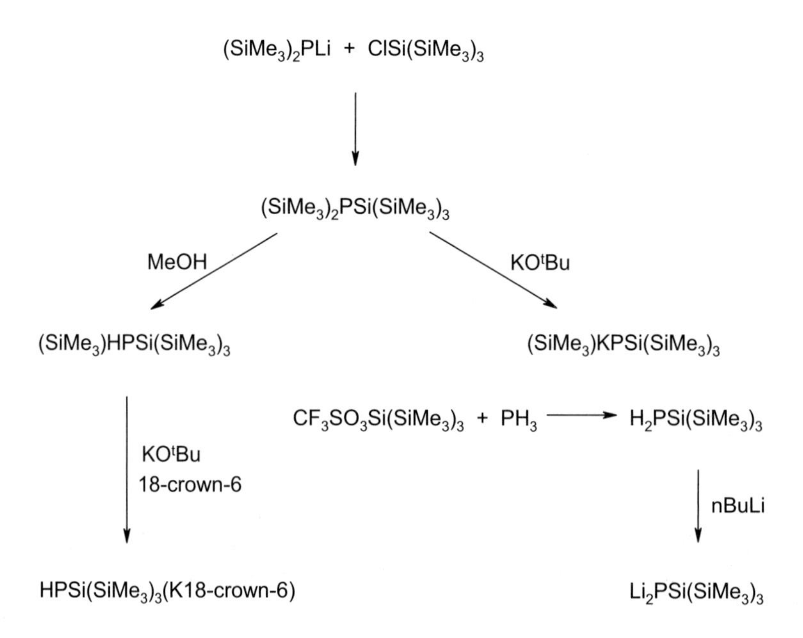

Scheme 10.1. Preparation of hypersilylmonophosphanes.

Reactions of **8** and **11** with *tert*-BuOK in toluene in the presence of 18-crown-6 proceed by cleavage of the P-$SiMe_3$ bond, giving $(Me_3Si)KP–Si(SiMe_3)_3$ (**10**, $\delta(^{31}P) = -352.7$ ppm) and $KHPSi(SiMe_3)_3$ (**12**, $\delta(^{31}P) = -352.2$ ppm, $^1J(P,H) = 127.7$Hz).

The oxidative coupling of **12** with $BrCH_2CH_2Br$ gives the diphosphane $(Me_3Si)_3SiHP–PHSi(SiMe_3)_3$ (**14**). **14** can also be prepared quantitatively from **9** and $^tBu_2Hg$. The $^{31}P\{^1H\}$ NMR spectrum consists of two multiplets centered at –201.8 and –211.8ppm (Figure 10.12) which originate from *meso*- and *rac*-**14** and are the XX' parts of an AA'XX' spin system (A = H, X = P). $^{31}P^{31}P$ coupling constants of –254.1 Hz were obtained for the d,l form of **14**, and –82.6Hz for the meso form by a simulation of the spectra. The large difference between the two coupling constants can be explained by the dihedral angle between the two phosphorus lone pairs. In the meso form, the dihedral angle is 180° if the hypersilyl groups are in *anti* position, which is reasonable because of their large size. Such an arrangement is known to give a small absolute value for the PP coupling constant.[14]

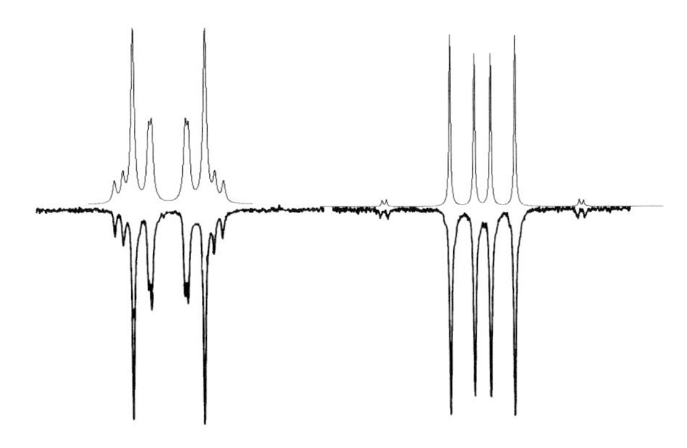

Figure 10.12. Calculated and experimental $^{31}P\{^1H\}$ NMR spectra of **14** with the XX' multiplets of two AA'XX' spin systems. Left: meso (centered at –201.8 ppm); right: d,l (centered at –211.8 ppm).

10.6 Conclusions

The outcome of the reaction of oligosilyl-substituted phosphanes with *tert*-BuOK depends on the steric demand of the silyl groups. A phosphanide anion is formed by cleavage of the Si-P bond if the size of the silyl group allows attack at the Si-P bond. If both $SiMe_3$ groups and oligosilyl groups are attached to the phosphorus atom, the smaller $SiMe_3$ group is invariably cleaved, which can be used for the preparation of oligosilyl-substituted phosphanide anions.

Reactions of $P(SiMe_3)_3$ with *tert*-BuOK have previously been investigated by Uhlig.[19] If the size of the oligosilyl groups is increased, *tert*-BuOK cleaves an Si-Si bond. The resulting silyl anion is not stable and rearranges into a phosphanide anion and a silylene. In the case of the P_7 anion, the silylene then inserts into a P-P bond forming novel heptaphosphanide anions.

10.7 Acknowledgement

We thank Dr. Guido Kickelbick (Vienna University of Technology) and Dr. Trixie Wagner for their help with the X-ray structure analysis of **7**.

10.8 References

[1] S. M. Kauzlarich (Ed.), *Chemistry, Structure and Bonding of Zintl Phases and Ions*, VCH, Weinheim, 1996.

[2] M. Baudler, *Angew. Chem.* **1982**, *94*, 520; *Angew. Chem. Int. Ed. Engl.* **1982**, *21*, 492; M. Baudler, *Angew. Chem.* **1987**, *99*, 429; *Angew. Chem. Int. Ed. Engl.* **1987**, *26*, 419.

[3] W. Hönle, H. G. v. Schnering, *Z. Anorg. Allg. Chem.* **1978**, *440*, 171; G. Fritz, W. Hölderich, *Naturwiss.* **1975**, *62*, 573.

[4] C. Mujica, D. Weber, H. G. v. Schnering, *Z. Naturforsch. B* **1986**, *41*, 991.

[5] I. Kovacs, G. Baum, G. Fritz, D. Fenske, N. Wiberg, H. Schuster, K. Karaghiosoff, *Z. Anorg. Allg. Chem.* **1993**, *619*, 453.

[6] G. Fritz, J. Härer, K. H. Schneider, *Z. Anorg. Allg. Chem.* **1982**, *487*, 44.

[7] G. Fritz, J. Härer, *Z. Anorg. Allg. Chem.* **1983**, *504*, 23.

[8] S. Charles, J. C. Fettinger, B. W. Eichhorn, *J. Am. Chem. Soc.* **1995**, *117*, 5303.

[9] H. Schmidbaur, A. Bauer, *Phosphorus, Sulfur, Silicon* **1995**, *102*, 217.

[10] H. Siegl, W. Krumlacher, K. Hassler, *Monatsh. Chem.* **1999**, *130*, 139.

[11] M. Baudler, R. Heumüller, K. Langerbeins, *Z. Anorg. Allg. Chem.* **1984**, *514*, 7.

[12] C. Marschner, *Eur. J. Inorg. Chem.* **1999**, 221.

[13] C. Marschner, private communication.

[14] M. Baudler, K. Glinka, *Chem. Rev.* **1993**, *93*, 1623.

[15] U. Englich, K. Hassler, K. Ruhland-Senge, F. Uhlig, *Inorg. Chem.*, **1998**, *37*, 3532.

11 Polysilanes: Formation, Bonding and Structure

Richard G. Jones[*]

11.1 Introduction

Polysilanes consist of catenated silicon atoms substituted with aryl groups, alkyl groups or other groups capable of withstanding the sometimes harsh conditions of the polymer synthetic procedure. Unlike carbon chain polymers, all the silicon atoms of the chain are capable of bearing substituents as a consequence of the greater Si-Si bond length and a lack of steric crowding resulting from the conformation usually adopted by the chain. Some typical examples of polysilanes are shown in Figure 11.1.

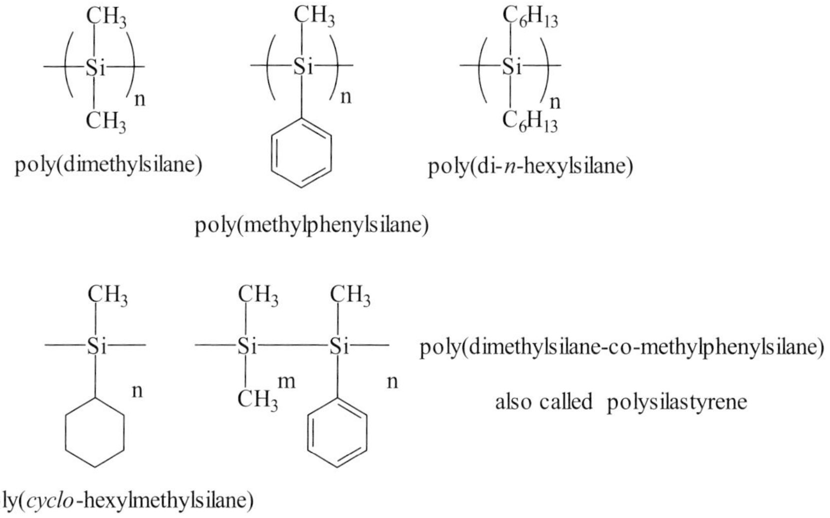

poly(dimethylsilane)

poly(methylphenylsilane)

poly(di-*n*-hexylsilane)

poly(*cyclo*-hexylmethylsilane)

poly(dimethylsilane-co-methylphenylsilane)

also called polysilastyrene

Figure 11.1. Typical polysilanes.

Many silicon-containing polymeric structures are of particular interest for their spectroscopic and electroactive properties and polysilanes are no exception. They absorb strongly between 290 nm and 400 nm and display strong narrow photoluminescence. They are photoconductive and conductive upon chemical doping. They have NLO properties and are photosensitive to UV light. Although it is only in recent years that the polymers are finding applications based on these properties, new possibilities are emerging regularly

[*] Address of the author: Centre for Materials Research, School of Physical Sciences, University of Kent, Canterbury, Kent CT2 7LD, UK

as polysilanes combined with carbon chain structures in block copolymers reveal aggregation capabilities that have the potential to enhance the properties of the parent polymers.

This chapter offers the newcomer to polysilanes a basic introduction to their synthesis, together with an understanding of the unique nature of the main chain bonding and the chain structures that gives rise to their special properties. Like all topics that are reaching maturity, there is far more that might be written than can be covered in a single chapter so copolymers will not be discussed. Likewise, applications will only receive mention so that polysilanes can be seen to take their rightful place in the context of materials for the future. The reader who wants to delve deeper into the subject is referred to a more comprehensive volume concerned with the full range of silicon-containing polymers that was published recently.[1]

11.2 Synthetic Procedures

The first polysilanes were synthesized more than 75 years ago by Kipping [2, 3] through Wurtz-type reductive dehalogenation of dihalodiorganosilanes. This procedure, which is alkali metal intensive, has in the last decade been better understood and has consequently acquired a degree of control that was not previously considered possible. It remains the easiest and most widely used route to a polysilane and it will be discussed later in this section after consideration has been given to three other methods of synthesis that have been researched in recent years.

11.2.1 Anionic Polymerization of Masked Disilenes

The first of these new synthetic methodologies, the anionic polymerization of masked disilenes developed by Sakurai and co-workers,[4–6] is elegant and its concept is unique. Almost uniquely, it offers access to a range of polyalkylsilane structures with narrow molecular weight distributions but involves some procedures that are so painstaking as to cause the principal researcher to caution the uninitiated who might wish to follow a similar path.

The overall procedures of the methodology are shown in Scheme 11.1. The first reaction depicts the synthesis of masked disilenes of representative structures and the second is a conventional anionic polymerization such as might be applied in the synthesis of polystyrene. Masked disilenes are so-called as they are to all intents and purposes disilenes stabilized by a biphenyl bridge. During the polymerization reaction the bridge is released. Product polymers have been reported to have molecular weights up to 50,000 with polydispersities between 1.5 and 2.0. Although these distributions are not particularly narrow, it is nevertheless a living polymerization and this feature can be exploited for the synthesis of block copolymers.[7, 8]

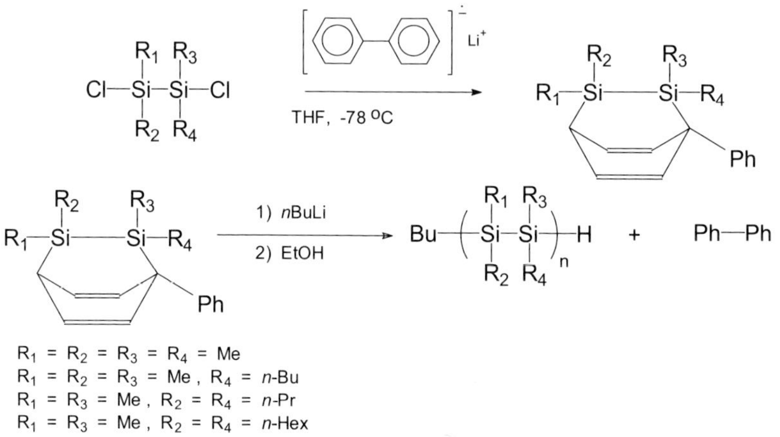

$R_1 = R_2 = R_3 = R_4 = Me$
$R_1 = R_2 = R_3 = Me$, $R_4 = n$-Bu
$R_1 = R_3 = Me$, $R_2 = R_4 = n$-Pr
$R_1 = R_3 = Me$, $R_2 = R_4 = n$-Hex

Scheme 11.1. Synthesis of polysilanes *via* the 'masked-disilene' methodology.

$Nu^- = BuMgBr : R = Bu$
$Nu^- = PhLi : R = Ph$

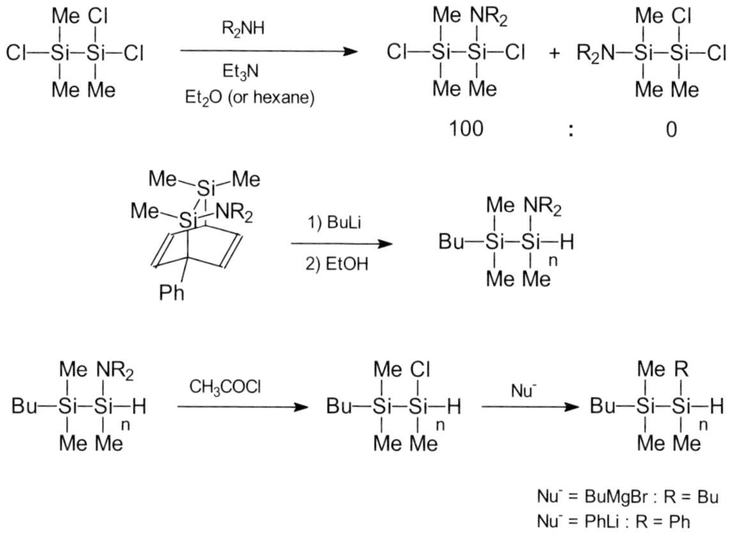

Scheme 11.2. The 'masked-disilene' methodology applied to the synthesis of dialkylamine-substituted polysilanes.

The synthesis of poly(1,1-dihexyl-2,2-dimethylsilane) as depicted in Scheme 11.1 was used to demonstrate the high order associated with these polymerizations.[5] The disilene units are incorporated exclusively in a head-to-tail fashion.

11.2.2 Anionic Ring-Opening Polymerisation

This method of synthesising polysilanes has so far found very limited applicability. As first demonstrated by Matyjaszewski and co-workers,[10] it requires the pre-synthesis of a strained silacycle. In general, these are not readily accessible; however, 1,2,3,4-tetramethyl-1,2,3,4-tetraphenylcyclotetrasilane can be synthesized from octaphenylcyclotetrasilane, after which the all-*trans* isomer was prepared by recrystallization of a hexane solution of stereoisomeric mixtures. The anionic polymerization reaction of this monomer is depicted in Scheme 11.3. The products had molecular weights up to 36,000, almost normal distributions, and were predominantly heterotactic.

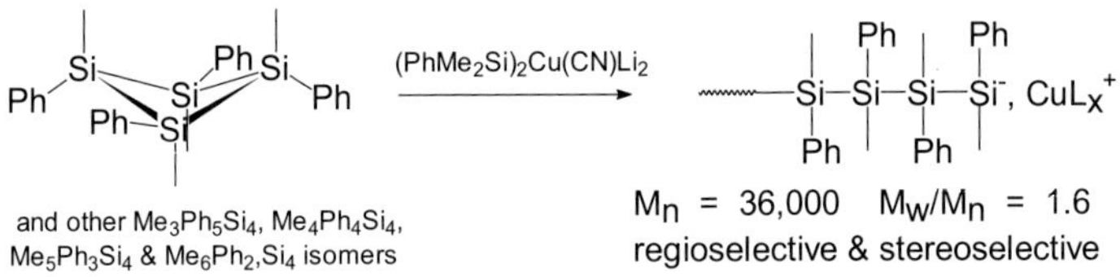

Scheme 11.3. Anionic ring-opening polymerization of cyclotetrasilanes.

Another anionic ring-opening polymerisation that can also be used for the synthesis of a poly(germasilane) as well as a polysilane has been described by Sakurai and co-workers [11] and is shown in Scheme 11.4. They are formed in greater than 40% yield and as well as having narrow molecular weight distributions (M_n = 17,000, M_w/M_n = 1.3) the polymers are highly ordered, as evidenced by NMR spectroscopy.

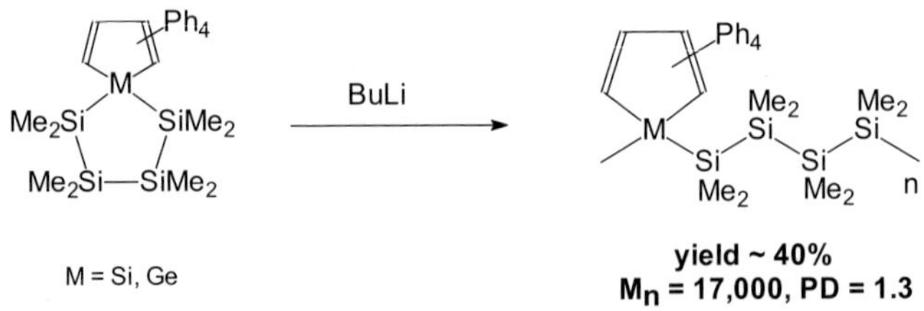

Scheme 11.4. Highly regulated polymers *via* anionic ring-opening polymerization.

11.2.3 Catalytic Dehydrocoupling of Hydrosilanes

The third method of synthesizing polysilanes that has received significant attention in recent years is one based on catalytic dehydrogenation using alkyl metallocenes of the type Cp_2MR_2 at room temperature. When this is applied to a primary silane, as depicted in Scheme 11.5, and in selected instances to secondary silanes, oligomers are readily formed.[12] Much of the effort has been directed towards extending the molecular weights of the products by varying the reaction conditions, catalyst, monomer, but degrees of polymerization no greater than about 50 have been obtained. This is possibly caused by growing chains disengaging from the catalyst at a particular average degree of polymerization for a reason that lies beyond the chemistry of the reaction. This will not be elaborated here but it is probably not dissimilar to influences that prevail in the Wurtz-type reductive-dehalogenation reaction to be discussed in the next sub-section.

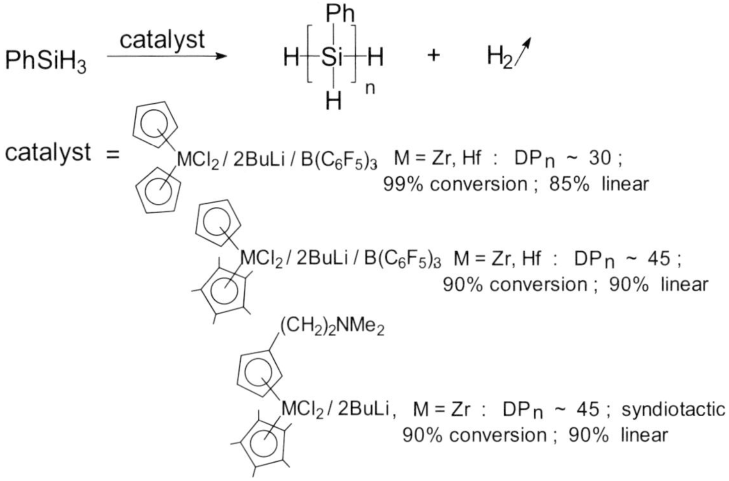

$$PhSiH_3 \xrightarrow{\text{catalyst}} H\!-\!\left[Si(Ph)(H)\right]_n\!-\!H + H_2\uparrow$$

catalyst = $Cp_2MCl_2 / 2BuLi / B(C_6F_5)_3$ M = Zr, Hf : DP_n ~ 30 ;
99% conversion ; 85% linear

$CpCp^{*}MCl_2 / 2BuLi / B(C_6F_5)_3$ M = Zr, Hf : DP_n ~ 45 ;
90% conversion ; 90% linear

$(CH_2)_2NMe_2$-substituted $CpCp^{*}MCl_2 / 2BuLi$, M = Zr : DP_n ~ 45 ; syndiotactic
90% conversion ; 90% linear

Scheme 11.5. The synthesis of polysilanes by the catalytic dehydrogenation of primary silanes.

11.2.4 The Wurtz Reductive Coupling Reaction

The synthesis of polysilanes by the Wurtz-type reductive-coupling reaction is shown in Scheme 11.6. It is applicable to the synthesis of any polysilane, the substituents of which can withstand the harsh reaction conditions. This limits the systems to ones with alkyl, aryl, silyl, fluoroalkyl, and ferrocenyl substituents but other structures are accessible through post-polymerization chemistry.[13]

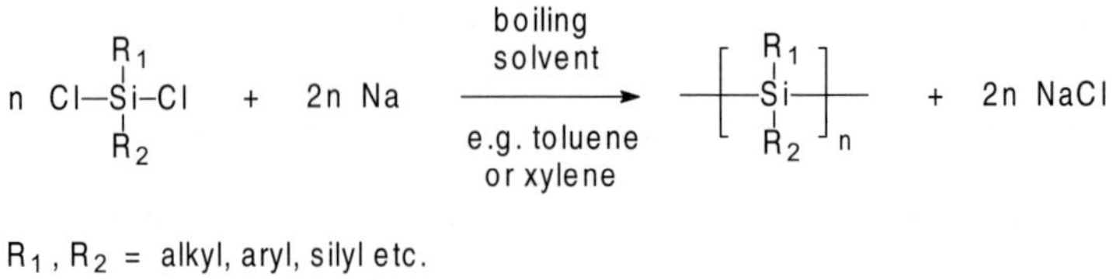

$$n\ \mathrm{Cl{-}Si(R_1)(R_2){-}Cl} + 2n\ \mathrm{Na} \xrightarrow[\text{e.g. toluene or xylene}]{\text{boiling solvent}} \left[\mathrm{Si(R_1)(R_2)}\right]_n + 2n\ \mathrm{NaCl}$$

R_1, R_2 = alkyl, aryl, silyl etc.

Scheme 11.6. Synthesis of polysilanes by the Wurtz-type reductive-coupling reaction.

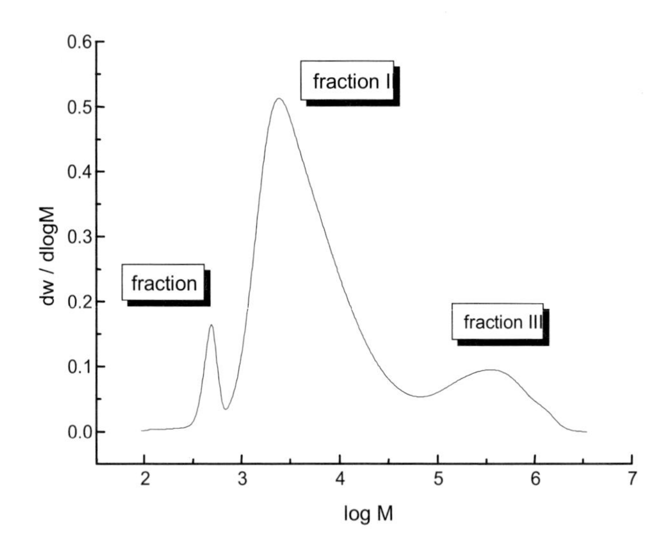

Figure 11.2. Molecular weight distribution of poly(methylphenylsilane) prepared by a Wurtz reductive-coupling in boiling toluene (Reprinted with permission from Organometallics, **1998**, *17*, 59. Copyright 1998 American Chemical Society).

The reaction has conventionally been carried out in boiling toluene using sodium as the reducing agent. It is highly exothermic and although at such an elevated temperature the reaction rates are high, yields can be quite variable. Another detraction is that the molecular weight distributions of polymers that result under these conditions of reaction are polymodal consisting of a narrow low molecular weight fraction ($M_{r,w} < 1{,}000$), a dominant intermediate fraction ($1{,}000 < M_{r,w} < 50{,}000$) with a polydispersity index of about 10, and a lesser high molecular weight fraction ($M_{r,w} \sim 10^6$) characterized by a polydispersity index of about 1.5. A typical distribution for poly(methylphenylsilane), PMPS, is shown in Figure 11.2.[14] Although the first of these fractions is easily removed using standard polymer isolation techniques, the other two fractions are inseparable other than by fractionation.

Given the generality of the Wurtz-type reductive-coupling to the synthesis of polysilanes, recent years have seen a substantial drive to understanding the mechanism of the reaction with a view to optimization of the procedures.[15] The reaction has been investigated at high, intermediate, and low temperatures. Solvent effects have been studied, as have homogeneous reaction systems and a range of alternative heterogeneous systems. It is now generally accepted that the lowest molecular weight fraction arises in part from an end-biting reaction that leads to the formation of cyclopentasilane in the early stages of polymerization, and in part from a back-biting degradation reaction leading to cyclohexasilane when the reaction is carried out at elevated temperatures such as that of boiling toluene. The latter reaction is considered to be the reason why polymer molecules that comprise the dominant intermediate molecular weight fraction are prevented from further growth, i.e. they become disengaged from the alkali metal surface as a consequence of the back-biting reaction. The reaction is thermally activated and so it follows that the lower the temperature at which the polymerization can be conducted, the higher will be the molecular weights attained within this fraction. This assertion is borne out by observations of the synthesis of PMPS in THF at ambient temperature. In this system no high molecular weight fraction is formed but the average molecular weights of the intermediate fraction are seen to increase steadily with time.[14, 15] This procedure, if necessary assisted by ultrasound,[16] has now been proved to work effectively for the production of other polysilanes in substantial yield, in particular polyalkylsilanes.[16, 17]

The underlying reason for the formation of the highest molecular weight fraction at elevated temperatures is intriguing, but first it is necessary to consider the overall progress of the reaction as depicted in Figure 11.3 for the synthesis of PMPS in boiling toluene. There are two clear phases of reaction. The first phase, lasting about 5 minutes, is one in which the precursor dichlorodiorganosilane is rapidly consumed, as would be consistent with a chain reaction. In this time all three of the product fractions appear. An acceptable molecular mechanism that accounts for this phase of reaction has been proposed by Matyjaszewski [18] in which silyl anions, radical anions, and radicals are all intermediates. In the second phase, lasting for the next 20 minutes or thereabouts, the oligomeric fraction is depleted to the advantage of the other two fractions. This can be explained by accepting that the lowest molecular weight fraction contains linear as well as cyclic oligomers and that the former can be involved in what are effectively single-stage condensation reactions onto the other two fractions. However, it is clear that molecular mechanism offers no explanation for the formation of two distinct polymeric fractions. In order to explain this phenomenon it is first necessary to understand a little of the structure of polysilanes.

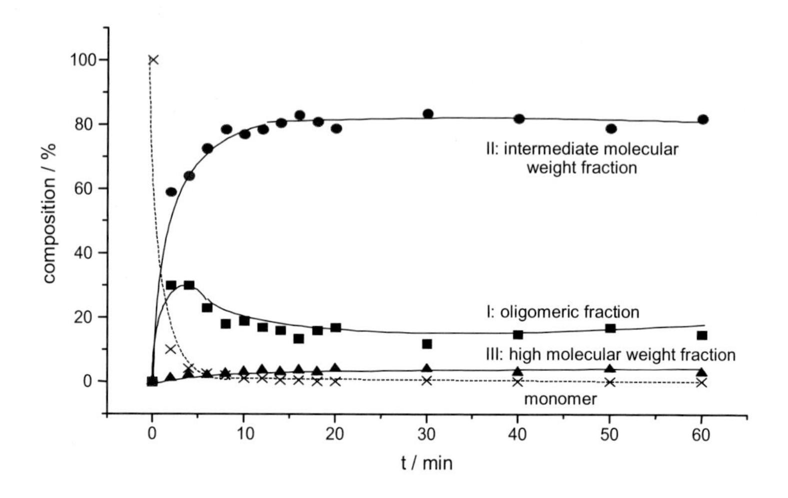

Figure 11.3. Variation of composition of reaction mixture with time for a conventional Wurtz synthesis in boiling toluene (Reprinted with permission from ref. [1], Copyright 2000 Kluwer Academic Publishers).

Fujiki and co-workers [19] have recently shown that many polysilanes, including PMPS, adopt a helical conformation which in the absence of internal or external preventative influences switch screw-sense, M to P or P to M, and pitch at intervals along the chain (see Section 11.4). These switch points represent structural kinks in the chain and within a polymer molecule in solution they move freely up and down the chain. For a polymer molecule growing on an alkali metal surface it is presumed that the kinks enter the chain at the free end and that as the chain grows it can accommodate an increasing number of them, this being predetermined in an average sense by a stability that correlates with a length of sequence for which a particular screw sense can be maintained. Figure 11.4 shows a representation of the kinks and the role they are considered to play in the disengagement of growing polymer chains from the alkali metal surface (included as an inset to the figure is a representation of helical screw-sense reversal that constitutes a kink). It is assumed that the translation of a kink to the metal surface facilitates the back-biting reaction and the concomitant disengagement of the polymer molecule.

At any given temperature these events would occur with a defined probability (P_i) that is at its greatest when the chain is only of sufficient length to accommodate one kink. As chain length increases this probability decreases, explaining the characteristic shape of the distribution of the intermediate molecular weight fraction. However, as the chain length increases the probability of adjacent kinks meeting and undergoing mutual annihilation also increases. This results in an abnormal reduction in the probability of the kink closest to the alkali metal surface ever reaching it and the polymerization progressively taking on the attributes of a living polymerization. This possibility explains the narrow distribution, high molecular weight fraction that is formed when Wurtz-type polymerisations are conducted at elevated temperatures. Theoretical molecular weight distributions detailed in a mathematical model elaborated by McLeish et al.[20] are in excellent agreement with the experimentally observed distributions.

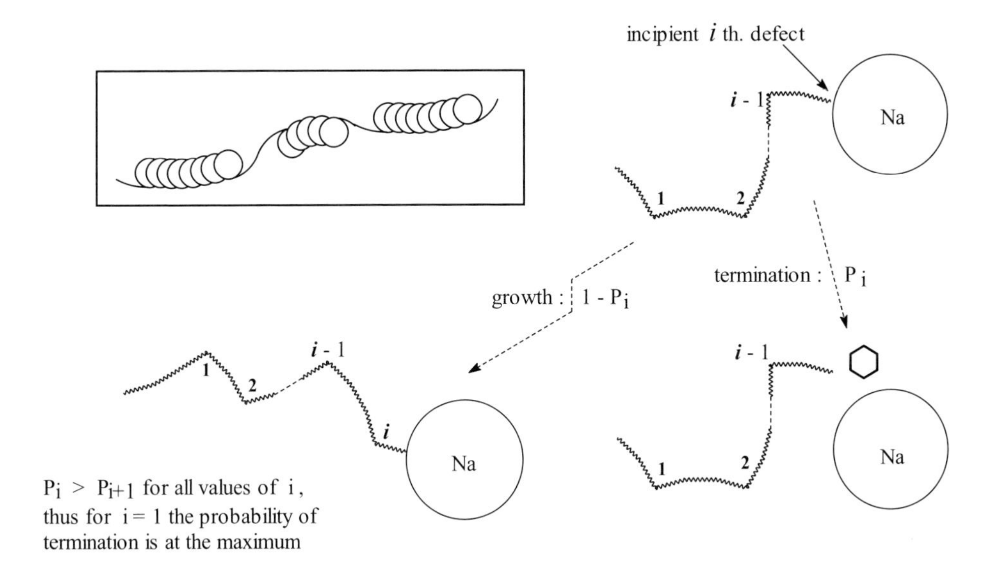

Figure 11.4. Schematic of an alkali metal mediated Wurtz-type synthesis of a polysilane. *Inset*: Representation of a helical structure showing screw-sense reversals.

11.3 Bonding

The spectroscopic and electroactive properties of polysilanes touched upon in the introduction arise from their unique chain structure. Underlying the helical conformations already alluded to in the previous section is the phenomenon of σ-conjugation represented in Figure 11.5. Included for purposes of comparison is a representation of the more familiar π-conjugation as found in a linear polyene. The greater size of the sp^3 bonding orbitals of the silicon atom facilitates a significant geminal overlap that simply does not occur in the case of carbon. As a consequence, electronic delocalization through the σ-bonded framework of a polysilane is a reality, whilst for a carbon chain polymer this can only be realized within the π-bonded framework of a polyene. Whereas each silicon atom contributes two orbitals to the conjugated backbone whilst each carbon atom of a polyene contributes only one, it is easily reasoned from Figure 11.5 that disilane is analogous to ethene, trisilane to butadiene, and tetrasilane to hexatriene, etc. Because the resonance integrals β_{vic} and β_{gem} of a polysilane are different just as those of alternate bonds in a real polyene with alternating shorter and longer bonds are different, the energy gap between HOMO and LUMO tends to a finite value, rather than to zero, with increasing chain length.

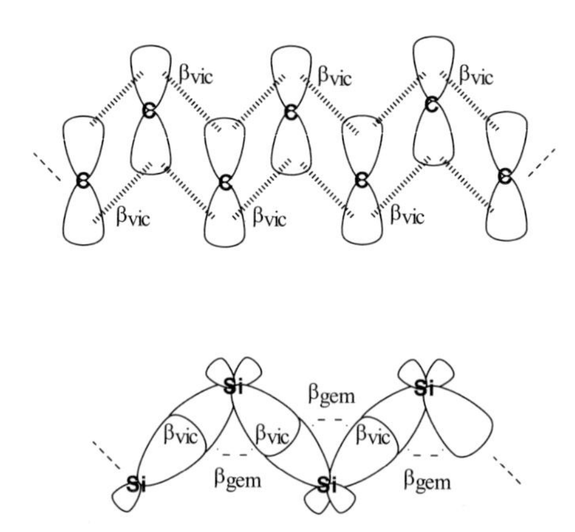

Figure 11.5. Orbital diagrams of a σ-conjugated linear polysilane (lower) and a π-conjugated polyene (upper).

The above conclusions follow from the so-called Sandorfy C model.[21] This model ignores the other two sp^3 hybrid orbitals that are used to bond the substituents and the latter are assumed not to interact with the backbone orbitals. This is adequate for describing the ground-state properties since most polysilane substituents are so much more electronegative than the silicon atoms that the orbitals that bind them to the chain are lower in energy and for the most part reside on the substituents. The corresponding antibonding orbitals are located mainly on the silicon atoms but they are unoccupied and are therefore not needed to describe the ground state. None of the occupied bonding orbitals have a node across a Si-Si bond but all of the unoccupied antibonding orbitals do. Just as the HOMO of the π system of a linear polyene has nodes between every alternate pair of carbon atoms, so the HOMO of a linear polysilane has nodes at every silicon atom. Likewise, whereas the LUMO of the linear polyene has nodes between the carbon atoms where they were not found in the HOMO and none where they were, so the LUMO of a polysilane has nodal planes between every silicon atom. In accordance with the Sandorfy C model, Figure 11.6 illustrates how the HOMO of the polysilane chain can be viewed as being composed from the $3p_x$ orbitals of the silicon atoms, whilst the LUMO has a large contribution from the silicon 3s orbitals. Consistent with this view of the ground state, the xy nodal plane at each silicon atom in the HOMO contains the orthogonal and therefore non-interacting $3p_y$ and $3p_z$ orbitals that contribute to the bonding of the substituents. In contrast, the LUMO, with its large contribution from silicon 3s orbitals, interacts quite strongly with the σ^* orbitals of the lateral bonds.

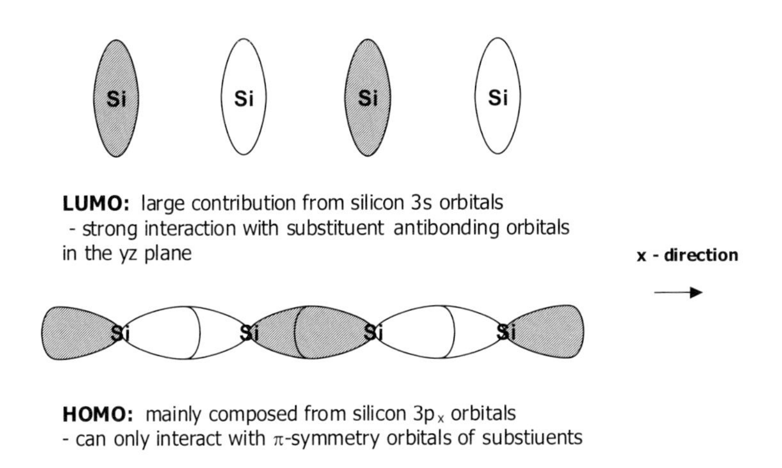

Figure 11.6. Molecular orbital representations of the HOMO and LUMO of a linear polysilane based on the Sandorfy C model.

The problem with the Sandorfy C model is that the resonance integrals (β_{vic} and β_{gem}) do not change on rotation about an Si-Si bond and all conclusions that can be drawn are independent of molecular conformation. This is inconsistent with experimental observation. Typically, the second ionization potential of peralkylated tetrasilanes shows that the energy of the gauche conformation is approximately 0.5 eV greater than the energy of the transoid conformation.[22] To account for this and related observations a non linear orbital interaction topology is required in which the magnitude and even the sign of β_{vic} changes as the dihedral angle changes from 0° (*syn*) to 180° (*anti*). Such a topology resembles a ladder and the so-called "ladder C" model [23] satisfactorily accounts for the ground-state electronic structure of polysilanes. It is from such considerations that an understanding of the absorption and emission spectroscopy and the related electroactive properties of polysilanes has developed.

11.4 Structure

Poly(arylmethylsilane)s absorb at longer wavelengths than poly(alkylsilane)s and poly(diarylsilane)s absorb at wavelengths greater than 400 nm. These shifts are too large to be explained by σ-π mixing and in the case of the poly(diarylsilane)s, based on a variety of supporting evidence, it was concluded that the explanation lay in extended sequences of silicon atoms in all-*trans* conformation enforced by the bulk of the substituents.[24] In addition, spectroscopic studies had until recently indicated that most polysilanes adopt an ordered, usually all-*trans* conformation under given conditions and some other conformation under other conditions. Typically, the position of the absorption maximum associated with the main chain σ-σ^* transition in poly(di-*n*-alkylsilane)s in solution is sensitive to temperature (thermochromism) and to solvent (solvatochromism). Thus, there is a discontinuous wavelength shift over relatively narrow temperature or compositional ranges indicating that two conformational isomers must coexist in solution.[25] Such transformations were explained as coil-to-rod transitions, the ordered all-*trans* conformation representing the rod-like form.

One of the first signs that the above model was an oversimplification came from the recognition that poly(di-*n*-hexylsilane) in its low temperature form (presumed to be an all-*trans* conformation even in the solid state) is crystalline but that its high temperature form is a liquid crystal-like *mesophase* in which crystallinity is lost but some ordering remains.[25, 26] This polymer in its low temperature form and its lower alkyl homologues that are not thermochromic are now understood to adopt helical conformations and it is the order within the helices that gives rise to the crystallinity.[27–29]

Fluorescence spectroscopy in combination with circular dichroism (CD), optical rotatory dispersion, X-ray crystallography, UV and NMR spectroscopy of the main chain is a powerful probe for identifying helical conformation, uniformity, and rigidity in polymers. In recent years, these techniques have been applied extensively to investigate the structures of polysilanes in both the solid state and in solution and it is now clear that after electronic structure main chain helicity is the principal determinant of the properties of polysilanes. In this final section it is intended to elaborate on some of the notions underlying this assertion but for a detailed understanding the reader is referred to the recently published and very excellent review of the subject by Fujiki.[30]

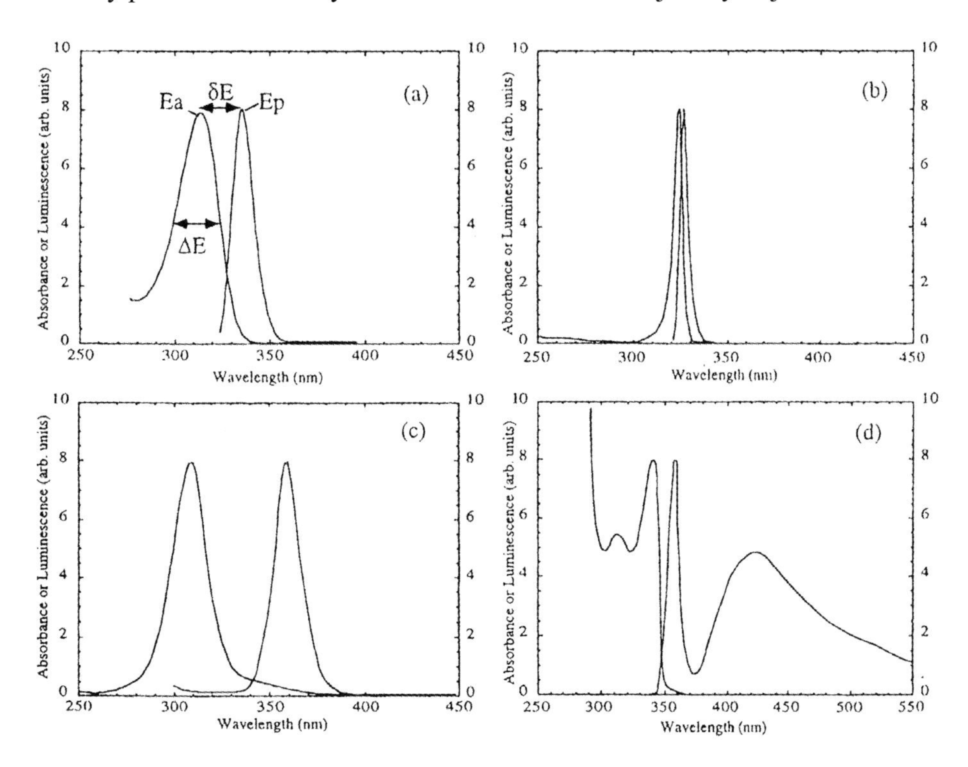

Figure 11.7. Absorption and luminescence spectra of polysilanes: (a) typical spectra in which E_a and E_p are the peaks of absorption and emission respectively, and δE and ΔE are respectively the Stokes shift and the width at half height of the absorption band; (b), (c) and (d) depict the spectra of a single screw-sense rod-like polysilane, poly(di-*n*-butylsilane), and PMPS, respectively (Reprinted from ref.[1]. Copyright 2000 Kluwer Academic Publishers).

Figure 11.7 depicts typical absorption and luminescence spectra for polysilanes. Absorption peak energies are within the range 3 to 4 eV and are determined mainly by backbone conformation. Photoluminescent efficiencies are high (10% to 50%) and the apparent Stokes shift increases with the breadth of the absorption spectrum. Whenever a single screw-sense helical structure pertains, the absorption and emission bands are narrow and the Stokes shift is small. Such a polymer is poly[decyl-(*S*)-2-methylbutylsilane], the chiral center within the (S)-2-methylbutyl substituent determining the screw sense. Such polymers are rod-like. The spectra of poly(di-*n*-butylsilane), for which two phases coexist over a limited temperature range, are broader. The Stokes shift is accordingly greater and originates from energy transfer between the two phases, which have different backbone conformations, within the excited state. Ignoring the short-wavelength absorption that arises from the ππ* transition of the phenyl substituent and the broad, long-wavelength emission, it is seen that the spectra of PMPS have features that accord more with those of poly[decyl-(S)-2-methylbutylsilane] than those of poly(di-*n*-butylsilane). The long-wavelength emission is known to be due to branching defects arising from the presence of trace amounts of trichlorosilane impurities in the synthesis and is not seen if the polymer is prepared from high purity dichloromethyl-phenylsilane.[31]

It has been shown from correlations of the σσ* UV spectroscopic parameters with the viscosity index (the α parameter of the Mark–Houwink equation) for a range of poly(dialkylsilane)s and poly(alkylarylsilane)s in THF solution at 30 °C that the peak intensities increase exponentially as α tends towards the maximum attainable value for an ideal rigid rod polymer. There is also a concomitant narrowing of peaks.[32] From this and corresponding data for a series of optically active polysilanes, the extent of σ-conjugation in the main chain is seen to be closely connected to the overall shapes of the molecules in solution. These range from shrink coil through random coil to stiff and rigid rod-like.

Structural dependence on temperature is now best illustrated by example. Optically inactive poly(hexyl-2-methylpropylsilane) has an α parameter of 1.29 in THF indicating a rigid rod-like structure. Furthermore, in isooctane solution at ambient temperature the peaks of the UV absorption and fluorescence spectra are narrow, they mirror each other, and the Stokes shift is only 3 nm. These are indicative of an almost homogeneous photoexcited energy state with minimal structural variation in the main chain. Notwithstanding this conclusion, when the solution is cooled to –80 °C, the absorptivity increases, the peaks narrow, and the Stokes shift drops to 2.2 nm. It is thus assumed that the polymer attains a perfectly extended rod shape at this temperature.[30]

If the above conclusions are applied to address the abrupt thermochromism of poly(dihexylsilane) not only in the solid state but also in solution, the conclusion is that the increase in UV absorption intensity does indeed result from a coil-to-rod transition but that the red shift arises from a helix-to-helix transition of the main chain. Recent calculations have indicated that the absorption maximum gradually red shifts as the helical dihedral angle changes from 60° to 180°.[33–35]

Although polysilane main chain conformations can range from shrink coiled through to rigid rod-like structures, it is now generally accepted that most of them (including PMPS) tend to adopt helical main chain structures regardless of substituents, physical state or temperature.[34, 35] The helices may be tight with approximately *gauche* conformations or loose with approximately *ortho* or transoid conformations. However, most polysilanes do not show Cotton effects in their CD spectra, a characteristic that would indicate a preferential screw-sense. Thus, they contain equal numbers of P and M segments, i.e. they are internal racemates and optically inactive. However, if the main chain helicity is driven by a chiral chemical influence, enantiomeric excesses of either the P or M screw sense are the result. Such an effect is shown in Figure 11.8 for the chiral substituted homopolymer and copolymer depicted alongside.[19] Positive Cotton effects at the absorption maxima are observed in both cases and there is little to distinguish the extent of the chiral influence within the homopolymer from that within the copolymer. Although it is not depicted, if the chiral substituent is changed from the (*S*) to the (*R*) isomer, the sign of the CD band changes accordingly, indicating that the polymers adopt the opposite screw sense.

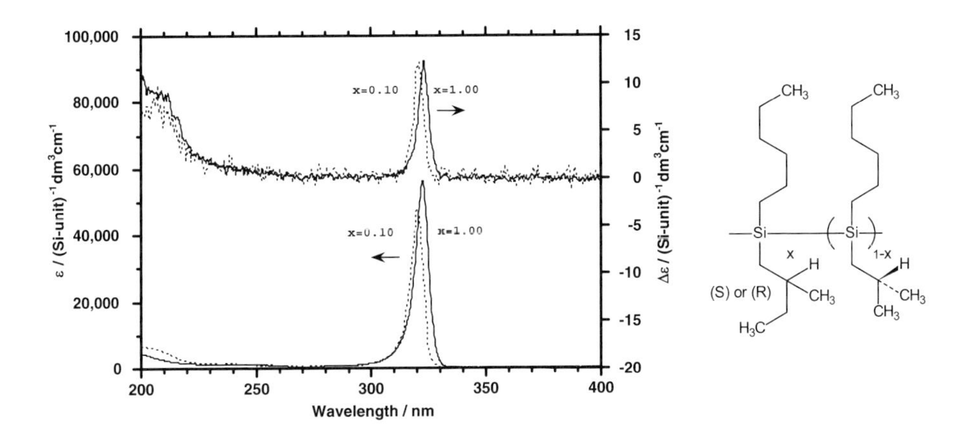

Figure 11.8. UV and CD spectra of polysilane copolymers in isooctane at –5 °C: solid lines x = 1; dotted lines x = 0.1

Figures 11.9 and 11.10, relating to PMPS and poly(methyl-*n*-hexylsilane) (PMHS) respectively, show that the chiral influence on main chain helicity of polysilanes need not be an internal influence.[36] In each case, the upper plots depict the UV spectra of the relevant polymer in THF solution and in the chiral solvent, (*S*)-2-methylbutylbenzyl ether. The lower plots depict the CD spectra overlaid on the corresponding UV spectra for solutions of the polymers in the chiral solvent. From the absorption spectra, a very slight change in solution-phase conformation is evident for PMPS on switching solvents but not so in the case of PMHS. However, both CD spectra display positive Cotton effects consistent with the main chains adopting enantiomeric excesses in the chiral solvent. In the case of PMPS the excess is only 8% but for PMHS it is 80%.

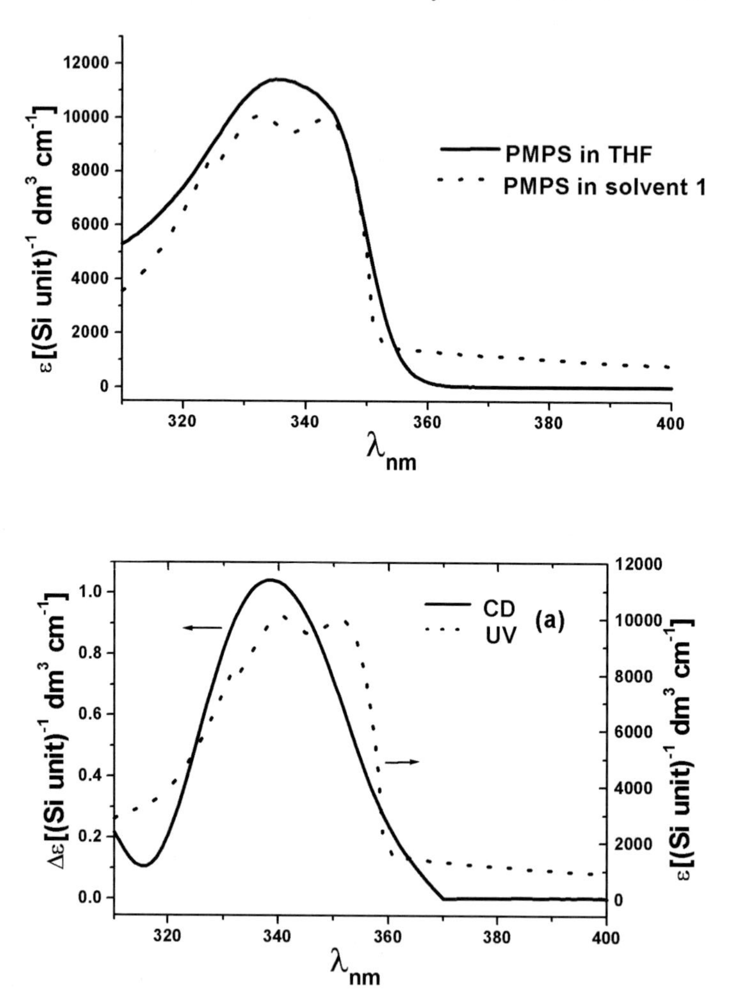

Figure 11.9. UV spectra of PMPS in THF and (*S*)-2-methylbutylbenzyl ether (top) and overlaid UV and CD spectra in (*S*)-2-methylbutylbenzyl ether (bottom) at ambient temperature.

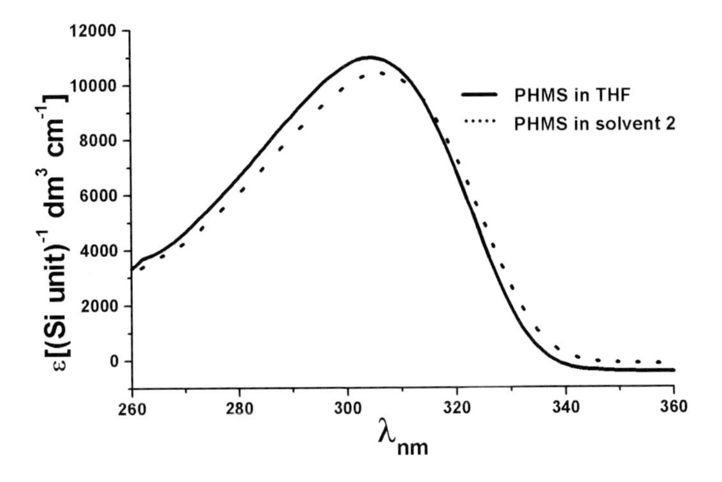

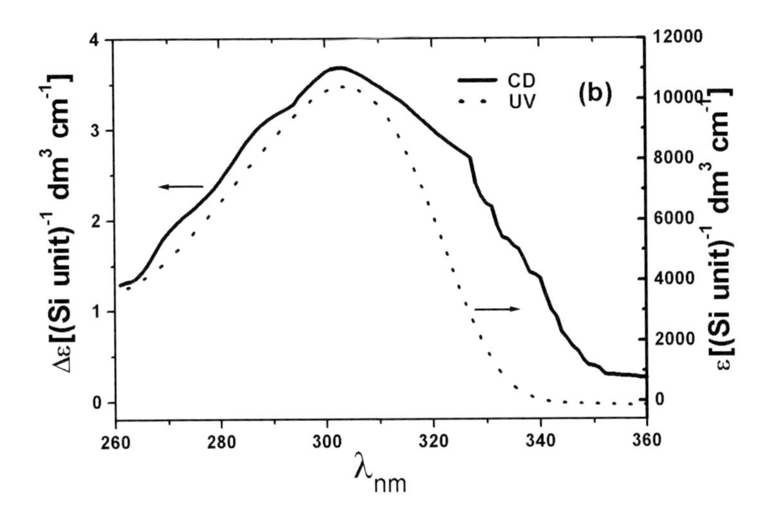

Figure 11.10. UV spectra of PMHS in THF and (*S*)-2-methylbutylbenzyl ether (top) and overlaid UV and CD spectra in (*S*)-2-methylbutylbenzyl ether (bottom) at ambient temperature.

11.5 Conclusion

The various physical properties indicated in the introduction arise from the unique bonding of the main chain of polysilanes. The consequent potential for exploitation of these properties is greatly dependent upon the conformation that the polymers adopt. Conformation influences the way in which they pack in the solid state and therefore the magnitude of the electroactive effects they display. Poly(*n*-hexylphenylsilane) has recently been shown to have a one-dimensional intrachain hole mobility of 0.23 $cm^2 V^{-1} s^{-1}$, which is comparable to the values determined for the better known π-conjugated carbon chain hole transport polymers.[37] Conformation also influences the ability of polymers to self-assemble. Recently, carefully designed amphiphilic block copolymers in which the hydrophobic component is a polysilane have been shown to assemble into vesicles in aqueous dispersion [38] or to be capable of being formed into shell-crosslinked micelles.[39] Such possibilities can be traced back to the unique chain structures of polysilanes and following from the greater understanding of these structures that has been gained in recent years will be rewards afforded by the effort to tailor them for the purposes of optimization in a comprehensive range of applications.

11.6 Acknowledgements

I should like to express my gratitude to the Japan Chemical Innovation Institute under whose management research effort into polysilanes at the University of Kent was sustained over ten years as part of the Industrial Science and Technology Frontier Programme supported by the Japanese New Energy and Industrial Technology Development Organisation (NEDO).

11.7 References

[1] R. G. Jones, W. Ando, J. Chojnowski, (Eds.), *Silicon-based Polymers: The Science and Technology of their Synthesis and Applications*, Kluwer Academic Publishers, Dordrecht, **2000**.
[2] F. Kipping, *J. Chem. Soc.* **1921**, *19*, 830.
[3] F. Kipping, *J. Chem. Soc.* **1924**, *125*, 2291.
[4] K. Sakamoto, K. Obata, H. Hirata, M. Nakajima, H. Sakurai, *J. Am. Chem. Soc.* **1989**, *111*, 7641.
[5] K. Sakamoto, M. Yoshida, H. Sakurai, *Macromolecules* **1990**, *23*, 4494.

[6] (a) H. Sakurai, K. Sakamoto, Y. Funada, M. Yoshida in P. Wisian-Neilson, H. R. Allcock, K. J. Wynne (Eds.), *Inorganic and Organometallic Polymers II, Advanced Materials and Intermediates*, ACS Symposium Series 572, **1994**, Chapter 2; (b) H. Sakurai, in N. Auner and J. Weis (Eds.), *Organosilicon Chemistry from Molecules to Materials*, VCH, Weinheim, Germany, **1994**, 285–292; (c) H. Sakurai in K. Hatada, T. Kitayama, O. Vogl (Eds.), *Macromolecular Design of Polymeric Materials*, Marcel Dekker, Inc., New York, **1997**, Chapter 27.
[7] K. Sakamoto, K. Obata, H. Hirata, M. Nakajima H. Sakurai, *J. Am. Chem. Soc,* **1989**, *111*, 7641.
[8] H. Sakurai, *Polym. Prepr.* **1990**, *31*, 230.
[9] H.Sakurai, M. Yoshida, in Ref. [1], pp 375–379.
[10] (a) M. Cypryk, J. Chrusciel, E. Fossum, K. Matyjaszewski, *Makromol. Chem. Macromol. Symp.* **1993**, *73*, 167; (b) E. Fossum, K. Matyjaszewski, *Makromolecules* **1995**, *28*, 1618; (c) M. Cypryk, Y. Gupta, K. Matyjaszewski, *J. Am. Chem. Soc.* **1991**, *113*, 1046.
[11] (a) T. Sanji, T. Sakai, C. Kabuto, H. Sakurai, *J. Am. Chem. Soc.* **1998**, *120*, 4552; (b) H. Sakurai, T Sanji, T. Sakai, H. Hanao, *Phosphorus, Sulfur, and Silicon* **1997**, *124*, 173.
[12] G. M. Gray, J. Y. Corey in Ref. [1], pp 401–418 and references therein.
[13] M. J. Went, H. Sakurai, T. Sanji, *ibid.*, pp 419-437 and references therein.
[14] R. G. Jones, W. K. C. Wong, S. J. Holder, *Organometallics* **1998**, *17*, 59.
[15] R.G. Jones, S. J. Holder in Ref. [1], pp 401–418 and references therein.
[16] R. West, *private communication.*
[17] S. J. Holder, M. Achilleos, R. G. Jones, *unpublished results.*
[18] K. Matyjaszewski, *Polym. Prepr. (Am. Chem. Soc., Div. Polym. Chem.)* **1987**, *28*, 224.
[19] M. Fujiki, J. R. Koe in Ref. [1] pp 643–665 and references therein.
[20] T. C. B. McLeish, R. G. Jones, S. J. Holder, *Macromolecules* **2002**, *35*, 548–554.
[21] C. Sandorfy, *Can. J. Chem.* **1955**, *33*, 1337.
[22] R. Imhof, D. Antic, D. E. David, J. Michl, *J. Phys. Chem. A* **1997**, *101*, 4579.
[23] H. S. Plitt, J. Michl, *Chem. Phys. Lett.* **1992**, *198*, 400.
[24] P. M. Cotts, R. D. Miller, R. Sooriyakumaran in J. M. Zeigler, F. W. G. Fearon (Eds.), *Silicon-Based Polymer Science*, Adv. Chem. Ser. 224, **1990**, pp 397–412.
[25] F. C. Schilling, F. A. Bovey, A. J. Lovinger, J. M. Zeigler, *ibid.*, pp 341–378.
[26] (a) P. Weber, D. Guillon, A. Skoulios, R. D. Miller, *J. Phys. France* **1989**, *50*, 795; (b) A. J. Lovinger, F. C. Schilling, F. A. Bovey, J. M. Zeigler, *Macromolecules* **1986**, *19*, 2660.

[27] R. D. Miller, J. Rabolt, R. Sooriyakumaran, W. Fleming, G.N. Fickes, B. L. Farmer, H. Kuzmany in M. Zeldin, K. J. Wynne, H. R. Alcock (Eds.), ACS Symposium Series; Amercian Chemical Society, Washington, DC, **1988**, Vol. *360*, pp 43–60.
[28] A. J. Lovinger, D. D. Davis, F. C. Schilling, F. A. Bovey, J. M. Zeigler, *Polym. Commun.* **1989**, *30*, 356.
[29] F. C. Schilling, A. J. Lovinger, J. M. Zeigler, D. D. Davis, F. A. Bovey, *Macromolecules* **1989**, *22*, 3055.
[30] M. Fujiki, *Macromol. Rapid. Commun.* **2001**, *22*, 539 and references therein.
[31] Y. Xu, T. Fujino, H. Naito, K. Oka, T. Dohmaru, *Chem. Letters* **1998**, 299.
[32] M. Fujiki, *J. Am. Chem. Soc.* **1996**, *118*, 7424.
[33] H. Teramae, K. Takeda, *J. Am. Chem. Soc.* **1989**, *111*, 128.
[34] R. Zink, T. F. Magnera, J. Michl, *J. Phys. Chem. A* **2000**, *104*, 3829.
[35] C. H. Ottosson, J. Michl, *J. Phys. Chem. A* **2000**, *104*, 3367.
[36] P. Dellaportas, R. G. Jones, S. J. Holder, *Macromol. Rapid Commun.* **2002**, *23*, 99
[37] F. C. Grozema, L. D. A. Siebbeles, J. M. Warman, S. Seki, S. Seiichi, U. Scherff, *Adv. Mater.* **2002**, *14*, 228.
[38] N. A. J. M. Sommerdijk, S. J. Holder, R. C. Hiorns, R. G. Jones, R. J. M. Nolte, *Macromolecules* **2000**, *33*, 8289.
[39] T. Sanji, Y. Nakatsuka, F. Kitayama, H. Sakurai, *Chem. Commun.* **1999**, 2201.

12 Phase Behavior of *n*-Alkyl-Substituted Polysilanes

Christian Mueller*, Christine Peter, Holger Frey[1], and Claudia Schmidt[2]

12.1 Introduction

Polysilanes are high-molecular-weight polymers with a pure silicon chain as backbone and organic side chains. The σ-conjugation of the silicon backbone leads to interesting properties, such as strong UV absorption and photoconductivity.[1–3] Poly(di-*n*-alkylsilanes) with *n*-alkyl side chains longer than ethyl form mesophases, in which the polymers are hexagonally packed in two dimensions but have highly flexible backbones. Such condis (conformationally disordered) crystals,[4] characterized by a dynamic exchange between different conformations, are also known for other polymers, for example, polysiloxanes and polyphosphazenes. A clear proof of the backbone mobility in these mesophases is provided by solid-state NMR spectroscopy.[5–7]

While the mesophase structures seem to bc similar for all poly(di-*n*-alkylsilanes), the crystal structures depend on the length of the side chains. In the following, symmetrically substituted poly(di-*n*-alkylsilanes) will be denoted by PDxS, with x specifying the length of the side chains, and asymmetrically substituted polysilanes with side chains of length x and y at each silicon atom will be called PxyS. PD6S was found to have a planar all-*trans* (2_1 helical) backbone conformation,[8, 9] whereas PD4S and PD5S reportedly have a 7_3 helical structure.[10, 11] Other structures, some still under dispute, have been reported for homologues with longer side chains.[8] Over the years, it has become obvious that most polysilanes can exist in more than one crystalline form.[8,12–18] Since the physical properties, which form the basis of all applications, depend strongly on structural details, the goal of our investigations was to achieve a better understanding of this polymorphism. Only a comprehension of how a variation of the chemical structure (in this case the length of the side chains) affects the phase structure, and hence other properties, can lead to a tailored synthesis of functional materials.

In the following, we will first report on the mobility of the alkyl side chains in polysilanes with short side chains up to length 6. The analogy between

---

* Address of the authors: Institut für Makromolekulare Chemie, Stefan-Meier-Straße 31, und Freiburger Materialforschungszentrum, Stefan-Meier-Straße 21, Albert-Ludwigs-Universität, D-79104 Freiburg, Germany

1 Present address: Institut für Organische Chemie, Johannes-Gutenberg-Universität, Duesbergweg 10–14, D-55128 Mainz, Germany

2 Present address: Fakultät für Naturwissenschaften, Department Chemie, Universität Paderborn, Warburger Str. 100, D-33098 Paderborn, Germany

polysilanes and polycarbosilanes will be pointed out. Then, the polymorphism of PD4S and PD5S at elevated pressure, known from reports on piezochromic effects,[19 20] and the phase behavior of PD8S and PD10S, as revealed by our DSC, NMR, and pVT experiments, will be discussed.[21]

12.2 Materials and Methods

12.2.1 Polymer Materials

Polysilanes with protonated or α-deuterated *n*-alkyl side chains were prepared from the respective dialkyldichlorosilanes by Wurtz-type coupling.[1,21] The products, high-molecular-weight polymers with number average molecular weights between 180,000 and 400,000 g/mol and polydispersities of about 1.5, were obtained by fractionated precipitation from toluene. Additional samples were prepared by dissolving small amounts of polymer in cyclohexanol and letting them slowly crystallize from the boiling solvent.

12.2.2 Measurements

12.2.2.1 Differential Scanning Calorimetry (DSC). DSC measurements were carried out with a Perkin Elmer DSC 7 at heating rates between 0.1 and 40 K/min.

12.2.2.2 Pressure-Volume-Temperature (pVT) Experiments. pVT experiments were performed with a Gnomix apparatus using the confining-fluid (Hg) method. Temperatures from 30–400 °C and pressures between 10 and 200 MPa were accessible.

12.2.2.3 Solid-State NMR Spectroscopy. The NMR experiments were carried out with a Bruker MSL 300 spectrometer at resonance frequencies of 46.073 and 59.625 MHz for ^{2}H and ^{29}Si, respectively. ^{2}H spectra were obtained using the quadrupole echo pulse sequence with 90° pulses of 3 μs. ^{29}Si signals in the crystalline phase were recorded using cross-polarization with a contact time of 2.5 ms and a recycling delay of 4 s. In the mesophase, either cross-polarization or direct excitation of the ^{29}Si nuclei with a single pulse, followed by acquisition under proton decoupling was used. Chemical shifts were calibrated using Q_8M_8 (12.4 ppm relative to TMS); the values reported are relative to TMS.

12.3 Results and Discussion

12.3.1 Side-Chain Motion in the Crystalline Phase

12.3.1.1 PD4S, PD5S, and PD6S. A typical example of the changes in the ^{2}H NMR line shape when passing through the phase transition from the crystalline state to the mesophase is shown in Figure 12.1 for a sample of PD4S. The wide Pake pattern at lower temperatures is the spectrum of the crystalline phase. It is replaced by the motionally narrowed Pake pattern of the mesophase at higher temperatures. The broad transition regime of several tens of K is typical of mesomorphic polymers. Above the glass transition temperature a narrow isotropic peak from a spurious amorphous fraction is observed. The high crystallinity of the samples discussed here is due to the special preparation technique using slow crystallization from boiling cyclohexanol.

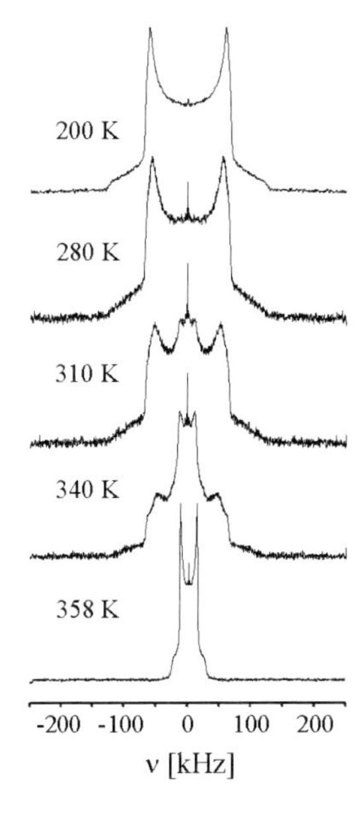

Figure 12.1. ^{2}H NMR spectra of PD4S, showing the transition from the crystalline phase at low temperature to the mesophase at high temperature.

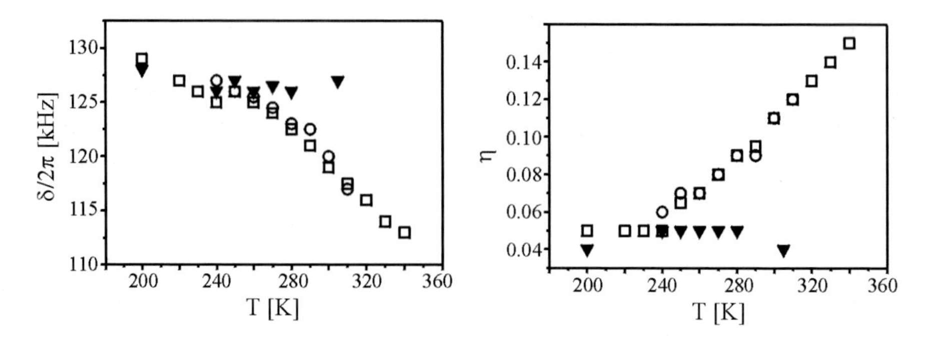

Figure 12.2. ^{2}H NMR line-shape parameters δ and η of highly crystalline PD4S (squares), PD5S (circles), and PD6S (triangles).

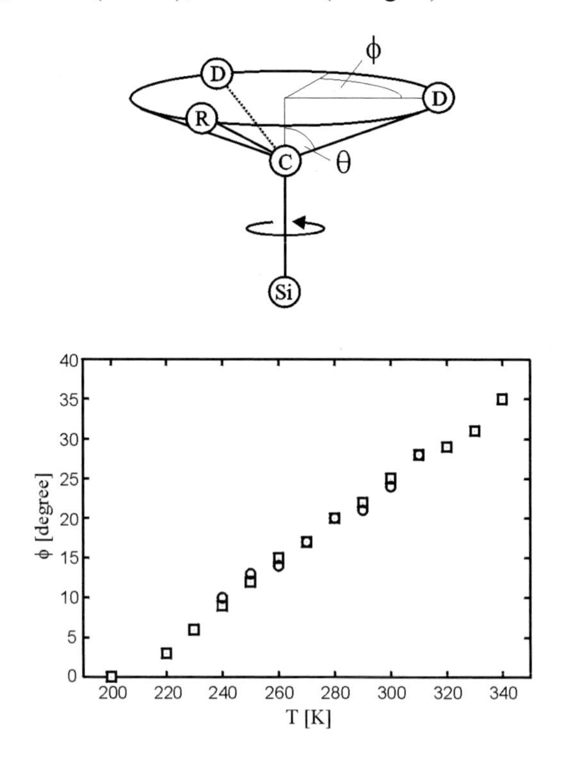

Figure 12.3. Top: Jump model for the motion of the α-CD_2 group in the crystalline phase; bottom: jump angle ϕ as a function of temperature for PD4S (squares) and PD5S (circles).

A closer inspection of the spectra obtained from the crystalline phase of PD4S reveals subtle changes in the line shape, which result from an increasing mobility of the side chains as the temperature is raised. The quadrupole splitting decreases and the shoulders at the outer flanks of the singularities become more pronounced. A quantitative line-shape analysis by fitting of the spectra yields the quadrupole coupling δ and the asymmetry η of the electric field gradient. The temperature dependence of these line-shape parameters is shown in Figure 12.2. Only the polymers PD4S and PD5S, which form a 7_3 helical backbone, show substantial side-chain motion in the crystal, whereas PD6S, with its all-*trans* (2_1) backbone conformation, remains almost rigid. Hence, mobility of the side chains is clearly related to the phase structure.

The side-chain motion can be described by the simple jump model illustrated in Figure 12.3, which was developed for poly(diethylsiloxane).[22] The backbone remains rigid but the α-CD_2 group of the side chain is allowed to rotate about the Si-C bond. This rotation can be modeled as a jump of the deuterons by a jump angle ϕ on a cone with a fixed cone angle θ. The experimental line shapes were fitted using ϕ as a free parameter. The results depicted in Figure 12.3 show a linear increase of ϕ with increasing temperature for both PD4S and PD5S.

As the side chains begin to move, the silicon backbone stays essentially rigid. This becomes evident from an inspection of the anisotropic chemical shift parameters of ^{29}Si, compiled in Figure 12.4 for PD4S, PD5S, and PD6S. The shift parameters of the crystalline phase do not vary with temperature, showing that the silicon backbone remains immobile. Figure 12.4 also demonstrates once more the different backbone conformation of PD6S compared to PD4S or PD5S. In the mesophase, there is no difference in the shift parameters of the different polymers, indicating that the dynamical state in the mesophase is the same for all the polymers.

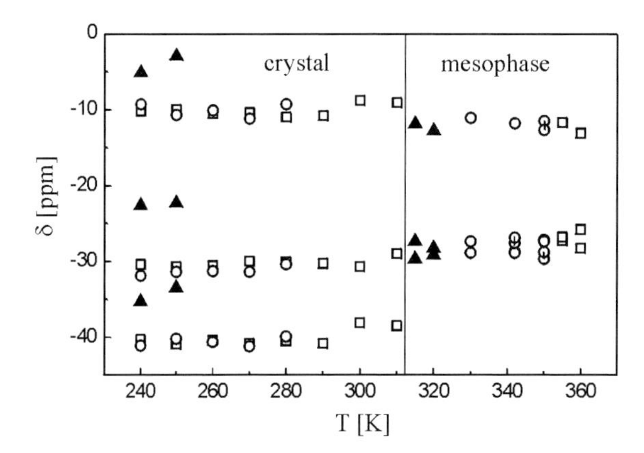

Figure 12.4. Principal values of the ^{29}Si chemical shift tensor (relative to TMS) for PD4S (squares), PD5S (circles), and PD6S (triangles) in the crystalline phase and in the mesophase.

12.3.1.2 Polycarbosilane. The ^{2}H NMR spectra of poly(di-*n*-butylsilylenemethylene),[23] depicted in Figure 12.5, are qualitatively similar to those of the polysilanes, showing clear evidence of a mesophase.[21, 24] However, the mesophase of polycarbosilane could not be obtained in pure form; it was always found to coexist with at least one other phase. The quadrupole splitting of the crystalline phase[21, 24] decreases with increasing temperature as seen for PD4S and PD5S, revealing mobile side chains. In the mesophase of polycarbosilane, the splitting is much smaller than that for the polysilanes. This is not surprising as the mesophase splitting is also related to the rapid conformational exchanges of the backbone, which will, of course, be different for the two classes of polymers.

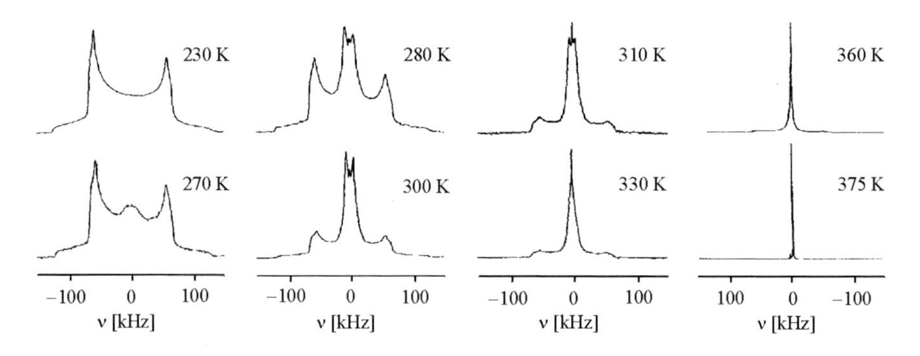

Figure 12.5. ^{2}H NMR spectra of poly(di-*n*-butylsilylenemethylene) deuterated in the α-position of the side chains.

12.3.2 Polymorphism of Polysilanes with Short Side Chains

The phase behavior of PD4S, PD5S, and PD6S appears simple at first glance since all three polymers exhibit only one transition in DSC experiments. This mesophase transition occurs at about 351, 338, and 318 K for PD4S, PD5S, and PD6S, respectively. Variations of up to 5 K from batch to batch have been observed.[7,21] However, previous reports on piezochromism indicate the existence of more than one crystalline modification, which led us to perform pVT investigations on polysilanes.[21]

A first result of our pVT experiments is that the volume change from the crystalline phase of PD6S (2_1 conformation) is rather large, whereas the "normal" transitions of PD4S and PD5S (7_3 conformation) show only a small volume change.[21,24] This is consistent with a less dense packing and the higher side-chain mobility of PD4S and PD5S described above. While pVT experiments on PD6S show no additional crystalline phases, for PD4S and PD5S a phase behavior much richer than previously thought was unraveled.

Figure 12.6 gives a summary of the transitions observed in our pVT investigations on these two polymers.

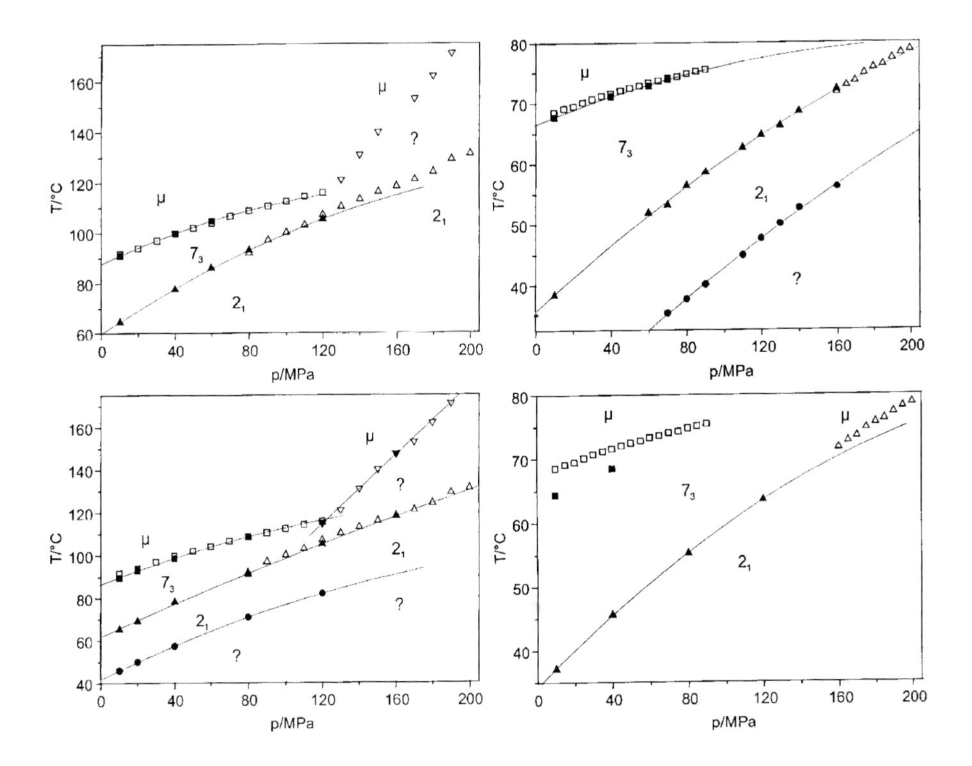

Figure 12.6. Phase transition temperatures of PD4S (left) and PD5S (right) determined by pVT experiments. Top: after isothermal crystallization; bottom: after isobaric crystallization.

First of all, for both PD4S and PD5S one additional phase can be obtained at high pressure by either isothermal or isobaric crystallization. This phase occurs at lower temperatures than the "normal" crystalline phase with its 7_3 backbone. Because of the relatively large volume change at the additional transition (from the new phase to the 7_3 structure or possibly to the mesophase) the new phase is assumed to be equivalent to the crystal phase of PD6S and is therefore labeled "2_1" in the diagrams of Figure 12.6. This additional phase is not the only new feature. Under certain conditions, by isobaric crystallization in the case of PD4S and isothermal crystallization in the case of PD5S, one more phase of unknown structure forms at low temperatures. Furthermore, PD4S reveals a complicated phase behavior at high pressures. The nature of the transition to the mesophase changes dramatically at about 120 MPa, as is evident from the different slopes for the mesophase transitions in Figure 12.6, showing yet another phase in this regime.

Experiments on polysilanes with asymmetric side chain substitution, P45S and P56S, showed no spectacular results, but the behavior of these polymers fits well into the overall picture. P45S has one transition at 319 K and a 7_3 backbone conformation;[8, 25, 26] pressure-induced additional phases were not observed. P56S shows two transitions[27] at 271 and 289 K, even at ambient pressure, indicating two different modifications, probably with 2_1 and 7_3 backbone conformations.

12.3.3 Phase Behavior of PD8S and PD10S

12.3.3.1 PD8S. The large variation in the phase transition temperatures[1, 16, 17, 21] and other physical properties such as UV absorption[16–18] as reported by different authors indicates a rich polymorphism of this polymer. Very recently, Chunwachirasiri et al.[17, 18] reported on five different crystalline or semicrystalline structures in quenched samples of PD8S, some of which occur only below the temperature range considered here. Our own DSC experiments on samples prepared in three different ways (denoted A–C) confirm that the phase behavior strongly depends on the conditions of sample preparation.

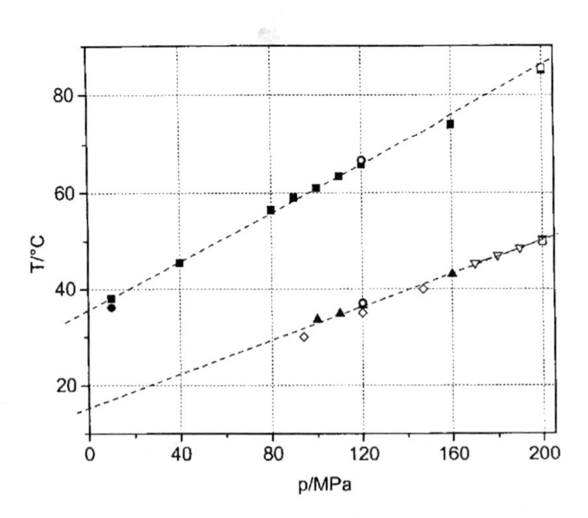

Figure 12.7. Phase transition temperatures of PD8S determined by pVT experiments under various conditions. Open triangles and squares, filled circle: after isobaric crystallization; filled triangles and squares, open circles: after isothermal crystallization; open diamonds: transition pressures after isothermal crystallization. The dashed lines are linear fits of the filled squares and triangles, respectively.

Precipitation from toluene at room temperature yields samples with only one transition at 285–286 K upon heating (type A). These samples are sticky solids at room temperature, in agreement with their low phase-transition temperatures. Type A is probably the type of material that most authors have obtained and investigated. Its ^{29}Si MAS spectra below the transition temperature show one major peak at $\delta = -28.5$ ppm together with a much smaller peak at $\delta = -20.8$ ppm.[21] These peaks disappear simultaneously at the transition to the mesophase; whether or not they belong to one structure is not clear. The chemical shift value of the smaller peak indicates a transoid structure similar to that of PD6S.

When PD8S is precipitated from toluene at high temperature, the first DSC heating scan (after prior cooling) yields the well-known transition at about 283 K plus an additional one at about 304 K, which, to our knowledge, has not been reported before. This second transition is no longer present in subsequent heating scans. The existence of two distinct crystalline forms with different transition temperatures to the mesophase is corroborated by extensive pVT experiments, applying different isobaric or isothermal crystallization procedures on a sample of type A.[21] The results, summarized in Figure 12.7, give evidence of only two distinct crystalline phases, no matter how the sample was treated. The phase with the lower transition temperature to the mesophase is obtained by rapid crystallization at low temperatures, that is, when the sample is quenched, even at moderate cooling rates, as in a typical DSC experiment. The second phase, which has a higher transition temperature, is obtained only when the sample is allowed to crystallize slowly at elevated temperatures. Both phases can be induced by the application of pressure. They correspond to the two phases found at ambient pressure in type A and B samples. This is proven both by the extrapolation of the data in Figure 12.7 to low pressure and by a DSC experiment carried out with a sample that had been slowly crystallized under pressure. This pressure-crystallized sample shows the same transition temperatures as a type B sample. Of the two crystalline forms, the slowly crystallizing one is the thermodynamically stable phase. Due to the slow kinetics of its formation, estimated to be slower than the formation of the other phase by a factor of 0.02, it is rarely obtained.

Finally, a third type of sample (C) was obtained by slow crystallization from boiling cyclohexanol. This procedure yields a white powder at room temperature. The sample shows three DSC transitions at 283, 313 (shoulder), and 321 K. In the pVT experiments, starting from a type A sample, we found no evidence of a crystalline form with such a high transition temperature. Interestingly, Miller and Michl also report a phase-transition temperature of 320 K for a PD8S sample with a transoid backbone structure.[1]

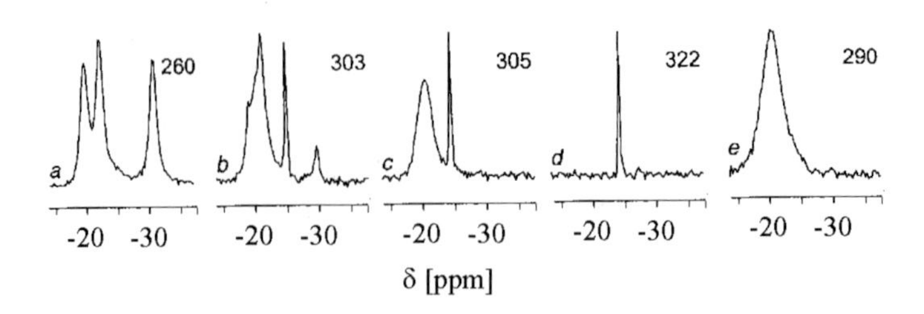

Figure 12.8. ^{29}Si MAS spectra of PD10S at different temperatures (in K). Spectra a–d were obtained upon heating after rapid cooling from the mesophase, spectrum *e* after slow cooling from the mesophase.

12.3.3.2 PD10S. Our conclusions on the phase behavior of PD10S based on DSC and solid-state NMR measurements have been reported before.[15] We assigned the two DSC peaks observed at about 304 and 322 K to the transitions of two different crystalline phases, labeled I and II, to the mesophase. Phase I is normally obtained at moderate cooling rates, whereas phase II is obtained only when the sample is annealed above 304 K or for samples prepared by crystallization from boiling cyclohexanol. ^{29}Si MAS NMR spectra are shown in Figure 12.8. Phase I (spectrum a) exhibits three narrow peaks at $\delta = -19.7$, –22.5, and –30.8 ppm, while phase II (spectrum e) has one broad signal at $\delta =$ –20.5 ppm, reminiscent of the spectrum of PD6S. The fact that phase I exhibits three peaks left some doubt about it being a pure phase. Based on X-ray investigations, Chunwachirasiri et al.[17] suggest a crystal structure of phase I with a two-chain orthorhombic unit cell and a backbone structure with $T_nD_-T_{n'}D_+$ sequences (T: transoid with dihedral angle of 170°, D: deviant with dihedral angle of 145°, n and n': small integer numbers) that includes considerable conformational disorder. This structure may have three basic Si environments with Si atoms surrounded by two T linkages, a T and a D_+ (or D_-), or even a D_+ and D_-, consistent with the three peaks seen in the NMR spectrum. The TG_+TG_- structure of the backbone suggested earlier[8] can be ruled out for phase I.[17]

In addition to the two structures mentioned here, three more crystalline phases have been observed by Chunwachirasiri et al. in samples quenched from the mesophase at 343 K to 243 K.[17] These authors also reported a fundamental difference between the structures of polysilanes with short and long side chains. The crystalline phases of PD10S have a biradial structure, that is, the side-chains are sticking out from the backbone in only two directions, in contrast to the tetraradial structure with interdigitated side chains from neighboring polymer molecules found for PD6S and other homologues with short side chains. The structures of PD8S are intermediate forms, with biradial structure but a more open packing of the side chains.

12.4 Conclusions

During the last few years, substantial progress has been made in understanding the phase behavior of polysilanes. Seemingly contradictory results from different groups could be understood once the complexity of the phase behavior of some polysilanes had been realized. A variety of phases with different chain packings, different backbone conformations, or different side-chain orders exists, not only for polysilanes with longer side chains but, as the pVT experiments discussed here demonstrate, also for PD4S and PD5S. Although the results reported here[21] and the investigations by Chunwachirasiri et al.[17,18] have shed more light on the polymorphism of alkyl-substituted polysilanes, many puzzles remain to be solved.

12.5 Acknowledgements

We thank Beate Gloderer, Alfred Hasenhindl, and Marie-France Quincy for help with the synthesis, the NMR experiments, and the preparation of this manuscript, respectively.

12.6 References

[1] R. Miller, J. Michl, *Chem. Rev.* **1989**, *89*, 1359.
[2] R. West, *J. Organomet. Chem.* **1986**, *300*, 327.
[3] R. West, in *Comprehensive Organometallic Chemistry II*, (Ed.: A. G. Davies), Pergamon, Oxford, 1994, p. 77.
[4] B. Wunderlich, J. Grebowicz, *Adv. Polym. Sci.* **1984**, *60–61*, 1.
[5] F. C. Schilling, F. A. Bovey, A. J. Lovinger, J. M. Zeigler, *Macromolecules* **1986**, *19*, 2660.
[6] G. C. Gobbi, W. W. Fleming, R. Sooriyakumaran, R. D. Miller, *J. Am. Chem. Soc.* **1986**, *108*, 5624.
[7] C. Mueller, C. Schmidt, H. Frey, *Macromolecules* **1996**, *29*, 3320.
[8] E. K. KariKari, A. J. Greso, B. L. Farmer, R. D. Miller, J. F. Rabolt, *Macromolecules* **1993**, *26*, 3937.
[9] S. S. Patnaik, B. L. Farmer, *Polymer* **1993**, *33*, 4443.
[10] R. D. Miller, B. L. Farmer, W. Fleming, R. Sooriyakumaran, J. Rabolt, *J. Am. Chem. Soc.* **1987**, *109*, 2509.
[11] F. C. Schilling, A. J. Lovinger, J. M. Zeigler, D. D. Davis, F. A. Bovey, *Macromolecules* **1989**, *22*, 3055.

[12] J. F. Rabolt, D. Hofer, R. D. Miller, G. N. Fickes, *Macromolecules* **1986**, *19*, 611.
[13] E. K. Karikari, B. L. Farmer, C. L. Hoffmann, J. F. Rabolt, *Macromolecules* **1994**, *27*, 7185.
[14] S. S. Bukalov, L. A. Leites, R. West, T. Asuke, *Macromolecules* **1996**, *29*, 907.
[15] C. Mueller, H. Frey, C. Schmidt, *Monatsh. f. Chem.* **1999**, *130*, 175.
[16] T. Kanai, H. Ishibashi, Y. Hayashi, T. Ogawa, S. Furukawa, R. West, T. Dohmaru, K. Oka, *Chem. Lett.* **2000**, 650.
[17] W. Chunwachirasiri, R. West, M. J. Winokur, *Macromolecules* **2000**, *33*, 9720.
[18] W. Chunwachirasiri, I. Kanaglekar, G. H. Lee, R. West, M. J. Winokur, *Synth. Met.* **2001**, *119*, 31.
[19] K. Song, H. Kuzmany, G. M. Wallraff, R. D. Miller, J. F. Rabolt, *Macromolecules* **1990**, *23*, 3870.
[20] K. Song, R. D. Miller, G. M. Wallraff, J. F. Rabolt, *Macromolecules* **1991**, *24*, 4084.
[21] C. Mueller, *Struktur und Dynamik mesomorpher, n-alkylsubstituierter Polysilane und Polysilylenmethylene*, Ph.D. Thesis, Universität Freiburg, 2001.
[22] V. M. Litvinov, V. Macho, H. W. Spiess, *Acta Polym.* **1997**, *48*, 471.
[23] F. Koopmann, H. Frey, *Macromolecules* **1996**, *29*, 3701.
[24] C. Mueller, C. Schmidt, F. Koopmann, H. Frey, in *Organosilicon Chemistry IV – From Molecules to Materials*, (Eds.: N. Auner, J. Weis), Wiley-VCH, Weinheim, 2000, p. 55.
[25] B. Klemann, R. West, J. A. Koutsky, *Macromolecules* **1993**, *26*, 1042.
[26] B. M. Klemann, R. West, J. A. Koutsky, *Macromolecules* **1996**, *29*, 198.
[27] G. Wallraff, M. Baier, R. Miller, J. Rabolt, V. Hallmark, P. Cotts, P. Shukla, *Polym. Prep.* **1990**, *31*, 245.

13 Structural and Electronic Systematics in Zintl Phases of the Tetrels

Reinhard Nesper*

Zintl phases are made up of metal ions and semi-metal cationic components. They involve both ionic and covalent interactions, which occur in discrete regions in their crystal structures. Covalent forces bind the Zintl anions if they do not occur as monoatomic entities, as oligo- or polymeric groups consisting exclusively of the semimetal component(s).[1–5]
In general, simple counting rules allow for a rationalization of the dependences of electron number and crystal structures.[6–10] For compounds of main group elements like $A_aX_x\square_d$ ($\square$ represents defects which are found in a given structure compared to a defect-free reference structure) Eq. (1) holds:[11]

$$E = (8\text{–}N)(a + x) + N^{*}d + E_c' + E_c'' \tag{1}$$

(*E* = valence electron number per formula unit; N = average bond number per semi-metal atom in a chosen defect-free reference structure; N* = additional electrons to be introduced upon formation of one defect in the reference structure, e.g. a defect in the diamond structure type localizes four electrons to form four lone electron pairs in the neighborhood of the defect; E_c = electrons located in electron-poor clusters which, for example, may be determined according to Wade-Mingos rules[12] or according to the 18e rule; E_c' = electron number in electron-poor cluster(s) composed of semi-metal atoms; E_c'' = electron number in electron-poor cluster(s) composed of metal atoms). By definition, Zintl phases do not contain metal-centered occupied electronic states; for these cases (1) may be rewritten as:[13]

$$E = (8\text{–}N)x + N^{*}d + E_c' \tag{2}$$

(E_c' = electron number in electron-poor clusters composed of semi-metal atoms). Eq. (2) states that bonds only occur between semi-metal atoms. This implies a complete formal electron-transfer concept for relating crystal and electronic structures. Such a complete formal charge transfer has proven to be a very reliable concept applicable not only to semiconducting Zintl phases but also if metallic conductivity occurs.[14–16] A graphical rationalization of this concept is given in Figure 13.1: the butterfly-shaped X_4^{6-} Zintl anion bears the

---

* Address of the author: Laboratory of Inorganic Chemistry, ETH Hönggerberg HCI, CH-8093 Zürich, Switzerland.

formal charge (6–) if X = Si, Ge. Such large formal charges (resulting in fairly large effective charges) must always be stabilized by a surrounding cation matrix. Such strong coulomb fields cannot be guaranteed outside of crystal lattices. Consequently, many attempts to isolate or react highly charged Zintl anions have resulted in redox and decomposition reactions. Inspection of the geometric distribution of valence states, for example by the electron localization function ELF,[17–20] shows that occupied bonding and nonbonding states are always centered at the semi-metal sites (yellow iso-ELF surfaces ELF = 0.8 in Figure 13.1). This means that the relationship between geometric structures and valence electron numbers can only be understood in simple terms if metal-centered states are not hosting valence electrons. For good reasons, it should be noted that such an understanding of charge transfer is completely independent of any definition of effective charge distributions. This should thus be emphazised that the cations coordinate the valence electron regions which are centered at the Zintl anions but do not take part in the covalent bonding patterns. This means that mainly ionic interactions are present between cations and Zintl anions.

If there is an energy gap (E_G) between the occupied semi-metal centered states (non-bonding and bonding; valence bands) and the unoccupied antibonding metal-centered states, then semiconductor (E_G< 3eV) or isolator (E_G > 3eV) behavior results. In general, but not exclusively, Zintl phases belong to the group of the semiconductors.

If the semi-metal components are either silicon or germanium, very frequently their Zintl phases exhibit the following characteristics:[21,22]

1. large cations group to trigonal prismens which are centered by X and have bond angles X-X-X close to 120° (Figure 13.2);
2. large cations envelope bridging parts of Zintl anions;
3. small cations of the same or a higher charge coordinate higher charged terminal groups of the Zintl anions because they generate a stronger coulomb field and thus introduce a better charge stabilization (Figure 13.2);
4. the Zintl anions are very frequently planar or contain planar parts; this tendency increases with increasing negative charge;
5. planar parts of different Zintl anions stack in a ecliptic manner with typical distances of 420 and 490 pm; these separations still allow for covalent interactions;
6. if 4. and 5. hold, then anisotropic and/or low-dimensional metallic conductivities arise along the stacking direction of the Zintl anions.

Despite the very frequent occurrence of planar Zintl anions, there is no indication at all that aromatic systems occur.

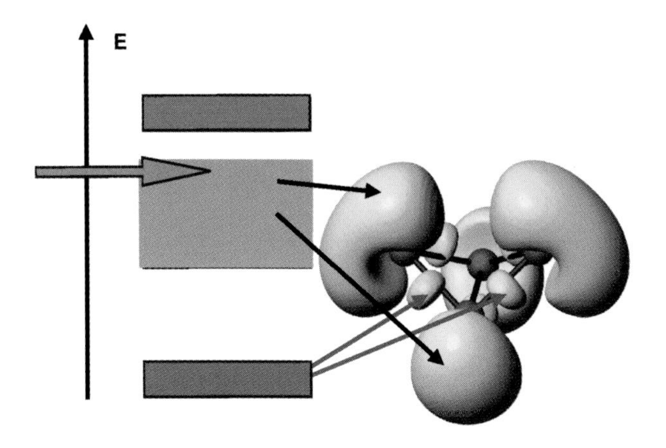

Figure 13.1. Butterfly-shaped X_4^{6-} Zintl anion.

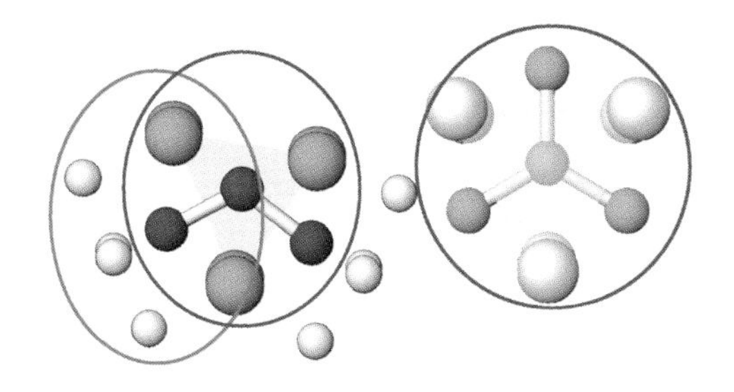

Figure 13.2. Arrangements in Zintl phases.

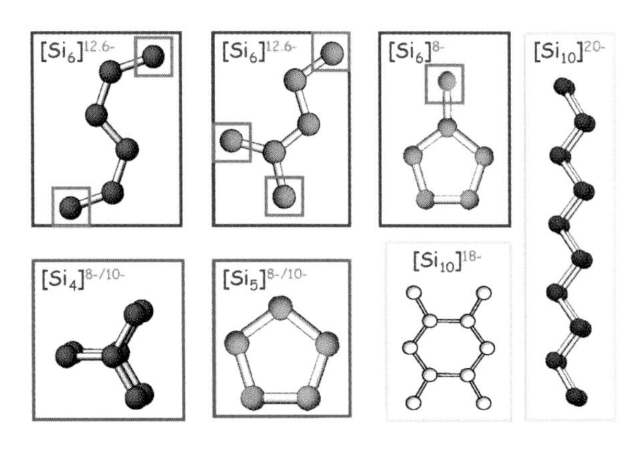

Figure 13.3. different geometries of Zintl anions.

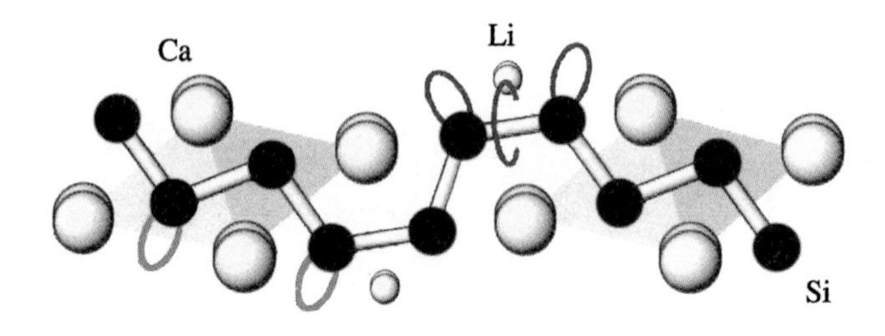

Figure 13.4. Effect of different cations on a chain segment.

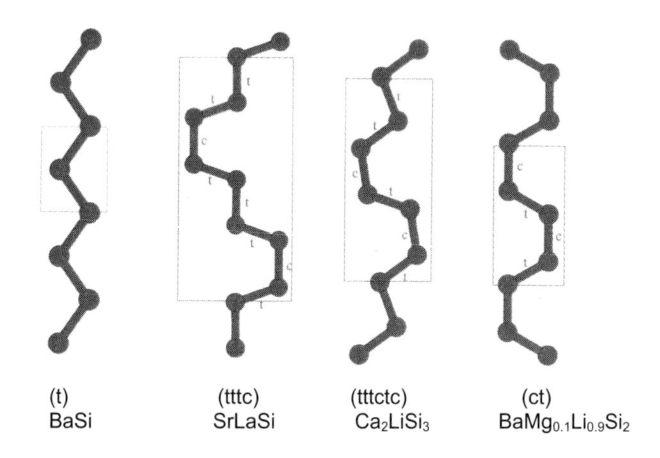

Figure 13.5. Different chain conformations found in Zintl phases.

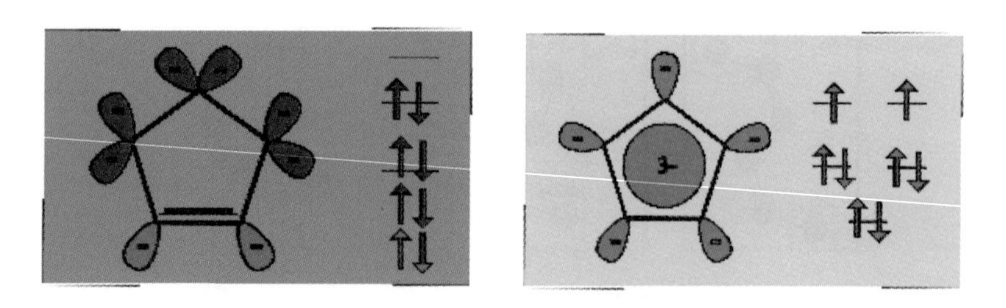

Figure 13.6. Two possible isomers of $[X_5]^{8-}$.

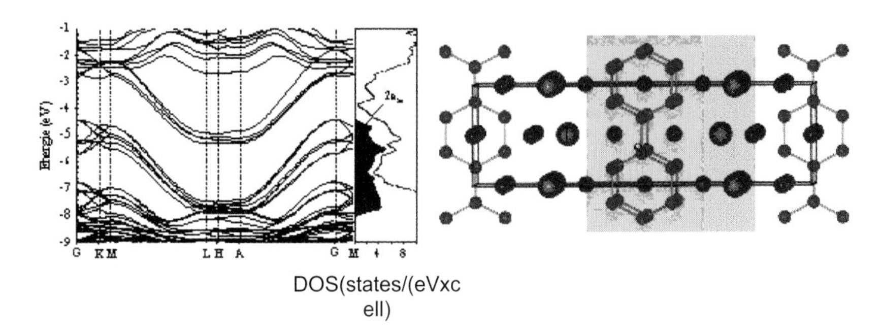

Figure 13.7. Eclipsed stack of transannularly linked Si6 rings forming a one-dimensional planar Zintl anion (polysilaphenylide).

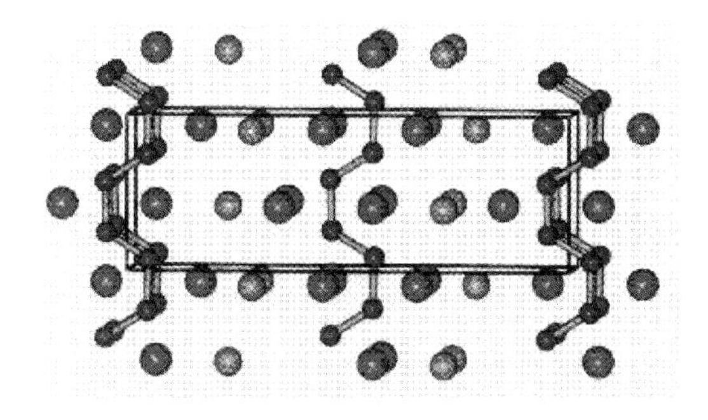

Figure 13.8. Structure of Ba_2X_2I.

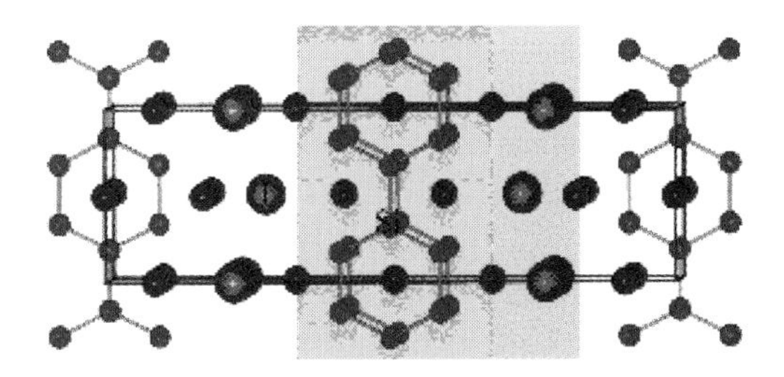

Figure 13.9. Structure of $Sr_5Si_6I_2(SiI_2)$.

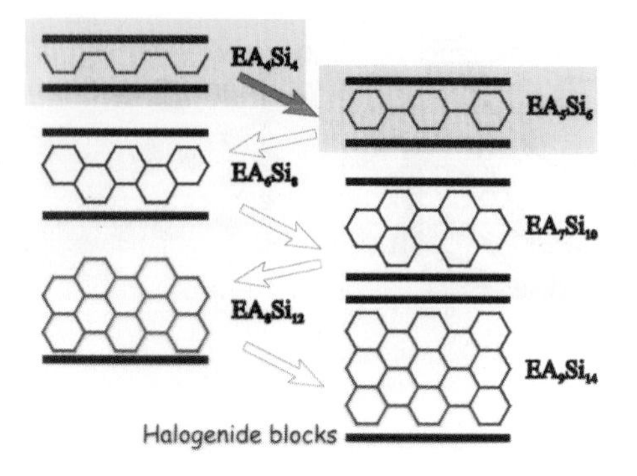

Figure 13.10. En route to hypothetical "graphite-like" silicon.

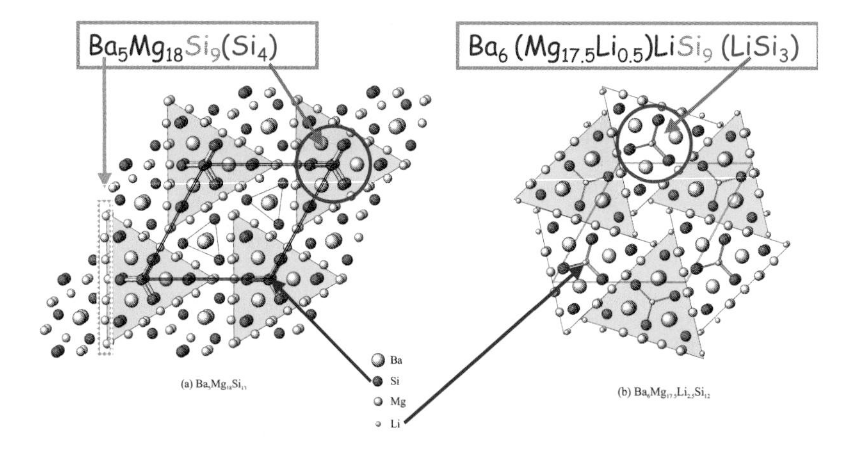

Figure 13.11. Crystal structures of $Ba_5Mg_{18}Si_9(Si_4)$ (left) and $Ba_9(Mg_{17.5}Li_{0.5})ZrSi_9(LiSi_3)$ (right).

Only in selected cases do these rules apply for stannides. As there are no carbides with sufficiently large carbanions, comparison is not possible here.

The electrostatic action of the cations on the very soft Zintl anions is considerable and determines the crystal structure formations in a pivotal way. Figure 12.3 shows Zintl anions with the same or similar charge but of different geometry. Short-range interactions surely give rise to the formation of different forms of anions, but general structural requirements (packing, etc.) may also be of importance.

Figure 13.4 shows the influence of different cations on a chain segment: small cations like Li^+ and Mg^{2+} favor a *cis*-conformation, because that results in

better contacts to the free electron pairs of the chain atoms. Figure 12.5 presents a few examples of different chain conformations found in Zintl phases.[14, 21]
As mentioned before, small cations also coordinate anionic end-groups, which means that such cations may exhibit different multidirectional behavior in a structure. Thus, the frequency of end-groups $(1b)X^{-}$ [23] is a function of the concentration of such cations. Complex cations like $[M_6O]^{10-}$ (M = Ca, Sr, Ba), which have also been found in Zintl phases, tend to act similarly. If the Zintl anions are viewed as soft geometrical sensors, then the coulomb fields generated by the complex cations should be similar to those of the above-discussed small or highly charged cations.

Besides the cation-anion interactions, the electron distribution in the Zintl anion itself plays a structure-directing role. In Figure 13.6, two possible isomers of $[X_5]^{8-}$ are shown, both of which have partially occupied π^* systems. In one case, a localized double bond is formed (left), in the other a spatially well delocalized π-system results with double bond contributions all over the anion. In general, the latter is only possible in planar systems and is preferentially found in the Zintl phases discussed here (cf. Figure 12.6). It should be emphasized that – according to our present stage of knowledge – there are no magic electron numbers and no aromaticity in these anions. The driving force for a certain electronic filling is simply that the Zintl anions receive a global charge, which is as large as possible without overloading local semi-metal centers. In other words, the preference for planar arrangements in highly charged Zintl anions stems from the need to achieve a stable charge distribution. Consequently, the filling of the π^*-systems must vary depending on the charge, concentration, and distribution of cations.[14–16, 21]

In Figure 13.7 an eclipsed stack of transannularly linked Si_6-rings forming a one-dimensional planar Zintl anion (poly-sila-phenylide) is shown. Inspection of the electronic band structure reveals that the high dispersion of bands crossing the Fermi level is due to the π^*–π^* overlap of consecutive anion parts in the stack.[24] To the right side of the band structure, the corresponding density of states is presented, where the black-filled regions arise from just these π^* bands. Thus, Zintl phases of this general kind, i.e. those with one-dimensional eclipsed stacking of planar Zintl anions, are always metals with a one-dimensional characteristic and the conduction electrons reside in π^* bands.

Comparing the structure of Ba_2X_2I ($Ba_xX_4I_2$; X = Si, Ge) in Figure 13.8 with the structure of $Sr_5Si_6I_2(SiI_2)$ in Figure 13.9 exemplifies two linking stages of Zintl anions which belong to a series as shown in Figure 13.10.

The upper left situation in Figure 13.10 is the least condensed member and the hypothetical "graphite-like silicon" the most highly condensed one. Based on the above described understanding of charge stabilization and planarity, a "graphite-like silicon" is highly unlikely because there would not be any charges to force planarity upon the system. However, a triel anion, for example formed by aluminum or gallium, could stabilize by adapting such a geometry. At present, we do not know how far the condensation series for tetrel elements can extend towards the graphite structure, because there are many parameters (synthesis, cation, types and combinations, etc.) which may be investigated.[25]

Ba_2X_2I may be written as $Ba_4X_4I_2$ = $(Ba_3X_4)(BaI_2)$ (X = Si, Ge) and consequently $Sr_5Si_6I_2$ as $(Sr_4Si_6)\cdot(SrI_2)$. Inspection of the crystal structures shows that these compounds comprise real double salts of Zintl phases and halogenide salts, and it is important to note that in such compounds a completely new way of separating Zintl anions has been found. This is due to the fact that a suitable cation coordination remains on introduction of the halogenide components. The two above-mentioned double salts in particular exhibit a clearcut separation of Zintl and halogenide into different layers. This leads directly to an electronic separation of the Zintl regions into quantum layers and in ideal cases to quasi-two-dimensional electron gases. It should be emphazised that such quantum layer systems – contrary to many others formed by different techniques – are generated by self-organization during the formation of homogeneous phases.[25–27]

Figure 13.11 depicts the crystal structures of $Ba_5Mg_{18}Si_9(Si_4)$ (left) and of $Ba_6(Mg_{17.5}Li_{0.5})LiSi_9(LiSi_3)$ for comparison.[24] Surprisingly, the general structural pattern is very similar, i.e. both structures contain similar building blocks, which have been emphazised by the yellow triangles, but the Zintl anions are very different. We assume that this is due to relatively consistent mutual coordination schemes as given above under points 1. to 4. and as shown in Figure 13.2. However, it should also be mentioned that there are only isolated full-shell $[Si^{4-}]$ anions in $Ba_6(Mg_{17.5}Li_{0.5})LiSi_9(LiSi_3)$, for which the same arguments as for the polynuclear Zintl anions need not necessarily be valid. A deeper understanding may arise here from future theoretical investigations.[28]

Zintl phases have always been understood within the framework of an elegant and simple scheme, the so-called Zintl-Klemm concept.[1–12] However, only Zintl compounds with semiconducting properties were assumed to fulfil this concept. In the last decade, important steps have been taken towards an expansion of the Zintl-Klemm concept into the metallic regime while still separating Zintl compounds quite clearly from intermetallic phases. This is due to the fact that in intermetallics the clearcut separation of components at which valence states are centered and those at which they are not, is lost.

However, despite an extremely fruitful and long period of expanding the family of Zintl phases and their theoretical understanding there are still very few data on their properties. Furthermore, just a few reactions in which Zintl phases are used as precursors and may even react in a topochemical way under preservation of at least parts of the original structure have been investigated. Such reactions may be very interesting for future applications of Zintl phases of the tetrels.

Acknowledgement

The Swiss National Science Foundation has funded our research on Zintl phases for more than a decade in a very generous way.

References

[1] E. Zintl, W. Dullenkopf, *Z. Phys. Chem.* **1932**, *B16*, 183.
[2] E. Zintl, *Angew. Chem.* **1939**, *52*, 1.
[3] W. Klemm, *Trab. Reun. Int. React. Solidos 3rd*, **1956**, *1*, 447.
[4] E. Mooser, W.B. Pearson, *Phys. Rev.* **1956**, *101*, 1608.
[5] W. Klemm, FIAT-Review of German Science, Naturforschung und Medizin in Deutschland 1939–1946, Anorganische Chemie Teil IV, 26, 106.
[6] W. Klemm, *Proc. Chem. Soc. London* **1959**, 329.
[7] H. O. Grimm, A. Sommerfeld, *Z. Phys.* **1926**, *36*, 36.
[8] H. G. von Schnering, Angew. Chem. **1981**, *93*, 44; *Angew. Chem. Int. Ed. Engl.* **1981**, *20*, 33.
[9] H. G. von Schnering, Rheinisch-Westfälische Akademie der Wissenschaften, Vorträge N325, S./, Westdeutscher Verlag, Opladen.
[10] H. G. von Schnering, W. Hönle, *Chem. Rev.* **1988**, *88*, 243.
[11] R. Nesper, H. G. von Schnering, *Tschermaks mineralog. Petrograph. Mitt.* **1983**, *32*, 195
[12] R. Nesper, *Prog. Solid State Chem.* **1990**, *20*, 1.
[13] K. Wade, *Adv. Inorg. Chem. Radiochem.* **1976**, *18*, 1.
[14] R. Nesper, A. Currao, S. Wengert, in *Organosilicon Chemistry II, From Molecules to Materials*, (Eds.: N. Auner, J. Weiss), VCH, Weinheim, FRG, **1995**.
[15] A. Currao, R. Nesper, *Angew. Chem.* **1998**, *110*, 843; *Angew. Chem. Int. Ed.* **1998**, *37*, 841.
[16] R. Nesper, A. Currao, S. Wengert, *Chem. Eur. J.* **1998**, *4*, 2251.
[17] A. D. Becke, K. E. Edgecombe, *J. Chem. Phys.* **1990**, *92*, 5397.
[18] U. Häussermann, S. Wengert, P. Hofmann, A. Savin, O. Jepsen, R. Nesper, *Angew. Chem.* **1994**, *106*, 2147; *Angew. Chem. Int. Ed.* **1994**, *33*, 2069.

[19] U. Häussermann, S. Wengert, R. Nesper, *Angew. Chem.* **1994**, *106*, 2151; *Angew. Chem. Int. Ed.* **1994**, *33*, 2073.
[20] A. Savin, R. Nesper, S. Wengert, T. F. Fässler, *Angew. Chem.* **1997**, *109*, 1892; *Angew. Chem. Int. Ed.* **1997**, *36*, 808.
[21] A. Currao, J. Curda, R. Nesper, *Z. Anorg. Allg. Chem.* 622 **(1996)** 85.
[22] R. Nesper, CHIMIA **2001**, *55*, 787.
[23] The formulation (*n*b) indicates an *n*-valent component in a cluster or polymer structure.
[24] S. Wengert, Dissertation Nr. 12070, ETH-Zürich **1997**.
[25] S. Wengert, H. Willems, R. Nesper, in preparation.
[26] S. Wengert, H. Willems, R. Nesper, *Chem. Eur. J.* **2001**, *7*, 3209.
[27] S. Wengert, R. Nesper, *J. Solid State Chem.* **2000**, *152*, 460.
[28] F. Zürcher, S. Wengert, R. Nesper, *Inorg. Chem.* **1999**, *38*, 4567.

14 Zintl Phases MSi_2 (M = Ca, Eu, Sr, Ba) at Very High Pressure

Jürgen Evers and Gilbert Oehlinger*

14.1 Introduction

In recent decades, the static high pressures that can be generated for the synthesis and study of materials have increased by a factor of approximately 50. Now, at a pressure range which is hard to imagine, materials with exciting structures and properties are being both prepared and investigated. This huge development has been possible due to the great improvements made in the diamond anvil cell technique. The pressure in the Earth´s interior is about 3500000 bar = 3500 kbar = 3.5 Mbar.[1] However, the highest pressure now achieved in a diamond anvil cell is 5.5 Mbar,[1] and it is envisaged that the upper limit for pressure generation with diamonds is approximately 7.5 Mbar.[2] At this pressure, metallization of diamond, accompanied by a large volume decrease, is believed to start.

Changes in the internal energy (U) of a material depend on the variables of state temperature (T) and pressure (p): $\Delta U = T{\cdot}\Delta S - p{\cdot}\Delta V$ (S: entropy and V: volume). By adjusting the temperature dependent product $T{\cdot}\Delta S$ the state of aggregation can be changed, e.g. by melting or by evaporation. Nevertheless, the pressure dependent product $- p{\cdot}\Delta V$ plays a more dominant role, since pressure can be increased over a much greater range than temperature. Table 14.1 summarizes the effects that can be achieved for a material with a volume decrease of $\Delta V = -20$ cm^3/mole, if one increases the pressure from 5000 bar to 500000 bar and then to 5000000 bar.

Applying pressures of 5000 and 50000 bar results in relativly small effects, such as the squeezing of materials or the bending of bonds. Such effects can also be achieved by adjusting the temperature at 1 bar. However, on reaching a pressure of the order of approximately 500000 bar, one enters a totally new field: new materials are built up with new bonds and new electronic levels.

* Address of the authors: Department Chemie, Universität München, Butenandtstr. 5-13, D-81377 München, Germany.

Table 14.1. Effects achieved by increasing pressure from 5000, to 50000, and then to 500000 bar.

p (bar)	$-p\cdot\Delta V$ (kcal/mol)	Effects
5 000	2	Squeezing of materials
50 000	20	Bending of bonds
500 000	200	Creating of new bonds with new electronic levels

As a highlight, the results of very high pressure research on nitrogen are mentioned here. Due to its triple bond, the N_2 molecule is one of the most stable diatomic molecules. Therefore, it possesses a very high bond energy. At low temperature and pressure, nitrogen is condensable into solid structures consisting of N_2 molecules. Nitrogen in such structures is an isolator with a large band gap. In 1985, McMahan and LeSar [3] predicted that at very high pressure, the triple bond in molecular nitrogen could be broken, creating a mono atomic arrangement of nitrogen. Here, every nitrogen atom is covalently bonded to three other nitrogen atoms. Such structures have already been observed at ambient pressure for the elements phosphorus, arsenic, antimony, and bismuth, the chemical relatives of nitrogen in the same main group V of the periodic table. According to the prediction, this transformation should take place in a pressure range between 500 and 940 kbar. Looking at the bond energies, it becomes apparent that mono atomic nitrogen must be a high energy material: the triple bond has an energy of 226 kcal/mol, the single bond one of 38 kcal/mol.[4] The difference is 188 kcal/mol, approximately the energy presented in Table 14.1 for 500000 bar. Recently, Mao et al. performed the transformation of molecular into mono atomic nitrogen *in situ* in a diamond anvil cell.[5] At 980 kbar and 300 K the transformation starts. It is observable with a light microscope looking through the diamond anvils. At the beginning, darkening is seen, then a color change from red to dark occurs at 1500 kbar. The dark phase is a semiconductor with a band gap of 0.4 eV at very high pressure. At ambient pressure, a band gap of 0.4 eV would result in a silvery white color, as is observed for germanium with a band gap of 0.7 eV. It is quite remarkable that after breaking the triple bond in nitrogen an amorphic phase is obtained under several pressure conditions. Here an energy barrier kinetically hinders the crystallization of the mono atomic nitrogen phase. Some other highlights of very high pressure research are also briefly mentioned. Molecular CO_2 is transformed at very high pressure into a solid structure which shows close analogy to the tridymit structure of SiO_2 at ambient pressure;[6] the isolator iron(II) oxide FeO becomes a metallic alloy under pressure;[7] due to a change of the 6s-electron into a 5d-band, the main group metal cesium becomes a transition metal at high pressure,[8] crown sulfur, S_8 – a yellow, transparent isolator at ambient pressure – transforms into a supraconducting material at very high pressure,[9] and a condensate of the composition $Ar(H_2)_2$ crystallizes at high pressure in the structure of the intermetallic Laves phase $MgZn_2$.[10]

14.2 Materials Studied at Very High Pressure

In the following, a research report is presented on compounds which belong to another class of intermetallic phases: the Zintl phases MSi_2 (M = Ca, Eu, Sr, Ba). In these phases, the number of covalently bonded silicon atoms can be derived by simple valence rules. In an ionic interpretation, the above Zintl phases are formulated as M^{2+} $(Si^-)_2$. Negatively charged Si^- ions with five valence electrons show structural analogies to the elements of main group V. Therefore, in the silicon sublattice of the MSi_2 compounds, each Si atom is connected to three neighbors, as is also observed in the solid-state structures of P, As, Sb, and Bi and now additionally in the structure of mono atomic N as well. One might regard such Zintl phases MSi_2 as mere curiosities of solid-state research laboratories, but the Zintl phase MgB_2 (ionic formulation $Mg^{2+}(B^-)_2$) recently prepared at high pressure, is a supraconductor with a critical temperature Tc = 39 K.[11] This is the highest value yet observed for an intermetallic compound. Therefore, one can speculate on a technical application of MgB_2, e.g. in supraconducting magnetic coils. The hithero practically applied intermetallic materials Nb_3Sn and NbTi only have critical temperatures near 20 K. In MgB_2 with the AlB_2 structure,[12] negatively charged B^- ions with four valence electrons, as in carbon, have three boron neighbors in planar graphite-like layers. Planar groups of four boron atoms with equidistant bonds and equal bond angles of 120° build up these layers, which are separated by Mg^{2+} counterions in order to compensate the negative charge. The four phases investigated in this project, $CaSi_2$-II,[13–15] $EuSi_2$-I,[16–18] $SrSi_2$-II,[19, 20] and $BaSi_2$-III [21–23] crystallize in structures that show some similarities to MgB_2. In these disilicides, each Si^- ion also has three neighbors of the same kind. In addition, $CaSi_2$-II and $SrSi_2$-II are supraconductors, albeit with a much lower critical temperature [13, 24] than MgB_2. Europium disilicide contains Eu^{2+} cations, with seven unpaired 4f electrons, and shows interesting magnetic ordering properties.[17, 18] The Roman numerals at the end of the four MSi_2 chemical formulae are used to distinguish between normal and high-pressure phases. "I" is used to characterize a normal-pressure phase, "II" and "III" are used for high-pressure phases. The high pressure phases $CaSi_2$-II, $SrSi_2$-II, and $BaSi_2$-III were prepared in a belt apparatus at 40 kbar and 1000 °C and were then quenched to ambient conditions as metastable phases.[23] At temperatures lower than 400 °C at 1 bar, these phases can be handled as "normal" materials. As is well known, at 1 bar diamond is metastable with respect to graphite. On heating diamond to 1500°C at 1 bar, it transforms exothermically into the thermodynamically stable phase graphite. The four MSi_2 Zintl phases (M = Ca, Eu, Sr, Ba) in their network structures were used as starting materials in this very high pressure investigation. Sometimes it is less laborious if one "starts" the investigation with a material, which is "dense" even at 1 bar. The phases $CaSi_2$-II,[13–15] $EuSi_2$-I [16–18] and $SrSi_2$-II [19, 20] crystallize in the tetragonal α-$ThSi_2$ structure.[25] Compared to MgB_2 the α-$ThSi_2$ structure also contains planar groups of four atoms. However, in contrast to the diboride, such planar groups of four atoms are rotated 90° with respect to each other in the disilicide in order to build up a three-dimensional net instead of a layer. Also, the symmetry of the planar group of four atoms in the disilicide with the α-$ThSi_2$ structure is lower than that in the diboride. The planar groups of four Si atoms in the disilicide consist of one short and two longer Si-Si bonds and have two different $Si^{Si}Si$ valence angles. The different bond distances and angles are a consequence of the axial ratio c/a, which, in these three compounds has a value of approximately 3.15, a 10% lower value than the ideal one of $c/a = 2 \cdot \sqrt{3} = 3.464$. With such an axial ratio c/a and a positional parameter of $z_{Si} = 5/12$, the Si-Si bonds are of equal length and the $Si^{Si}Si$ valence angles are 120°. The Si-Si bond distance in $CaSi_2$-II at 2.299(1) Å indicates one of the shortest Si-Si bonds to have been determined by modern single-crystal techniques at ambient pressure.[15] In the cavities of the silicon network, the M^{2+} counterions compensate the negative charge of the Si^- net. Figure 14.1 shows this remarkable structure. Laves describes the silicon sublattice of the α-$ThSi_2$ structure as a "three dimensional graphite analogue",[26] and according to calculations by Hoffmann and Kertesz [27] it is a hypothetical carbon structure with a density between those of graphite and diamond.

Up to 40 kbar and 1000 °C, the α-$ThSi_2$ structure is stable for the disilicides of Ca, Eu, and Sr. The metallic radius can be increased from that of calcium (r_{Ca} = 1.974 Å) [28] up to that of strontium (r_{Sr} = 2.151 Å).[28] However, for barium disilicide (r_{Ba} = 2.243 Å) [28] a polymorph with the cubic $SrSi_2$ structure is obtained in the belt apparatus.

This structure also consists of a three-dimensional three-connected silicon net. Here, however, four Si^- ions built up pyramidal groups with equal bond distances and equal bond angles of 118.1°. The top of the pyramid is 0.36 Å higher than its base. It is interesting that in its ideal version the three-dimensional three-connected net of the $SrSi_2$ structure shows a very high topological symmetry. According to A. F. Wells,[29] this net is the three-connected analogue of the four-connected diamond net. Figure 14.2 shows the three-dimensional, three-connected net of the $SrSi_2$ structure. Among phases with the $SrSi_2$ structure, no supraconducting material has yet been identified.

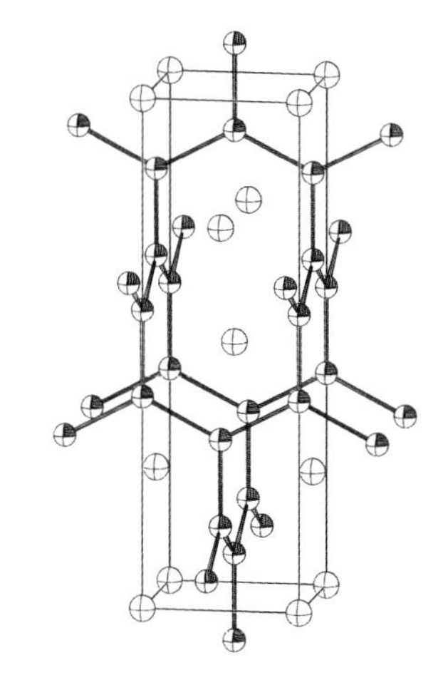

Single-Crystal Data [15, 20, 23]

	Si-Si (Å) (1 bar)	
$CaSi_2$-II	2.299(1) 1x,	2.400(1) 2x
$EuSi_2$-I	2.306(8) 1x,	2.421(4) 2x
$SrSi_2$-II	2.330(2) 1x,	2.489(2) 2x

Figure 14.1. Three-dimensional, three-connected silicon net of the α-$ThSi_2$ structure for $CaSi_2$-II, $EuSi_2$-I, and $SrSi_2$-II. The net consists of planar groups of four Si atoms with one short and two longer Si-Si bonds. The distances are given above. In the cavities of the net, M^{2+} counterions compensate the negative charge of the silicon net.

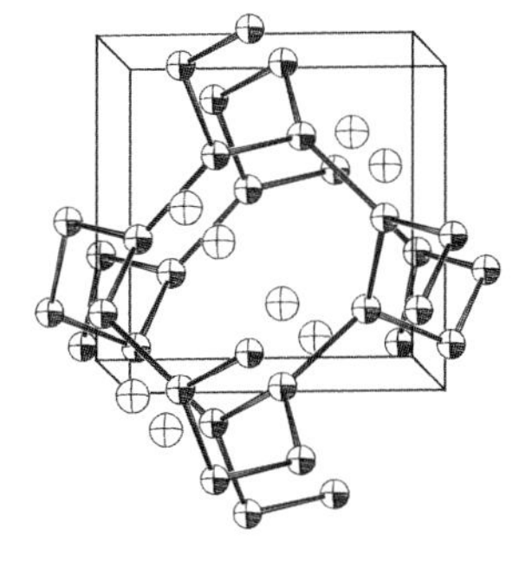

Single-Crystal Data [23]

	Si-Si (Å)(1 bar)
$BaSi_2$-III	2.447(1) 3x

Figure 14.2. The three-dimensional, three-connected silicon net of the $SrSi_2$ structure for $BaSi_2$-III. The net consists of pyramidal groups of four Si atoms with equal bond distances and equal bond angles (118.1°). In the cavities of the net, Ba^{2+} counter ions compensate the negative charge of the silicon net.

14.3 Apparatus and Experimental Technique

The pressure calibration was performed by adding 20% silver powder to the samples of the Zintl phases MSi_2 (M = Ca, Eu, Sr, Ba) at ambient pressure and determining the specific volumes V/V_o from the lattice parameter of silver at variable pressure. Specific volumes V/V_o of silver are tabulated as a function of pressure up to approximately 1 Mbar.[30] The diamonds used in the anvil cell had culets of 0.3 mm. A maximum pressure of approximately 600 kbar was then created. For 0.2 mm culets, a calibration with silver up to 1.04 Mbar was performed. The gaskets were made of stainless steel with a hole of 130 µm (0.3 mm culet) or 90 µm (0.2 mm culet). No pressure-transmitting material was applied. The diamond anvils were used in a lever apparatus which amplifies the pressure by a ratio of 5:1.[31] Heating was performed with a CO_2 laser at a wavelength λ = 10.6 µm. The diamond anvils were thermally isolated with Al_2O_3 discs. In the high-pressure diffractometer filtered Mo-K_α or Ag-K_α X-ray radiation was used to aquire diffractograms in the Θ range 1 to 12.5° with a typical increment of 0.03°. Counting times were between 600 and 900 seconds per increment. The X-ray microcollimators used had diameters between 110 and 80 µm.

14.4 Results

In Tables 14.2, 14.3, and 14.4, the pressure (kbar), the tetragonal axes *a* (Å) and *c* (Å), the axial ratio *c*/*a* and the specific volumes V/V_o for $CaSi_2$-II, $EuSi_2$-I and $SrSi_2$-II with the α-$ThSi_2$ structure are summarized. In Table 14.5, the data for $BaSi_2$-III with the $SrSi_2$ structure are presented. The experimental results in these four tables were obtained without laser heating.

Heating experiments with a CO_2 laser – at first only up to 700°C – showed only the sequence of reflections for the α-$ThSi_2$ structure for $EuSi_2$-I up to 500 kbar. As a first heating candidate, europium(II) disilicide was chosen, since here the change of one 4f electron of Eu into a d-band results in a change to valence three. $Eu^{3+}(Si^-)_2$ should be as stable as the corresponding 4f disilicide $Gd^{3+}(Si^-)_2$. $GdSi_2$ [16] crystallizes at ambient pressure in an orthorhombic variant of the α-$ThSi_2$ structure with a 13% lower volume than $EuSi_2$ with divalent europium. The decrease in volume on changing from divalent to trivalent should promote the transformation through the product term $-p{\cdot}\Delta V$.

In Table 14.2, 14.3, and 14.4 it is shown that for the phases $CaSi_2$-II, $EuSi_2$-I, and $SrSi_2$-II up to 400 kbar specific volumes with approximately V/V_o = 0.72 are obtained. This means that in comparison to the volumes at ambient pressure, the volumes at 400 kbar are squeezed by approximately 28%.

Squeezing in the diamond anvil cell at 400 kbar is more efficient than in the belt apparatus at 40 kbar. The metastable quenched high-pressure phases MSi_2 (M = Ca, Sr, Ba) [23] obtained in the belt apparatus showed only a volume decrease of approximately 10%.

Table 14.2. $CaSi_2$-II, α-$ThSi_2$ structure.

Pressure (kbar)	a(Å)	c(Å)	c/a	V/V_o
0.001	4.283(3)	13.53(1)	3.159(5)	1.000
58(5)	4.14(2)	13.40(4)	3.24(2)	0.926(5)
83	4.08	13.37	3.28	0.897
123	4.04	13.36	3.31	0.879
157	3.97	13.28	3.35	0.844
180	3.92	13.22	3.37	0.819
200	3.89	13.14	3.38	0.802
215	3.86	13.12	3.40	0.788
227	3.83	13.08	3.42	0.774
255	3.82	13.02	3.41	0.766
271	3.79	13.01	3.43	0.754
284	3.78	12.96	3.43	0.747
302	3.76	12.94	3.44	0.738
333	3.73	12.87	3.45	0.722
360	3.71	12.87	3.47	0.714

Table 14.3. $EuSi_2$-I, α-$ThSi_2$ structure.

Pressure (kbar)	a(Å)	c(Å)	c/a	V/V_o
0.001	4.304(3)	13.66(1)	3.174 (5)	1.000
8(5)	4.27(2)	13.62(4)	3.19(2)	0.981(5)
35	4.16	13.54	3.25	0.925
137	4.02	13.49	3.36	0.862
183	3.96	13.46	3.40	0.834
250	3.90	13.29	3.41	0.799
278	3.86	13.27	3.44	0.781
354	3.81	13.21	3.47	0.758
375	3.78	13.14	3.48	0.742
470	3.73	13.11	3.51	0.721

Table 14.4. $SrSi_2$-II, α-$ThSi_2$ structure.

Pressure (kbar)	a(Å)	c(Å)	c/a	V/V_o
0.001	4.438(3)	13.83(1)	3.116(5)	1.000
68(5)	4.31(2)	13.61(4)	3.16(2)	0.928(5)
138	4.18	13.53	3.24	0.868
186	4.10	13.45	3.28	0.830
208	4.06	13.40	3.30	0.811
275	4.00	13.30	3.33	0.781
330	3.95	13.24	3.35	0.758
375	3.92	13.18	3.36	0.744
405	3.89	13.16	3.38	0.731

Table 14.5. $BaSi_2$-III, $SrSi_2$ Structure.

Pressure (kbar)	a(Å)	V/V_o
0.001	6.7154(3)	1.000
34	6.61(2)	0.954(4)
63	6.55	0.928
80	6.50	0.907
97	6.48	0.897
143	6.39	0.862
178	6.34	0.842
224	6.24	0.803
297	6.16	0.772
335	6.12	0.757
368	6.07	0.739
420	6.01	0.717

At ambient pressure, $EuSi_2$-I has an axial ratio c/a = 3.17. This value deviates only slightly from that in the quenched high-pressure phases $CaSi_2$-II with c/a = 3.16 and $SrSi_2$-II with c/a = 3.12 (see Table 14.2–14.4). As already mentioned, the deviation of the axial ratio from its ideal value of $c/a = 2\cdot\sqrt{3}$ = 3.464 is responsible for the distortion in the planar groups of four Si atoms within the net of the α-$ThSi_2$ structure (Figure 14.1). In Table 14.2, it is shown for $CaSi_2$-II that the axial ratios increase with increasing pressure. At 83 kbar e.g. the axial ratio c/a is 3.28, and at 200 kbar it is 3.38. At 360 kbar, it reaches a value of 3.47, i.e. approximately the ideal value of 3.464.

Recently, a high-pressure investigation of $CaSi_2$ in a diamond anvil cell has been published.[32] Nitrogen was used as a pressure-transmitting material and the diamond anvil cell was calibrated by the ruby luminescence method.

However, this investigation was only performed up to 178 kbar with a specific volume of $V/V_o = 0.826$. Here also, an increase in the axial ratio with increasing pressure was observed, e.g. from $c/a = 3.17$ at 6.1 kbar to $c/a = 3.32$ at 178 kbar. Inspection of Table 14.2 shows that our c/a values are slightly higher ($CaSi_2$-II, e.g. at 180 kbar, $c/a = 3.37(2)$). The small deviations are probably the result of the two different methods of high-pressure calibration. In Table 14.6, the bond distances and bond angles are summarized for $CaSi_2$-II at three pressures applied in our investigation (83, 200, 360 kbar) and are compared with those obtained by a single-crystal investigation at ambient pressure.[15]

Table 14.6. Axial ratio c/a, Si-Si bond lengths (Å) and $Si^{Si}Si$ bond angles (°) for $CaSi_2$-II with increasing pressure (kbar).

Pressure (kbar)	c/a	Si-Si bond lengths (Å) 1x	2x	$Si^{Si}Si$ valence angles (°) 1x	2x
0.001	3.16	2.299(1)	2.400(1)	126.3(1)	116.8(1)
83	3.28	2.27(2)	2.30(2)	124(2)	118(2)
200	3.38	2.19(2)	2.23(2)	122(2)	119(2)
360	3.47	2.14(2)	2.14(2)	120	120

In Table 14.6, it is shown that in $CaSi_2$-II the differences in bond distances and $Si^{Si}Si$ bond angles become smaller with increasing pressure. At 360 kbar, the ideal arrangement of Si atoms with equidistant bonds of 2.14 Å and bond angles of 120° is obtained. Density functional theory (DFT) calculations confirm the experimental results. In addition, from DFT calculations it is evident that at much smaller specific volumes than $V/V_o = 0.71$ (Table 14.2, 360 kbar) the axial ratio increases to higher values than 3.46.

It is interesting that for strontium disilicide with the α-$ThSi_2$ structure, with a larger metallic radius than for calcium or europium, the axial ratio $c/a = 3.46$ is obtained only at a much higher pressure than 400 kbar. Inspection of Table 14.4 shows that in $SrSi_2$-II at 405 kbar, the axial ratio is $c/a = 3.38$. This value is achieved for the smaller $CaSi_2$-II at a much lower pressure of 200 kbar (Table 14.2). With increasing metallic radius, it becomes more and more difficult to compress the α-$ThSi_2$ structure. From this fact, it can now be readily understood, as to why for $BaSi_2$,with the large barium cations, only the $SrSi_2$ structure and not the α-$ThSi_2$ structure is obtained in the belt apparatus up to 40 kbar.

The cubic $SrSi_2$ structure shows no characteristic axial ratio from which a change in the pyramidal group of Si atoms can be deduced. For $BaSi_2$-III in diamond anvil cell experiments up to 420 kbar, only the decreases in both the cubic lattice parameters and the specific volumes V/V_o are apparent (Table

14.5). At 1 bar, the Si-Si bond distance in this phase is 2.447(1) Å (3x) (Figure 14.2) and the $Si^{Si}Si$ bond angle is 118.1(1)° (3x). These data result from the positional parameter x_{Si} = 0.4191(1) determined in a single-crystal investigation at ambient pressure.[23] DFT calculations show that in $BaSi_2$-III up to 420 kbar, the minima in total energy are also obtained for the positional parameter x_S = 0.419. Therefore, the pyramidal groups of Si atoms in $BaSi_2$-III do not become planar with increasing pressure up to 420 kbar. In addition, with this positional parameter and the lattice parameter (6.01 Å, Table 14.5) the Si-Si bond distance of 2.19 Å in $BaSi_2$-III at 420 kbar is obtained.

The short Si-Si bond distances for $CaSi_2$-II, $EuSi_2$-I, $SrSi_2$-II, and $BaSi_2$-III at very high pressures of approximately 400 kbar can also be estimated from the specific volumes V/V_o at very high pressure. These volumes are the cubic power of specific lengths L/L_o: $V/V_o = (L/L_o)^3$. Therefore, from the cubic root of the specific volumes, the specific lengths can be calculated: $(V/V_o)^{1/3} = L/L_o$. For the four investigated phases $CaSi_2$-II, $EuSi_2$-I, $SrSi_2$-II, and $BaSi_2$-III from the lowest V/V_o values (0.714, 0.721, 0.731, 0.717, respectively (Table 14.2–14.5)), L/L_o values are calculated (0.893, 0.897, 0.901, 0.895, respectively). With the averaged L_o values (= Si-Si bond distances) from the single-crystal investigations at 1 bar (2.366, 2.383, 2.436, 2.447 Å, respectively, Figure 14.1–14.2) one may calculate rough estimates of the Si-Si bond distance in $CaSi_2$-II at 360 kbar (2.11 Å), in $EuSi_2$-I at 470 kbar (2.14 Å), in $SrSi_2$-II at 405 kbar (2.19 Å), and in $BaSi_2$-III at 420 kbar (2.19 Å). The average value of these four Si-Si bond distances is 2.16 Å at approximately 400 kbar.

Such a value deviates only slightly from 2.14 Å, which is tabulated for the length of an Si-Si double bond at 1 bar.[4] In a double bond, each Si atom has four valence electrons in bonding orbitals. In the Si^- ions of the Zintl phases MSi_2 (M = Ca, Eu, Sr, Ba), however, there are five valence electrons, three in bonding and two in non-bonding orbitals. It would appear that the two Si electrons in the non-bonding orbitals with large volume requirements, as well as the repulsive effects due to the negative charges, create problems. Therefore, with increasing pressure in the range above 600 kbar the limit of stability of Si-Si bonds in the four Zintl phases MSi_2 (M = Ca, Eu, Sr, Ba) with three-connected Si atoms is reached. The quality of the diffractograms obtained at such very high pressures decreases: reflections are increasingly broadened until they disappear totally: amorphization starts as was also mentioned for the very high pressure study on nitrogen.[5]

14.5 Summary and Outlook

On increasing the pressure up to about 400 kbar, in a diamond anvil cell, the Zintl phases MSi_2 (M = Ca, Eu, Sr, Ba) are compressed to approximately 28%

of their volumes at 1 bar. In the three phases with the α-$ThSi_2$ structure ($CaSi_2$-II, $EuSi_2$-I, $SrSi_2$-II), the axial ratios c/a (approximately 3.15 at 1 bar) increase to near their ideal value ($2 \cdot \sqrt{3} = 3.464$). Therefore, the planar group of four Si atoms becomes more symmetrical with increasing pressure. At 360 kbar, the Si-Si bond distances in $CaSi_2$-II are equal (2.14 Å, 3x), as are the bond angles (120°, 3x). For $SrSi_2$-II, with the larger strontium cation in comparison to calcium, much higher pressure is needed to squeeze the planar group of four Si atoms into a more symmetrical arrangement. It is interesting that both $CaSi_2$-II and $SrSi_2$-II phases with the α-$ThSi_2$ structure are supraconductors, probably due to the sp^2 hybridization of the silicon atoms. $EuSi_2$-I, which also has an sp^2 hybridization of the silicon atoms, becomes magnetically ordered at very low temperatures and is, therefore, not supraconducting. The behavior of $BaSi_2$-III, with the $SrSi_2$ structure, is different from that of $CaSi_2$-II, $EuSi_2$-I, and $SrSi_2$-II. According to DFT calculations, the groups of four Si atoms in barium disilicide remain in their pyramidal arrangement (probably with sp^3 hybridization of the Si atoms), even at 420 kbar. Here, no supraconducting property is found.

At very high pressures of 400 kbar, short Si-Si bond distances of approximately 2.16 Å are observed for the four Zintl phases MSi_2 (M = Ca, Eu, Sr, Ba). This short bond distance is comparable to that of an Si-Si double bond at 1 bar, even though the bonding must be quite different. In an Si-Si double bond between neutral Si atoms, four valence electrons permit, at a distance of 2.14 Å, a quite good overlap of orbitals to form one σ and one π bond. In the Zintl phases MSi_2 (M = Ca, Eu, Sr, Ba) with Si^- ions, however, at 2.16 Å the limit of the stability of a single bond is reached. Here, with the five valence electrons (three in bonding, two in non-bonding orbitals), the two electrons in the non-bonding ("free") orbitals seem to create problems at very high pressures. As is well-known, non-bonding orbitals have large spatial requirements. In addition, the repulsive effects due to the negative charges become very important, as the Si-Si distances decrease drastically to 2.16 Å. The energy input from the product $-p \cdot \Delta V$ is estimated to be approximately 100 kcal/mol at 400 kbar. This energy is sufficient to break Si-Si bonds between the Si^- ions, but due to kinetic hindrance it is insufficient to induce crystallization in a new structure with Si-Si coordination higher than three, as already observed in the Zintl phases MSi_2 (M = Ca, Eu, Sr, Ba). Si-Si coordination higher than three should produce two stabilizing effects. Firstly, it could lead to an increase in the Si-Si distance, even at high pressures, e.g. from 2.16 to 2.21 Å. Secondly, it could activate the electrons in the non-bonding orbitals to create new bonds to more than three silicon neighbors.

Creating new bonds in materials with exciting structures needs a lot of energy from the term $-p \cdot \Delta V$ as is indicated in Table 14.1. A possible candidate for such a new structure could be an intermetallic Laves phase with the $MgCu_2$

structure [33] with a Si-Si coordination of six. A transformation from the ternary Zintl phase EuPdSi (with the LaIrSi structure,[34] a ternary variant of the $SrSi_2$ structure) into the intermetallic Laves phase with the $MgZn_2$ structure has already been observed in quenched samples obtained in belt experiments at 40 kbar and 1000 °C.[35] However, for the binary Zintl phases MSi_2 (M = Ca, Eu, Sr, Ba) the transformation pressure must be much higher. According to DFT calculations, for $CaSi_2$ a pressure of approximately 100–200 kbar should be sufficient to transform the α-$ThSi_2$ structure into the $MgCu_2$ structure. However, the diamond anvil cell would have to be heated with a laser, and one would first have to solve the kinetic problems which hinder the induction of crystallization.

14.6 References

[1] J. A. Xu, H. K. Mao, P. M. Bell, *Science* **1986**, *232*, 1404.

[2] A. L. Ruoff, H. Luo, Y. Vohra, *J. Appl. Phys.* **1991**, *69*, 6413.

[3] A. K. McMahan, R. LeSar, *Phys. Rev. Lett.* **1985**, *54*, 1929.

[4] *Holleman-Wiberg*, Lehrbuch der Anorganischen Chemie, (Ed.: N. Wiberg), Walther de Gruyter, Berlin, New York, **1995**, p. 141, p. 887.

[5] M. I. Eremets, R. J. Hemley, H. K. Mao, E. Gregoryanz, *Nature* **2001**, *411*, 170.

[6] C. S. Yoo, H. Cynn, F. Gygi, G. Galli, V. Ioata , *Phys. Rev. Lett.* **1999**, *83*, 5527.

[7] E. Knittel, R. Jeanloz, *Geophys. Res. Lett.* **1986**, *13*, 1541.

[8] S. G. Louie, M. L. Cohen, *Phys. Rev.* **1974**, *B10*, 3237.

[9] V. V. Struzkhin, R. J. Hemley, H. K. Mao, Y. A. Timofeev, *Nature* **1997**, *390*, 382.

[10] P. Loubeyre, R. LeToullec, J. P. Pinceaux, *Phys. Rev. Lett.* **1994**, *72*, 1360.

[11] J. N. Nagamatsu, N. Nakagawa, T. Muranaka, Y. Zenitani, J. Akimitsu, *Nature* **2001**, *410*, 63.

[12] *Pearson's Handbook of Crystallographic Data for Intermetallic Phases*, (Ed.: P. Villars & L. D. Calvert), American Society for Metals, Metals Park, OH 44073, **1985**.

[13] D. B. McWhan, V. B. Compton, M. Silverman, J. R. Soulen, *J. Less-Common Met.* **1967**, *12*, 75.

[14] J. Evers, *J. Solid State Chem.* **1979**, *28*, 369.

[15] J. Evers, G. Oehlinger, A. Weiss, *Z. Naturforsch.* **1982**, *B37*, 1487.

[16] J. A. Perri, I. Binder, B. Post, *J. Phys. Chem.* **1959**, *63*.

[17] J. Evers, G. Oehlinger, A. Weiss, F. Hulliger, *J. Less-Comm. Met.* **1983**, *90*, L19.

[18] S. Labroo, N. Ali, *J. Appl. Phys.***1990**, *67*, 4811.
[19] J. Evers, *J., Solid State Chem.* **1978**, *24*, 199.
[20] J. Evers, G. Oehlinger, A. Weiss, *Z. Naturforsch.* **1983**, *B38*, 899.
[21] J. Evers, G. Oehlinger, A. Weiss, *Angew. Chem.* **1978**, *90*, 562; Angew. Chem. Int. Ed. Engl. **1978**, *17*, 538.
[22] J. Evers, *J. Solid State Chem.* **1980**, *32* 77.
[23] J. Evers, *Habilitationsschrift,* **1982**, Universität München.
[24] J. Evers, G. Oehlinger, H. R. Ott, *J. Less-Common Met.* **1980**, *69*, 389.
[25] G. Brauer, A. Mitius, *Z. Anorg. Allg. Chem.* **1942**, *249*, 325.
[26] F. Laves, *Theory of Alloy Phases*, American Society of Metals, Cleveland, Ohio, **1955**, 169.
[27] M. Kertesz, R. Hoffmann, *J. Solid State Chem.* **1984**, *54*, 313.
[28] W. B. Pearson, *The Crystal Chemistry and Physics of Metals and Alloys*, Wiley-Interscience, New York, London, Sydney, Toronto, **1972**, p. 151.
[29] A. F. Wells, *Structural Inorganic Chemistry*, Clarendon Press, Oxford, **1984**, p. 90.
[30] H. K. Mao, P. M. Bell, J. W. Shaver, J. D. Steinberg, *J. Appl. Phys.* **1978**, *49*, 3276.
[31] J. Evers, G. Oehlinger, G. Sextl, *Eur. J. Solid State and Inorg. Chem., Special Issue, Chemistry Under Extreme Conditions*, **1997**, *34*, 773.
[32] P. Bordet, M. Affronte, S. Sanfilippo, M. Nunez-Regueiro, O. Laborde, G. L. Olcese, A. Palenzoa, S. LeFloch, D. Levy, M. Hanfland *Phys. Rev. B* **2000**, *62*, 11392.
[33] J. B. Friauf, *J. Am. Chem. Soc.* **1927**, *49*, 3107.
[34] K. Klepp, E. Parthé, *Acta Crystallogr.* **1982**, *B38*, 1105.
[35] B. Sendlinger, *Thesis*, Ludwig-Maximilians-Universität München, **1993**.

15 Silicon- and Germanium-Based Sheet Polymers and Zintl Phases

Martin S. Brandt*, Günther Vogg and Martin Stutzmann

15.1 Introduction

Depending on their coordination in the backbone network, group IV elements can form extended, covalently bonded structures of different dimensionality. As the most commonly known example, the fourfold coordination of the sp^3 hybridized atoms leads to the three-dimensional (3D) crystalline solids diamond, c-Si, c-Ge, and α-Sn with their well-known semiconducting properties. On the other hand, linear (1D) polymer chains $(XR_2)_n$ with X = C, Si, Ge, Sn are based on a twofold coordination of the backbone atoms and are of great importance in organic and inorganic polymer chemistry.[1]

The classical layered 2D structure of group IV elements is graphite, although this is a special case due to the sp^2 hybridization of the C atoms. In contrast, layered graphite fluoride $(CF)_n$ is purely based on sp^3 hybridized carbon,[2] whereas both hybridizations are present in layered graphite oxide.[3] For the heavier group IV elements, the formation of extended 2D structures with a pure threefold backbone coordination between sp^3 hybridized atoms has so far only been reported for Si in the form of polysilyne $(SiH)_n$ ("layered polysilane" [4]) or, in the case of higher oxygen contents, as oxypolysilyne, also known as siloxene.[5] These crystalline layered polymers are formed by a backbone of weakly interacting, puckered Si 2D layers similar to the {111} layers of c-Si.[6] The single layers are stabilized through termination of the free valences pointing out of the layer plane by hydrogen or hydroxide groups.

Silicon sheet polymers were formed for the first time by Wöhler in 1863, who studied the effect of acids on "Kieselcalcium" $CaSi_2$. Treating $CaSi_2$ with ice-cold concentrated aqueous HCl, he obtained a yellow compound which he called "Silicon".[5] Kautský later modified the reaction conditions and obtained a gray-green product he called "siloxene".[7] The fact that the latter compound could easily be modified by various substitution reactions [8] led to considerable scientific interest in this material in the following decades.[9] From chemical substitution experiments, Kautsky concluded that siloxene obtained according to his modified reaction conditions consists of planes of Si_6 rings connected to each other by linear oxygen bridges.[8] However, this structure has never been observed by means of X-ray diffraction (XRD); Kautsky siloxene is rather

---

* Address of the authors: Walter Schottky Institut, Technische Universität München, Am Coulombwall 3, D-85748 Garching, Germany
E-mail: mbrandt@physik.tu-muenchen.de

found to be amorphous. Indeed, the bond angles of oxygen bridges are typically well below 180°, so that the insertion of O atoms into the Si layers leads to stress and ultimately to the destruction of the planar structure. Due to the high luminescence intensity of Kautsky siloxene and the possibility of shifting the luminescence wavelength through substitution, this material is certainly of considerable interest on its own.[10, 11] Nevertheless, here we focus on sheet polymers with a threefold homoatomic coordination, which is not found in ideal Kautsky siloxene.

In contrast, the transformation of $CaSi_2$ according to Wöhler leads to crystalline sheet polymers with a threefold coordinated Si backbone, as first shown by Weiss et al.[6] This well-defined structure has advantages compared to the amorphous Kautsky material, such as the possibility of forming high quality epitaxial thin films which are accessible to the usual structural analysis. In this contribution, we will summarize the present state of knowledge on sheet polymers, in particular with respect to their formation and physical properties. Special attention is given to the growth of epitaxial thin films of the layered Zintl phases of Ca with Si and Ge and the formation of Ge and SiGe alloy sheet polymers, which have been achieved during the last years. More details on specific topics can be found in the literature cited below. For a thorough introduction to layered Zintl phases and sheet polymers, the reader is referred to ref.[12]. Earlier reviews on silicon sheet polymers can be found, e.g., in the Gmelin series [13, 14] or in refs.[9, 15, 16].

15.2 Chemical Transformation of the Layered Zintl Phases of Ca with Si and Ge

The only method known to date for the formation of Si or Ge sheet polymers is the topochemical transformation of the layered Zintl phases $CaSi_2$ or $CaGe_2$ in aqueous HCl solutions.[5, 17] According to the 8-*N* rule of Zintl phases,[18] the strong heteropolar bonding character in $CaSi_2$ and $CaGe_2$ leads to the formation of puckered $(Si^-)_n$ and $(Ge^-)_n$ polyanion layers, which are separated from each other by planar monolayers of Ca^{2+}.[19, 20] The corresponding crystal structure is shown in Figure 16.1. In spite of the heteropolar bonding, $CaSi_2$ is a metal.[21–23] During a topochemical transformation, the Ca^{2+} ions are removed without destroying the polyanion structure, which then forms the backbone of the resulting sheet polymer. The structures of the prototype sheet polymers obtained, siloxene $(Si_2HOH)_n$ and polygermyne $(GeH)_n$, are also shown in Figure 16.1.[6, 17] The preserved puckered layers of the group IV atoms are terminated by hydrogen or hydroxyl groups bonded to each backbone atom. To date, all attempts to obtain such sheet polymers by polymerization have only led to the formation of statistical network polymers, which, on average, do exhibit a threefold coordination, but form a 3D amorphous structure rather than stacked layers.[24–26] Thus, only the topochemical transformation of layered Zintl phases can be used to obtain sheet polymers. Nevertheless, the exact stoichiometry of the polymers obtained and their properties depend on the specific reactions taking place. We will start our discussion of the topochemical reactions of the layered Zintl phases with the well-known reactions of $CaSi_2$ and then turn to the reactions of $CaGe_2$ investigated more recently.

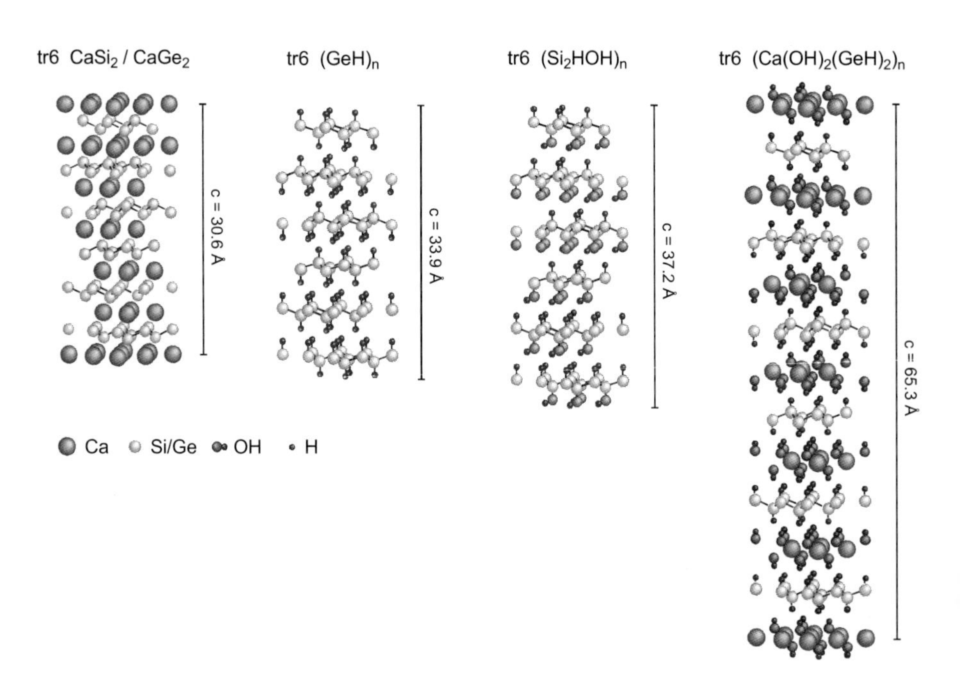

Figure 15.1. Structures of the trigonal-rhombohedral tr6 modifications of the Zintl phases $CaSi_2$ and $CaGe_2$, of the sheet polymers polygermyne $(GeH)_n$ and siloxene $(Si_2HOH)_n$ and of the polygermyne calcium dihydroxide intercalation compound $(Ca(OH)_2(GeH)_2)_n$. In each case, a full unit cell in the *c* direction is shown.

The simplest reaction to remove Ca from $CaSi_2$ would be by exposure to pure Cl_2:

$$(CaSi_2)_n + n\,Cl_2 \rightarrow (Si)_{2n} + n\,CaCl_2 \qquad (1)$$

e.g. by treating a suspension of $CaSi_2$ and CCl_4 with gaseous Cl_2 according to ref.[27]. However, the fourth valences of the Si atoms would then form highly reactive dangling bonds after the transformation. It is therefore not surprising that the 2D "active silicon" obtained according to Eq. (1) is unstable, highly reactive, and can hardly be handled or characterized.[28, 29] In contrast, much more stable sheet polymers are formed by exposure of $CaSi_2$ to aqueous HCl. The overall reaction in this case is:

$$(CaSi_2)_n + 2n\,HCl + xn\,H_2O \rightarrow (Si_2H_{2-x}(OH)_x)_n + n\,CaCl_2 + xn\,H_2 \qquad (2)$$

The parameter x in Eq. (2) describes the degree of hydrolysis that takes place during the transformation. In accordance with the results from other groups,[30,31] we typically find a hydrolysis corresponding to $x \approx 1$ over a broad range of reaction conditions (temperature, acid concentration). The polymer $(Si_2HOH)_n$ obtained for $x = 1$, with H and OH groups alternately bonded to the Si atoms, is called Wöhler siloxene and its ideal structure is shown in Figure 15.1. A lowering of the reaction temperature to below –20 °C has been reported to suppress hydrolysis and to lead to the formation of the oxygen-free sheet polymer polysilyne $(SiH)_n$.[32, 33] However, we have not been able to confirm this result, possibly due to the very fast spontaneous oxidation of this material upon exposure to air.

It is surprising that, although $CaGe_2$ has been known since 1944,[20] the analogous topochemical transformation into Ge sheet polymers has only been studied recently.[17] In contrast to Eq. (2), we do not observe a significant hydrolysis during the reaction of $CaGe_2$ with aqueous HCl, even at reaction temperatures of 0 °C or slightly above. The overall reaction is instead:

$$(CaGe_2)_n + 2n\,HCl \rightarrow (GeH)_{2n} + n\,CaCl_2 \qquad (3)$$

Therefore, the Ge sheet polymer obtained is pure polygermyne $(GeH)_n$ (Figure 15.1). In contrast to $(SiH)_n$, $(GeH)_n$ is stable when exposed to the ambient atmosphere, indicating a significant difference between the oxidation behavior of Ge and Si sheet polymers, reminiscent of the different oxidation behavior of Ge and Si surfaces. The stronger tendency of Si to undergo hydrolysis in sheet polymers can be rationalized by comparing the relevant binding energies involved. Whereas the Si-O bond (8.0 eV) is significantly stronger than the Ge-O bond (6.6 eV), the Si-H bond (3.0 eV) is slightly weaker than the Ge-H bond (3.2 eV). Only after prolonged reaction times are there indications of the formation of a partially OH-substituted germoxene $(Ge_2H_{2-x}(OH)_x)_n$, albeit with significantly less structural integrity than siloxene.[34]

Si and Ge are known to be completely miscible,[35] forming homogeneous SiGe alloys over the entire composition range. The successful topochemical transformation of $CaGe_2$ and the growth of $Ca(Si_{1-x}Ge_x)_2$ alloy Zintl phases [36] therefore suggests that $Si_{1-x}Ge_x$ alloy sheet polymers should be obtainable in the same way as pure Si and Ge polymers. Indeed, such polygermanosilynes $(Si_{1-x}Ge_x\,H_{1-y}(OH)_y)_n$ have been obtained over the entire composition range.[37] The different oxidation behavior of Si and Ge leads to a characteristic dependence of the OH concentration in the polygermanosilynes as a function of the Ge content x: For $x < 0.5$, the OH concentration is approximately constant. For $x > 0.5$, it decreases linearly to eventually become 0 in pure polygermyne. Fourier-transform infrared spectroscopy (FTIR) and thermal effusion experiments (thermally programmed desorption, TPD) show that the hydroxyl groups are preferentially bonded to Si in the backbone rather than to Ge.[37] We therefore observe the overall topochemical transformation reactions

$$(Ca(Si_{1-x}Ge_x)_2)_n + 2n\,HCl + n\,H_2O \rightarrow ((Si_{1-x}Ge_x)_2\,HOH)_n + n\,CaCl_2 + n\,H_2 \quad (4)$$

for $x < 0.5$ and

$$(Ca(Si_{1-x}Ge_x)_2)_n + 2n\,HCl + 2(1-x)n\,H_2O \rightarrow ((SiOH)_{1-x}(GeH)_x)_{2n} + n\,CaCl_2 + 2(1-x)n\,H_2 \quad (5)$$

for $x > 0.5$.

While the overall stoichiometry of the different polymers and the chemical reactions leading to their formation are now known in most cases, no information concerning the kinetics of the topochemical process has been available. Topochemical reactions are typically characterized by extended reaction zones and the formation of product layers separating the reactants, which have to be overcome by diffusion. Consequently, no generally valid theoretical model exists for the description of such solid-state reactions.[38] However, the pronounced changes in the physical properties when metallic Zintl phases are transformed into semiconducting polymers can easily be used to monitor the reaction kinetics.[39] As an example, Figure 16.2 shows that the transformation according to Eq. 2 proceeds with a constant reaction rate along the Si layers. This indicates that the overall transformation is limited by the reaction rate and that diffusion along the siloxene layers formed does not limit the transformation kinetics, even over macroscopic distances. A more complete analysis of the reaction kinetics, as well as of the anisotropy of the reaction rate, is given in ref.[39].

Thus far, we have reviewed the reaction of $CaSi_2$, $CaGe_2$, and the $Ca(Si_{1-x}Ge_x)_2$ alloys with aqueous HCl solutions. We now turn to the reaction of these Zintl phases with pure water. $CaSi_2$ reacts with moisture or oxygen only

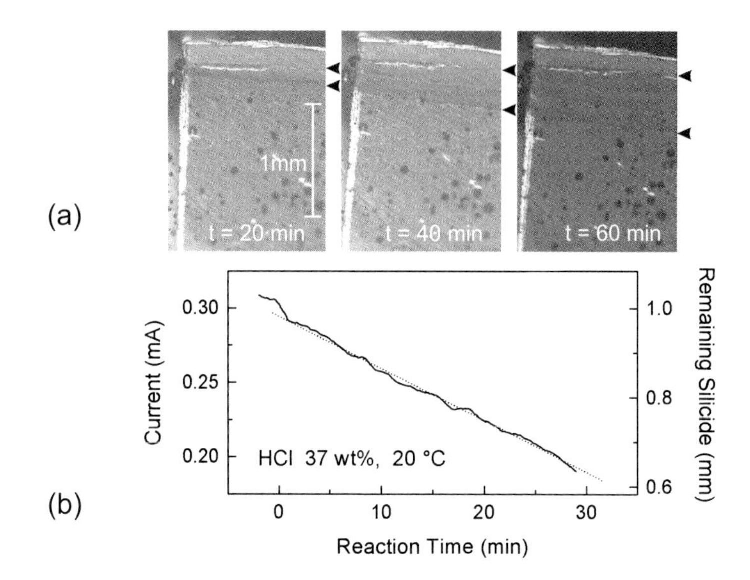

Figure 15.2. Time dependence of the propagation of the topochemical transformation according to Eq. 2 of $CaSi_2$ along the layers of the $(Si^-)_n$ polyanions. (a) The surface of an epitaxial layer of $CaSi_2$ on a (111)-oriented Si substrate is photographed through a covering epoxy layer, which ensures that the $CaSi_2$ is only in contact with the aqueous HCl solution at the top of the photographs. The yellow material is the siloxane formed, the metallic green material is the remaining $CaSi_2$. The linear progression with time of the reaction zone indicated by the lower arrow is also evident from the measurement of the sample conductivity as a function of reaction time shown in (b).

very slowly. Even the heated material exposed to H_2O vapor shows no significant decomposition.[5] However, a slow reaction in hot water is reported [40] as

$$(CaSi_2)_n + 6n\,H_2O \rightarrow n\,Ca(OH)_2 + 2n\,SiO_2 + 5n\,H_2 \tag{6}$$

It is obvious that this reaction destroys the original layered crystal structure by the formation of SiO_2. In contrast, $CaGe_2$ and $Ca(Si_{1-x}Ge_x)_2$ alloys with a high Ge concentration x are much more sensitive to the attack of H_2O [36] and react rapidly under ambient atmosphere, forming new layered Ca-Si-Ge-O-H compounds. Pure thin films of $CaGe_2$ transform in moist air or in water according to

$$(CaGe_2)_n + 2n\,H_2O \rightarrow (Ca(OH)_2\,(GeH)_2)_n \tag{7}$$

Analysis of the structural and vibrational properties of the resulting material indicate that the reaction product contains the $(GeH)_n$ layers of polygermyne.[41] However, in contrast to the transformation in aqueous HCl, Ca cannot be removed, especially when the reaction is performed in air. Rather, the Ca layers are transformed into $(Ca(OH)_2)_n$ layers, which are intercalated between the $(GeH)_n$ layers. We therefore call this material "polygermyne calcium dihydroxide intercalation compound" $(Ca(OH)_2(GeH)_2)_n$. Its structure is also shown in Figure 15.1.

The time needed for thin films of $CaGe_2$ to form this intercalation compound is of the order of several hours at ambient conditions. In contrast, $Ca(Si_{1-x}Ge_x)_2$ alloys react significantly faster, with thin films of CaSiGe being decomposed in less than 60 s. In the reaction of $Ca(Si_{1-x}Ge_x)_2$ with H_2O, the formation of $(GeH)_n$ layers competes with the formation of SiO_2. Indeed, crystalline layered polygermanosilyne calcium dihydroxide intercalation compounds have been found only for $0.7 < x < 1$.[41] At lower Ge contents x, the reaction products become amorphous.

15.3 Preparation and Properties of Si and Ge Sheet Polymers

15.3.1 Reactive Deposition Epitaxy and Topotactic Transformation

We have already noted above that the only known way to obtain Si- and Ge-based sheet polymers is by transformation of the corresponding layered Zintl phases. Bulk $CaSi_2$ or $CaGe_2$ can be prepared by several methods which, however, are rendered difficult by the incongruent melting of the material, which can lead to the formation of several different Ca silicides and germanides.[21, 42, 43] An alternative route, which is, moreover, widely compatible with existing semiconductor technology, is the preparation of thin films on Si or Ge substrates.[36, 44–46] For the growth of silicides on Si, solid-phase epitaxy (SPE) is commonly used. In this method, the silicides are formed from a metallic film that is deposited on the cold substrate and then subjected to a thermal anneal. Although this method has been shown to work for $CaSi_2$ as well,[47] the comparatively high vapor pressure of Ca severely limits the obtainable film thickness. We have, therefore, used the alternative method of reactive deposition epitaxy (RDE) and optimized this process for the growth of Si- and Ge-based layered Zintl phases.[36, 44–46]. In RDE of $CaSi_2$, the metal is evaporated onto a heated reactive substrate which, at the same time, functions as the source for Si. Since Si is diffusing faster through the $CaSi_2$ layer than Ca, the growth zone is at the sample surface.[48] For thick $CaSi_2$, this results in an erosion at the Si/$CaSi_2$ interface, limiting the maximum silicide thickness that can be grown by RDE to about 2 μm. On the other hand, the fast diffusion of Si prohibits the formation of other, Ca-rich silicides at the Si/$CaSi_2$ interface, leading to atomically abrupt interfaces. We have used this method to grow $CaSi_2$ on Si, $CaGe_2$ on Ge, and $Ca(Si_{1-x}Ge_x)_2$ on $Si_{1-x}Ge_x$ epilayers, which were grown by molecular beam epitaxy (MBE) on Si. The technique, as well as the optimum growth parameters to achieve laterally homogenous layers, are discussed in detail in refs.[36, 45, 46].

Growth of silicide layers on substrates with different crystallographic orientations showed that the puckered Si layers in epitaxial $CaSi_2$ films grow parallel to the {111} planes of the substrate, a behavior called templating.[45] On Si(111), the $CaSi_2$ layers grow parallel to the surfaces, while on Si(110), the $CaSi_2$ planes are inclined at an angle of 35.5° with respect to the substrate surface (Figure 15.3). In the latter case, twinning can be suppressed by using off-axis cut substrates. Due to the much higher electrical conductivity of the sheet polymers parallel to the layers,[49] the latter orientation will be beneficial for the realization of optoelectronic devices where current flow across the Si/sheet polymer interface is required.

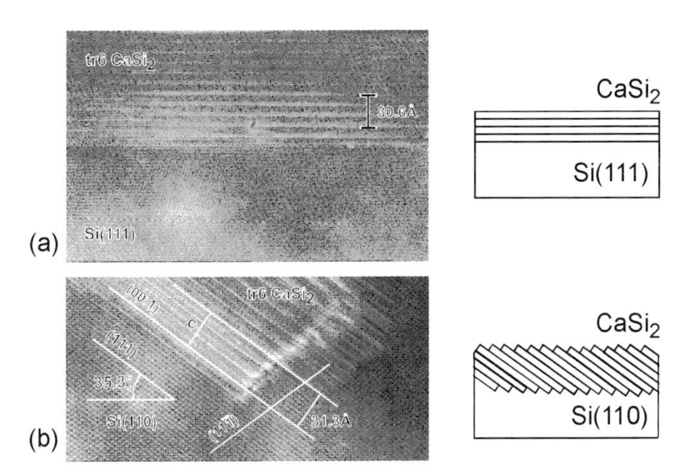

Figure 15.3. High-resolution transmission electron micrographs (TEM) of the interface of $CaSi_2$ and Si for $CaSi_2$ films grown on Si(111) surfaces (a) and Si(110) surfaces (b). The diagrams on the right-hand side indicate the orientation of the Si layers in the $CaSi_2$ formed with respect to the substrate.

We have analyzed the structure and the chemical composition of the Zintl phases, as well as of the sheet polymers, with the help of X-ray diffraction (XRD),[17, 34, 37, 36, 41, 45, 46, 50] scanning electron microscopy (SEM),[45, 46, 51] energy dispersive X-ray analysis (EDX),[17, 34, 36, 37] elastic recoil detection analysis (ERD),[46] nuclear magnetic resonance (NMR),[16, 52] thermally programmed desorption (TPD),[37, 51] and vibrational spectroscopy (Raman, infrared absorption and neutron scattering).[53, 54] The strained epitaxial layers of $CaSi_2$ have also been used to study the elastic properties of this Zintl phase.[50] As the dominant polytype in all epitaxial Zintl phase layers, we have identified the trigonal rhombohedral tr6 modification. However, other polytypes are also formed in the thin films, usually simultaneously with the tr6 modification.[36, 45, 46] The tr3 modification of $CaSi_2$ was found to grow on top of thick $CaSi_2$ films,[45] while at the Ge/$CaGe_2$ interface, a thin hexagonal h2 $CaGe_2$ modification is formed.[46] The reason for the formation of a particular polytype in the layered Zintl phases is still unclear, but probably can be attributed to contamination with impurities or to strain.[50] X-ray diffraction clearly shows that, upon topochemical transformation, the stacking sequence or polytype of the Zintl phase is preserved in the sheet polymer. Using reciprocal space maps, in particular the formation of tr6 siloxene [45] and tr6 polygermyne [12, 17] has been shown. The presence of a stacking sequence in the sheet polymers demonstrates the high structural quality of the films obtained. It also shows that the term "topotactic" transformation should be used rather than topochemical, since the relative crystallographic orientations of the thin film and the substrate are preserved.[55]

15.3.2 Vibrational Properties

The large concentration of H- and OH-ligands present in the sheet polymers leads to various easily observed local vibrational modes, which have been used for the structural and compositional analysis of these materials.[56, 57] Isotopically substituted polymers can be obtained by the use of DCl in D_2O or HCl in $H_2{}^{18}O$ for the topochemical transformation and are very helpful in the assignment of the different modes. It is beyond the scope of this brief review to discuss the large number of different local vibrational modes in detail.[53] Instead, we would like to briefly discuss two particularly interesting issues which can be addressed by vibrational spectroscopy of sheet polymers: the question of bonding between the polymer sheets and the use of these polymers as model substances for Si surfaces.

The spontaneous cleavage of bulk $CaSi_2$ during topochemical transformation yielding siloxene platelets clearly shows that bonding between adjacent polymer sheets is weak as would be expected. From Figure 15.1, it is obvious that interlayer bonding can either be through van der Waals forces or through hydrogen bonds. In infrared absorption spectroscopy, the O-H stretching vibrations are observed between 3200 and 3600 cm^{-1}. Isolated OH groups give rise to a narrow line at 3595 cm^{-1}. In contrast, OH groups participating in a hydrogen bond absorb in a broad frequency band centered at around 3400 cm^{-1}. Taking into account the rather different oscillator strengths of these OH groups,[58] we find that typically only 10% of the hydroxyl groups participate in hydrogen bonds, while the vast majority of the ligands remain isolated. However, because of the much higher strength of hydrogen bonds, we conclude that both van der Waals forces and hydrogen bonds contribute significantly to the interlayer bonding.

At lower frequencies, the Si-H bending mode and the Si phonon modes are found. In crystalline Si, these modes do not couple due to their different symmetries. However, in the sheet polymers, both the bending mode and the phonon with the Si atoms vibrating in-plane are two-fold degenerate (type E) and undergo a Fermi resonance. This is clearly seen in Raman spectroscopy, where the Si phonon appears to harden under H-D substitution from 515 to 575 cm^{-1}.[54] Such behavior has already been predicted for Si surfaces.[59] However, only the chlorine chemistry available for H-D substitution in sheet polymers has enabled the experimental observation of this mode coupling, while it still eludes observation on Si surfaces, where the necessary fluorine chemistry complicates the H-D substitution.[60] This mode coupling is also observed in neutron scattering from siloxene,[54] which clearly shows that sheet polymers can be used as a convenient model system for the properties of Si surfaces, even with bulk-sensitive methods.

15.3.3 Band Structure and Luminescence

The most prominent physical property of the Si and Ge sheet polymers is their intense visible photoluminescence (PL), which can reach efficiencies of up to 10% at room temperature.[61] Crystalline Si and Ge only show very weak infrared luminescence due to the indirect character of their electronic band structure.[62] Linear chain polymers like polysilane or polygermane, on the other hand, exhibit a much stronger luminescence, albeit in the near-UV spectral range.[63] In order to obtain Si- or Ge-based materials which show strong photo- and electroluminescence and which are compatible with existing microelectronics technology, various approaches for a physical or chemical modification of crystalline Si or Ge have been followed, such as the use of superlattices for band structure folding, incorporation of impurities, introduction of disorder or quantum confinement.[64–67] Band structure calculations show that the dimensional reduction from 3D crystals to pure H-terminated 2D sheet polymers results in a near degeneration of the direct and indirect band gaps. While the theoretical predictions disagree as regards whether polysilyne $(SiH)_n$ has a direct or an indirect band gap, a direct band gap is unanimously predicted for siloxene $(Si_2HOH)_n$.[67–71] Recent calculations find the same for polygermyne $(GeH)_n$.[51, 71]

The luminescence properties of siloxene have now been studied in great detail. A typical siloxene PL spectrum is shown at the bottom of Figure 15.4 (b). It has a maximum in the yellow-green spectral range at around 2.4 eV. At low temperatures, a radiative lifetime of 10 ns and a polarization memory were observed.[72] These properties and the small Stokes shift between the photoluminescence and its excitation spectra [73] provide experimental evidence for the hypothesis that siloxene does indeed have a direct band gap. Details of the excited states in siloxene leading to the luminescence have also been obtained from measurements of optically detected magnetic resonance (ODMR).[74–76]

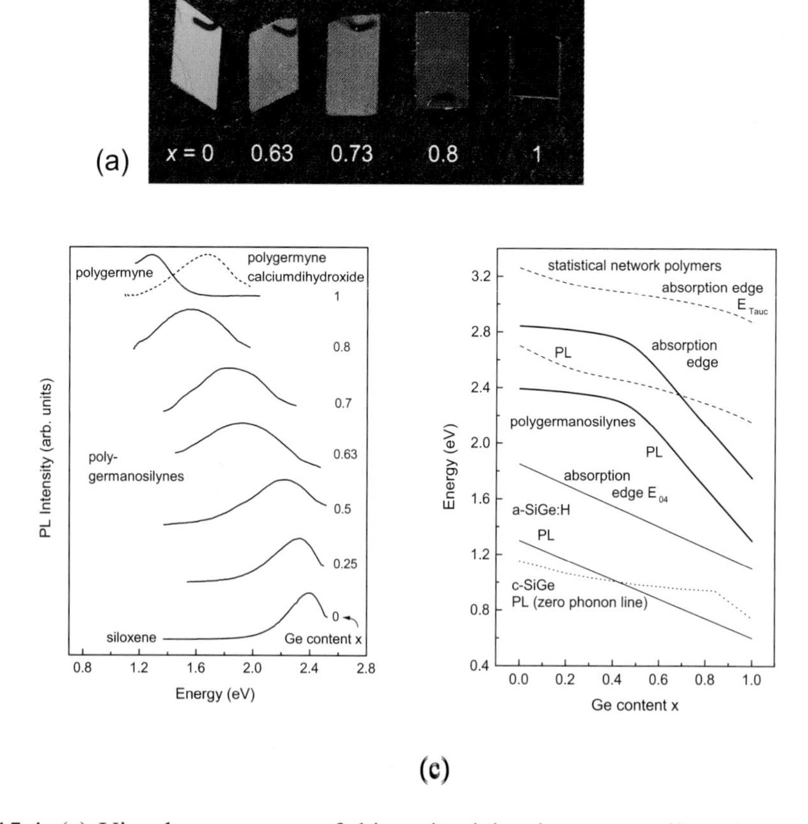

Figure 15.4. (a) Visual appearance of thin epitaxial polygermanosilyne layers as a function of the Ge content *x*. (b) Comparison of the photoluminescence spectra of the different sheet polymers as a function of the Ge content *x*. All spectra have been measured at room temperature and have been normalized. (c) Comparison of the dependences of the photoluminescence energies and the absorption edges of several $Si_{1-x}Ge_x$ compounds on the Ge content x. [26, 37, 62, 77]

Polygermyne has a strong infrared luminescence at 1.3 eV (Figure 15.4 (a) top), which can also be excited with a small Stokes shift.[17, 71] Figure 15.4 (b) shows that the photoluminescence peak of the polygermanosilynes is continuously tunable between 1.3 and 2.4 eV.[37] The continuous variation of the band gap of SiGe alloy sheet polymers is also quite obvious from Figure 15.4 (a), which shows the visual appearance of some of the polymers with different Ge content x. A broadening of the luminescence spectra in the alloy sheet polymers due to compositional disorder and lower crystalline quality is also seen in Figure 15.4 (b). In contrast to siloxene and polygermyne, no band structure calculations have yet been performed for alloy sheet polymers. Similarly, the influence of ligand substitution on the band structure of polygermyne and the polygermanosilynes has not been studied, neither theoretically nor experimentally.

The dependences of the optical absorption edges and of the luminescence maxima of various $Si_{1-x}Ge_x$ alloys on the Ge content are compared to that of the polygermanosilynes in Figure 16.4 (c). Apart from in the case of crystalline $Si_{1-x}Ge_x$ alloys, where the characteristic dependence of the zero phonon luminescence line on alloy composition is caused by the fact that the conduction band minima in Si and Ge occur in different directions of the Brillouin zone,[62] a more or less linear dependence of the optical band gap and of the luminescence energy is typically observed. This is in marked contrast to the dependence observed for the polygermanosilynes. Here, the optical properties are more or less constant for $x < 0.5$, but exhibit a strong decrease with increasing Ge content x for $x > 0.5$. The shift for $x > 0.5$ is compatible with a linear decrease of the characteristic optical energies from the values expected for pure $(SiH)_n$ to those observed for pure $(GeH)_n$.[69] It is possible that the deviation from this linear behavior for samples with lower Ge content is caused by the well known red-shift induced by the hydroxyl groups attached to the Si atoms.[8, 68]

The polygermyne calcium dihydroxide intercalation compound also shows an intense photoluminescence, which is shifted with respect to that of pure polygermyne to the red spectral region at 1.7 eV due to the intercalated $Ca(OH)_2$ layers (Figure 15.4 (b)). Optical experiments indicate that this compound also has a direct band gap.[41] The amorphization that takes place during the reaction of $Ca(Si_{1-x}Ge_x)_2$ alloys with water leads to a strong decrease in the photoluminescence intensity even for moderate Si concentrations. The large difference between the optical absorption and the luminescence observed in these compounds suggests that this luminescence originates at defects. No band structure calculations have yet been performed for the intercalation compounds.

The photoluminescence line width of siloxene of about 0.3 eV is the same in crystalline platelets obtained from bulk $CaSi_2$ and in epitaxial thin films, despite the considerably higher structural quality of the latter.[71, 73] Possible origins of this line width include inhomogeneities in the termination with H- or OH-ligands, stacking faults or a strong electron-phonon coupling. The large line width has probably prohibited the observation of stimulated emission from siloxene until now. To study optical gain in siloxene, we have grown $CaSi_2$ layers on silicon-on-insulator substrates, which were topochemically transformed into 500 nm thick siloxene layers on 400 nm thick SiO_2 buffer layers. SiO_2 rather than Si is needed below the siloxene film to ensure optical waveguiding. However, using a frequency-doubled Ti:sapphire laser for excitation, no gain narrowing could be observed on these structures for pulse energies up to 1 nJ and pulse lengths of 2 ps. In addition to the large line width of the PL, scattering caused by surface and interface roughness could lead to an increase of the pumping level required for stimulated emission.

15.4 Future Directions

15.4.1 Sheet Polymers Containing Other Group IV Elements

We have described above the preparation and properties of $Si_{1-x}Ge_x$ sheet polymers over the entire compositional range. This motivates a discussion concerning the possibilities of forming sheet polymers containing other group IV atoms such as C or Sn in the backbone, summarized in Table 16.1.

In marked contrast to Si and the heavier group IV atoms, carbon is able to form multiple bonds. As a consequence, CaC_2 does not form a layered structure similar to that of $CaSi_2$, but modifications containing triply-bonded monomeric $(C^-)_2$ units.[78] Therefore, C-based sheet polymers with sp^3 hybridization of the carbon backbone atoms are only obtained directly from graphite. In this case, the sp^2 hybridization of carbon can be changed in favor of an sp^3 hybridization by stabilizing each C atom by a strongly electronegative ligand such as fluorine. The resulting product is the sheet polymer graphite fluoride $(CF)_n$, which is isotypic to $(GeH)_n$, containing similar buckled backbone layers.[2] In contrast, the formation of C sheet polymers with less electronegative ligands is energetically unfavorable, and, in particular, purely H-terminated $(CH)_n$ does not exist. Instead, an O- and H-containing carbon sheet polymer called graphite oxide or graphite acid can be prepared by treating graphite with strong aqueous oxidizing agents.[3] However, since the electronegativity of OH groups is not sufficient for a general sp^3 hybridization of the C atoms, the layer structure of this compound is characterized by the simultaneous presence of C=C double bonds, C-O-C bridges, and some OH termination.[3] Graphite oxide with different colors has been reported, with some samples exhibiting strong yellow to red luminescence.[79]

Table 15.1. Compilation of the Ca Zintl phases with layered polyanions and of typical sheet polymers based on an sp^3 hybridized backbone for the different group IV atoms. The arrows indicate the route for obtaining the polymers.

2D polymer	Graphite fluoride $(CF)_n$ ↑		Siloxene $(Si_2HOH)_n$ ↑	Polygermano-silynes $(Si_{1-x}Ge_xO_yH_z)_n$ ↑	Poly-germyne $(GeH)_n$ ↑			
Zintl-phase with 2D polyanion	↑		$CaSi_2$ ↑	$Ca(Si_{1-x}Ge_x)_2$ ↑	$CaGe_2$ ↑	$Ca(Ge_{1-x}Sn_x)_2$ with $x \leq 0.25$ ↑		
Element or alloy	C	SiC	Si	$Si_{1-x}Ge_x$	Ge	$Ge_{1-x}Sn_x$	Sn	Pb

To a certain degree, the method of preparing sheet polymers by the formation of layered Zintl phases should be applicable to $Ge_{1-x}Sn_x$ and $Si_{1-x}C_x$ alloys. Pure $CaSn_2$ does not exist,[80] but $Ca(Ge_{1-x}Sn_x)_2$ alloys have been prepared as bulk material under equilibrium conditions for $x \leq 0.25$, though no topochemical transformation was attempted.[81] Epitaxial films of Zintl phases with even higher Sn content may probably be grown by non-equilibrium methods. Epitaxial thin films of α-$Ge_{1-x}Sn_x$ have been grown up to $x = 0.3$,[82] which might be used for RDE growth of the corresponding Zintl phases. Similarly, $Si_{1-x}C_x$ alloys with carbon concentrations up to $x = 0.2$ have been grown using non-equilibrium techniques.[83] However, no layered Zintl phases other than $CaSi_2$ have been reported in the Ca-Si-C system. In particular, attempts to obtain bulk CaSiC have led to phase separations into $CaSi_2$ and CaC_2.[84] In view of the different structure and chemical behavior of CaC_2, it would be surprising if SiC alloy sheet polymers with C concentrations above a couple of percent could be obtained via the Zintl phase route, but $Ge_{1-x}Sn_x$ sheet polymers with Sn contents of at least up to 0.25 can most probably be formed.

Broadening the range of elements used in sheet polymers further, one might even speculate about the possibility of forming group III–V compound sheet polymers such as $(GaAsH_2)_n$. Attempts to form a CaGaAs Zintl phase have, however, failed also due to phase separation,[84] so that the formation of such compound polymers, if stable at all, does not appear to be possible via the Zintl phase route.

15.4.2 Physics and Chemistry of Sheet Polymers

We finally point out some issues concerning the physics and chemistry of sheet polymers and Zintl phases which could be interesting research topics in the near future.

For the existing SiGe sheet polymers, some rather fundamental investigations of the electronic band structure remain to be performed. For example, the layered structure should allow the direct determination of the complete band structure by angular resolved photoemission spectroscopy or inverse photoemission. The anisotropy of the dielectric constants leading to optical activity has been reported,[85] but has not been studied systematically as a function of wavelength, e.g. by ellipsometry. Also, for the realization of optical gain and stimulated emission, it appears necessary to identify the origin of the broad luminescence line width and to find ways of reducing it. Furthermore, as mentioned above, no calculations have been performed to predict the band structure of the polygermanosilyne alloys and of the intercalation compounds, nor to study the effect of ligand substitution on the band structure of these polymers. Finally, superlattices of Si and Ge sheet polymers might be formed from ordered $(CaSi_2)_n(CaGe_2)_m$ structures, which would have to be grown with atomic layer precision e.g. by molecular beam epitaxy.

Concerning electronic transport in thc sheet polymers, the situation is similar. Both siloxene and polygermyne are intrinsic semiconductors.[17, 49, 51] As would be expected from the crystallographic structure, the conductivity of siloxene is very anisotropic.[12, 49] The photoconductivity of siloxene is large and dominated by bimolecular recombination, which shows that the material has very few electronically active defects such as dangling bonds.[49, 75] However, for the realization of optoelectronic devices, doping of the sheet polymers is required. Substitutional doping with group III or group V atoms will most probably not lead to any doping effect, since these atoms will be incorporated into the backbone with their preferred threefold coordination. Substitutional doping with atoms of group II or group VI elements could be more promising, but might also lead to an unwanted destruction of the polymer layers during topochemical transformation. Therefore, intercalation after the formation of the polymers with group I atoms such as lithium, which is a known interstitial donor in silicon,[85] or group VII atoms such as iodine should be attempted first. Without doping, the first siloxene diodes exhibiting an asymmetric current-voltage characteristic have been fabricated using p-type Si and Ca, which have very different work functions, as contacts for charge injection.[12, 44] Further development of such devices is required for the realization of electroluminescence.[87]

Turning to the chemistry of the sheet polymers, we first have to note that the substitution reactions reported for Kautsky siloxene [8] do not appear to lead to ligand substitution in Wöhler siloxene. The origin of this behavior is unclear at the moment. A significant concern with respect to the application of sheet polymers is the easy oxidation, in particular of siloxene and the Si-rich polygermanosilynes. This is due to the energy of about 1 eV gained when an oxygen atom from an OH-group is incorporated into an Si-Si bond of the backbone structure.[68] While exposure to ambient oxygen can be limited by appropriate encapsulation, the incorporation of O from the hydroxyl groups is an inherent problem of siloxene and can only be solved by the use of other ligands. Finally, the weak interlayer bonding, which leads to an easy cleavage of siloxene, could be strengthened by suitable cross-linking.

Knowledge of the topochemistry of the Zintl phases of Ca with Si and Ge known is still rather limited. New approaches could include the use of different sources of chlorine or of protic solvents in the topochemical transformation of the layered Zintl phases. A first attempt in this direction has involved the use of aqueous solutions of $CoCl_2$ for the removal of Ca.[88] A different approach to broaden the topochemistry is the transformation of Zintl phases with different structures of the group IV polyanions, leading to Si or Ge compounds with other backbone structures.[15] An example of this is the reaction of CaSi with HCl, which leads to the formation of linear polysilane chain polymers.[89] However, the Zintl phases of Si and Ge have an even richer structural chemistry, containing polyanions in the form of tetrahedra or planar Si rings, which could lead to Si compounds of great interest from both a chemical and a physical point of view.[15, 90]

The discovery of carbon nanotubes showed that sheet polymers do not necessarily have to exist in planar form.[91] Recent calculations have shown that the Si- and Ge-based sheet polymers discussed here should also be able to form such nanotubes, retaining the sp^3 hybridization, threefold homoatomic coordination, and H- and OH-termination. In contrast to the carbon nanotubes, these tubular structures are expected to remain semiconducting, but should show a dependence of the band gap on the diameter of the tube.[92] Various techniques aimed at forming free-standing Si quantum wires are being pursued, but no hollow, tubular structures have yet been grown. It will be a challenging endeavor to identify the type of templating necessary to form such siloxene or polygermyne tubes.

15.5 Conclusion

We have summarized the present knowledge about the layered Zintl phases of Ca with Si and Ge and about the sheet polymers obtained from them. In particular, we have been able to demonstrate in recent years the high quality epitaxial growth of $CaSi_2$, $CaGe_2$, and $Ca(Si_{1-x}Ge_x)_2$ alloys on crystalline Si and

Ge substrates, for the first time synthesizing the alloy Zintl phases. By topotactic transformation of these Zintl phases, a variety of formerly unknown sheet polymer compounds was obtained, most notably polygermyne and the alloy polygermanosilynes. With the alloy sheet polymers, we now have Si- and Ge-based materials available which show a strong visible photoluminescence tunable from 2.4 to 1.3 eV. While the variation of the luminescence through ligand substitution in sheet polymers has been known for a long time, we have shown that substitution of backbone atoms can also be used to tune the optical properties of these substances. Nevertheless, the issues raised in the last section ensure that sheet polymers will be an interesting and rewarding research field for years to come.

15.6 Acknowledgements

The authors are very grateful to the coordinators Prof. P. Jutzi and Prof. U. Schubert for the excellent organization of the Schwerpunktprogramm and to the referees for their constructive criticism. The authors gratefully acknowledge the cooperation with Z. Hajnal (Universität Paderborn) and M. Albrecht (Universität Erlangen-Nürnberg) on electronic band structure calculations and transmission electron microscopy, respectively. M.S.B. and M. St. would also like to thank their diploma students Thies Puchert, Günther Vogg, Lutz Höppel and Lex Meyer and their Ph.D. students Nikta Zamanzadeh-Hanebuth and Günther Vogg for their enthusiasm and dedicated work on this project.

15.7 References

[1] I. Manners, *Angew. Chem. Int. Ed. Engl.* **1996**, *35*, 1602.
[2] Y. Kita, N. Watanabe, Y. Fujii, *J. Am. Chem. Soc.* **1979**, *101*, 3832.
[3] H. He, J. Klinowski, M. Forster, A. Lerf, *Chem. Phys. Lett.* **1998**, *287*, 53.
[4] G. Schott, *Z. Chemie (Leipzig)* **1962**, *6/7*, 194.
[5] F. Wöhler, *Liebigs. Ann.* **1863**, *127*, 257.
[6] A. Weiss, G. Beil, H. Meyer, *Z. Naturforsch.* **1979**, *35b*, 25.
[7] H. Kautsky, *Z. anorg. Chem.* **1921**, *117*, 209.
[8] H. Kautsky, G. Herzberg, *Z. anorg. allg. Chem.* **1924**, *139*, 135.
[9] E. F. Hengge, A. Kleewein, in R. Corriu, P. Jutzi (Eds.) *Tailor-made Silicon-Oxygen Compounds*, Vieweg, Braunschweig, **1996**, 89.
[10] E. Hengge, *Fortschr. Chem. Forsch.* **1967**, *9*, 145.
[11] see Chapter 16, herein.

[12] G. Vogg, Ph.D. thesis, Technische Universität München, **2001**.
[13] *Gmelin Handbuch der anorganischen Chemie, Silizium B*, Verlag Chemie, Weinheim, **1959**, 591.
[14] *Gmelin Handbook of Inorganic Chemistry, Silicon B1*, Springer, Berlin, **1982**, 231.
[15] W. Hönle, U. Dettlaff-Weglikowska, S. Finkbeiner, A. Molassioti-Dohms, J. Weber, in Ref. [9], p. 99.
[16] M. S. Brandt, T. Puchert, M. Stutzmann, in Ref. [9], p. 117.
[17] G. Vogg, M. S. Brandt, M. Stutzmann, *Adv. Mater.* **2000**, *12*, 1278.
[18] H. Schäfer, B. Eisenmann, W. Müller, *Angew. Chem. Int. Ed. Engl.* **1973**, *12*, 694.
[19] J. Böhm, O. Hassel, *Z. anorg. allg. Chem.* **1927**, *160*, 152.
[20] H. J. Wallbaum, *Naturwissenschaften* **1944**, *32*, 76.
[21] T. Hirano, *J. Less-Common Metals* **1991**, *167*, 329.
[22] E. Kulatov, H. Nakayama, H. Ohta, *J. Phys. Condens. Matter* **1997**, *9*, 10159.
[23] R. Würz, M. Schmidt, A. Schöpke, W. Fuhs, *Appl. Surf. Sci.* **2002**, *190*, 437.
[24] P. A. Bianconi, F. C. Schilling, T. W. Weidman, *Macromolecules* **1989**, *22*, 1697.
[25] W. J. Szymanski, G. T. Visscher, P. A. Bianconi, *Macromolecules* **1993**, *26*, 869.
[26] H. Kishida, H. Tachibana, M. Matsumoto, Y. Tokura, *Appl. Phys. Lett.* **1994**, *65*, 1358.
[27] E. Bonitz, *Chem. Ber.* **1961**, *94*, 220.
[28] H. Kautsky, *Z. Naturforsch.* **1952**, *7b*, 174.
[29] E. Bonitz, *Angew. Chem. Int. Ed. Engl.* **1966**, *5*, 462.
[30] U. Dettlaff-Weglikowska, W. Hönle, A. Molassioti-Dohms, S. Finkbeiner, J. Weber, *Phys. Rev. B* **1997**, *56*, 13132.
[31] J.R. Dahn, B. M. Way, E. Fuller, *Phys. Rev. B* **1993**, *48*, 17872.
[32] J. He, J. S. Tse, D. D. Klug, K. F. Preston, *J. Mater. Chem.* **1998**, *8*, 705.
[33] S. Yamanaka, H. Matsu-ura, M. Ishikawa, *Mat. Res. Bull.* **1996**, *31*, 307.
[34] G. Vogg, M. S. Brandt, M. Stutzmann, in N. Auner, J. Weis (Eds.), *Organosilicon Chemistry V*, Wiley-VCH, Weinheim, **2003**, in print.
[35] M. Rubenstein, *J. Cryst. Growth* **1972**, *17*, 149.
[36] G. Vogg, C. Miesner, M. S. Brandt, M. Stutzmann, G. Abstreiter, *J. Cryst. Growth* **2001**, *223*, 573.
[37] G. Vogg, A. J. P. Meyer, C. Miesner, M. S. Brandt, M. Stutzmann, *Appl. Phys. Lett.* **2001**, *78*, 3956.
[38] H. R. Oswald, A. Reller, *Ber. Bunsenges. Phys. Chem.* **1986**, *90*, 671.
[39] G. Vogg, M. S. Brandt, M. Stutzmann, *Chem. Mat.* **2003**, *15*, 910.

[40] S. Abe, H. Nakayama, T. Nishino, S. Iida, *Appl. Surf. Sci.* **1997**, *113/114*, 562.
[41] G. Vogg, L. J. P. Meyer, C. Miesner, M. S. Brandt. M. Stutzmann, *Chem. Monthly* **2001**, *132*, 1225.
[42] J. Evers, G. Oehlinger, A. Weiss, *J. Solid State Chem.* **1977**, *20*, 173.
[43] S. Yamanaka, F. Suehiro, K. Sasaki, M. Hattori, *Physica B* **1981**, *105*, 230.
[44] G. Vogg, N. Zamanzadeh-Hanebuth, M. S. Brandt, M. Stutzmann, M. Albrecht, *Chem. Monthly* **1999**, *130*, 79.
[45] G. Vogg, M. S. Brandt, M. Stutzmann, M. Albrecht, *J. Cryst. Growth* **1999**, *203*, 570.
[46] G. Vogg, M. S. Brandt, M. Stutzmann, I. Genchev, A. Bergmaier, L. Görgens, G. Dollinger, *J. Cryst. Growth* **2000**, *212*, 148.
[47] J. F. Morar, M. Wittmer, *J. Vac. Sci. Technol. A* **1988**, *6*, 1340.
[48] M. A. Nicolet, S. S. Lau, *Formation and Characterization of Transition Metal Silicides*, in *VLSI Electronics Vol. 6* (Ed.: N. G. Einspruch), Academic Press, New York, **1983**.
[49] M. S. Brandt, T. Puchert, M. Stutzmann, *Solid State Commun.* **1997**, *102*, 365.
[50] G. Vogg, M. S. Brandt, M. Stutzmann, p*hys. stat. sol. (a)* **2001**, *185*, 213.
[51] G. Vogg, M. S. Brandt, L. J. P. Meyer, M. Stutzmann, Z. Hajnal, B. Szücs, T. Frauenheim, *Mat. Res. Symp. Proc.* **2001**, *667*, G3.13.
[52] M. S. Brandt, S. E. Ready, J. B. Boyce, *Appl. Phys. Lett.* **1997**, *70*, 188.
[53] N. Zamanzadeh-Hanebuth, M. S. Brandt, M. Stutzmann, *J. Non-Cryst. Solids* **1998**, *227-230*, 503.
[54] M. S. Brandt, L. Höppel, N. Zamanzadeh-Hanebuth, G. Vogg, M. Stutzmann, p*hys. stat. sol. (b)* **1999**, *215*, 409.
[55] F. K. Lotgering, *J. Inorg. Nucl. Chem.* **1959**, *9*, 113.
[56] H. Ubara, T. Imura, A. Hiraki, I. Hirabayashi, K. Morigaki, *J. Non-Cryst. Solids* **1983**, *59&60*, 641.
[57] H. D. Fuchs, M. Stutzmann, M. S. Brandt, M. Rosenbauer, J. Weber, A. Breitschwerdt, P. Deak, M. Cardona, *Phys. Rev. B* **1993**, *48*, 8172.
[58] R. Laenen, C. Rauscher, *J. Chem. Phys.* **1997**, *107*, 9759.
[59] B. J. Min, Y. H. Lee, C. Z. Wang, C. T. Chan, K. M. Ho, *Phys. Rev. B* **1992**, *46*, 9677.
[60] I. P. Ipatova, O. P. Chikalova-Luzina, K. Hess, p*hys. stat. sol. (b)* **1999**, *212*, 287.
[61] M. Stutzmann, M. S. Brandt, M. Rosenbauer, H. D. Fuchs, S. Finkenbeiner, J. Weber, P. Deak, *J. Luminesc.* **1993**, *57*, 321.
[62] J. Weber, M. I. Alonso, *Phys. Rev. B* **1989**, *40*, 5683.
[63] R. D. Miller, J. Michl, *Chem. Rev.* **1989**, *89*, 1359.

[64] M. S. Brandt, H. D. Fuchs, A. Höpner, M. Rosenbauer, M. Stutzmann, J. Weber, M. Cardona, H. J. Queisser, *Mat. Res. Soc. Symp. Proc.* **1992**, *262*, 849.
[65] S. S. Iyer, Y. H. Xie, *Science* **1993**, *260*, 40.
[66] L. Brus, *J. Phys. Chem.* **1994**, *98*, 3575.
[67] K. Takeda, K. Shirashi, *Comments Cond. Mat. Phys.* **1997**, *18*, 91.
[68] P. Deak, M. Rosenbauer, M. Stutzmann, J. Weber, M. S. Brandt, *Phys. Rev. Lett.* **1992**, *69*, 2531.
[69] C. G. Van de Walle, J. E. Northrup, *Phys. Rev. Lett.* **1993**, *70*, 1116.
[70] K. Takeda, K. Shiraishi, *Phys. Rev. B* **1997**, *56*, 14985.
[71] Z. Hajnal, G. Vogg, A. J. P. Meyer, B. Szücs, M. S. Brandt, T. Frauenheim, *Phys. Rev. B* **2001**, *64*, 33311.
[72] I. Hirabayashi, K. Morigaki, S. Yamanaka, *J. Non-Cryst. Solids* **1983**, *59&60*, 645.
[73] M. Stutzmann, M. S. Brandt, M. Rosenbauer, J. Weber, H. D. Fuchs, *Phys. Rev. B* **1993**, *47*, 4806.
[74] M. S. Brandt, M. Stutzmann, *Solid State Commun.* **1995**, *93*, 473.
[75] M. S. Brandt, M. Stutzmann, in Z. C. Feng, R. Tsu (Eds.) *Porous silicon*, World Scientific, Singapore, **1994**, 417.
[76] H. Pioch, J. U. von Schütz, H. C. Wolf, *Chem. Phys. Lett.* **1997**, *277*, 89.
[77] R. Carius, in H. Fritzsche (Ed.) *Amorphous Silicon and Related Materials Vol. B*, World Scientific, Singapore, **1988**, 939.
[78] M. A. Bredig, *J. Phys. Chem.* **1942**, *46*, 801.
[79] H. Thiele, *Z. anorg. allg. Chem.* **1930**, *190*, 145.
[80] R. D. Hoffmann, D. Kussmann, U. C. Rodewald, R. Pöttgen, C. Rosenhahn, B. D. Mosel, *Z. Naturforsch.* **1999**, *54b*, 709.
[81] A. K. Ganguli, J. D. Corbett, *J. Solid State Chem.* **1993**, *107*, 480.
[82] P. R. Pukite, A. Harwit, S. S. Iyer, *Appl. Phys. Lett.* **1989**, *54*, 2142.
[83] H. Rücker, M. Methfessel, E. Bugiel, H. J. Osten, *Phys. Rev. Lett.* **1994**, *72*, 3578.
[84] J. Evers, private communication.
[85] H. Kautsky, H. Zocher, *Z. Phys.* **1922**, *9*, 267.
[86] K. Winer, R. A. Street, *J. Appl. Phys.* **1989**, *65*, 2272
[87] M. Rosenbauer, M. Stutzmann, *J. Appl. Phys.* **1997**, *82*, 4520.
[88] H. Nakano, S. Yamanaka, M. Nawate, S. Honda, *J. Mater. Sci.* **1994**, *29*, 1324.
[89] P. Royen, C. Rocktäschel, *Z. anorg. allg. Chem.* **1966**, *346*, 279.
[90] see Chapter 13, herein.
[91] M. S. Dresselhaus, G. Dresselhaus, P. C. Eklund, *Science of Fullerenes and Carbon Nanotubes*, Academic Press, San Diego, **1996**.
[92] see Chapter 17, herein.

16 Kautsky-Siloxene Analogous Monomers and Oligomers

Harald Stüger*

16.1 Introduction

Because of its indirect band structure, crystalline silicon does not show visible light emission at room temperature. Low-dimensional silicon polymers such as polysilanes, or silicon polymers with sheet-like structures, however, have a direct band structure. They attract considerable attention in solid-state physics mainly because of their outstanding luminescence properties.[1–3] Especially the two-dimensional sheet polymers such as siloxene $(Si_6O_3H_6)_n$ or polysilyne $(SiH)_n$ are promising candidates for technological application because of their higher mechanical stability and higher conductivity. Since $(SiH)_n$, accessible from $(SiBr)_n$ and $LiAlH_4$ or, more recently, from $CaSi_2$ and HCl,[4,5] is highly unstable and shows luminescence only in the UV part of the spectrum, siloxene with its strong room temperature photoluminescence in the green or yellow has been most thoroughly investigated. Research on siloxene has been further intensified by the recent suggestion that the efficient visible luminescence observed in porous silicon, a phenomenon of considerable current interest, is mainly due to the presence of siloxene species on the surface of the porous silicon particles.[6]

A prerequisite for an understanding of the luminescence properties of siloxene is to understand its solid-state structure. In the following section, the present knowledge about the structure of siloxene is briefly reviewed, and this is followed by a short summary of current studies aimed at rebuilding siloxene-like structures from well-defined molecular precursors to obtain structurally well defined sub-elements of the siloxene lattice. Particular attention will be devoted to the question as to whether the fluorescence properties of siloxene can be matched by selectively prepared siloxene sub-units or substituted cyclopolysilanes.

16.2 Synthesis and Structure of Siloxene

When $CaSi_2$ is reacted with HCl in a topochemical reaction, insoluble solid

---

* Address of the author: Institute of Inorganic Chemistry, Graz University of Technology, Stremayrgasse 16, A-8010 Graz, Austria

siloxene is obtained (Eq. 1).

$$3\ CaSi_2 + 6\ HCl + 3\ H_2O \rightarrow Si_6O_3H_6 + 3\ CaCl_2 + 3\ H_2 \qquad (1)$$

Two different experimental methods are available yielding siloxenes with different properties. The original Wöhler method,[7] later optimized by Hönle et al.,[5] affords a crystalline yellow product with green luminescence ($\lambda_{max,em}$ = 560 nm), while siloxene prepared according to Kautsky [8] is a grayish-green, X-ray amorphous solid showing blue luminescence ($\lambda_{max,em}$ = 500 nm). It was frequently argued that different structures might be responsible for this different behavior.

There are three commonly proposed structures for siloxene (Figure 16.1): (A) the planar modification, resembling the 111-double layers in crystalline silicon, consists of two-dimensional corrugated silicon layers with three Si-Si bonds per silicon and alternating OH or H substituents; (B) the ring structure consisting of Si_6 rings, which are linked by oxygen bridges to form planes, and (C) the chain structure containing Si chains interconnected by oxygen bridges.[9] In structures (B) and (C) the fourth valence of the silicon atoms is passivated by a hydrogen atom.

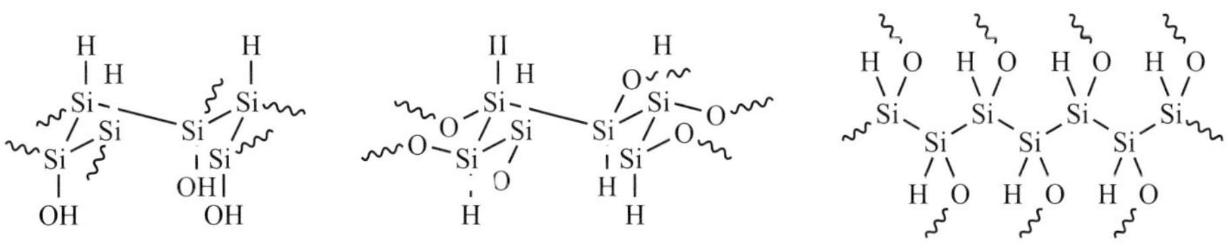

A B C

Figure 16.1. Structural models for siloxene.

Single crystals of Wöhler siloxene can be obtained either by growing epitaxial $CaSi_2$ on Si and transforming the layer to siloxene by treatment with HCl [2] or from $CaSi_2$ single crystals and HCl.[5, 10] As a consequence, it is now well established from X-ray diffraction studies that siloxene freshly prepared by the Wöhler method predominately consists of the planar modification (A).[4, 5, 10] This assignment is also supported by ^{29}Si NMR [11] and vibrational spectroscopy.[12] There is no experimental proof for the existence of the ring (B) or chain structure (C). However, it has been argued that siloxene may undergo structural transformations upon heat treatment[13] or further oxidation,[14] and that polymers with structures related to the modifications (B) and (C) might be generated due to the stepwise insertion of oxygen into the Si_6

rings, although experimental evidence for this interpretation is very limeted.

The exact structure of Kautsky siloxene has not been determined so far, mainly because of its X-ray amorphous character and the inhomogeneous composition of the material. X-ray emission and infrared spectra confirm that *Kautsky*-siloxene samples are mixtures of several modifications with a predominant contribution of the ring structure (B).[12, 15] An idealized structure according to model (B) is therefore not unlikely, as has already been proposed by Kautsky mainly on the basis of the substitution chemistry of the material.[16]

16.3 Siloxene-like Polymers from Molecular Precursors

One of the basic properties of siloxenes is their general insolubility in organic solvents, a fact that strongly impedes physical and structural characterization. As a result, the question arose as to whether structurally better defined siloxene-like polymers with improved solubility can be assembled in a stepwise manner starting from appropriate molecular precursors, and whether the properties of siloxene, such as the intense photoluminescence, can be matched. We thus attempted to rebuild the proposed structure of *Kautsky*-siloxene by the controlled hydrolysis of cyclic or linear oligosilanes bearing hydrolytically labile substituents followed by the thermal condensation to polymeric siloxanes.[17] The general route is outlined in Scheme 16.1.

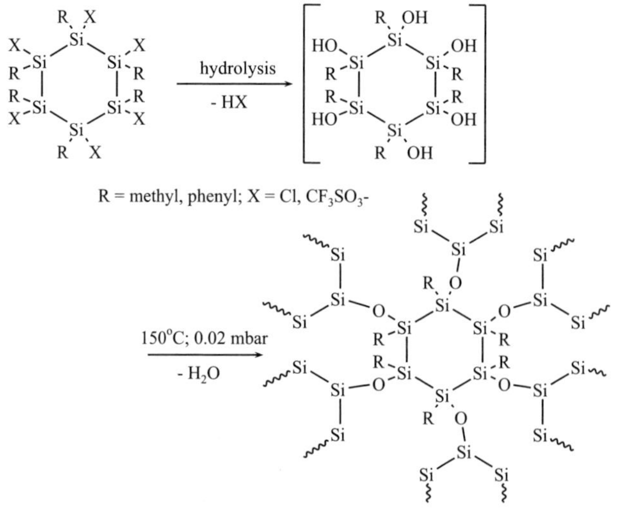

hydrolysis

- HX

R = methyl, phenyl; X = Cl, CF_3SO_3-

150°C; 0.02 mbar

- H_2O

Scheme 16.1. General reaction scheme for the synthesis of polymers with siloxene-like structures.

The resulting polymers exhibit properties closely resembling those of siloxene. The materials are insoluble in organic solvents, their IR spectra exhibit characteristic bands near 3400 cm^{-1} [ν (O-H)] and 1050 cm^{-1} [ν_{as} (Si-O-Si)], and in some cases intense photoluminescence is observed. Depending on the starting materials the fluorescence maxima of the obtained products range from 400 to 550 nm. The fluorescence spectra of selected examples are shown in Figure 16.2.

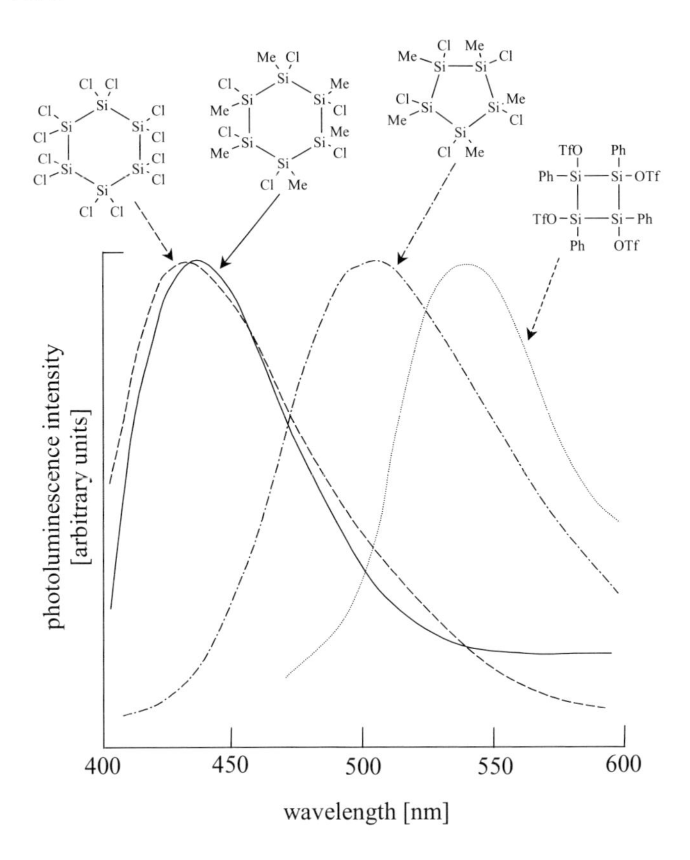

Figure 16.2. Normalized fluorescence spectra of selected siloxene-like polymers (the molecular precursors used for polymer preparation are indicated above the spectra).

Criteria for fluorescence are the presence of cyclosilanyl sub-units and polymeric multidimensional structures,[18] as clearly indicated by the diminishing fluorescence on going from polymer **1** to **4** in Scheme 16.2. While the precursor of polymer **1** has six Si-Cl functionalities per cyclohexasilanyl unit, which is likely to result in the formation of highly cross-linked materials upon hydrolysis and condensation, the degree of cross-linking in polymer **2** is strongly reduced, because only three hydrolytically labile substituents per precursor molecule are present. Polymer **3** most probably has a linear structure, and polymer **4**, derived from an open-chained precursor, does not exhibit fluorescence at all.

Since only polymers derived from cyclic starting materials, which are likely to have siloxene-like structures, exhibit color and fluorescence, the polysilane ring seems to be essential for the exceptional optical properties. This is in agreement with the original idea assuming the cyclosilane ring to be the chromophore responsible for the photoluminescence of siloxenes.

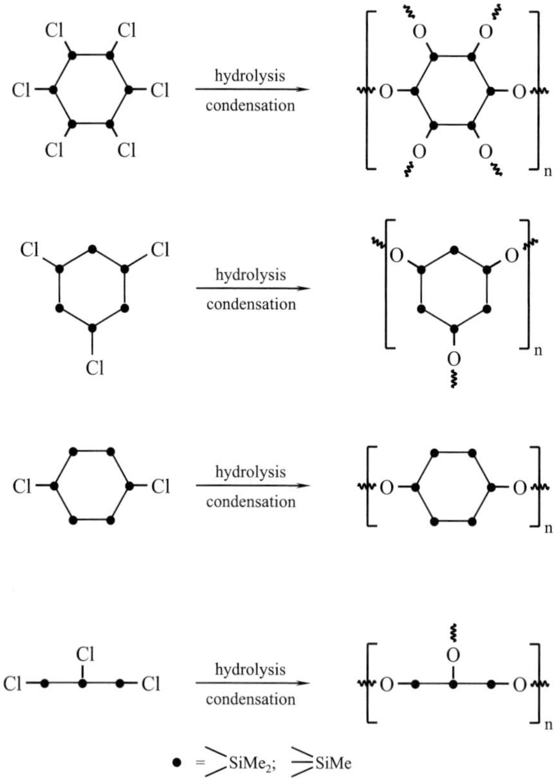

hydrolysis / condensation

Polymer **1**: yellow solid, highly fluorescent, $\lambda_{max,\ em}$ = 436 nm

Polymer **2**: white solid, weak fluorescence, $\lambda_{max,\ em}$ = 420 nm

Polymer **3**: white solid, non-fluorescent

Polymer **4**: white solid, non-fluorescent

• = $SiMe_2$; SiMe

Scheme 16.2. Possible structures and properties of hydrolysis products derived from chloropermethylpolysilanes of different constitutions.†

† In the following the common notation will be used in which a dot represents a Si atom with methyl groups attached to bring the total coordination number to four.

16.4 Synthesis and Fluorescence of Molecular Models

If the luminescence of siloxenes is a molecular rather than a solid-state property, as has been argued in an earlier study,[12] smaller sub-units of siloxene such as cyclohexasilane derivatives of the general formula $Si_6Me_n(OSiR_3)_{6-n}$ bearing siloxy groups attached to the polysilane ring should represent suitable models. While numerous cyclosilanes bearing organic side groups have been prepared in the past, synthetic routes to functional cyclopolysilanes are limited to a very few examples so far. The following section covers only selected aspects of cyclohexasilane chemistry relevant to the scope of this article. For a more general overview the interested reader is referred to several reviews on cyclopolysilanes.[19]

16.4.1 Synthesis of Methylsiloxycyclohexasilanes $Si_6Me_n(OSiR_3)_{6-n}$

Si_6Me_{12} can be conveniently prepared from Me_2SiCl_2 with sodium/potassium alloy in THF in the presence of an equilibrating catalyst, which causes depolymerization of the initially formed permethylpolysilane (Eq. 2).[20]

$$Me_2SiCl_2 \xrightarrow{\text{Na/K, THF}} \underset{>80\%}{(Me_2Si)_6} + \underset{<10\%}{(Me_2Si)_5} + \underset{\text{traces}}{(Me_2Si)_7} \qquad (2)$$

The methyl groups in Si_6Me_{12} can be substituted by halogen without destruction of the ring structure. With $SbCl_5$, either one, two or three of the methyl groups can be replaced by chlorine depending on the stoichiometric ratio of the reactants.[21] In the case of $Si_6Me_{10}Cl_2$, the 1,3- and 1,4-isomers are produced exclusively (Scheme 15.3).

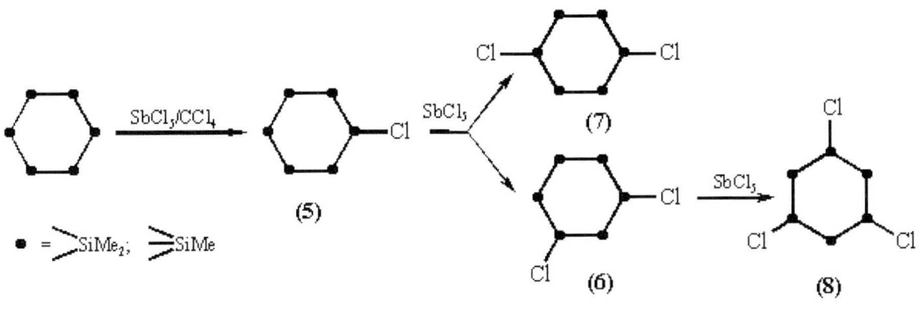

Scheme 3. Partial chlorination of Si_6Me_{12}.

Separation of **6** and **7** can easily be accomplished by hydrolysis of the isomeric mixture, distillation of the hydrolysis products 1,3-dihydroxydecamethylcyclohexasilane and decamethyl-7-oxa-1,2,3,4,5,6-hexasilanorbornane, followed by rechlorination of the Si-O bonds with acetyl chloride.[22] 1,2-Dichlorodecamethylcyclohexasilane can be synthesized systematically from 1,2-diphenyltetramethyldisilane by several Si-Si bond-formation and selective substituent exchange steps.[23] 1,2,3,4,5,6-Hexachlorohexamethylcyclohexasilane (**9**) is prepared as shown in Eq. 3.[24]

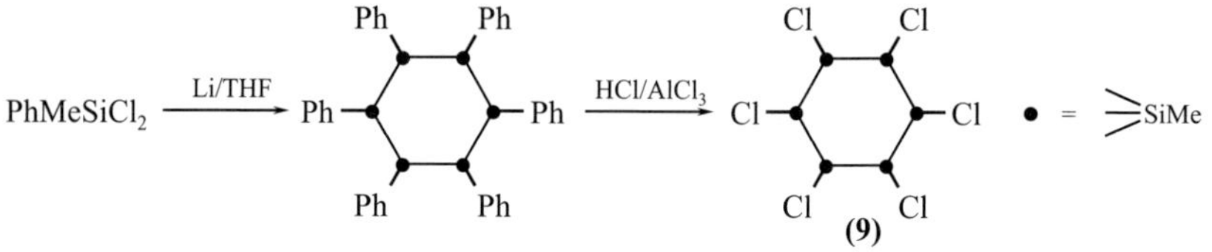

$$PhMeSiCl_2 \xrightarrow{Li/THF} Si_6Me_6Ph_6 \xrightarrow{HCl/AlCl_3} Si_6Me_6Cl_6\ \textbf{(9)} \qquad \bullet = \text{SiMe} \quad (3)$$

Reactions of the partially chlorinated cyclohexasilanes **5–8** with various nucleophiles such as H^-, $[(CO)_2CpFe]^-$, RS^- or R^-, afford the corresponding substitution products.[25] The siloxy-substituted cyclohexasilanes **10–13** can be prepared in a similar way by reacting **5–8** with lithium silanolates according to Scheme 15.4.[26]

The di- and trisubstituted derivatives **11–13** were obtained as statistical mixtures of *cis/trans* isomers. Isolation of the least soluble isomer was achieved by crystallization. Single-crystal X-ray diffraction studies of **11** and **12** [27] reveal that the cyclohexasilane ring adapts a slightly distorted chair conformation with the bulky $OSiMe_2{}^tBu$ groups in equatorial sites in order to minimize non-bonding interactions (Figure 16.3).[28]

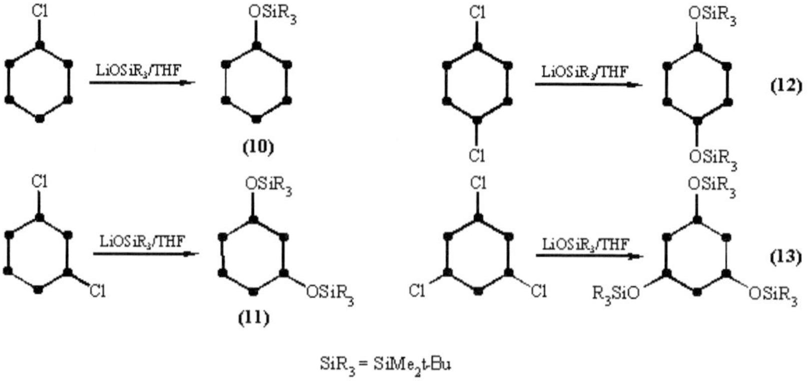

LiOSiR₃/THF → (10), (11), (12), (13)

$SiR_3 = SiMe_2t\text{-}Bu$

Scheme 16.4. Synthesis of siloxy-substituted cyclohexasilanes.

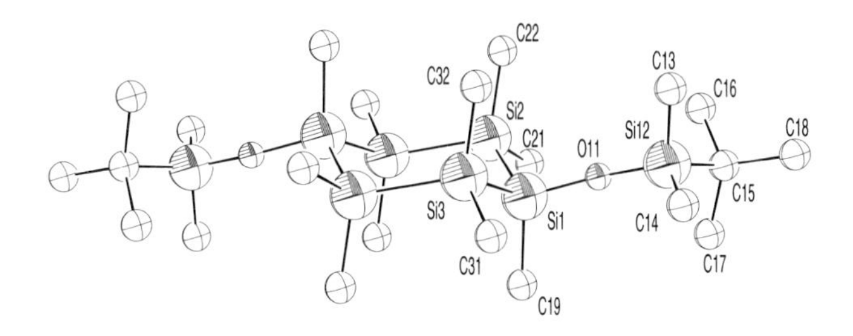

Figure 16.3. Crystal structure of compound **12**.

16.4.2 Absorption and Photoluminescence Spectra of Methylsiloxycyclohexasilanes

UV absorption and fluorescence emission spectra of compounds **10–13** are shown in Figure 15.4. All spectra show relatively weak absorption bands at the low energy side, which are not present in the spectrum of Si_6Me_{12} and exhibit pronounced bathochromic shifts as the number of siloxy groups increases.

When isomerically pure samples of **10–13** are excited at the wavelength of the first UV maximum around 290 nm, distinct photoluminescence is observed. Several overlapping fluorescence bands appear at around 340 nm, the positions of which are not influenced markedly by the number of the siloxy groups attached to the cyclohexasilane ring (see Figure 16.4).

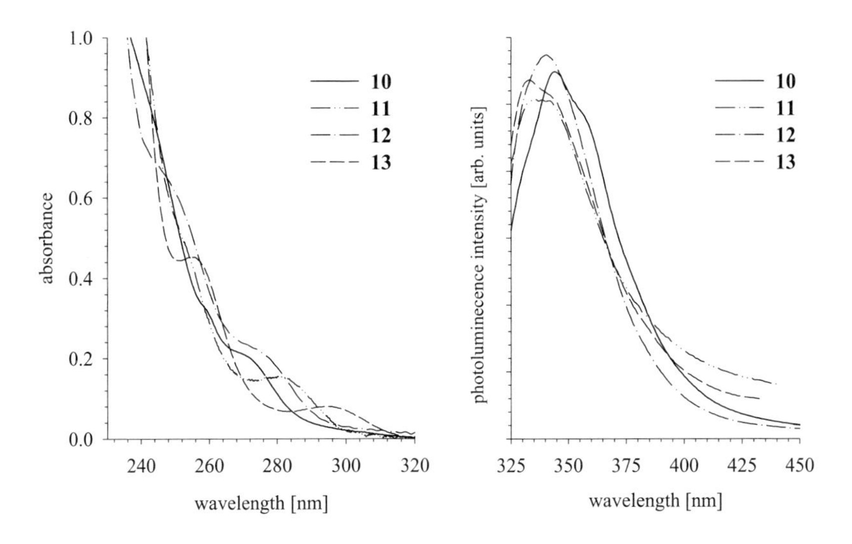

Figure 16.4. Absorption and photoemission spectra of **10–13** in cyclohexane ($c = 5 \cdot 10^{-4}$ mol/L, $\lambda_{ex} = 288$nm).

Although the trends appearing in the luminescence spectra of **5**–**8** and related compounds cannot yet be interpreted straightforwardly, the fluorescence intensity of siloxy-substituted cyclohexasilanes turns out to be considerably higher compared to that of the corresponding phenyl derivatives, while the fluorescence of Si_6Me_{12} is rather weak (Figure 16.5). Intense photoluminescence, therefore, can be traced back to the smallest subunit of *Kautsky* siloxene, that is the cyclohexasilane ring linked to other silyl groups by oxygen bridges.

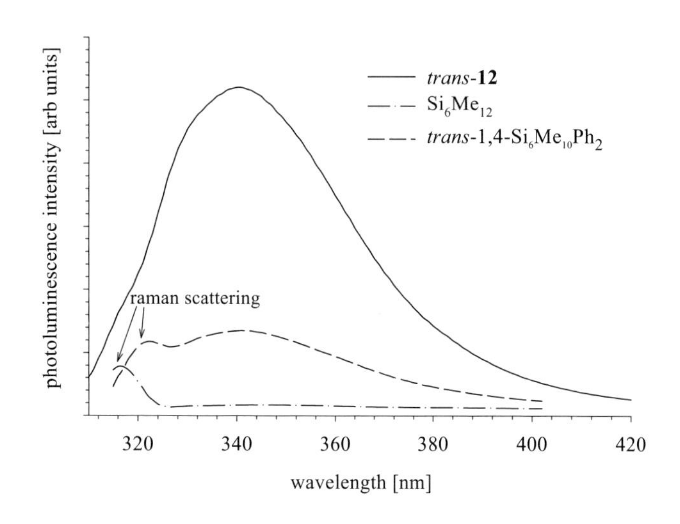

Figure 16.5. Normalized photoemission spectra of *trans*-**12**, Si_6Me_{12} ($c = 5 \cdot 10^{-4}$ mol/L, $\lambda_{ex} = 288$nm) and *trans*-1,4-$Si_6Me_{10}Ph_2$ ($c = 1 \cdot 10^{-3}$ mol/L; $\lambda_{ex} = 291$nm)

16.4.3 Studies Concerning Methylaminocyclohexasilanes $Si_6Me_n(NR_2)_{6-n}$

In order to clarify whether the presence of oxygen is a prerequisite for the pronounced fluorescence of the cyclopolysilane derivatives **10**–**13**, the photoluminescence behavior of aminocyclohexasilanes $Si_6Me_{6-n}(NR_2)_n$ was also studied. Aminosilanes are most commonly prepared from chlorosilanes and amines or alkali metal amides.[28] When chloropermethylcyclohexasilanes are reacted with NH_3, the corresponding amino derivatives are obtained. As shown in Scheme 15.5, the reaction of **5** with ammonia cleanly affords $Si_6Me_{11}(NH_2)$ (**14**), which can be further silylated yielding the silazanes **15** and **16** after lithiation with *n*-BuLi. **16** is also easily accessible directly from **14** by thermal condensation. The air-stable compound **17** is obtained in good yield from the reaction of **10** with $LiN(SiMe_3)_2$.

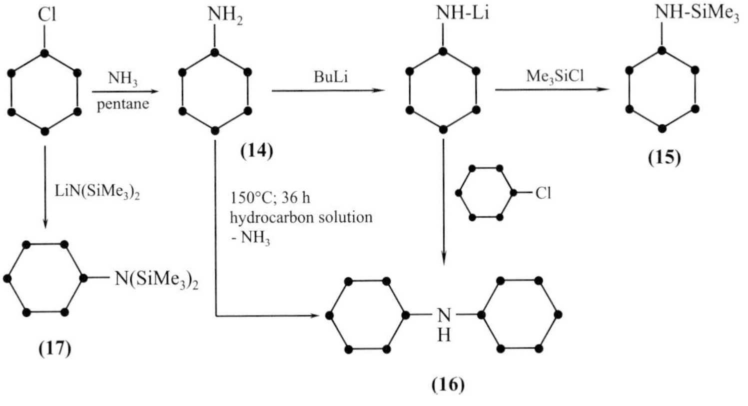

Scheme 16.5. Synthesis of aminocyclohexasilanes **14**–**17**.

The crystal structure of **14** again exhibits the cyclohexasilanyl ring in a chair conformation. Unlike the bulky hetero substituents in the structures of **6** and **7**, however, the small NH_2 group occupies an axial position with an unusually long Si-N bond distance of 183.4 pm. This extraordinarily long Si-N bond length implies a relatively weak Si-N bond, what is manifested by the sensitivity of **14** towards thermal condensation.

1,4-Diaminodecamethylcyclohexasilane **18**, which can be prepared from **7** and Na/NH_3 according to Scheme 16.6, undergoes inter- and intramolecular condensation even at room temperature when stored for several days. As a consequence, silazane polymers with sheet-like structures similar to the polymers **1** and **2** depicted in Scheme 16.2 might be accessible from **8** or **9** under very mild and controlled conditions.

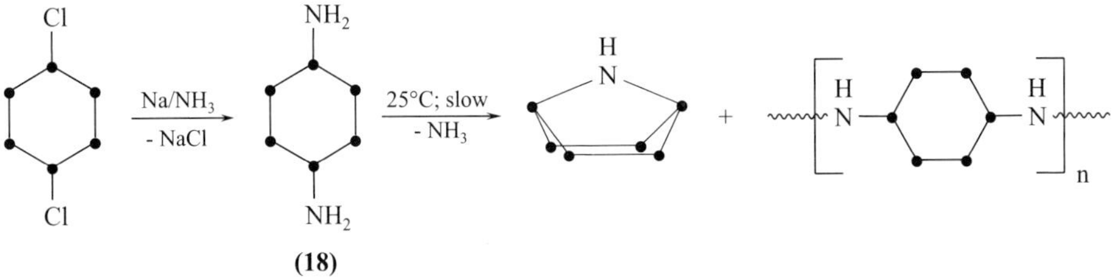

Scheme 16.6. Synthesis and condensation of 1,4- $Si_6Me_{10}(NH_2)_2$.

Remarkably, the UV absorption and emission spectra of **15**–**17** do not show any photoluminescence above 300 nm. The absorption spectra of **16** and **17** also lack the weak, low-energy absorption band near 270 nm present in the absorption spectrum of the corresponding siloxy derivative **10**.

16.5 Summary

It has been demonstrated that polymeric materials with siloxene-like properties, including intense photoluminescence, can be synthesized by the hydrolysis and thermal condensation of partially functionalized oligosilanes bearing hydrolytically labile substituents. However, only polymers obtained from cyclic starting materials exhibit color and fluorescence. Well-defined monomeric model substances $Si_6R_{6-n}(OSiR_3)_n$ bearing siloxy groups on the cyclohexasilane ring were also found to exhibit significantly increased luminescence intensities, while the corresponding aminocyclohexasilane $Si_6R_{6-n}(NR_2)_n$ do not show any photoluminescence at all. Cyclosilanyl subunits in the polymer lattice and the presence of oxygen-containing substituents are therefore the essential criteria for the observed photoluminescence behavior of the molecular and polymeric materials investigated in this study.

16.6 References

[1] R. D. Miller, M. Baier, A. F. Diaz, E. J. Ginsburg, G. M. Wallraff, *Pure Appl. Chem.* **1992,** *14*, 1291.
[2] M. S. Brandt, T. Puchert, M. Stutzmann in *Tailor-Made Silicon-Oxygen Compounds,* (Eds.: R. Corriu, P. Jutzi), Vieweg, Wiesbaden, 1994, p. 117.
[3] M. Stutzmann, J. Weber, M. S. Brandt, H. D. Fuchs, M. Rosenbauer, P. Deak, A. Höpner, A. Breitschwerdt, *Adv. Solid State Phys.* **1992**, *32,* 179.
[4] J. R. Dahn, B. M. Way, E. Fuller, *Phys. Rev. B* **1993**, *48*, 17872.
[5] U. Dettlaf-Weglikowska, W. Hönle, A. Molassioti-Dohms, S. Finkbeiner, J. Weber, *Phys. Rev. B: Condens. Mater.* **1997**, *56*, 13132.
[6] H. D. Fuchs, M. Stutzmann, M. S. Brandt, M. Rosenbauer, J. Weber, A. Breitschwerdt, P. Deak, M. Cardona, *Phys. Rev. B* **1993**, *48*, 8172.
[7] F. Wöhler, *Liebigs Ann. Chem.* **1863**, *127*, 257.
[8] H. Kautsky, G. Herzberg, *Z. Anorg. Allg. Chem.* **1924**, *139*, 135.
[9] H. Kautsky, W. Vogell, F. Oeters, *Z. Naturforsch. B* **1955**, *10*, 597.
[10] A. Weiss, G. Beil, H. Meyer, *Z Naturforsch. B* **1979**, *34*, 25.
[11] M. S. Brandt, S. E. Ready, J. B. Boyce, *Appl. Phys. Lett.* **1997**, *70*, 188.
[12] H. D. Fuchs, M. Stutzmann, M. S. Brandt, M. Rosenbauer, J. Weber, A. Breitschwerdt, P. Deak, M. Cardona, *Phys. Rev. B* **1993**, *48,* 8172.
[13] P. Deak, M. Rosenbauer, M. Stutzmann, J. Weber, M. S. Brandt, *Phys. Rev. Lett.* **1992**, *69*, 2531.

[14] W. Hönle, U. Dettlaf-Weglikowska, S. Finkbeiner, A. Molassioti-Dohms in *Tailor-Made Silicon-Oxygen Compounds,* (Eds.: R. Corriu, P. Jutzi), Vieweg, Wiesbaden, 1994, p. 99.

[15] E. Z. Kurmaev, S. N. Shamin, D. L. Ederer, U. Dettlaf-Weglikowska, J. Weber, *J. Mater. Res.* **1999**, *14*, 1235.

[16] Reviews on the chemical properties of siloxene: E. Hengge: *Fortsch. Chem. Forsch.* **1967**, *9*, 145. E. Hengge, *Top. Current Chem.* **1974**, *51*, 1. *Gmelin Handbook of Inorganic Chemistry*, *Silicon* Vol. B1, Springer, Berlin, 1982, p. 252f.

[17] A. Kleewein, H. Stüger, *Monatsh. Chem.* **1999**, *130*, 69

[18] A. Kleewein, H. Stüger, S. Tasch, G. Leising in *Organosilicon Chemistry 4,* (Eds.: N. Auner, J. Weis), Verlag Chemie, Weinheim, 2000, p. 389.

[19] R. West in *Compreh. Organomet. Chem.* Vol. 2, (Eds.: G. Wilkinson, F. G. A. Stone, E. W. Abel), Pergamon Press, Oxford, 1982, p. 365. E. Hengge, K. Hassler in *The Chemistry of Inorganic Homo- and Heterocycles*, Vol. 1, (Eds.: I. Haiduc, D. B. Sowerby), Academic Press, London, 1987, p. 191. R. West, *Pure Appl. Chem.* **1982**, *54*, 1041. E. Hengge, R. Janoschek, *Chem. Rev.* **1995**, *95,* 1495. E. Hengge, H. Stüger in *The Chemistry of Organosilicon Compounds,* Vol. 2, (Eds.: S. Patai, Z. Rappoport), Wiley, 1998, p. 2177.

[20] L. F. Brough, R. West, *J. Organomet. Chem.* **1980**, *145*, 139; E. Carberry, R. West, *J. Am. Chem. Soc.* **1969**, *91*, 5440.

[21] E. Hengge, M. Eibl, *J. Organomet. Chem.* **1992**, *428*, 335; M. Eibl, U. Katzenbeisser, E. Hengge, *J. Organomet. Chem.* **1993**, *444*, 29.

[22] A. Spielberger, P. Gspaltl, H. Siegl, E. Hengge, K. Gruber, *J. Organomet. Chem.* **1995**, *499*, 241.

[23] W. Uhlig, *J. Organomet. Chem.* **1993**, *452*, C6.

[24] E. Hengge, W. Kalchauer, F. Schrank, *Monatsh. Chem.* **1986**, *117*, 1399; S. Chen, L. David, K. Haller, C. Wadsworth, R. West, *Organometallics* **1983**, *2*, 409.

[25] E. Hengge, M. Eibl, *Organometallics* **1991**, *10*, 3185; F. Uhlig, B. Stadelmann, A. Zechmann, P. Lassacher, H. Stueger, E. Hengge, *Phosphorus, Sulfur, Silicon Relat. Elem.* **1994**, *90*, 29.

[26] K. Renger, A. Kleewein, H. Stueger, *Phosphorus, Sulfur, Silicon Relat. Elem.* **2001**, *168*, 449.

[27] K. Renger, *Dissertation*, Technische Universität Graz, 2001.

[28] D. A. Armitage in *The Silicon–Heteroatom Bond,* (Eds.: S. Patai, Z. Rappoport), Wiley, Chichester, 1991, p.365.

17 Silicon-based Nanotubes: A Theoretical Investigation

Th. Köhler[#], G. Seifert[*], Th. Frauenheim[#]

17.1 Introduction

Since the discovery of fullerenes by Kroto [1] in 1985, following studies of the role of carbon within interstellar dust, and their successful synthesis by Kraetschmer,[2] nanotechnology has progressed rapidly to become a major field of innovative technical applications. Fullerenes are closed-cage carbon molecules built-up of pentagonal and hexagonal rings, the most prominent example being the "soccer-ball" molecule C_{60}. A few years later in 1991, Iijima [3] discovered a new fullerene-related form of carbon, carbon nanotubules, by analyzing material deposited on the cathode during arc-evaporation of graphite for the synthesis of fullerenes. The first structures reported by Iijima were invariably multi-walled nanotubes (MWNT), comprising a central tubule of nanometric diameter surrounded by graphene layers separated by about 3.4 Å. It took less than two years until single-walled nanotubes (SWNT) became available from another route of production. The addition of metal catalysts, such as Fe, Co, or Ni to the dc arc process under He atmosphere yields extremely fine and uniform single-walled tubes.[4] The pioneering work of E. Smalley's group has led to a route, involving the condensation of a laser-evaporated carbon cobalt nickel mixture at 1200 °C. Using this technique, the diameter of the nanotubes can be influenced by the process parameters, the tubes being self-assembled into ropes of a two-dimensional triangular lattice forming bundles of a few hundred of tubes. This was the breakthrough for producing fibers consisting of CNTs. In the latter, the carbon nanotubes exhibit an extraordinarily high Young's modulus of about 1.2 TPa, which indicates an axial stiffness comparable to that of diamond, thus favoring the use of CNTs for tailoring novel robust materials.

Electronically, the carbon nanotubes can have either metallic or semiconducting properties depending upon their chirality and diameter. The electronic gap is also controllable within a range of ~0–1 eV. This gives rise to possible metal–semiconductor or semiconductor–semiconductor junctions for use in nanoelectronic devices. The possibility of integrating carbon nanotubes into logic circuits was demonstrated in 2001.[5] Another property of CNTs, the ability of electron field emission, has reached technological relevance.

---

* Address of the authors: Technische Universität Dresden, Physikalische Chemie, D-01062 Dresden, Germany

Universität Paderborn, Theoretische Physik, Warburger Str. 100, D-33098 Paderborn, Germany

Together with the production of large macroscopically aligned and defect-reduced nanotube bundles, the first light-emitting devices have been successfully produced.[6] The potential for display application is enhanced by the high packing density of tubes in aligned bundles and the resulting high emission current density. A further improvement could be achieved by taking advantage of the recently discovered nanohorns.[7] A tip instead of a round, spherical cap at the end of the tube will increase the emission considerably.
Based on the analogy between chemically equivalent layered structures, it might even be possible to create phosphorus (P) NTs [8] or, as one goal of this work, to propose stable silicon-based tubular structures. The search for possible technical applications of each tubular system, or a combination of them, is a wide active field of research in nanotube science which is greatly extending the science of the fundamental tube – the carbon nanotube.

For all further considerations of nanotube properties, some common concepts are discussed in this chapter. This discussion is based on the description given in refs. [9] and [10]. For a further and deeper understanding, we refer to these references and furthermore to a review by Rao et al. [11] and two monographs.[10,12]

17.1.1 Stuctural Classification of Carbon Nanotubes

Many of the structural and physical properties of single-walled nanotubes can easily be understood based on a the picture of a strip of a two-dimensional graphite sheet rolled into a cylindrical form. Adapted to this symmetry, we will use the real space unit vectors $\mathbf{a}_1$, $\mathbf{a}_2$ of a hexagonal lattice in x,y coordinates:

$$\mathbf{a}_1 = \left(\frac{\sqrt{3}}{2}, \frac{1}{2}\right)a \quad , \quad \mathbf{a}_2 = \left(\frac{\sqrt{3}}{2}, -\frac{1}{2}\right)a \ ,$$

where $a = |\mathbf{a}_1| = |\mathbf{a}_2| = a_{C\text{-}C} \times \sqrt{3} = 2.46$ Å is the in-plane lattice constant of graphite, while $a_{C\text{-}C}$ denotes the bond length between sp^2-bonded carbon atoms in order to define the basis vectors of a nanotube.

Two orthogonal vectors, see Figure 17.1, are necessary to describe the nanotube unit cell completely, the so-called "chirality" vector $\mathbf{C_h}$ and the "translational" vector $\mathbf{T}$ with the inner scalar product $\mathbf{C_h}*\mathbf{T} = 0$ (*). The very common (n,m)-notation for characterizing the nanotubes will be introduced by expressing $\mathbf{C_h}$ in terms of $\mathbf{a}_1$, $\mathbf{a}_2$

$$\mathbf{C_h} = n\,\mathbf{a}_1 + m\,\mathbf{a}_2 \equiv: (n, m) \quad (n, m \text{ integer}; 0 \leq |m| \leq n).$$

Analogously, the translational vector **T** will be expressed as:

$$\mathbf{T} = t_1\,\mathbf{a}_1 + t_2\,\mathbf{a}_2 \qquad \gcd(t_1, t_2) = 1\,, \quad t_1 = \frac{2m+n}{d_R}\,, \quad t_2 = -\frac{2n+m}{d_R}\,,$$

where gcd() denotes the largest common divisor; $d_R = \gcd(2n+m, 2m+n)$, while the integer factors t_1, t_2 guarantee that condition (*) is met. Another often used quantity is the chiral angle θ, defined by the relationship

$$\cos\theta = (\mathbf{C}_\mathbf{h} * \mathbf{a}_1) / (|\mathbf{C}_\mathbf{h}||\mathbf{a}_1|) = \frac{2n+m}{2\sqrt{(n^2+m^2+mn)}}\,, \quad 0 < \theta \le \frac{\pi}{6}.$$

During the historical development of nanotube research, special notations for the limiting cases have been favored. Thus, we distinguish between so-called “zig-zag” tubes (n,0) $\theta = 0$ and “armchair” tubes (n,n) $\theta = 30°$. All other nanotubes are described as “chiral”, (n,m) $n \neq m$.
Within this framework, a lot of geometrical properties can easily be written in terms of n and m. The diameter D of the tube is given by the circumferential length $L = |\mathbf{C}_\mathbf{h}|$ simply as

$$D = L/\pi = a\,\sqrt{(n^2+m^2+mn)}\,/\pi$$

and the length of the translational vector $T = |\mathbf{T}|$ becomes

$$T = \sqrt{3}\,L/d_R\,.$$

To obtain further the number N of hexagons per unit cell of a nanotube we have to divide the unit cell area $|\mathbf{C}_\mathbf{h} \times \mathbf{T}|$ by the area of one hexagon of our two-dimensional graphite sheet $|\mathbf{a}_1 \times \mathbf{a}_2|$, which gives

$$N = \frac{2(m^2+n^2+nm)}{d_R}.$$

Thus, there are 2N atoms within the unit cell.

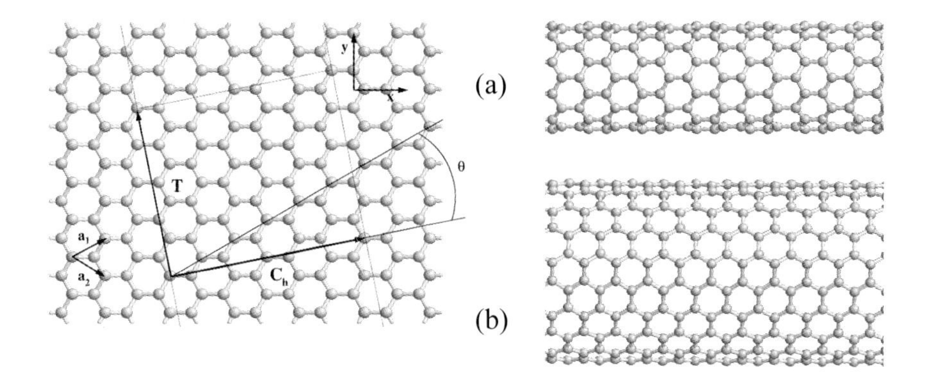

Figure17.1. Illustration of the basic unit vector system defining the nanotube. The limiting cases of nanotubes are shown on the right-hand side, (n,0) "zig-zag" (a) and (n,n) "armchair" (b), respectively.

17.1.2 Computational Methodology

A self-consistent charge density functional based tight binding (SCC-DFTB) method [13, 14] was used for the calculations of the geometries, electronic structure, and the molecular dynamics simulations. This method is based on the expansion of the Kohn–Sham density functional $E[\rho]$ in the local density approximation (LDA) to second order over the density fluctuations $\delta\rho$ around a suitable electron density, ρ_0:

$$E[\rho_0 + \delta\rho] = \sum_i^{occ} n_i \varepsilon_i^0 + E_{rep}[\rho_0] + E^{(2)}[\rho_0, \delta\rho]$$

The first term, called "band-structure" energy E_{BS}, represents the sum over occupied one-electron eigenstates of the system at the reference density ρ_0. The one- and two-center matrix elements of the associated Hamiltonian are calculated (rather than fitted) within LDA in an explicit minimal valence basis for neutral atoms embedded in a confinement potential. The second term, E_{rep}, is a repulsive energy, which includes double counting in Coulomb and exchange-correlation contributions to the reference Hamiltonian, as well as the internuclear repulsion. This term is approximated by universal pair potentials, which are fitted to the difference between the sum of E_{BS} and $E^{(2)}$ energy curves and potential energy curves from self-consistent field LDA calculations of dimers and crystals. Lastly, the third term represents second-order corrections to the Hartree and exchange-correlation energies due to the density fluctuations. This term is approximated through net atomic Mulliken charges, Δq, as

$$E^{(2)} = \frac{1}{2} \sum_{\alpha,\beta} \gamma_{\alpha\beta}(U_\alpha, U_\beta, R_{\alpha\beta}) \Delta q_\alpha \Delta q_\beta$$

where the summation extends over all atom pairs in the system. $R_{\alpha\beta}$ stands for interatomic distance, U_α is the chemical hardness, determined parameter-free within LDA, and $\gamma_{\alpha\beta}$ describes the Madelung-like distance–dependence. The variation of the density functional with respect to the Kohn–Sham orbital expansion coefficients results in a set of secular equations that are solved self-consistently for the Mulliken charges.

Interatomic forces for molecular dynamics can be calculated from analytical expressions for the gradients of the total energy at the atomic positions.[15] All integral tables and repulsive parameters have been determined for carbon previously.[13] For a review of the DFTB methodology and its application, see ref. [14].

17.2 Silicon-based Nanotubes – Structure and Properties

Presently, there is growing interest in one-dimensional (1D) silicon structures as possible elements of nanoelectronic devices. However, there are as yet no efficient ways for synthesizing such 1D Si systems. This is certainly partly due to deficiencies in the understanding of their structural, energetic, and electronic properties. Recently, two popular models describing Si nanowires as either bulk-like fourfold-coordinated cores surrounded by a possibly reconstructed surface with a large fraction of threefold coordinated atoms [16] or as hollow fullerene-like structures,[17] have been discussed in the literature. The first 1D Si structures synthesized have been clearly characterized as nanowires consisting of a bulk-like core while the surface is probably coated with SiO_x.[18] Thus, in contrast to CNTs the corresponding 1D Si structures are not hollow. This behavior can easily be explained in terms of the inability of Si to form strong π-bonds and thus to favor stabilization of threefold (sp^2) coordination, this being the basis of the stability of 2D graphite, 1D CNTs or fullerene-like structures of carbon. The overlap of the p-π carbon valence orbitals is about ten times greater than that for silicon at corresponding equilibrium distances, i.e. any significant π bonding contribution in silicon structures can be excluded. Therefore, any report about the hypothetical existence of Si nanotubes seems to be doubtful.[19]

17.2.1 Silicide and Silane Nanotubes

However, as a starting point for the creation of new silicon–based tubular systems, there are indeed layered silicon systems among the inorganic crystalline modifications of the alkaline earth metal silicides. These consist of silicon layers separated by the alkaline earth metal ions.[20] In one of the most prominent representatives, $CaSi_2$, each silicon atom is threefold coordinated by other silicons, as described in detail in the contribution of M. S. Brandt et al. (Chapter 15 herein). Formally, the silicon atoms are negatively charged in these silicides, Si^-. Thus, silicon becomes isoelectronic with phosphorus, and phosphorus forms a layer structure too, known as black phosphorus (b-P). As in b-P the silicon layers in silicides are not ideally planar – like in graphene –, but are puckered. On the other hand, the structure greatly resembles a graphene layer. From recent calculations [21] we know that nanotubes of phosphorus corresponding to the b-P layer structure are indeed stable, and Bi tubes have recently been synthesized.[22] Based on the analogy between silicide and b-P layers, it might become possible to form new tubular modifications of silicon. Furthermore, using silicides for the synthesis of polymeric silane SiH, also showing the puckered layer structure (Figure 17.2), would expand the number of candidates possible for Si-based nanotube design.

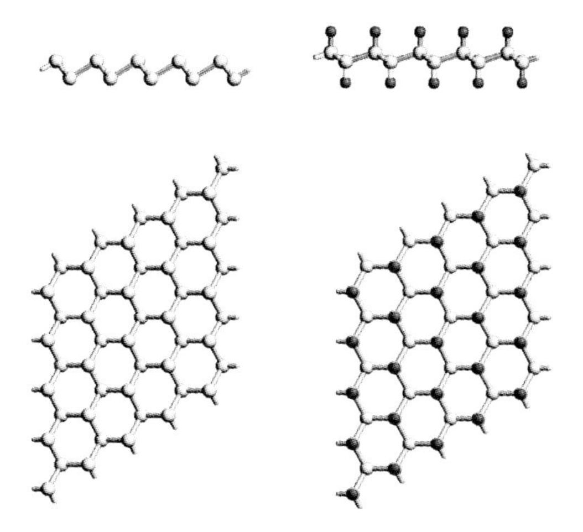

Figure 17.2. The structures (side view, upper figure; top view, below) of a silicide (left) and silane (right), as predicted by DFTB calculations. From the side view, the puckered structure of the layer is clearly visible. (Taken from ref. [24].)

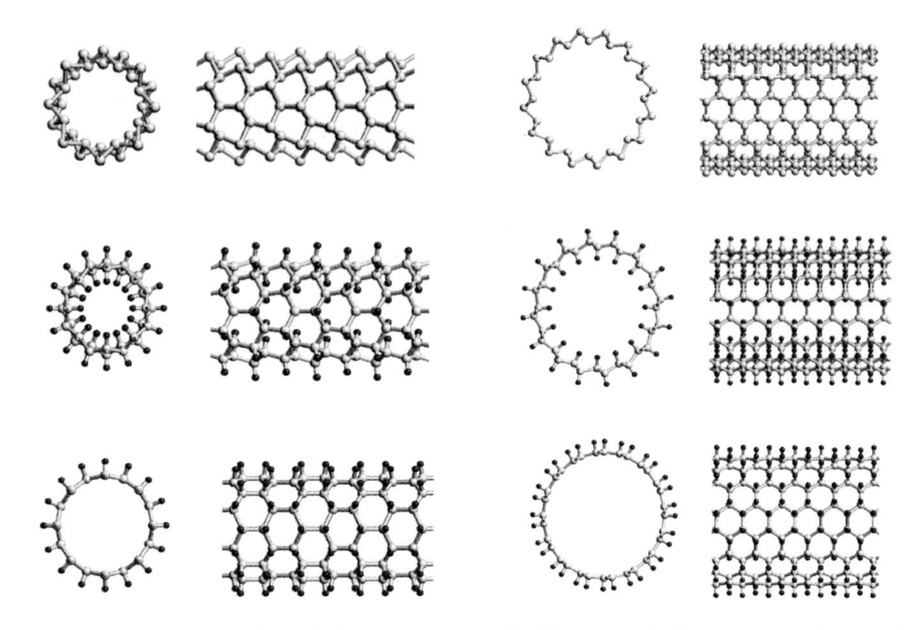

Figure 17.3. Examples of the structures of silicide and silane nanotubes with (8,0) (left panel) and (8,8) (right panel) chirality. From top to bottom are shown pure silicide (Si^-) and the silanes (SiH-io) and (SiH-sf), in each case in top and side views (taken from ref. [24]).

Experimentally, it is known [20] that $CaSi_2$ consists of hexagonal puckered layers with an Si-Si bond length of 2.41 Å and an Si-Si-Si bond angle of 104.8°. Nearly the same values are obtained for the P-P bond distance and for the P-P-P angle in the b-P system, confirming the structural similarity of the two systems. The puckered Si layers are separated by planar layers of Ca with a Ca-Si distance of 3.82 Å. Since we are mainly interested in the layer properties we assumed a complete ionic model concerning the Ca-Si interaction and considered the Si layer only with one additional electron per Si atom in correspondence to Ca^{2+} and Si^-. The positive countercharge was treated as a homogeneously distributed background charge. As a result of our DFTB calculations, we predict for silicides a puckered layer structure as a stable configuration with bond lengths of 2.49 Å and bond angles of 96°. This is in reasonably good agreement with the experimental values for $CaSi_2$ quoted above. A similar stable puckered configuration was determined for silane (SiH), with an Si-Si bond length of about 2.34 Å and an Si-H distance of 1.51 Å, these being very close to the bond lengths in disilane (Si_2H_6).

In order to study the energetic viability of the corresponding Si^- and SiH nanotubes, as well as to determine their electronic and mechanical properties, a series of calculations was performed, in which initial-guess tubular structures were fully relaxed with respect to atomic positions and tube cell length. An initial configuration of the nanotube was constructed by folding a flat 2D graphene-like sheet of Si with an Si-Si bond distance of 2.49 Å. We started from a flat sheet in which the high symmetry of the hexagonal structures was disrupted by small random displacements of atoms so as not to bias the structure of the nanotube and to accelerate convergence of the conjugate-gradient technique used. For each nanotube thus constructed, a set of structural relaxation calculations had to be performed, each one imposing a different axial stress on the tube, with the aim of finding the atomic configuration and lattice parameter of minimum energy. We considered both limiting cases of nanotubes: zigzag nanotubes (n,0) and armchair tubes (n,n) with $n \in \{6,10\}$.[23] Figure 17.3 illustrates the minimum–energy structures of silicide (Si^-) and silane (SiH) nanotubes for the case n = 8.

The interesting structures can be understood in terms of conventional graphitic CNTs by replacing the flat hexagons by puckered rings similar to the silicide and silane layer structure.

The two chosen examples are structural representatives for all tubes studied within the considered series. The tubes only differ in their diameter D and in the number of rings on their circumference. In comparison to recently predicted phosphorus tubes there is no qualitative structural difference, as expected from the conclusion drawn by chemical analogy. Even though the silicide tubes are charged and the counterions have not yet been considered, the results indicate the possible existence of stable tubular silicon modifications. This is strongly supported by analyzing the strain (curvature) energy of the tubes, obtained by calculation of the energy difference between the tube and the corresponding planar layer structure. The predicted $1/D^2$ behavior of the strain energy, as expected from elasticity theory, is clearly visible in Figure 17.4. All values converge roughly as $1/D^2$ towards a strain energy of zero for the planar reference system with increasing mean tube diameter D. It should be noted that for a given diameter the strain energy per atom is only slightly larger than that of pure carbon nanotubes [26] and even smaller than that of the corresponding P tubes.[8]

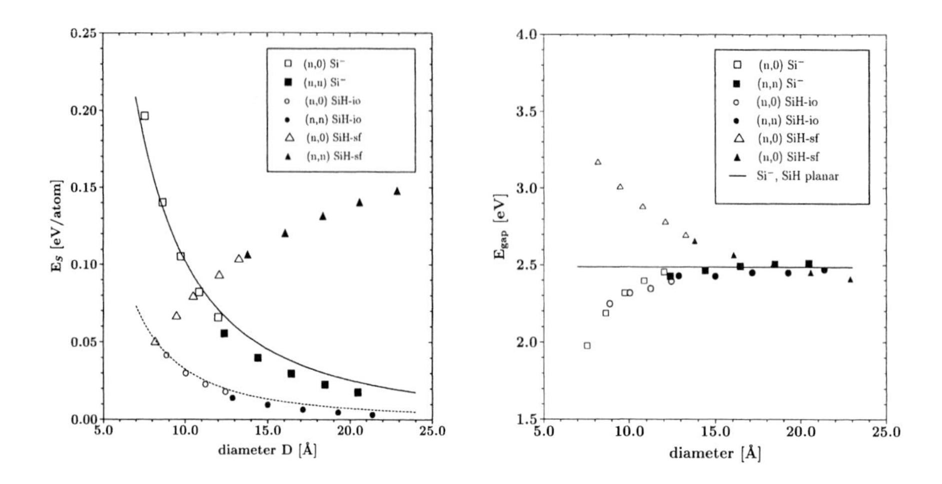

Figure 17.4. Strain energies (left) and gap sizes (right) as functions of mean diameter are shown for all calculated (n,0) and (n,n) silicide and silane nanotubes. The gap sizes of planar reference structures are symbolized by a single line due to nearly equal gap widths of 2.49 eV and 2.50 eV for planar Si^- and SiH, respectively. (Taken from ref. [24].)

Some interesting behavior can be discussed in the case of silane tubes. Besides the two limiting (n,0) and (n,n) cases, we can bind all hydrogen atoms either on the exterior surface of the tube (SiH-sf) or alternating between the inside and outside (SiH-io), as shown in Figure 17.3. Surprisingly, it turned out that for small diameters the SiH-sf tubes are almost isoenergetic with the SiH-io configurations. However, compared to the SiH-io tubes, the SiH-sf tubes show the opposite strain behavior to what might be expected – becoming less stable with increasing diameter. This can be understood quite easily. With increasing diameter of the tube and simultaneously decreasing curvature, the Si-Si-H bond angle is forced towards 90°, whereas in the SiH-io case the ideal tetrahedral angle of 109.5° for fourfold coordinated silicon can be established. A flat 2D SiH layer with all hydrogens bound on one side would not be stable at all, but it could bend spontaneously to tubular structures and favor the formation of SiH-sf tubes with small diameters.

Analyzing the electronic properties of the investigated systems, the flat silicide and the SiH sheets, as well as all nanotubes considered here, were found to be semiconducting. Within our DFTB method, we obtain band gaps of 2.49 eV and 2.50 eV for Si^- and SiH layers, respectively, agreeing quite nicely with that obtained by other calculations [27] (2.48 eV) and experimental results [28] (2.50 eV) for layered SiH. As shown in Figure 17.4. the gap size rapidly tends towards the value for flat sheets as the mean tube diameter D increases.

For the SiH-sf nanotubes, the behaviour is opposite, and the gap shrinks down to the value for the flat sheet.

Additionally, the mechanical properties of silicide and silane nanotubes have been obtained by DFTB calculations. The data clearly indicate that Si-based nanotubes are less stiff than other nanotubes hitherto considered, such as those made of P, BN, or C. Compared with typical values of Young's moduli of about 1.2 TPa for CNTs, the values for Si-based tubes are of the order of the bulk modulus of crystalline silicon, 98 GPa, predicted with the same theoretical method. Depending on tube diameter and type, the Young's moduli vary between 55 and 80 GPa.

17.2.2 Siloxene Nanotubes

Siloxenes and their derivatives have been known for a long time in chemistry,[29,30] but they have attracted remarkable attention in recent years as possible active structures in the strong visible luminescence of porous silicon.[31] Siloxenes may be characterized as "silicon-backbone materials", where each Silicon atom has only three nearest-neighbor silicon atoms. The structure of the "silicon back bone", a buckled layer of silicon rings, is closely related to that of the starting material of the siloxene synthesis, the above discussed $CaSi_2$. $CaSi_2$ is the starting material for the so-called Wöhler synthesis,[29] a topochemical reaction, in which the Ca ions are washed out and hydroxyl groups as well as hydrogen atoms are bound to the silicon leading to an ideal composition $Si_6H_3(OH)_3$. An image of the planar "back bone" structure is shown in Figure 17.5.

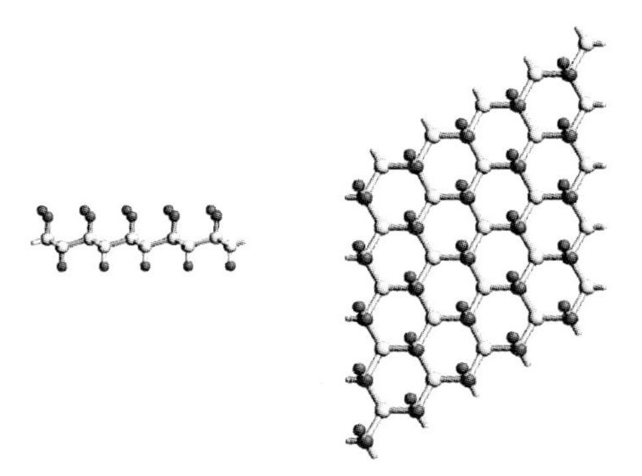

Figure 17.5. Structure (side and top views) of a siloxene layer as predicted by our DFTB method. The OH-groups above and the hydrogens below the silicon "backbone" layer can be seen clearly, as can the puckered structure. (Taken from ref. [23].).

Viewed from above the layer, the structure clearly resembles a graphene-like layer, although it is puckered in a similar way as in the silicide, silane, and phosphorus cases. Thus, the question arises as to whether siloxenes may form stable tubular structures as predicted for hypothetical silicides, silanes, and black phosphorus nanotubes. The experimentally synthesized [32] siloxene consists of hexagonal puckered layers, in which the Si-Si bond distance is 2.34 Å, the Si-H bond length is 1.54 Å, and the Si-O bond length is 1.60 Å. The Si-O-H bond angle is 115°.

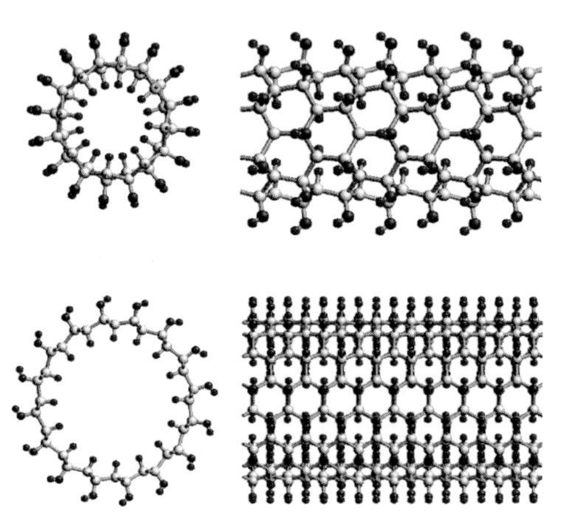

Figure 17.6. Hypothetical structures of (8,0) and (8,8) siloxene nanotubes. On the left-hand side are views down the tubular axes. The right-hand views clearly show the OH-groups bound on the outside and the hydrogens pointing towards the interior of the tube (taken from ref. [23]).

Within our DFTB method stable, puckered layers are predicted with respective bond lengths of 2.34, 1.51, and 1.63 Å. The Si-O-H bond angle exactly matches the experimental value.

In order to determine the energetic viability of the corresponding tubes, we performed similar optimizations of tubular structures by using conjugate-gradient techniques as described for silicides and silanes. We constructed various sized nanotubes by folding 2D sheets of puckered siloxene layer following the notation as zig-zag (n,0) and armchair (n,n) tubes with $n \in \{6,20\}$.[23] Figure 17. 6 illustrates the minimum-energy atomistic structures obtained for siloxene (8,0) and (8,8) nanotubes. The siloxene tube structures can be understood in terms of those of conventional graphitic CNTs by simply replacing the flat hexagons present in the latter by cyclohexane-like rings, just in the same way as the layered structure of siloxene is related to that of graphene. The structural details for all other siloxene tubes are similar, the only differences arising from the changing diameter, as reflected by the number of hexagon-like rings on the circumference of the tube.

The stability and possible routes for the synthesis of these hypothetical tubular siloxene structures are strongly supported by the calculated strain energy as a function of tube diameter. The energy difference between a tube and the corresponding layer structure converges in proportion to $1/D^2$ towards the value of the reference structure (strain energy zero) as the diameter D increases; see Figure 17.7.

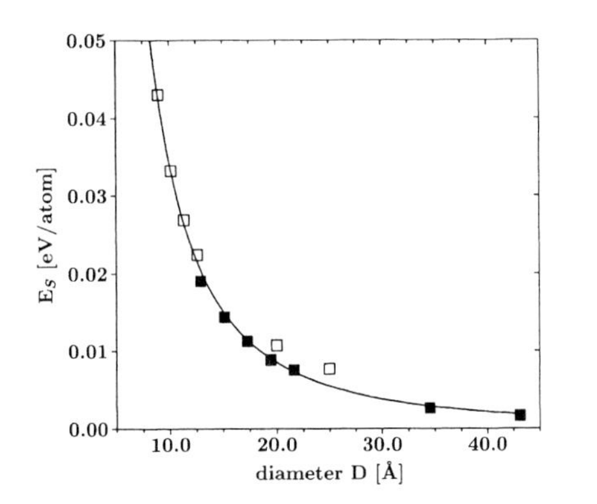

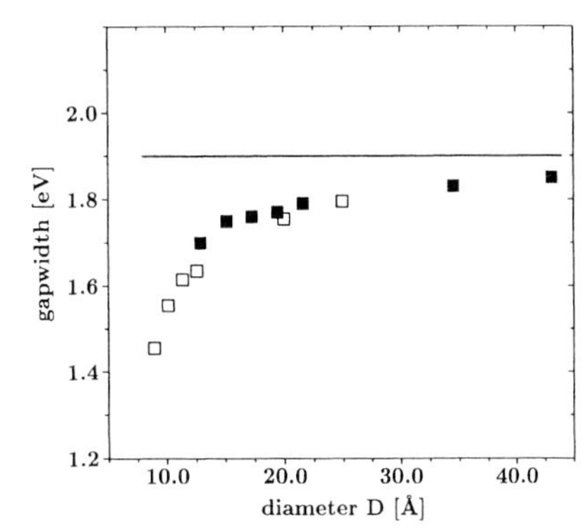

Figure 17.7. Strain energies and gap sizes as a function of the tube diameter. The (n,0) tubes are denoted by open squares, while the (n,n) tubes are represented by closed squares. The gap width of a corresponding siloxene monolayer is drawn as a horizontal line at 1.9 eV; for an explanation, see reference [23]. (Note: This gap width is usually underestimated within DFTB as in LDA in general!).

Electronically, the flat sheet as well as all the nanotubes considered here were determined to be semiconducting. In Figure 17.7, the gap sizes of nanotubes are shown as a function of the mean diameter. The gap sizes grow from about 1.5 eV for the smallest (n,0) nanotube towards the value for flat siloxene sheets (1.9 eV) as the tube diameter increases, as in the case of silicides and silanes discussed above.

The mechanical properties were again investigated by analyzing Young's moduli. The moduli, simply for reasons of the topology of the sp^3 hybridized Si(111) backbone sheets, are controlled by the stiffness of the Si-Si chemical bonds along the tube and thus reflect similar mechanical properties to those of crystalline silicon with a bulk modulus of 98 GPa. Depending on tube diameter and chirality, the Young's moduli vary between 70 and 80 GPa.

17.2.3 Potential Applications – Why Germyne Nanotubes?

In this section, we speculate on possible applications of all up to now theoretically predicted silicon-based tubular structures. However, the discussion will become more complete if we involve germanium-based tubes, too. It is not necessary within this chapter to essentially repeat all things written for the above discussed silicon-based tubes. We refer the reader to ref. [33]. The basic idea was to apply the similarities in the structural and electronic properties of silicon and germanium in order to support the possible existence of germyne tubular structures. As is well known, germanium also forms a diamond-like cubic structure with a tetrahedrally sp^3 hybridized bonding configuration. Germanium is a semiconductor with an indirect gap, and we find with calcium digermanide ($CaGe_2$) the ability to form Zintl phases similar to $CaSi_2$.

The chemical analogy between these group IV elements has very recently been extended to the synthesis of germanium sheet polymers.[34] The polygermynes were prepared by a topochemical reaction from calcium digermanide, similar to the classical siloxene synthesis from calcium disilicide. However, whereas the siloxenes contain OH groups as ligands, the polygermynes are completely saturated only by hydrogen atoms.

The experimental success in preparing germanium-based (111) layered backbone structures has enabled us to apply our theoretical investigations to study the formation of possible germyne (GeH) tubular systems based on chemical analogy. Polygermynes consist of hexagonal GeH layers, which are quite well separated from one another. Within DFTB, we obtain stable puckered GeH layers with a lattice constant of 4.15 Å. DFT-LDA calculations and experimental results give in-plane lattice constants of 3.94 Å [35] and 3.98 Å [34] respectively. The calculated Ge-Ge bond length is 2.49 Å and the Ge-H bonds length 1.56 Å, which compare quite well with the values of 2.39 and 1.54 Å, respectively, obtained from DFT-LDA calculations. Considering these arguments, we have performed similar series of calculations to construct germyne nanotubes and we find them to be stable, showing the now familiar dependences of strain energy and gap width on the tube diameter.

The germyne GeH sheet polymers show strong photoluminescence at 1.4 eV and a direct band gap close to this energy is predicted. The GeH nanotubes have been determined to be semiconducting in our calculations, too. Similarly to the siloxene and silane nanotubes the germyne tubes show a dependence of the gap size on diameter, as illustrated in Figure 17.8. The gap size grows from about 1.1 eV for the smallest (n,0) tube towards the value for germyne sheets (1.35 eV). Both types of nanotubes, the zig-zag (n,0) and armchair (n,n) tubes, possess a direct band gap. In conclusion, such tubes are expected to show interesting photoluminescence properties. Due to the size-dependent direct band gap their photoluminescence might be tuned in the near-infrared (IR).

Furthermore, taking into account the predicted gap properties of the silane and siloxene nanotubes, the group IV nanotubes could span a *photoluminescence over a wide wavelength range* from near-IR (narrow germyne tubes) to nearly blue light emitting large silane tubes. The tunable gap sizes make silicon- and germanium-based nanotubes attractive as potential materials for optoelectronic applications.

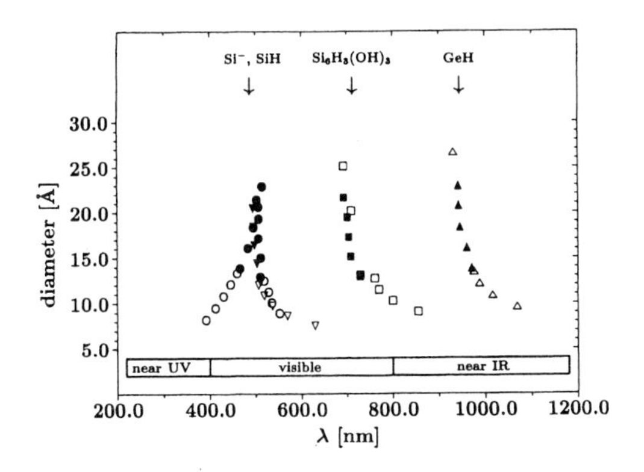

Figure 17.8. Illustration of the wide wavelength range of tunable gap sizes of all the silicon- and germanium-based nanotubes discussed herein for possible photoluminescence applications in optoelectronic nanodevices.

17.3 Conclusions

As mentioned in the introduction, the tubular materials are fascinating new structures with a considerable potential for various applications. Most of the nanotubes investigated to date have been carbon-based nanotubes. In this chapter, we have discussed the properties of various nanotubes with a silicon or germanium backbone structure. Most of these nanotubes are still hypothetical, "made" only in the computer. We proposed the stability of phosphorus nanotubes,[8] and about a year later Li et al. [22] reported the successful synthesis of nanotubes from bismuth, which is isoelectronic with phosphorus. The stability of NbS_2 nanotubes was also proposed in 2000,[36] and these were synthesized by Nath and Rao in 2001.[37]

Therefore, there are good chances that these tubular structures can be synthesized and thereby extend the manifold of already existing nanotubes with interesting potential for application. Another promising means of designing properties and functionality of tubular systems has recently been opened by the use of simple nanotubes as templates for novel nanocomposite materials.[38]

The nanotechnology for creating new classes of artificial nanostructured materials is still in its infancy.

17.4 Acknowledgement

This work has been supported by the German Israeli Foundation (GIF).

17.5 References

[1] H. W. Kroto et al., *Nature* **1985**, *318*, 162.
[2] W. Krätschmer, L. D. Lamb, K. Forstiropoulos, D. R. Huffman, *Nature* **1990**, *347*, 354.
[3] S. Iijima, *Nature* **1991**, *354*, 56.
[4] A. Thess, R. Lee, P. Nikolaev, H. Dai, P. Petit, J. Robert, C. Xu, Y. H. Lee, S. G. Kim, A. G. Rinzler, D. T. Colbert, G. E. Scuseria, D. Tomanek, J. E. Fischer, R. E. Smalley, *Science* **1996**, *273*, 483.
[5] P. C. Collins, M. S. Arnold, P. Avouris, *Science* **2001**, *292*, 706.
[6] W. B. Choi, D. S. Chung, J. H. Kang, H. Y. Kim, Y. W. Jin, I. T. Han, Y. H. Lee, J. E. Jung, N. S. Lee, G. S. Park, J. M. Kim, *Appl. Phys. Lett.* **1999**, *75*, 3129.
[7] A. Krishnan, E. Dujardin, M. M. J. Treacy, J. Hugdahl, S. Lynum, T. W. Ebbesen, *Nature* **1997**, *388*, 451.
[8] G. Seifert, E. Hernandez, *Chem. Phys. Lett.* **2000**, *318*, 355.
[9] M.S. Dresselhaus, G. Dresselhaus, Ph. Avouris (Eds.), *Carbon Nanotubes*, Springer-Verlag, Berlin, 2001.
[10] R. Saito, G. Dresselhaus, M. S. Dresselhaus, *Physical Properties of Carbon Nanotubes*, Imperial College Press, London, 1998.
[11] C. N. R. Rao, B. C. Satiskumar, A. Govindaraj, M. Nath, *Chem. Phys. Chem.* **2001**, *2*, 78.
[12] P. J. F. Harris, *Carbon Nanotubes and Related Structures*, Cambridge University Press, 2000.
[13] D. Porezag, Th. Frauenheim, Th. Köhler, G. Seifert, R. Kaschner, *Phys. Rev.* **1995**, *B51*, 12947.
[14] Th. Frauenheim, G. Seifert, M. Elstner, T. A. Niehaus, C. Köhler, M. Amkreutz, M. Sternberg, Z. Hajnal, A. di Carlo, S. Suhai, *J. Phys.* **2002**, *C14*, 3015.
[15] Th. Frauenheim, G. Seifert, M. Elstner, Z. Hajnal, G. Jungnickel, S. Suhai, R. Scholz, *Phys. Stat. Sol. (B)* **2000**, *227*, 41.
[16] M. Menon, E. Richter, *Phys. Rev. Lett.* **1999**, *83*, 792.

[17] B. Marsen, K. Sattler, *Phys. Rev.* **1999**, *B 60*, 11593.
[18] N. Wang, Y. H. Tang, Y. F. Zhang, C. S. Lee, S. T. Lee, *Phys. Rev.* **1998**, *B58*, R16024.
[19] S. B. Fagan, R. J. Baierle, R. Mota, A. J. R. da Silva, A. Razzio, *Phys. Rev.* **2000**, *B61*, 9994.
[20] K. H. Janzon, H. Schäfer, A. Weiss, *Z. Anorg. Allg. Chem.* **1970**, *372*, 87.
[21] M. E. Spahr, P. Bitterli, R. Nesper, F. Kruumeick, H. U. Nissen, *Angew. Chem. Int. Ed. Engl.* **1998**, *37*, 1263.
[22] Y. Li, J. Wang, Z. Deng, Y. Wu, X. Sun, D. Yu, P.Yang, *J. Am. Chem. Soc.* **2001**, *123*, 9904.
[23] G. Seifert, Th. Frauenheim, Th. Köhler, H. M. Urbassek, *Phys. Stat. Sol (B)* **2001**, *225*, 393.
[24] G. Seifert, Th. Köhler, H. M. Urbassek, E. Hernandez, Th. Frauenheim, *Phys. Rev.* **2001**, *B 63*, 193409.
[25] G. Seifert, Th. Köhler, Th. Frauenheim, *Appl. Phys. Lett.* **2000**, *77*, 1313.
[26] E. Hernandez, C. Goze, P. Bernier, A. Rubio, *Phys. Rev. Lett.* **1998**, *80*, 4502.
[27] K. Takeda, K. Shiraishi, *Phys. Rev. B.* **1989**, *39*, 11028.
[28] U. Dettlaff-Weglikowska, W. Hönle, A. Molassioti-Dohms, S. Finkbeiner, J. Weber, *Phys. Rev. B* **1997**, *56*, 13132.
[29] F. Wöhler, *Ann. Chem. Pharm.* **1863**, *127*, 257.
[30] H. Kautsky, *Z. Anorg. Allg. Chem.* **1921**, *117*, 209.
[31] P. Deak, M. Rosenbauer, M. Stutzmann, J. Weber, M. Brandt, *Phys. Lett.* **1992**, *69*, 2531; see also contribution of M. S. Brandt et al. (Chapter 15).
[32] A. Weiss, G. Beil, H. Meyer, *Z. Naturforsch.* **1979**, *B 35*, 25.
[33] G. Seifert, Th. Köhler, Z. Hajnal, Th. Frauenheim, *Solid State Commun.* **2001**, *119*, 653.
[34] G. Vogg, M. S. Brandt, M. Stutzmann, *Adv. Mat.* **2000**, *12*; see also contribution of M. S. Brandt et al. (Chapter 15).
[35] Z. Hajnal, G. Vogg, L. J. P. Meyer, B. Szücs, M. S. Brandt, Th. Frauenheim, *Phys. Rev.* **2001**, *B64*, 33311.
[36] G. Seifert, H. Terrones, M. Terrones, Th. Frauenheim, *Solid State Commun.* **2000**, *115*, 635.
[37] M. Nath, C. N. R. Rao, *J. Am. Chem. Soc.* **2001**, *123*, 4841.
[38] T. Seeger, Th. Köhler, Th. Frauenheim, N. Grobert, M. Rühle, M. Terrones, G.Seifert, *Chem. Commun.* **2002**, 34.

18 Structure and Reactivity of Solid SiO

Ulrich Schubert*, Thomas Wieder†

18.1 Introduction

While gaseous SiO is the most abundant silicon oxide in the universe, solid SiO does not exist naturally on earth. It was first prepared by Potter in 1905 by reduction of SiO_2 with carbon, silicon or SiC.[1] Today, several tons of solid SiO are produced industrially each year by comproportionation of silica and silicon at low pressure (10^{-3}–10^{-4} mbar) and high temperatures (1250–1400 °C). The gaseous SiO formed under these conditions is then condensed at colder surfaces. However, SiO is always present whenever silica or silicates are reduced at high temperatures, such as in the carbothermal reduction of SiO_2 (production of SiC) or in blast furnaces. The local structure at the interface between SiO_2 films on elemental Si, or the nature of nanocrystalline SiO_x particles,[2] is also related to that of solid SiO.

When gaseous SiO is condensed at cold surfaces, different forms of SiO are formed, depending on the temperature of the surface.[3] Among them, only black SiO with a coke-like appearance is commercialized. Other forms are black or yellow glasses, black fibers or yellow powders, depending on the temperature of the condensation surface.

The main commercial use of solid SiO is as a vapor-deposition material for the production of SiO_x thin films for optical or electronic applications (antireflective coatings, interference filters, beam-splitters, decorative coatings, dielectric layers, isolation layers, electrodes, thin-film capacitors, thin-film transistors, etc.), for diffusion barrier layers on polymer foils or for surface protection layers.[4] Other uses for SiO have been proposed, such as the substitution of elemental silicon in the Müller-Rochow process for the production of organosilicon halides,[5] because solid SiO can be produced at lower temperatures than elemental silicon.

* Address: Institute of Materials Chemistry, Vienna University of Technology, Getreidemarkt 9/165, A-1060 Wien, Austria

† Address: Fachbereich Material- und Geowissenschaften, Technische Universität Darmstadt, Petersenstr. 23, D-64287 Darmstadt, Germany

18.2 Structure of Solid SiO

All known modifications of solid SiO are X-ray amorphous. The structure of solid SiO must therefore be deduced by combinations of physical-chemical methods. Several structural models, some of them at variance, were proposed and controversially debated over several decades. An overview is given in ref. [6]. Part of the controversy is probably due to the fact that samples with different thermal histories were used (which was not specified in most papers), which almost certainly had different (local) structures.

There is general agreement that the silicon atoms in solid SiO are tetrahedrally coordinated. This implies that the structure of SiO is composed of five structural units, i.e. $Si(Si_{4-x}O_x)$ (x = 0–4). The stoichiometry of SiO requires, as a side condition, that 50% of the bonds must be Si-Si and 50% Si-O. The various models differ in the proportion and distribution of the different units. One limiting structural model is based on the assumption of a complete disproportionation into small particles of elemental silicon and SiO_2,[7] where only $Si(Si_4)$ and $Si(O_4)$ units would be present. The other extreme would be a statistical distribution of the $Si(Si_{4-x}O_x)$ units (6.25% $Si(Si_4)$, 25% $Si(Si_3O)$, 37.5% $Si(Si_2O_2)$, 25% $Si(SiO_3)$ and 6.25% $Si(O_4)$).[8] Several intermediate models have been proposed, such as one based on chains and rings,[9] and another, according to which silica and silicon regions are connected by a suboxide interphase.[10] In this latter model, the proportions of the suboxide units $Si(Si_3O)$, $Si(Si_2O_2)$, and $Si(SiO_3)$ depend on the relative sizes of the silica and silicon regions.

The notion that SiO is to some extent disproportionated, i.e. that the structure is intermediate between a complete disproportionation and a statistical distribution of the $Si(Si_{4-x}O_x)$ units, is currently generally accepted. However, it is still unclear as to what extent this disproportionation occurs.

To shed light on the nature of at least one modification of solid SiO, structural investigations by a combination of spectroscopic methods (T. Wieder et al.) as well as reactivity studies (U. Schubert et al.) were recently performed on samples with a well-defined thermal history and clearly specified macroscopic properties, i.e. with the commercially available Patinal® (Merck KGaA, Darmstadt; particle size < 0.044 mm). Thus, all experiments of the two groups discussed in the remainder of this article refer to Patinal.

Diffraction experiments with X-rays or neutrons, as well as high-resolution transition electron microscopy, showed that SiO (Patinal) is not crystalline on a length scale > 1 nm.[6] However, areas with a different contrast were observed by HRTEM, indicating a local heterogeneity. From in situ crystallization experiments, the size of these areas can be estimated to be 1–2 nm.

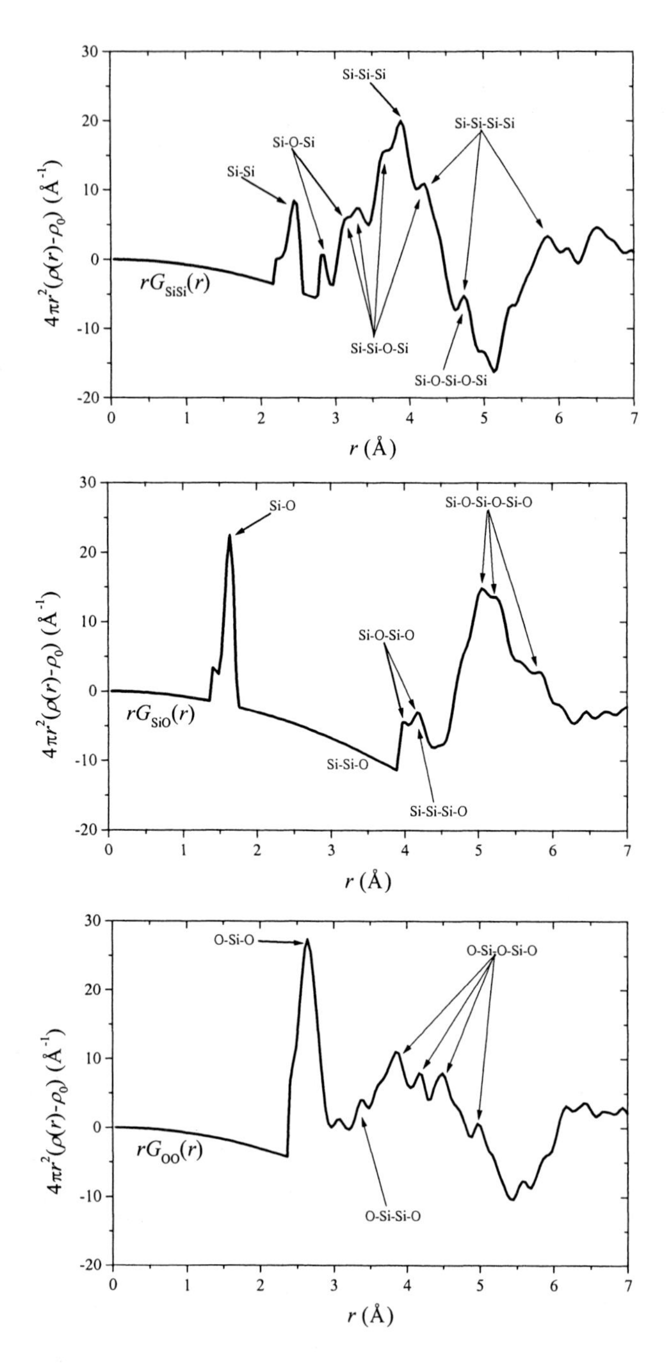

Figure 18.1. The partial radial distribution functions $G_{Si\text{-}Si}(r)$, $G_{Si\text{-}O}(r)$ and $G_{O\text{-}O}(r)$.

Further information was obtained from the radial distribution function $\rho(r)$, which is the probability of finding another atom at a distance r from the reference atom. Normally, the Si-O, Si-Si and O-O distances cannot be separated, and the interpretation of $\rho(r)$ is therefore difficult. With three different types of radiation (X-ray, neutrons, electrons), a separation of the three partial radial distribution functions is possible. In Figure 18.1, the pair distribution functions $G_{Si\text{-}Si}(r)$, $G_{Si\text{-}O}(r)$ and $G_{O\text{-}O}(r)$ of SiO are shown,[6] where G is the probability of finding an element j in a distance r from atom i.

Figure 18.1 shows two important features with regard to the structure of solid SiO. First, the $G_{ij}(r)$ cannot be reproduced by a superposition of the corresponding $G_{ij}(r)$ of SiO_2 and Si. This proves that SiO is not just a mixture of (very small) SiO_2 and Si particles. Second, the peaks corresponding to connected $Si(O_4)$ and $Si(Si_4)$ tetrahedra are high, and those assigned to $Si(O_{4-x}Si_x)$ (x = 1–3) are low. In particular, the very low abundance of Si-Si-O units indicates a spatial separation of the Si-Si and Si-O bonds.

The local environment of the silicon atoms, i.e. the nature and number of nearest neighbors, was probed by ELNES spectroscopy (energy-loss near-edge structure, a section of the EELS spectrum) at the Si K-edge, XPS (X-ray photoelectron spectroscopy), solid-state ^{29}Si NMR spectroscopy, and electron spin resonance spectroscopy. Two overlapping bands at ΔE = 1842 and 1847 eV with relative intensities of 0.37 and 0.7 are observed in the ELNS spectrum, which are assigned to $Si(Si_4)$ and $Si(O_4)$ units, respectively. Curve fitting shows that these units are predominant. A similar result was obtained by XPS (Figure 18.2). It was concluded that < 25% of the silicon atoms belong to $Si(O_{4-x}Si_x)$ (x = 1–3) units.[6]

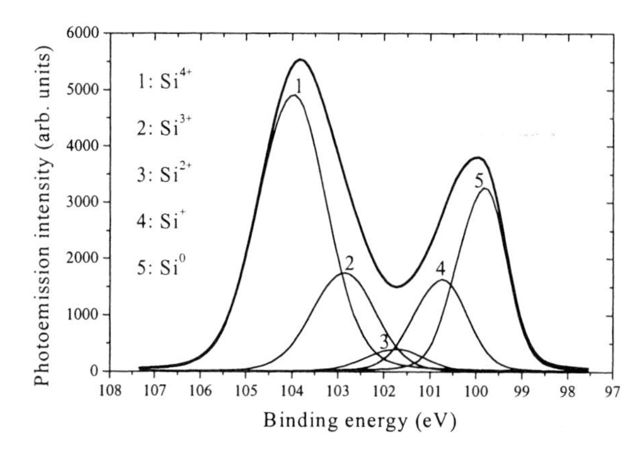

Figure 18.2. XPS spectrum fitted by five peaks.

Two resonances are found in the ^{29}Si MAS NMR spectrum of solid SiO, at $\delta = -69$ and -109 ppm.[11,12] The former signal has a lower intensity than the latter. The ^{29}Si chemical shifts expected for $Si(Si_4)$ and $Si(O_4)$ units are approximately –53 and –110 ppm, respectively. ESR measurements indicated the presence of silicon-based radicals, predominantly in the $Si(Si_4)$ regions. The density of the paramagnetic centers is about $2.2 \cdot 10^{19}$ cm^{-3}.[11] Since unpaired electrons from broken bonds give rise to large local magnetic fields, nuclei close to these unpaired electrons (within the so-called "wipe-out radius") will not contribute to the observed NMR signal. This wipe-out radius was calculated to be about 1.1 nm and accounts for the fact that no $Si(Si_4)$ resonance is observed in ^{29}Si NMR, if the dimensions of the silicon regions are of this magnitude.

In summary, all spectroscopic investigations of SiO (Patinal) clearly show that there are no extended areas of Si or SiO_2. The size of the Si and SiO_2 areas is 0.5–2.5 nm. This implies that a large percentage of the silicon atoms is in the thin interphase surrounding and interconnecting the Si and SiO_2 areas (Figure 3).

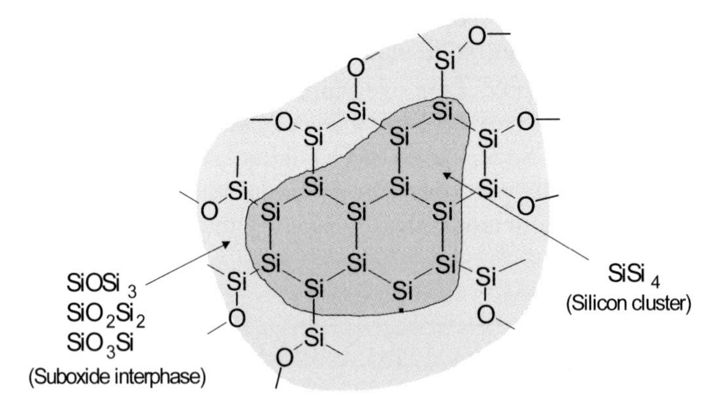

Figure 18.3. A two-dimensional structural model for solid SiO (Patinal) (see text).

This model accounts for the previously observed differences, which probably originated from different degrees of (nano)disproportionation and phase separation. On condensing molecular (gaseous) SiO onto cold surfaces, cyclic compounds have been identified, in which the silicon atoms are two-coordinate (two oxygen atoms). The further steps are hypothetical, but some "nano-disproportionation" must occur, which starts to separate Si-O and Si-Si bonds.

18.3 Reactions with Solid SiO

Although SiO is produced on an industrial scale, only few a reactions of solid SiO are known. The most obvious reaction is that with air or O_2, which should produce SiO_2. The rate of this reaction appears to depend on the SiO modification and on the presence of a passivating silica layer on the surface of SiO. Thus, while some authors have reported freshly prepared SiO to be pyrophoric,[3,13] others note that exhaustive oxidation cannot be achieved even at high temperatures.[13,14]

When SiO (Patinal) was heated in pure O_2, thermogravimetric analysis showed a slow weight increase above 1000 °C, but even at 1500 °C oxidation was not exhaustive. When a sample was kept at 1300 °C in air for 48 h, crystalline Si and SiO_2 (cristobalite) were observed by XRD. Thus, disproportionation into Si and SiO_2 is much faster than oxidation at this temperature. The conclusion from these experiments is that solid-state reactions of SiO must be carried out at temperatures below 1000 °C to exclude the possibility that the observed products originate from initial oxidation and/or disproportionation reactions.[15]

Other oxidants for SiO include water vapor (leading to SiO_2 and H_2 at 600 °C), chlorine (leading to SiO_2 and $SiCl_4$ at 800 °C) and SO_2 (leading to SiO_2 and S at 800 °C).[16] SiO has even been used to obtain elemental metals (Eq. 1).

$$\mathrm{SiO + MO + 2\,CaO \xrightarrow{1300\ ^\circ C} Ca_2SiO_4 + M \quad (M = Mg, Zn)} \tag{1}$$

A variation of this reaction was found when solid SiO (Patinal) was reacted with CuI, AgI, AgCl or AgF to give the silicon tetrahalides SiX_4 and elemental silver or copper (Eq. 2).[17] These products are also obtained when elemental silicon is employed instead of SiO.[18]

$$\mathrm{2\,SiO + 4\,MX \rightarrow SiX_4 + SiO_2 + 4\,M \quad (M = Cu, Ag;\ X = I, Cl, F)} \tag{2}$$

Mixtures of fluorine- or chlorine-substituted mono- and oligosilanes were obtained when SiO was reacted with aqueous HF [1, 19] and HCl gas,[20] respectively. In the case of the HCl reaction SiO_2 was claimed to be a by-product. SiO was similarly reacted with organic halides in the presence of catalysts to give organosilicon chlorides, similar the Müller-Rochow reaction of elemental silicon.[5]

One of the few straightforward reactions of SiO_2 leading to molecular compounds is its reaction with alkali metal hydroxide and ethylene glycol, by which Laine et al. obtained monomeric and dimeric pentacoordinate complexes of the composition $M[Si(OCH_2CH_2O)_2OCH_2CH_2OH]$ or $M_2[Si_2(OCH_2CH_2O)_5]$ (M = Li, Na, K, Cs) with two chelating and one monodentate glycolate ligands per silicon.[21] When solid SiO was reacted with a solution of lithium or sodium ethylene glycolate in ethylene glycol at $T > 140$ °C, the same compounds were obtained (Eq. 3).[22] Interestingly, elemental silicon gives the same products under the same conditions. As in the case of SiO, gas evolution was observed. While the reaction of elemental silicon with alkali metal glycolates has not previously been investigated, reactions of Si with monoalcohols in the presence of metal alkoxides are known to give tetraalkoxysilanes.[23] The same result was obtained in the reaction with catechol (Eq. 4).

Thus, the reactions with alkali metal glycolates or with catechol do not differentiate between SiO, elemental Si or SiO_2 (except that silicon is oxidized in the former two cases).

SiO_2 / SiO / Si + Li / $HOCH_2CH_2OH$ ⟶ Li^+ $[Si(OCH_2CH_2O)_2OCH_2CH_2OH]^-$ (+ H_2)

SiO_2 / SiO / Si + Na / $HOCH_2CH_2OH$ ⟶ 2 Na^+ $[Si_2(OCH_2CH_2O)_5]^{2-}$ (+ H_2)

(3)

SiO_2 / SiO / Si + $C_6H_4(OH)_2$ / NH_3 / H_2O ⟶ 2 NH_4^+ $[Si(O_2C_6H_4)_3]^{2-}$ (+ H_2)

(4)

Although oxidation reactions of solid SiO have been known for a long time, its reduction by elemental metals has only recently been investigated. When a stoichiometric mixture of SiO and Mg was heated to 700 °C under argon, silicon nanoparticles (11 nm) and MgO were observed by XRD.[15] However, the reaction is more complex than it may appear. When a 1:1 mixture of Mg and SiO was heated and X-ray diffractograms were taken in situ, Mg_2Si and MgO were observed between 300 °C and 600 °C. At $T > 600$ °C, the Zintl phase Mg_2Si was completely converted into elemental Si and MgO (Eq. 5). This suggests the following sequence of reactions:

$$\begin{array}{lll} < 300\ ^\circ C: & 3\ Mg + SiO \longrightarrow & Mg_2Si + MgO \\ 300\text{–}600\ ^\circ C: & Mg_2Si + 2\ SiO \longrightarrow & 3\ Si + 2\ MgO \\ \hline \text{overall:} & Mg + SiO \longrightarrow & Si + MgO \end{array} \quad (5)$$

Thus, Mg as the reductant gives Mg_2Si, while Mg_2Si as the reductant gives elemental silicon.

The two-stage reaction of SiO with Mg was not observed in the reduction with other elemental metals, although the final products were mostly analogous. The alkali metals Li, Na, K, and Rb did react with SiO, although the products (possibly silicides) could not be identified with certainty due to their extreme sensitivity. In the reaction mixtures of solid SiO with strontium or titanium, SrSi, $SrSi_2$ and Ti_5Si_3 were identified by XRD. Reaction with aluminum did not result in the formation of a silicide but of elemental silicon instead.[24] It should be noted that elemental silicon forms silicides with Mg, Ca, Sr, Ti and the alkali metals, for example, but not with Al.

Reaction of solid SiO with calcium allowed again a deeper insight into the overall reaction: reaction in a 1:3 ratio at 750 °C resulted in the formation of CaO and Ca_3SiO.[24] However, Ca_3SiO is not stable at higher temperatures. Upon heating to 900 °C, this phase was no longer observed, and a mixture of CaSi, Ca_5Si_3, and CaO was obtained. The structure of Ca_3SiO (a distorted anti-perovskite structure), which was erroneously described in the literature as "cubic Ca_2Si", was determined from the powder diffraction data.

The metals which react with SiO have redox potentials smaller than –1.6 V, while those that do not react have less negative potentials. For example, zinc ($E^\circ_{Zn/Zn2+}$ –0.76 V) did not react. Although other factors certainly also influence the reduction behavior of SiO, this result allows the prediction that only compounds with a potential of less than -1.2 ± 0.4 V will be capable of reducing SiO.

18.4 Conclusions

The reactivity of Patinal, a well-defined form of solid SiO, is very similar to that of elemental silicon. Most reactions found for solid SiO are also observed for elemental silicon. In a few reactions, its reactivity resembles that of silica, and only a few reactions (reduction to Si, disproportionation with Mg_2Si) are unique.

The reactivity behavior is very consistent with the picture that emerges from physical-chemical investigations. The most probable structural model is that Patinal consists of Si and SiO_2 regions of 0.5–2.5 nm in diameter, which are connected by a thin interphase consisting of $Si(O_{4-x}Si_x)$ (x = 1–3) units. The volume fraction of these interphase units is probably less than 25%.

18. 5 References

[1] H. N. Potter, *Trans. Electrochem. Soc.* **1907**, *12*, 191.

[2] For example: G. Lucovsky, *J. Non-Cryst. Solids* **1998**, *227–230*, 1. H. Hofmeister, P. Ködderitzsch, U. Gösele, *Ber. Bunsenges.* **1997**, *101*, 1647.

[3] F. Stetter, M. Friz, *Chemiker Ztg.* **1973**, *97*, 138.

[4] H.-D. Klein, F. König in *Tailor-made Silicon-Oxygen Compounds*, (Eds.: R. Corriu, P. Jutzi), Vieweg, Braunschweig, Wiesbaden **1996**, p. 141.

[5] G. N. Bokerman, J. P. Cannady, C. S. Kuivila, *US Patent* 5,051,247 (1991), 5,120,520 (1992).

[6] A. Hohl, T. Wieder, P. A. van Aken, T. E. Weirich, G. Denninger, M. Vidal, S. Oswald, C. Deneke, J. Mayer, H. Fuess, *J Non-Cryst. Solids*, **2002**, in press.

[7] G. W. Brady, *J. Phys. Chem.* **1959**, *63*, 1119; C. F. George, P. d' Antonio, *J. Non-Cryst. Solids* **1979**, *34*, 323; G. Etherington, A. C. Wright, R. N. Sinclair in *The Structure of Non-Crystalline Materials*, (Eds. P. H. Gaskell, J. M. Parker, E. A. Davis), Taylor & Francis, London, 1983, p. 501.

[8] H. R. Philipp, *J. Phys. Chem. Solids* **1971**, *32*, 1935; *J. Non-Cryst. Solids* **1972**, *8–10*, 627.

[9] J. A. Yasaitis, R. Kaplow, *J. Appl. Phys.* **1972**, *53*, 995.

[10] R. J. Temkin, *J. Non-Cryst. Solids* **1975**, *17*, 215.

[11] R. Dupree, D. Holland, D. S. Williams, *Philos. Mag. B* **1984**, *50*, L13.

[12] B. Friede, M. Jansen, *J. Non-Cryst. Solids* **1996**, *204*, 202.

[13] H. H. Emons, P. Hellmond, *Z. Anorg. Allg. Chem.* **1967**, *355*, 265.

[14] A. Weiss, A. Weiss, *Z. Anorg. Allg. Chem.* **1954**, *276*, 95.

[15] E. Füglein, U. Schubert, *Chem. Mater.* **1999**, *11*, 865.

[16] E. Zintl, W. Bräuning, H. K. Grube, W. Krings, W. Morawietz, *Z. Anorg. Allg. Chem.* **1940**, *245*, 1.

[17] E. Biehl, U. Schubert, *Monatsh. Chem.* **2000**, *131*, 813.

[18] K. H. Lieser, H. Elias, H. W. Kohlschütter, *Z. Anorg. Allg. Chem.* **1961**, *313*, 200.

[19] P. L. Timms, C. S. G. Phillips, *Inorg. Chem.* **1964**, *3*, 606.

[20] G. N. Bokerman, J. P. Cannady, C. S. Kuivila, *US Patent* 5,051,248 (1991).

[21] R. M. Laine, K. Y. Blohowiak, T. R. Robinson, M. L. Hoppe, P. Nardi, J. Kampf, J. Uhm, *Nature* **1991**, *353*, 642. B. Herreros, S. W. Carr, J. Klinowski, *Science* **1994**, *263*, 1585. K. Y. Blohowiak, D. R. Treadwell, B. L. Mueller, M. L. Hoppe, S. Jouppi, P. Kansal, K. W. Chew, C. L. Scotto, F. Babonneau, J. Krampf, R. M. Laine, *Chem. Mater.* **1994**, *6*, 2177.

[22] W. Donhärl, I. Elhofer, P. Wiede, U. Schubert, *J. Chem. Soc., Dalton Trans.* **1998**, 2445.

[23] For example: W. Joch, A. Lenz, W. Rogler, *German Patent* 2,354,683 (1973); R. D. Yarwood *UK Patent* 2,140,814 (1984).

[24] E. Biehl, U. Schubert, F. Kubel, *New J. Chem.* **2001**, *25*, 994.

19 Si Nanocrystallites in SiO_x Films by Vapour Deposition and Thermal Processing

Herbert Hofmeister and Uwe Kahler*

19.1 Introduction

Materials consisting of Si nanocrystallites embedded in an Si suboxide matrix exhibit peculiar electrical, optical, and optoelectronic properties, which depend strongly on the Si nanocrystallite size as well as on the structural characteristics and dielectric properties of the surrounding medium.[1–9] One of the main issues in these studies has been attempted tuning of the photoluminescence emission wavelength by controlling the nanocrystallite size. Various techniques, like co-sputtering of Si and SiO_2, ion implantation of Si in SiO_2, reactive modes of chemical vapor deposition, laser ablation, and gas-phase evaporation have been employed for the production of Si/SiO_x nanoparticulate composites.[8, 10–17] Investigations of the structure, composition and properties of these materials have been directed towards the formation of nanosized Si crystallites during synthesis and/or subsequent processing,[18–20] the formation and/or variation of the SiO_2 matrix,[21–25] and the role and nature of Si–O bonds at the particle/matrix interface.[4, 6, 22, 26, 27] Physical vapor deposition of silicon monoxide, on the other hand, is widely used for thin-film formation aimed at exploiting the unique mechanical, chemical, and dielectric properties of such materials. However, despite the ease of making these films due to the high vapor pressure of SiO of 3.8×10^{-3} mbar at 1000 °C, little is known about their tendency to decompose into Si and SiO_2 at elevated temperatures. A better understanding of this behavior may enable controlled phase separation into composites of Si nanocrystallites embedded in an SiO_x matrix upon thermal processing. This requires a more detailed picture of the structure of SiO_x.

Since the first synthesis of silicon monoxide by Potter 1905,[28] the chemical state and atomic structure of this amorphous solid have been discussed controversially. The known binary phase diagrams of silicon and oxygen that

*

Address of the authors: Max Planck Institute of Microstructure Physics, Weinberg 2, D-06120 Halle, Germany include the existence of SiO [29, 30] predict stability of the compound only above 1000 °C. Below this temperature the material is found in a metastable state depending on the conditions of synthesis and storage. To explain the observed behavior, principally two different models have been developed. One is the random bond (RB) model,[31] considering a random network of statistically distributed $Si(Si_{4-i}O_i)$ tetrahedra with $i = 0$ to 4 denoting the number of oxygens bound to the central silicon. The other is the random mixture (RM) model,[32] considering only two phases, Si and SiO_2, with the corresponding tetrahedral units $Si(Si_4)$ and $Si(O_4)$ randomly dispersed down to the nanometer scale. Any silicon suboxide SiO_x between Si and SiO_2 is described in the RM model by varying proportions of $Si(Si_4)$ and $Si(O_4)$ (the straight dashed lines in Figure 19.1), and in the RB model by varying portions of $Si(Si_4)$, $Si(Si_3O)$, $Si(Si_2O_2)$, $Si(SiO_3)$, and $Si(O_4)$ (the solid curves in Figure 19.1). The restriction to $Si(Si_4)$ and $Si(O_4)$ constituents of the RM model is not realistic because of the non-negligible volume of an interfacial region where other tetrahedral units must also occur. This interphase material amounts to a significant fraction when the dispersion of Si and SiO_2 phases approaches linear dimensions of only a few nanometers. This is taken into account in several modified mixture models, which differ in how small these dimensions may be and in what fraction of elemental Si may occur.[33–35]

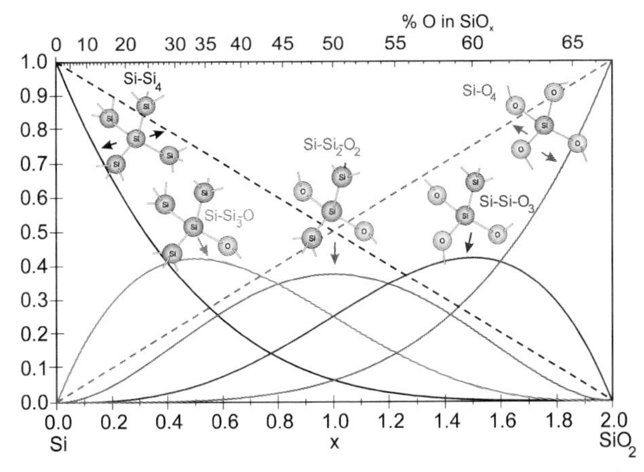

Figure 19.1. Representation of silicon suboxide composition in terms of the relative probabilities of RB (solid curves) and RM (dashed lines) model constituents $Si(Si_{4-i}O_i)$.

Regardless of the model, there is one feature in common, namely the assumption of various building blocks instead of only one SiO unit. Strictly speaking, all models consider a certain degree of disproportionation of 2 SiO into Si and SiO_2 and even in the most homogeneous distribution [31, 35] we must account for the initial stage of disproportionation.

It is well known that this initial stage may proceed to further disproportionation upon applying special heat treatment.[36, 37] This behavior opens a route to Si nanocrystallite formation provided that the scale of the phase separation may be confined appropriately. Because of the relatively high temperatures required for crystallization of Si in SiO_x,[38–41] the variation of the available volume fraction of Si is recommended as a more convenient control of crystallite size. Such control of the suboxide stoichiometry may be achieved by applying well-defined oxygen admission during the film fabrication. Based on the structural investigation of the source material used, we studied the disproportionation behavior and Si nanocrystallite formation in SiO_x films of various stoichiometry using mainly Fourier-transform infrared (FTIR) spectroscopy and high-resolution electron microscopy (HREM). The obtained results are related to model considerations concerning some principal aspects of the processes taking place. Details of the additionally studied photoluminescence emission of these films, in particular the evolution with crystallite size, have been published elsewhere.[39–41]

19.2 Fabrication of SiO_x Films

19.2.1 Characteristics of the SiO Source Material

The non-porous SiO powder of BALZERS (nominal purity 99.9%; grain size 0.2 to 0.7 mm) was studied by X-ray diffraction (XRD) and transmission electron microscopy (TEM) in the as-received state as well as after heating to about 1300 °C in high vacuum for 1 h to ensure that the material exhibits the expected behavior. Figure 19.2 shows the results of XRD measurements obtained with the copper Kα line of a PHILIPS X-ray diffractometer.

The positions and widths of the diffraction peaks of the as-received sample diffraction pattern are characteristic of an amorphous random network of tetrahedrally coordinated silicon and oxygen with broad maxima at scattering angles 2θ of 22.3, 52.2, and 69.2°. Three additional maxima at 28.4, 47.6, and 56.3° occur after heating, which are more narrow and correspond to {111}, {220}, and {311} lattice planes, of crystalline silicon. By using the Scherrer formula a mean size of Si nanocrystallites of 3.5 to 4.9 nm was calculated from the width of the diffraction peaks.

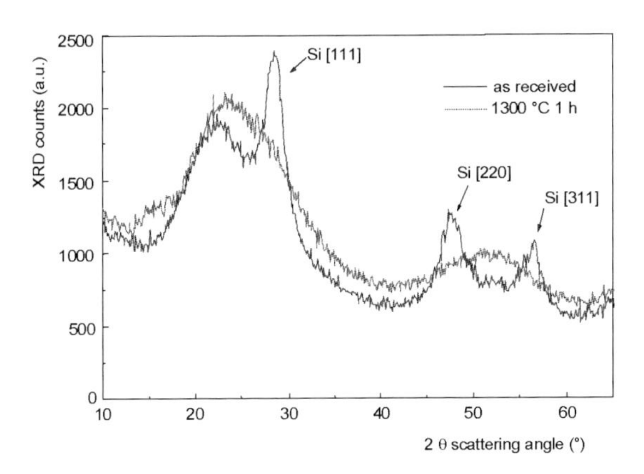

Figure 19.2. XRD measurement of Balzers SiO, as-received and upon 1 h heating to 1300 °C, with crystal lattice reflections indexed.

Electron microscopy investigations, using a JEM 4000 EX operating at 400 kV, were additionally performed on the same samples pulverized to appropriate size. They revealed a homogeneous appearance for the as-received sample, with the characteristic contrast features of a completely amorphous material. Upon heating, the sample contains densely arranged Si crystallites of the diamond cubic lattice type having sizes approximately ranging from 3 to 5 nm. XRD and HREM investigation unambiguously confirmed that the source material used is of homogeneous nature and has a completely amorphous structure. After heating to 1300 °C, the formation of Si nanocrystallites is observed. This result suggests that annealing at slightly lower temperatures may yield Si nanocrystallites in the size range below 5 nm which would be interesting because of their photoluminescence properties.

To get a more accurate picture of the behavior of SiO at elevated temperatures, we performed a differential scanning calorimetry (DSC) measurement up to 1100 °C. To this end, the SiO powder was heated in a Pt crucible at a rate of 20 K min^{-1} under flowing Ar gas (75 ml min^{-1}) in a DSC Pegasus 404C from NETZSCH. Figure 19.3 shows the heat flow rate plotted versus temperature in the range of 500 to 1100 °C. Exothermic effects may be recognized by negative deviations from the base line. At an extrapolated onset of about 800 °C, the beginning of crystallization can be seen. The maximum is situated at 997 °C. The crystallization enthalpy amounts to –178.7 J g^{-1}. From this measurement, it is clear that in pure SiO powder crystallization occurs between 800 and 1050 °C. Consequently, we routinely applied annealing at 1000 °C to our SiO_x films since one may expect complete crystallisation of the Si fraction at this temperature.

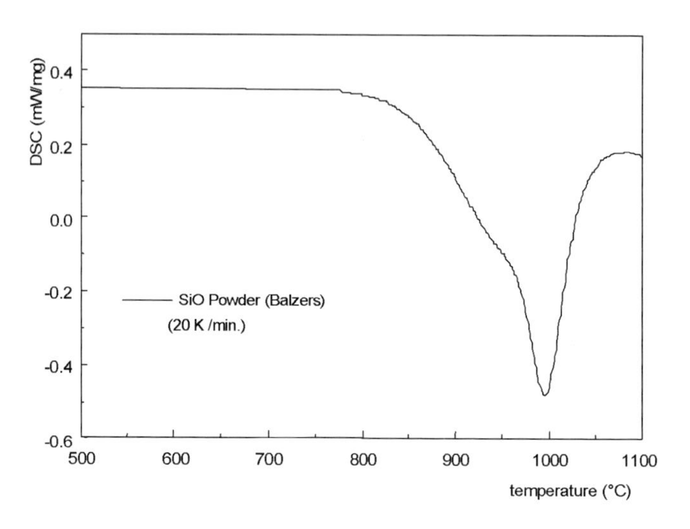

Figure 19.3. High-temperature DSC measurement of Balzers SiO revealing exothermic effects (crystallization) between 800 and 1050 °C.

19.2.2 Vapor Deposition and Thermal Processing of SiO_x Films

The SiO_x films we fabricated was done by physical vapor deposition of SiO with controlled oxygen admission in an oil-free vacuum apparatus at 1×10^{-7} mbar base pressure. The experimental set-up is shown schematically in Figure 19.4. Thermal evaporation of SiO from two Ta boats was directed to rotating 10 cm Si(100) wafer substrates heated by rear illumination with tungsten halide lamps. To achieve uniform film thickness over the large area of the Si wafer, the angle between the evaporators and substrate center was fixed at 53°, while the distance between the substrate plane and evaporator plane was kept at 20 cm. Routinely, films of 100 nm thickness were deposited this way. A defined oxygen partial pressure was adjusted by admitting high purity oxygen gas into the evaporation chamber. Incorporation of oxygen into the growing film is determined by the SiO evaporation rate, the residual gas pressure, and the residual gas composition on the one hand, and by the sticking coefficients of SiO and oxygen on the other. Since the oxygen sticking coefficient is temperature dependent, the substrate temperature also plays a role. Without oxygen admittance, a film stoichiometry parameter x as near as possible to 1 was achieved by setting the evaporation rate to 12 nm min^{-1} and the substrate temperature to 100 °C. By increasing the oxygen partial pressure (2×10^{-7} to 1×10^{-4} mbar), the additional incorporation of oxygen could be adjusted up to a stoichiometry of $x = 2$.

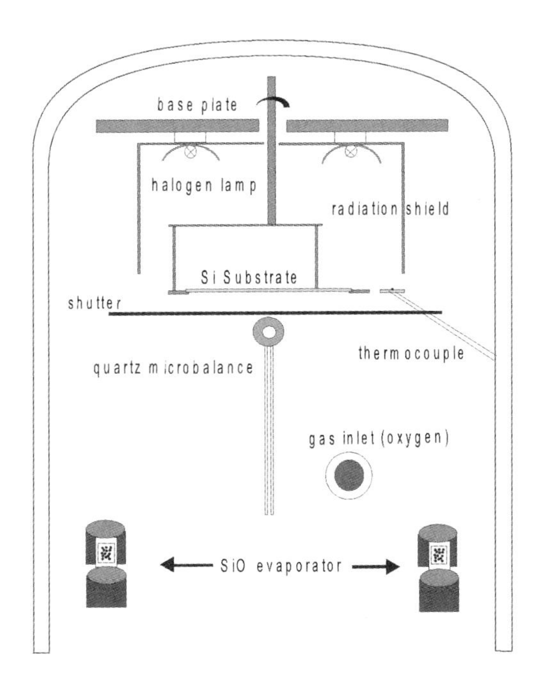

Figure 19.4. Schematic drawing of the experimental set-up for SiO_x film deposition.

After deposition, the films were annealed in a conventional quartz tube furnace. This heat treatment, performed in vacuum to avoid supplementary oxidation, was aimed at starting disproportionation of the Si suboxide and, based on the corresponding atomic processes, proceeding to a phase separation into amorphous Si and SiO_2. The latter process should enable the formation of Si nanocrystallites above 800 °C, provided that the Si regions are large enough. Even at lower temperatures, up to 400 °C, a certain restructuring of the slightly porous material is observed, accompanied by a removal of surface hydroxyl groups, which results in an increase in the density of the films.[42] For the phase separation study described in section 19.3.2, a vertical evaporation geometry was utilized yielding more dense films that do not contain hydroxyl groups, even though the film thickness becomes less homogeneous.[42]

19.3 Decomposition of SiO_x Films and Formation of Si Nanocrystallites

19.3.1 Initial Stoichiometry

Since the Si–O–Si stretching vibrations exhibit stoichiometry–dependent resonances in the infrared spectral region, one could use their maximum position as a measure of the oxygen content.[43] However, variations of the film density, due to a porosity caused by the deposition geometry,[41] as well as the degree of disproportionation may influence this position.[44] To get an as accurate as possible determination of the absolute oxygen content of the fabricated SiO_x films, we applied Rutherford back-scattering (RBS) analysis. This method does not require extensive calibration and is well-suited for direct quantification of thin-film measurements. It is based on the detection of energy losses of highly energetic ions (1.4 MeV He) bombarded onto and back-scattered from the surface of a sample of unknown composition. RBS enables a considerable depth resolution and is even used for thick ness evaluation. The stoichiometry of the SiO_x films could be determined with an accuracy of ± 1% by comparing the measured curves with calculations based on models of known composition. In this way, RBS measurements proved that under otherwise identical conditions the SiO_x film stoichiometry was a monotonic function of the oxygen partial pressure.[39, 40] This is demonstrated by Figure 19.5, where the results for films of various stoichiometry are plotted. RBS measurements also confirmed that no changes occurred in the overall composition of the films during subsequent heat treatment.

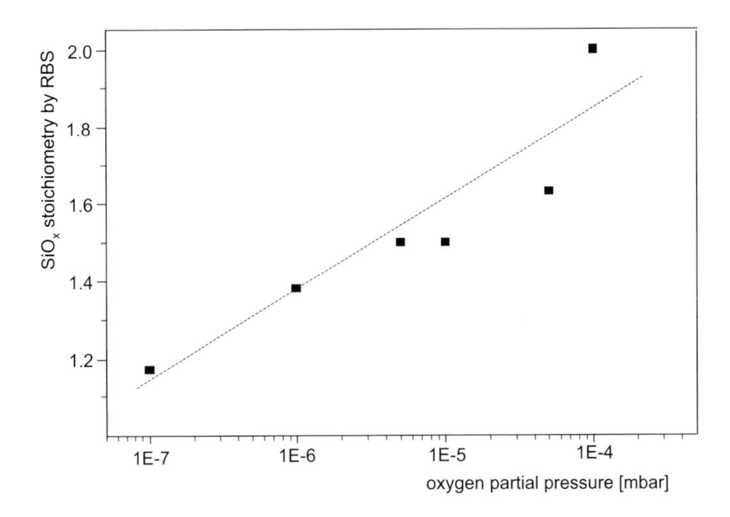

Figure 19.5. SiO_x film stoichiometry by RBS in dependence on the oxygen partial pressure (straight line is to aid the eye).

19.3.2 Disproportionation and Phase Separation

To monitor structural changes in the SiO_x films caused by annealing at elevated temperatures, we recorded their FTIR spectra in the range of the Si–O–Si stretching vibration. The spectra were acquired on a BRUKER IFS66v FTIR spectrometer equipped with an infrared detector. Recording was performed in 90° transmission geometry using a piece of plain wafer, subjected to the same thermal processing, as reference. Figure 19.6 shows such spectra for a series of samples of stoichiometry $x = 1$ annealed at temperatures between 300 and 1100 °C. Beginning at an annealing temperature of about 400 °C, a shift of the Si–O–Si stretching vibration maximum position to higher frequency is observed, indicating distinct changes in the environment of Si.[45] The as-deposited film shows a peak at 1024 cm^{-1}.

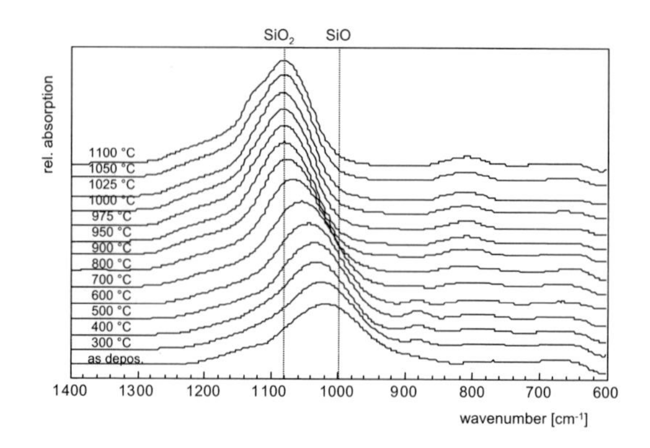

Figure 19.6. FTIR spectra of SiO_x films of stoichiometry $x = 1$ in dependence on the temperature of thermal processing for 1 h.[42]

Assuming no contribution of already disproportionated Si to the above vibration and a homogeneous distribution of oxygen throughout the remaining phase, one could correlate the maximum position to the stoichiometry of this phase.[46] This rough estimate yields values for the as-deposited films which agree fairly well with those determined by RBS. We introduce below a detailed model which relates the maximum position to a more realistic description of the degree of disproportionation. Upon annealing at 950 °C the peak is shifted to 1080 cm^{-1}, indicating a complete phase separation into Si and SiO_2. The peak shift is accompanied by an increase of peak height at the expense of the halfwidth, pointing to a more narrow distribution of the various Si–O coordinations. This behavior has been previously explained in terms of approaching the bonding situation of SiO_2 upon annealing.[45, 46]

The blue shift of the Si–O–Si stretching vibration peak upon thermal processing of SiO_x films clearly indicates that disproportionation may occur well below the temperature of Si nanocrystallite formation. To understand the reason for this peak shift a simple model was developed which correlates the maximum position with the degree of disproportionation, based on an atomic structure of SiO_x according to the RB model. However, it seems to be more appropriate not to consider isolated tetrahedral units $Si(Si_{4-i}O_i)$ with Si in the center, but clusters of two such tetrahedra which share a common oxygen for reasons of symmetry. Since oxygen is always bound to two Si nearest neighbors, regardless of stoichiometry, a change in vibration with x can only be simulated by considering the next-nearest neighbors as well, i.e. the three neighbors of the two nearest Si atoms. Seven configurations of such clusters are obtained by varying the number of next-nearest oxygens from zero to six as shown in Figure 19.7(a), where these clusters are numbered 1 to 7.

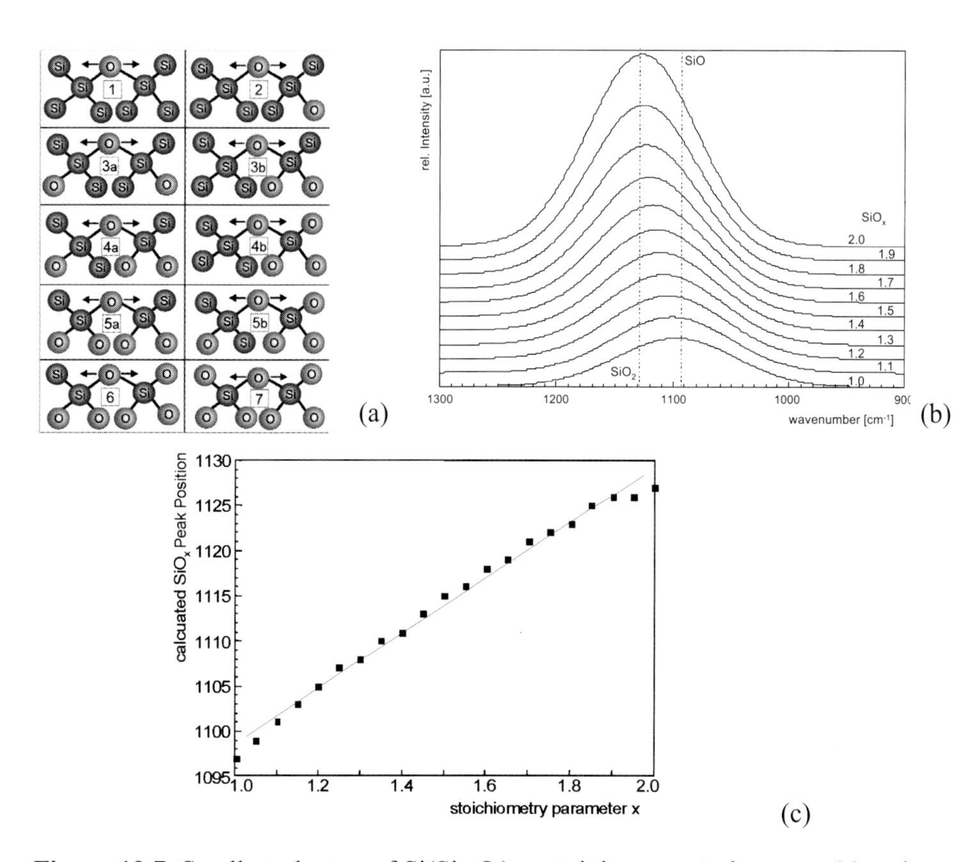

Figure 19.7. Smallest clusters of $Si(Si_{4-i}O_i)$ containing a central oxygen (a); calculated spectra for varying stoichiometry parameter x (b); Si–O–Si stretching vibration peak position of calculated spectra versus stoichiometry parameter x (c).

The bonding site of a certain oxygen at the silicon is unspecific, but the number of oxygens bonded to a silicon makes a difference in the spectra. Therefore, for clusters 3, 4, and 5, we must consider the corresponding modifications, which increases the number of clusters to ten. The probability of each of these clusters in SiO_x results from the product of probabilities of the tetrahedra involved according to the RB model. The sum of these values, subtracted from unity, gives the probability of all clusters having no central oxygen. Since these clusters do not exhibit Si–O–Si stretching vibrations, they do not contribute to the spectra under consideration. The stretching vibration frequency of each of these clusters is calculated by means of a semi-empirical molecular orbital program.[47, 48] More details of the simulation, such as the Gaussian broadening of single lines and the weighting of individual fractions according to the absorption cross-section of the clusters, are given elsewhere.[42] The effective dynamic charge of the clusters was estimated using the partial charges of the atoms involved. The spectra obtained after summation of all fractions for the stoichiometry changing from $x = 1$ to $x = 2$ are shown in Figure 19.7(b), and the essential parameters of the simulation are listed in Table 19.1.

Table 19.1. Calculated structural and vibrational characteristics of the clusters shown in Figure 19.7.

cluster	combined RBM probability	dynamic charge e*	bond angle Si–O–Si [°]	Si–O–Si stretching vibration [cm^{-1}]
1	$P_1 \times P_1$	0.643	118.7	1061
2	$P_1 \times P_2$	1.061	121.9	1072
3a	$P_2 \times P_2$	1.562	125.8	1086
3b	$P_1 \times P_3$	1.354	138.0	1126
4a	$P_2 \times P_3$	1.903	130.9	1103
4b	$P_1 \times P_4$	1.989	142.0	1144
5a	$P_3 \times P_3$	2.332	132.8	1105
5b	$P_1 \times P_3$	1.731	139.9	1135
6	$P_3 \times P_4$	2.783	134.3	1123
7	$P_4 \times P_4$	3.288	133.4	1129

There is a clear shift to higher frequencies, which is almost linearly dependent on the stoichiometry, as can be seen from the plot in Figure 19.7(c). The extent of the shift and the absolute position of the peaks do not agree completely with the experimental findings since the model is not yet sufficiently realistic. However, it allows a qualitative description of the stretching vibration peak shift in terms of restructuring effects during disproportionation. It can be used to calculate the stoichiometry of the oxygen-rich phase formed during thermal processing and to estimate the degree of disproportionation taking place.

The evaluation of rapid thermal annealing experiments with SiO_x films of $x = 1$ initial stoichiometry proves that during disproportionation two different processes occur,[42] one with a rather short time constant and one with a somewhat larger time constant. For the latter process, leading to complete disproportionation, an effective activation energy of 0.44 eV could be calculated. The first process is assumed to involve only bond switching in the close neighborhood of atoms, whereas in the second process the oxygen diffusion should be rate determining.

19.3.3 Formation of Si Nanocrystallites

As a consequence of the observed phase separation, Si nanocrystallites are expected to occur when the amorphous Si regions are large enough and the temperature is high enough to allow solid-phase crystallization. To this end, we investigated the annealed samples by high-resolution electron microscopy (HREM) using a JEM 4000 EX (JEOL) operating at 400 kV. A cross-section preparation technique was applied that enabled not only a characterization of the size, size distribution, and shape of the Si nanocrystallites, but also of their spatial distribution within the films. Formation of nanocrystallites was generally not observed below 900 °C, regardless of stoichiometry. Upon 1 h annealing at this temperature, relatively few Si nanocrystallites were found. Therefore, we chose 1 h annealing at 1000 °C to study the crystallization as a function of oxygen content. The nanocrystallites formed are uniformly distributed within the films and have approximately spherical shape. As an example, Figure 19.8(a) shows part of a HREM cross-section micrograph of a film with $x = 1.17$. The interface between single-crystalline substrate in the lower left and the SiO_x film is marked by a straight line and Si nanocrystallites are marked by arrows and circles. From the course of the lattice plane fringes imaged in these nanocrystallites, their orientation appears to be random. There is no preferred arrangement throughout the films. Frequently, the nanocrystallites exhibit planar lattice defects such as stacking faults and twin boundaries characteristic of solid-phase crystallization processes. Examples of cyclic and parallel arrangements of such defects (marked by arrows) are shown in Figures 19.8(b) and 19.8(c).

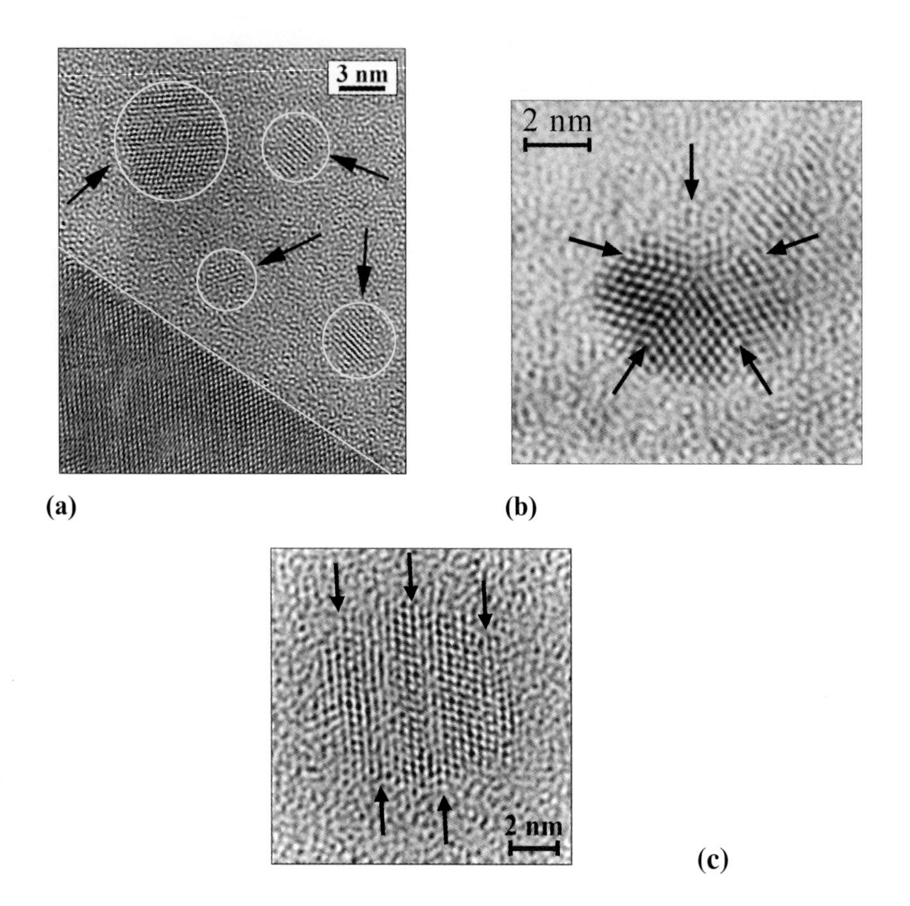

Figure 19.8. HREM images of Si nanocrystallites in an SiO_x film (x = 1.17) illustrating their random orientation (a) and examples of cyclic (b) and parallel (c) planar defects.

The size distributions of Si nanocrystallites determined for samples of various stoichiometry are represented in Figure 19.9. For equal annealing conditions, the mean nanocrystallite size is strongly correlated to the initial oxygen content of the SiO_x film. The size distribution maximum clearly shifts to smaller sizes with increasing x. Accordingly, the mean nanocrystallite size achieved upon 1 h annealing at 1000 °C drops from 4.3 nm to 3 nm with the stoichiometry parameter increasing from x = 1.17 to x = 1.63. The higher oxygen content not only prevents the formation of larger nanocrystallites, but also reduces their formation rate. Simultaneously, the size distribution becomes more narrow, which may be rationalized in terms of consumption of the available amorphous Si regions, because of their limited size, before the annealing is finished. The highest oxygen partial pressure applied resulted in a stoichiometry parameter x close to 2 for which only very few crystallites could be observed. Therefore, no accurate size distribution of this sample was determined.

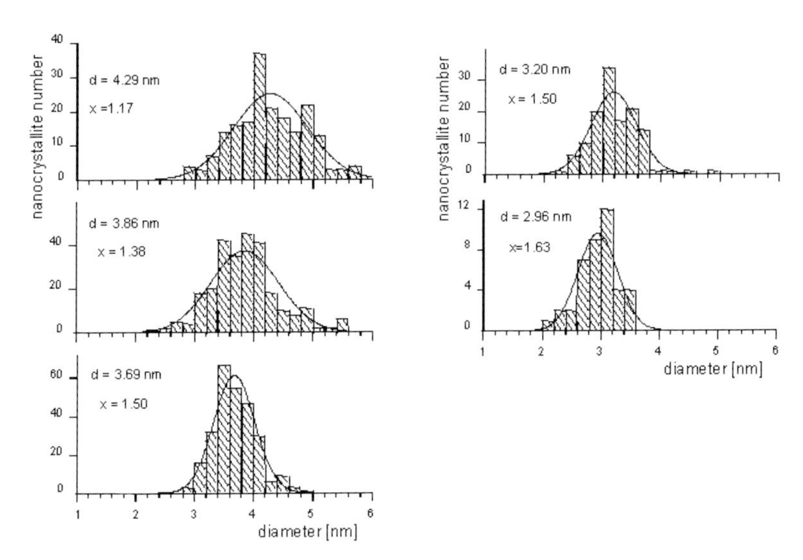

Figure 19.9. Size distribution of Si nanocrystallites in SiO_x films as a function of their oxygen content.

19.4 Conclusion

Starting from SiO vapor-deposited films, which may be regarded as nano-disproportionate materials, we have studied the evolution of further disproportionation and phase separation into a nano-particulate Si/SiO_x composite material; within which; at sufficiently high temperatures, Si nanocrystallites may form. The mean size and concentration of these nanocrystallites are mainly governed by the initial oxygen content of the films and the annealing temperature.

Their internal structure is determined by solid-phase crystallization processes.

19.5 Acknowledgements

The authors would like to express their thanks to R. Mattheis for performing the RBS measurements, S. Senz for the XRD measurements, J. Blumm for the DSC measurement, and O. Lichtenberger for the molecular orbital simulations.

6 References

[1] U. Gösele,V. Lehmann, *Materials Chemistry and Physics* **1995**, *40*, 253–259.
[2] H. Tamura, M. Rückschloss, T. Virschem, S. Veprek, *Thin Solid Films* **1995**, *255*, 92–95.
[3] H. Hofmeister, J. Dutta, H. Hofmann, *Phys. Rev. B* **1996**, *54*, 2856–2862.
[4] M. Zacharias, D. Dimova-Malinovska, M. Stutzmann, *Phil. Mag. B* **1996**, *73*, 799–816.
[5] M. H. Ludwig, A. Augustin, R. E. Hummel, T. Gross, *J. Appl. Phys.* **1996**, *80*, 5318–5324.
[6] T. Shimuzu-Iwayama, S. Nakao, K. Saitoh, *Nucl. Instrum. and Meth. B* **1996**, *120*, 97–100.
[7] A. G. Cullis, L. T. Canham, P. D. J. Calcott, *J. Appl. Phys.* **1997**, *82*, 909–965.
[8] Y. Wakayama, T. Inokuma, S. Hasegawa, *J. Cryst. Growth* **1998**, *183*, 124–130.
[9] M. A. Laguna, V. Paillard, B. Kohn, M. Ehbrecht, F. Huisken, G. Ledoux, R. Papoular, H. Hofmeister, *J. Luminesc.* **1999**, *80*, 223–228.
[10] S. Vijayalakshmi, F. Shen, H. Grebel, *Appl. Phys. Lett.* **1997**, *71*, 3332–3334.
[11] A. Dowd, R. G. Elliman, M. Samoc, B. Luther-Davies, *Appl. Phys. Lett.* **1999**, *74*, 239–241.
[12] T. Inokuma, Y. Wakayama, T. Muramoto, R. Aoki, Y. Kurata, S. Hasegawa, *J. Appl. Phys.* **1998**, *83*, 2228–2234.
[13] H. Seyfarth, R. Grötzschel, A. Markwitz, W. Matz, P. Nitzsche, L. Rebohle, *Thin Solid Films* **1998**, *330*, 202–205.
[14] S. Li, M. S. El-Shall, *Appl. Surf. Sci.* **1998**, *127–129*, 330–334.
[15] T. Makimura, Y. Kunii, N. Ono, K. Murakami, *Appl. Surf. Sci.* **1998**, *127–129*, 388–392.

[16] M. L. Bronsgerma, A. Polman, K. S. Min, E. Boer, T. Tambo, H. A. Atwater, *Appl. Phys. Lett.* **1998**, *72*, 2577–2579.
[17] Kenyon, P. F. Trwoga, C. W. Pitt, G. Rehm, *Appl. Phys. Lett.* **1998**, *73*, 523–525.
[18] I. T. Chang, F. Niu, D. Slimovici, C. Wildig, P. A. Leigh, P. J.Dobson, B. Cantor, *Mat. Sci. Forum* **1996**, *225–227*, 175–178.
[19] V. G. Baru, A. P. Chernushich, V. A. Luzanov, G. V. Stepanov, L.Yu. Zakharov, K. P. O'Donnel, I. V. Bradley, N. N. Melnik, *Appl. Phys. Lett.* **1996**, *69*, 4148–4150.
[20] T. Komoda, J. Weber, K. P. Homewood, P. L. F. Hemment, B. J. Sealy, *Nucl. Instrum. and Meth. B* **1996**, *120*, 93–96.
[21] S. Sato, H. Ono, S. Nozaki, H. Morisaki, *Nanostruct. Mat.* **1995**, *5*, 589–598.
[22] W. Calleja, C. Falcony, A. Torres, M. Aceves, R. Osorio, *Thin Solid Films* **1996**, *270*, 114–117.
[23] R. K. Soni, L. F. Fonseca, O. Resta, M. Buzaianu, S.Z. Weisz, *J. Luminesc.* **1999**, *83–84*, 187–191.
[24] K. Murakami, T. Suzuki, T. Makimura, M. Tamura, *Appl. Phys. A* **1999**, *69 (Suppl.)*, S13–S15.
[25] L. Patrone, D. Nelson, V. I. Safarov, S. Giorgio, M. Sentis, W. Marine, *Appl. Phys. A* **1999**, *69 (Suppl.)*, S217–221.
[26] L. N. Dihn, *Phys. Rev. B* **1996**, *54*, 5029–5037.
[27] A. J. Kenyon, P. F. Trwoga, C. W. Pitt, *J. Appl. Phys.* **1996**, *79*, 9291–9300.
[28] H. N. Potter, *Trans. Electrochem. Soc.* **1905**, *12*, 191–228.
[29] F. A. Shunk, *Constitution of Binary Alloys, Second Supplement*, McGraw-Hill, New York, **1969**.
[30] M. Nagamori, J.-A. Boivin, A. Claveau, *J. Non-Crystall. Solids* **1995**, *189*, 270–276.
[31] H. R. Phillipp, *J. Non-Crystall. Solids* **1972**, *8–10*, 627–632.
[32] R. J. Temkin, *J. Non-Crystall. Solids* **1975**, *17*, 215–230.
[33] W. Y. Ching, *Phys. Rev. B* **1982**, *26*, 6610–6621.
[34] C. Deneke, *Diploma Thesis*, Technical University Darmstadt, Darmstadt, **2000**.
[35] A. Hohl, T. Wieder, H. Fuess, *Z. Krist.* **2000**, *S18*, 66.
[36] I. T. H. Chang, B. Cantor, P. A. Leigh, P. A. Dobson, *Nanostructured Materials* **1995**, *6*, 835–838.
[37] L. You, C. L. Heng, S. Y. Ma, Z. C. Ma, W. H. Zong, Z. Wu, G. G. Qin, *J. Cryst. Growth* **2000**, *212*, 109–114.
[38] Y. Wakayama, T. Tagami, T. Inokuma, S. Hasegawa, S. I. Tanaka, *Recent*

Res. Devel. Crystal Growth Res. **1999**, *1*, 83–101.
[39] U. Kahler, H. Hofmeister, *Materials Science Forum* **2000**, *343–346*, 488–493.
[40] U. Kahler, H. Hofmeister, *Optical Materials* **2001**, *17*, 83–86.
[41] U. Kahler, H. Hofmeister, *Appl. Phys. A* **2002**, *74*, 13–17.
[42] U. Kahler, *Ph.D. Thesis*, University Halle-Wittenberg, Halle **2001**; ISBN 3-935316-27-5, Der Andere Verlag, Osnabrück, **2001**.
[43] J. Knights, R. Street, G. Lucovsky, *J. Non-Crystall. Solids* **1980**, *35–36*, 279–284.
[44] B. Hinds, F. Wang, D. Wolfe, C. Hinkle, G. Lucovsky, *J. Vac. Sci. Technol. B* **1998**, *16*, 2171–2176.
[45] T. Hirata, *J. Phys. Chem. Solids* **1997**, *58*, 1497–1501.
[46] D. V. Tsu, G. Lucovsky, B. N. Davidson, *Phys. Rev. B* **1989**, *40*, 1795–1805.
[47] J. Stewart, *J. Comp. Chem.* **1989**, *10*, 209–220.
[48] J. Stewart, *J. Comp. Aid. Mol. Des.* **1990**, *4*, 1–45.

20 Theoretical Treatment of Silicon Clusters

Alexander F. Sax*

20.1 Introduction

The study of small and intermediate-sized clusters has become an important research field because of the role clusters play in the explanation of the chemical and physical properties of matter on the way from molecules to solids.[1] Depending on their size, clusters can show reactivity and optical properties very different from those of molecules or solids. The great interest in silicon clusters stems mainly from the importance of silicon in microelectronics, but is also due in part to the photoluminescence properties of silicon clusters, which show some resemblance to the bright photoluminescence of porous silicon.[2] Silicon clusters are mainly produced in silicon-containing plasma as used in chemical vapor deposition processes. In these processes, gas-phase nucleation can lead to amorphous silicon films of poor quality and should be avoided.[3–6] On the other hand, controlled production of silicon clusters seems very suitable for the fabrication of nanostructured materials with a fine control on their structure, morphological, and functional properties.[7]

Theoretical investigations of bulk as well as surface properties of crystalline, porous, and amorphous solids are only possible with the help of theoretical models, which reduce the dimension of the problem to a manageable size. Very popular with solid-state physicists is the *periodic cluster approach*, where a cluster of appropriate size is periodically reproduced by applying periodic boundary conditions. When the surface of a crystal or a surface defect is considered, slabs formed by several layers of atoms are frequently used as models. The periodic cluster model is also used for the investigation of small silicon clusters. Because such calculations are generally performed in momentum space, very large unit cells have to be used.[8] In the *large cluster approach* a cluster of some hundred atoms treated like a molecule simulates the solid. In the case of covalent crystals such as silicon, hydrogen atoms are frequently used to saturate the unpaired electrons at the surface.

* Address of the author: Institut für Chemie, Karl-Franzens-Universität Graz, Strassoldogasse 10, A-8010 Graz, Austria

The description of defects is frequently achieved with the *embedded cluster approach*, where a small cluster, which includes the defect, is embedded in the surrounding bulk. This model is usefully applied when the bulk can be described with simpler methods than the embedded cluster. Embedding of a cluster can be done both with and without use of periodic boundary conditions. Therefore, the embedded cluster model can be seen merely as a modification of the first two models.[9]

In this article, only small gas-phase clusters will be discussed, as well as the use of clusters in the description of solids. Mesoscopic systems like nanoparticles will not be considered.

20.2 Small Silicon Clusters

Silicon clusters, both bare and hydrogenated, have been the focus of numerous experimental and theoretical investigations over the past decade (see, for example, refs. [10-13] and literature cited therein). The theoretical investigations comprise studies on the structure and stability, electronic and optical properties, and chemical reactivities of neutral and ionized clusters with up to 90 silicon atoms. Calculations have been performed with methods at very different levels of theory: classical model potentials,[14,15] tight-binding,[15,16] semi-empirical Hartree–Fock (HF),[12,17] all-electron HF and post-HF methods like Moller–Plesset perturbation theory of the fourth order,[18–20] generalized valence bond (GVB) wavefunctions,[21] density functional methods from local density approximation (LDA) [13,14] to generalized-gradient-approximations (GGA),[22–24] and Quantum Monte Carlo.[13,25]

20.2.1 Bare Silicon Clusters

The diatomic Si_2 molecule can be seen as the smallest bare silicon cluster, which has only one equilibrium structure. Starting with the Si_3 molecule all clusters have several possible structures; the most stable ones for Si_3 to Si_7 are: an isosceles triangle, a planar rhombus, a flattened trigonal bipyramid, an edge-capped trigonal bipyramid, and a tricapped tetrahedron, respectively.[18–20] For these clusters, the following construction scheme was found: each Si_n can be built from a smaller cluster Si_{n-1} by adding an Si atom at an appropriate edge- or face-capping site. For small clusters the edge-capped structures are favored; for intermediate clusters the face-capped clusters become comparable in energy. Si_7 is a tricapped tetrahedron and Si_{10} is a tetracapped octahedron. Compactness is however not the only criterion of stability; the tetragonal pyramid of Si_5 where the apex atom forms four bonds is less stable than the trigonal bipyramid where each of the two apex atoms forms only three bonds.

The cohesive energy, that is the binding energy per atom, calculated at the MP4 level, increases from 1.56 eV in Si_2 to 3.82 eV in Si_{10}, or from 33.7% to 82.5% of the bulk cohesive energy. This is surprising because, for metallic clusters, the corresponding fraction is much smaller. In silicon clusters, this is explained in terms of the formation of different kinds of bonds between silicon atoms in other than sp^3 hybridization states.

The stabilities of clusters can be better compared on the basis of incremental binding energies, that is the reaction energy of the reaction $Si_n \rightarrow Si_{n-1} + Si$. These energies show the existence of clusters which are especially stable, like Si_4, Si_6, Si_7 or Si_{10}. Grossmann and Mitáš [13] investigated silicon clusters with up to 20 silicon atoms using both LDA DFT and Monte Carlo methods. For $n \leq 10$, they compared their results with those of Raghavachari et al., and found good agreement in geometries for all methods used; LDA, however, overestimates the binding energy by 15 to 20%. The results of Grossmann and Mitáš show that Si_{13} and Si_{20} are also very stable.

For large cluster sizes, the problem of finding the global minimum becomes more and more important. Although the two possible ways of capping an existing structure lead to plausible starting geometries from where the minimum search can start, the number and the stability of all possible isomers is not known. It turns out that cluster isomers with high symmetry are often less stable than those with lower symmetry. For example, the dodecahedral Si_{20} is by 4 eV less stable than an elongated, stacked cluster with C_{3v} symmetry. In this cluster structure all atoms except for the two capping atoms are four-coordinate.

Studies on large clusters are therefore carried out either to investigate a special class of structures like elongated, stacked clusters or to investigate only clusters with magic numbers to find an explanation for their exceptional stability. Grossmann and Mitáš [13,25] in their Monte Carlo study of Si_n with $n \leq 50$ investigated elongated clusters, which are regarded as important in the construction of nanoscale wires. They found that these structures can be regarded as stacked, equilateral triangles in planes parallel to the *xy* plane with a common *z* axis and one or two capping atoms along the *z*-axis. The triangles are rotated by 60° with respect to their neighboring triangles. An Si_6 chair (D_{3d} symmetry) consists, according to this scheme, of two stacked triangles, labeled A and B, without capping atoms; an Si_{26} tube consists of 8 triangles or 4 AB pairs and two capping atoms. If the number of triangles is even, there is a pronounced trend toward chair-like six-membered rings (one AB pair), which can also be seen as building blocks, and can be stacked along an axis. Except for the capping atoms all silicon atoms are four-coordinate.

In the case of an odd number of triangles, the clusters differ considerably from their even counterparts because here the triangles no longer pair, i.e. there is no tendency to form six-membered rings, and the bonding and coordination of the atoms varies considerably. Clusters with an odd number of triangles and $n \geq 29$ have a significantly larger cohesive energy than smaller clusters with odd n. LDA was found to correctly predict the energy differences between the various isomers.

20.2.2 Hydrogenated Silicon Clusters

The optical band gap of silicon clusters increases with decreasing cluster size and all surface atoms with dangling bonds contribute to mid-gap electronic states (defects) which are assumed to quench photoluminescence. Saturation of the surface dangling bonds with hydrogen is the simplest way to eliminate defects, and therefore theoretical studies of the electronic structure of silicon clusters without defects are most easily performed on hydrogenated silicon clusters. The thermodynamic stability of these species is also of interest, because hydrogenated silicon clusters are formed in thermal and plasma CVD of silane.[3–6]

Theoretical studies on Si_nH_m have been carried out in a rather unsystematic way because of the great variety of possible structures. Such clusters are frequently investigated for cases in which the bare clusters show great stability.[23] In this study, bare and hydrogenated clusters with 5, 17, 29, 35, and 47 silicon atoms were investigated using DFT at the B3LYP level. A study on Si_6H_x clusters ($1 \leq x \leq 14$) showed that there are many structures which can be related to tetrahedral-bond-network clusters,[22] made of bent bonds, long bonds, and lone pairs, and polyhedral bonding network clusters, where triangular faces or tetrahedral interstices share electron pairs, and which are much more stable than hexasilaprismane.[26,27] A common feature of all these studies was that the selection of the hydrogenated clusters was either led by the stabilities of the parent bare clusters or motivated by a comparison with molecules such as hexasilaprismane. A systematic search for stable structures is only possible when techniques like simulated annealing or evolutionary algorithms are used.[28]

The question "What happens when one hydrogen atom is attached to a bare silicon cluster?" was investigated by Balamurugan and Prasad,[8] who calculated the influence of the hydrogen atom on ground-state geometry, total energy, and the first excited energy level in the series Si_nH, with $2 \leq n \leq 10$. Based on Car–Parinello molecular dynamics results using LDA-DFT, they state that hydrogen forms bonds to only one silicon atom in all clusters except Si_2H and Si_3H, where bridged bonds like those in the butterfly isomer of Si_2H_2 are found. However, these authors do not mention that for Si_2H two bridged structures exist, corresponding to local minima on the potential energy surfaces for the $^2A'$ and the $^2A''$ electronic states. These two local minima differ in the Si-Si bond length by about 10 pm but are nearly isoenergetic; the energy difference is only 2 kJ/mol.[29] For the $^2A'$ state, Kalcher and Sax also found a structure with a conventional Si-H bond and this structure is 1 kJ/mol lower in energy than the bridged structure; the saddle point connecting the two local minima is only 40 kJ/mol higher. These results were obtained with multi-reference CI methods. Using coupled cluster methods, Kalcher and Sax [30] likewise found for Si_3H on the $^2A'$ surface two local minima corresponding to a bridged structure and a structure with a conventional Si-H bond. Here, the bridged structure is 6 kJ/mol lower in energy, but the saddle point is only 24 kJ/mol above the bridged minimum. This means that such small clusters are extremely dynamic and it makes little sense to speak of such local minima as if they correspond to rigid structures.

20.3 Silicon Clusters in Solid State Investigations

Two quite different approaches are used to describe solids with the help of clusters. In the first approach a periodic cluster represents the solid, while in the second it is represented by a single large cluster, which is treated like a large molecule. An important difference lies in the way the electronic Schrödinger equation is solved: periodic cluster calculations are done in momentum space or *k* space, whereas large clusters are treated like molecules in real space. The theoretical methods used are of different levels of sophistication. They range from tight-binding (i.e. extended Hückel) and semi-empirical methods (see, for example, refs. [31–33]) to DFT-LDA coupled with various pseudopotentials;[34–36] GGA-DFT (generalized gradient approach) or quantum chemical *ab initio* methods are very rarely used.[31]

20.3.1 The Periodic Cluster Approach

In the periodic cluster approach, a supercell consisting of a small number of unit cells on which periodic boundary conditions are imposed represents a crystal. This simulates the translational symmetry of the supercell in an ideal crystal. For example, the unit cell for the diamond lattice contains 8 atoms, and so in solid state calculations a supercell made from $2^3 = 8$ unit cells with 64 atoms is frequently used; the next largest supercell is made from $3^3 = 27$ unit cells with 216 atoms, etc. The electronic spectrum of a supercell of carbon or silicon atoms contains not only bonding and antibonding molecular orbitals (MO) but also several MOs from the unsaturated electrons of the surface atoms. Due to their non-bonding character, these MOs lie in the energy gap between bonding and antibonding MOs. The surface dangling bonds are saturated by applying periodic boundary conditions, and the mid-gap energy levels disappear. A periodic cluster has a discretized electronic spectrum instead of continuous bands; the electronic structure is calculated mainly in k space, and the wavefunction is spanned by plane waves. The combination of LDA-DFT and LAPW (Linear Augmented Plane Waves) is frequently used for such calculations; for large supercells the much simpler tight-binding method is frequently used.

In the periodic cluster approach, the size of the supercell influences the number of orbitals and thus the discrete representation of the band structure. Using this model to describe localized defects such as vacancies introduces an additional difficulty because now the spatial extent of the vacancy wavefunction must be related to the size of the supercell. A 64-atom supercell was found to be too small to adequately describe the vacancy wavefunction,[36,37] the amplitude of the wave function of a monovacancy at the cell boundary is still $^1/_4$ of the maximum value. This means that the wavefunction has a considerable overlap with the vacancy wavefunctions of the neighboring cells. This artificial defect–defect interaction leads to a dispersion of about 0.8 eV for the vacancy states. When a 216-atom supercell is considered, the wavefunction amplitude at the cell boundary is only $^1/_8$ of the maximum value and the dispersion is less than 0.2 eV.[37] A supercell of at least this size was therefore recommended for a correct description of a monovacancy. Öğüt et al. found that only a cluster with 12 shells of silicon atoms around the vacancy contains about 95% of the squared vacancy wavefunction.[36] The literature does not give a clear answer as to how strongly the geometric structure of vacancies depends on the size of the supercell because frequently only incomplete geometry relaxation is allowed (see below).

20.3.2 The Large Cluster Approach

A large cluster treated like a large molecule is a very old model of a solid going back to Coulson,[38,39] and defect molecules are used to describe defects in solids.

When the central silicon atom in *neo*-Si_5H_{12} is removed, the four silyl groups in a tetrahedral arrangement constitute the simplest model of a monovacancy in silicon with only first-silicon neighbors. If X represents a monovacancy, then the spherical clusters XSi_4H_{12}, $XSi_{16}H_{36}$, $XSi_{32}H_{36}$, $XSi_{34}H_{36}$ are defect molecules with inclusion of the first-, second-, third-, and fourth-silicon neighbor shells. Such clusters have frequently been used to calculate the energies of the orbitals representing the vacancy, and the HOMO of the perfect cluster has been regarded as a good estimate of the band edge of the valence band. Comparison of the orbital energies of the two calculations showed, however, that the a_1 orbital of the vacancy may lie above the band edge in contrast to the results from calculations not based on the cluster model. The reason for this is that the orbitals of the perfect cluster cannot correctly describe bulk silicon due to both the small size of the clusters and the terminating hydrogen atoms. This means a non-physical interaction of the vacancy with the surface, similar to the spurious defect–defect interactions in the periodic cluster approach. This problem becomes even more pronounced when semi-empirical methods are used, where the influence of parameterization on the results must also be accounted for. Semi-empirical quantum chemistry programs have frequently been used because of their success in organic chemistry calculations without considering that the silicon parameters are somewhat less reliable than those for carbon, oxygen, and other "organic" elements. Good reviews of these problems are given in refs. [32,33]. All in all, the large cluster approach has never been the dominant method used for solid-state investigations.

In the last few years, however, the large cluster approach has again become attractive for investigations of defects in materials due to advances in electronic structure algorithms such as "real-space" methods.[40–43] One main advantage is the straightforward application to charged systems without the need to add neutralizing background charges, but real-space methods can also easily be adapted to parallel computers. Combined with efficient pseudopotential techniques, real-space methods are successfully used to investigate the physical properties of defects in materials.[31,36]

20.3.3 Methodic Problems

It has been shown that in both approaches the size of the model system is of great importance with regard to the reliability of the results. This is true for the geometric structure of vacancies as well as for their respective electronic properties. However, it would be wrong to believe that increasing the size of the supercell or the large cluster will guarantee relevant results without consideration of the methods used for the calculation of the electronic wavefunction and for the geometry relaxation. Öğüt et al. found, for example, a relaxation energy of 0.9 eV for the monovacancy in a $Si_{166}H_{124}$ cluster with DFT using the Ceperley–Alder exchange-correlation functional; this is larger than the 0.36 eV obtained from a 64-atom supercell calculation using the same exchange-correlation functional,[36] but much smaller than the tight-binding values of 1.9 eV [44] and 1.4 eV [45] obtained with 512- and 64-atom supercell calculations, respectively. Any energetic comparison of silicon clusters is only possible when a complete geometry optimization has been performed.

It is obvious that semi-empirical methods are especially attractive for time-consuming calculations such as geometry relaxations of large supercells. However, semi-empirical methods are also used when small supercells are relaxed, and even for such calculations complete geometry relaxation is frequently avoided. Roberson and Estreicher investigated neutral and charged vacancies and used hydrogenated silicon clusters up to $Si_{44}H_{42}$ to approximate the solid. Only the four nearest neighbors of the vacancy were allowed to relax. They found only a slight distortion of the original tetrahedral configuration. When the Si-Si bond length in bulk silicon is 2.35 Å, the distance between the four neighboring atoms is 3.84 Å. In the relaxed geometry the silicon atoms are only slightly displaced, and the authors find "bonds" between two pairs of silicon atoms with a bond length of 3.76 Å. This result is completely at variance with other investigations where complete geometry relaxation was allowed. Clark obtained bond lengths of 2.4 Å with a 64-atom supercell when complete geometry relaxation was allowed,[46] and similar results were obtained by Krüger and Sax with QM/MM methods.[47] It is, however, not clear whether the incomplete geometry relaxation or the semi-empirical PRDDO method used by Roberson and Estreicher is responsible for their results. Clark, as well as Krueger and Sax, used DFT methods for the geometry optimization. Hastings et al. [35] calculated the energy gain by successively removing silicon atoms starting from the monovacancy up to the hexavacancy. Using a 64-atom supercell and the PRDDO method, they optimized the geometries of the vacancies but allowed only the nearest neighbors to the vacancies to relax. These geometries were then used for subsequent single-point *ab initio* Hartree–Fock calculations. Bearing in mind the results of Roberson one must be doubtful of the reliability of these results.

20.3.4 QM/MM Techniques for Solid-State Calculations

Clark showed in his Ph.D. thesis [46] that for the monovacancy new bonds are formed in the first shell of neighboring atoms, and that major relaxation of the bulk occurs in the second shell and only minor changes in the third shell. This demonstrates nicely that the structure of local defects like monovacancies can be regarded as a relatively small part of the crystal where the atoms deviate considerably from their respective positions in a perfect crystal whereas the surrounding bulk is nearly undisturbed. The quantum chemical QM/MM hybrid method is one of several methods that can be used in the embedded cluster approach. In QM/MM methods, a large system is divided into a small part, the core, and a larger part, the bulk. Electron reorganization through bond breaking or bond formation may occur in the core, and therefore this part must be treated with quantum theoretical methods. The bulk is necessary to provide an elastic environment in which the core is embedded. It is mainly needed to prevent the core from collapsing or dissociating, and therefore a method of a much lower theoretical level can be used; for a review see ref. [48]. We chose this method for a treatment of multivacancies ranging from the monovacancy to various tetravacancies.[47,49] We used the density functional BP86 as a high-level method for the core and the semi-empirical method AM1 as a low-level method for the bulk; these methods were combined through Morokuma's subtractive QM/MM scheme ONIOM.[50]

The starting point for the calculation of the electronic structure of vacancies is a highly symmetrical distribution of unpaired electrons (dangling bonds) in the first silicon neighbor shell. For example, four tetrahedrally placed dangling bonds about a monovacancy give rise to a low-lying non-degenerate and a triply-degenerate orbital. In the electronic ground state, two electrons must be placed into the degenerate orbitals, which results in a degenerate electronic state. Through Jahn–Teller displacements of the atoms the degeneracy is lifted, but the direction of the displacements depends on which component of the degenerate orbital is occupied. In a divacancy, each of the removed silicon atoms is surrounded by three dangling bonds with local C_3 symmetry. From each set of three, we obtain a low-lying non-degenerate and a doubly-degenerate orbital, into which one electron can be placed. Linear combinations of the degenerate local orbitals give two molecular orbitals for the vacancy describing two types of bonding situations. In the first, a single bond is formed between two silicon atoms at each end of the vacancy. In the second, new bonds are formed between all atoms with dangling bonds, and one atom becomes pentacoordinate. A similar situation is found for the trivacancy, and for a tetravacancy where four silicon atoms in a zig-zag arrangement are removed. When a tetravacancy is formed by removing four silicon atoms in a pyramidal arrangement, three sets of three of dangling bonds with local C_3 symmetry are found. In the optimized structure of this vacancy, six silicon atoms with dangling bonds, together with three bulk silicon atoms, form a nine-membered ring.

A great advantage of QM/MM methods is that laborious geometry optimization need not be done for the whole system with the high-level method, but much work has still to be done to find the best combination of high- and low-level methods for solid-state investigations. Another advantage of QM/MM is that traditional quantum chemistry techniques can be applied to investigate electron excitations of the localized vacancy, which is treated as a molecule.

However, all caveats concerning spurious interactions of the vacancy with the cluster surface mentioned above also hold for QM/MM methods. One must be very careful with orbitals that are used for the construction of the excited states, because they may be considerably more extensive than the core. Of course, one can make the size of cores as large, as in the large cluster approach, but they will always be embedded in a larger bulk for which no quantum theoretical method need be applied. This should make QM/MM methods especially attractive for applications in materials science.

20.4 References

[1] P. Jena, B. K. Rao, S. N. Khanna (Eds.), *Physics and Chemistry of Small Clusters*, Plenum, New York, 1987.

[2] M. F. Jarrold, *Science* **1991**, *252*, 1085.

[3] M. T. Swihart, S. L. Girshick, *J. Phys. Chem.* **1999**, *103*, 64.

[4] S. L. Girshick, M. T. Swihart, S.-M. Suh, M. R. Mahajan, S. Nijhawan, *Proc. Electrochem. Soc.* **1999**, *98-23*, 215.

[5] U. R. Kortshagen, U. V. Bhandarkar, M. T. Swihart, S. L. Girshick, *Pure Appl. Chem.* **1999**, *7,* 1871.

[6] U. V. Bhandarkar, M. T. Swihart, S. L. Girshick, U. R. Kortshagen, *J. Phys. D* **2000**, *33*, 2731.

[7] S. Iannotta, P. Milani, *AIP Conf. Proc.* **2001**, *576*, 979.

[8] D. Balamurugan, R. Prasad, *Phys. Rev. B* **2001**, *64*, 205406.

[9] G. Pacchioni, P. S. Bagus, F. Parmigiani (Eds.), *NATO ASI Series, Series B*, Vol. 283: *Cluster Models for Surface and Bulk Phenomena*, Plenum, New York, 1992.

[10] G. A. Rechtsteiner, O. Hampe, M. F. Jarrold, *J. Phys. Chem. B* **2001**, *105*, 4188.

[11] H. Hiura, T. Kanayama, *Chem. Phys. Lett.* **2000**, *328*, 409.
[12] V. Meleshko, Yu. Morokov, V. Schweigert, *Chem. Phys. Lett.* **1999**, *300*, 118.
[13] J. C. Grossmann, L. Mitáš, *Phys. Rev. B* **1995**, *52*, 16735.
[14] D. W. Brenner, B. I. Dunlap, J. A. Harrison, J. W. Wintmire, R. C. Mowrey, D. H. Robertson, C. T. White, *Phys. Rev. B* **1991**, *44*, 3479.
[15] J. Pan, M. V. Ramakrishna, *Phys. Rev. B* **1994**, *50*, 15431.
[16] D. R. Alfonso, S. Wu, C. S. Jayanthi, E. Kaxirias, *Phys. Rev. B* **1999**, *59*, 7745.
[17] A. M. Mazzone, *Phys. Rev. B* **1996**, *54*, 5970.
[18] K. Raghavachari, V. Logovinsky, *Phys. Rev. Lett.* **1985**, *55*, 2853.
[19] K. Raghavachari, *J. Chem. Phys.* **1986**, *84,* 5672.
[20] K. Raghavachari, M. C. Rohlfing, *J. Chem. Phys.* **1988**, *89*, 2219.
[21] C. H. Patterson, R. P. Messmer, *Phys. Rev. B* **1990**, *42,* 7530.
[22] T. Miyazaki, T. Uda, I. Štich, K. Terakura, *Surf. Sci.* **1997**, *377-379*, 1046.
[23] Ş. Katircioğlu, Ş. Erkoç, *Physica E* **2001**, *9,* 314.
[24] K. Deng, J. Yang, L. Yuan, Q. Zhu, *Phys. Rev. A* **2000**, *62*, 045201.
[25] J. C. Grossmann, L. Mitáš, *Phys. Rev. Lett.* **1995**, *74*, 1323.
[26] S. Nagase, T. Kudo, M. J. Aoki, *J. Chem. Soc., Chem. Commun.* **1985**, 1121.
[27] A. Sax, R. Janoschek, *Angew. Chem. Int. Ed. Engl.* **1986,** *25* 651.
[28] B. Hartke, *J. Comput. Chem.* **1999**, *20*, 1752.
[29] J. Kalcher, A. F. Sax, *Chem. Phys. Lett.* **1993**, *215*, 601.
[30] J. Kalcher, A. F. Sax, *Chem. Phys. Lett.* **1996**, *259*, 165.
[31] M. A. Roberson, S. K. Estreicher, *Phys. Rev. B* **1994**, *49*, 17040
[32] P. Deák, L. C. Snyder, R. K. Singh, J. W. Corbett, *Phys. Rev. B* **1987**, *36*, 9612.
[33] P. Deák, L. C. Snyder, *Phys. Rev. B* **1987**, *36*, 9619.
[34] H. Seong, L. J. Lewis, *Phys. Rev. B* **1996**, *53*, 9791.
[35] J. L. Hastings, S. K. Estreicher, P. A. Fedders, *Phys. Rev. B* **1997**, *56*, 10215.
[36] S. Öğüt, H. Kim, J. R. Chelikowsky, *Phys. Rev. B.* **1997**, *56*, R11353.
[37] O. Sugino, A. Oshiyama, *Phys. Rev. Lett.* **1992**, *68*, 1858.
[38] C. A. Coulson, M. J. Kearsley, *Proc. Roy. Soc. A* **1957**, *241*, 433.
[39] T. Yamaguchi, *J. Phys. Soc. Jpn.* **1962**, *17* 1359.
[40] J. R. Chelikowsky, N. Troullier, Y. Saad, *Phys. Rev. Lett.* **1994**, *72*, 1240.
[41] E. L. Briggs, D. J. Sullivan, J. Bernholc, *Phys. Rev. B* **1995**, *52*, R5471.
[42] F. Gypi, G. Galli, *Phys. Rev. B* **1995**, *52*, R2229.
[43] G. Zumbach, N. A. Modine, E. Kaxiras, *Solid State Commun.* **1996**, *99*, 57.
[44] C. Z. Wang, C. T. Chan, K. M. Ho, *Phys. Rev. Lett.* **1991**, *66*, 189.
[45] E. G. Song, E. Kim, Y. H. Lee, Y. G. Hwang, *Phys. Rev. B* **1993**, *48*, 1486.
[46] http: //cmt.dur.ac.uk/sjc/thesis/thesis/thesis.html

[47] T. Krüger, A. F. Sax, *Phys. Rev. B* **2001**, *64*, 5201.
[48] J. Gao, *Rev. Comput. Chem.* **1993**, *7*, 119.
[49] T. Krüger, A. F. Sax, *Physica B* **2002**, *308-310,* 155.
[50] S. Dapprich, I. Komaromi, K. S. Byun, K. Morokuma, M. J. Frisch, *J. Mol. Struct. (Theochem)* **1999,** *461-462*, 1.

21 Isomers of Neutral Silicon Clusters

Rolf Schäfer*‡, Michael Rosemeyer#, Christian Herwig# and Jörg August Becker#

21.1 Introduction

Semiconductor clusters like Si_N have been the subject of many investigations concerning the evolution from molecules through clusters to solid materials, revealing a significant change in electronic, chemical, and structural properties with particle size.[1–4] It turned out that some of the most interesting systems are silicon clusters in the size range of N = 20–100 for which different structural types are observed. For Si_N^+ and Si_N^- clusters, such isomers were first identified by drift mobility measurements of silicon cluster ions in helium buffer gas by Jarrold's group.[5–7] The results were rationalized by classifying the isomers by their global structure, one type with spherical and the other with elongated shape. Since the different isomers could play a crucial role in the growth kinetics of solid semiconductors, it is important to gain more experimental information on the members of these isomer classes of semiconductor clusters. Mobility measurements are one of the few experimental methods allowing size-selective studies on structural properties, although they are limited to charged clusters.

Several quantum chemical calculations [8–10] support the idea of spherically and non-spherically shaped silicon cluster classes as well, but the distribution of isomers on the internal energy scale of the isomers is still unknown. There is theoretical evidence that at room temperature at least several isomers within each class have to be considered. Theoretical ab initio calculations and dynamical simulations for a temperature of 300 K show that the energy surfaces of the elongated silicon clusters are very flat and thermal fluctuations can cause bond-breaking and subsequent rebonding on a time scale of about 0.1 ps without a significant change in the overall shape.[11] The coexistence of several isomers is important for the isomerization rates of elongated to spherical clusters, especially if they affect the activation entropies. The activation energies for the isomerization of cluster cations obtained from drift mobility experiments have been discussed on the basis of the RRK

* Address of the authors: ‡Institut für Physikalische Chemie, Technische Universität Darmstadt, Petersenstraße 20, D-64287 Darmstadt, Germany
# Institut für Physikalische Chemie und Elektrochemie, Universität Hannover, Callinstraße 3-3A, D-30167 Hannover, Germany

(Rice–Ramsperger–Kassel) theory,[12] giving activation energies of 1.0–1.5 eV, but entropic effects were not considered in this analysis.
Since cluster ions and neutral clusters may have different geometrical structures, it was important to develop advanced experimental methods that offer new information on neutral silicon clusters.

21.2 Experimental

21.2.1 Preparation of Neutral Silicon Clusters

In order to investigate the properties of neutral silicon clusters, we set up a molecular beam experiment,[13, 14] in which the silicon clusters are generated in a laser vaporization cluster source. A small amount of material is evaporated from a silicon rod in the presence of helium with a nanosecond pulse of an Nd:YAG laser. By collisions with helium the silicon atoms are rapidly cooled down and start to form clusters. The mixture of helium atoms and silicon clusters is then expanded through a nozzle into a high-vacuum apparatus forming a molecular beam. The molecular beam is an appropriate starting point for a detailed investigation of neutral silicon clusters. Photoionization mass spectrometry allows one, for example, to determine the size distribution of the neutral clusters in the molecular beam. A schematic drawing of the laser vaporization cluster source together with a typical time-of-flight mass spectrum of neutral silicon clusters is shown in Figure 21.1.

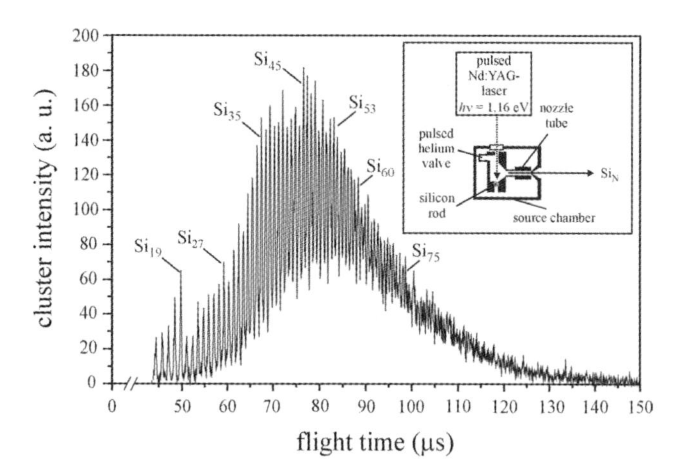

Figure 21.1. Photoionization time-of-flight mass spectrum of neutral silicon clusters. The clusters have been ionized with an excimer laser ($h\nu = 7.89$ eV). In the inset a schematic drawing of the pulsed laser vaporization cluster source is shown.

21.2.2 Calorimetric Studies on Silicon Clusters

The identification of different isomers requires the measurement of significant physical or chemical properties of the neutral silicon clusters, by which one can discriminate between possible existing cluster isomers. Since the various cluster isomers should have different thermodynamic properties, a measurement of the binding energies of silicon clusters should be adequate to identify different isomers. In order to measure the binding energies of neutral clusters, we have applied highly sensitive calorimetric techniques, which enable the detection of the heat of condensation during cluster deposition on a solid substrate. Such a device, based on a pyroelectric polymer foil,[15] permits one to probe a heat of condensation of a few nanojoules, which corresponds to the deposition of 10^8 silicon clusters with an average size of about 100 atoms.[13] If one takes the enhanced kinetic energy of the clusters in the molecular beam experiments into account, it is possible to determine the binding energies of the neutral silicon clusters from the measured heats of condensation.[16] Within the limits of sensitivity of the calorimeter, binding energies of neutral silicon clusters having a mean cluster size of $\overline{N}$ = 65–890 atoms, could be measured. The half-width of the investigated size distributions was typically 100 atoms.

21.2.3 Collision Cross-Sections of Silicon Clusters

An alternative method to investigate silicon cluster isomers is to study cluster–rare gas collisions, because one would expect the different shapes of the various isomers to influence their collision cross-sections. Therefore, we have investigated how efficiently the clusters are accelerated during the expansion at the end of a nozzle tube due to collisions with rare gas atoms by measuring the terminal velocity that the clusters attain in the molecular beam.[14] Depending on their collision cross-sections and masses, the clusters reach a lower velocity than the helium atoms, i.e. there is always a "velocity-slip" between clusters and carrier gas.[17] In order to estimate how the velocity slip can be affected by the shape of cluster isomers one can use a hard-sphere model to calculate the helium–cluster–collision numbers.[14] According to the hard-sphere model, the isomers having a less compact shape are faster than the corresponding compact, spherical-like clusters.

The starting point of the velocity measurement is provided by a system of two mechanical shutters that determine a time window during which the beam can pass through. The finishing point is determined by the pulsed ionization laser beam crossing the molecular beam. The ionized clusters are then counted with a time-of-flight mass spectrometer (TOF-MS). This is shown schematically in Figure 21.2 (a). The cluster intensity is measured as a function of the time of ionization. The time difference between the opening of the shutter system and the maximum of the cluster intensity allows one to calculate the mean cluster velocity, taking into account the distance between the shutter system and the point at which the ionization laser crosses the molecular beam.

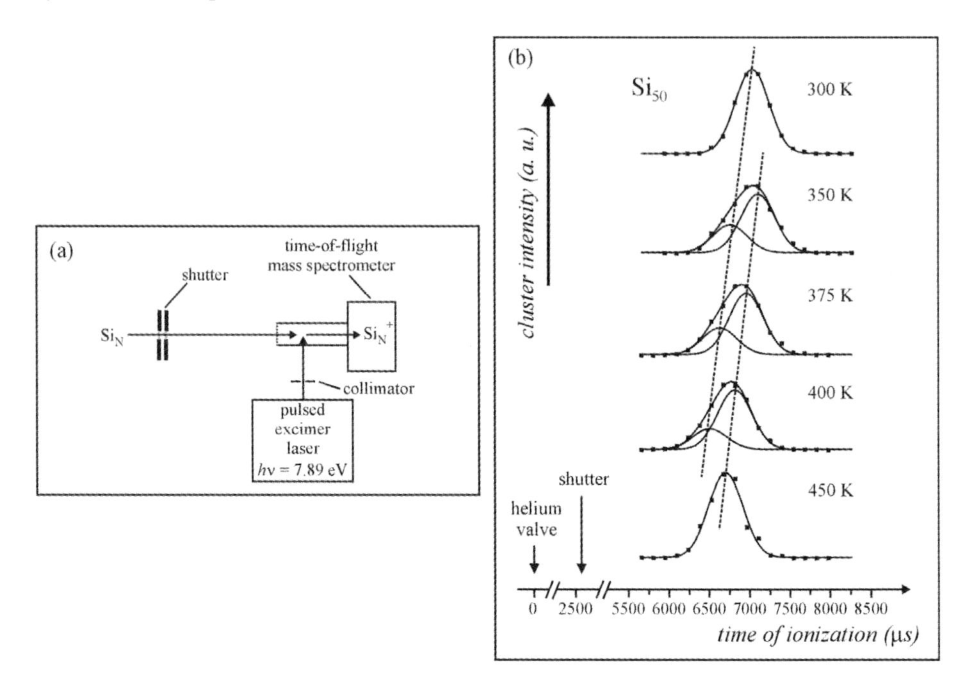

Figure 21.2. Schematic outline of the apparatus used for the measurement of cluster velocities (a) and typical cluster intensity profiles for different nozzle tube temperatures (b). The broken lines indicate the shift of the maxima of the intensity profiles with increasing nozzle tube temperature for the two different cluster isomers.

21.3 Evidence for Neutral Silicon Cluster Isomers from Calorimetric Studies

In order to investigate different neutral cluster size distributions, we also used a shutter system that allows one to select different regions of the cluster pulse for the calorimetric investigations.[13] Since the various cluster sizes are not homogeneously distributed over the total cluster pulse, this technique enables one to attain a powerful size selection for neutral clusters. The binding energies per atom of silicon clusters, e_{bin}, for different mean cluster sizes $\overline{N}$, determined with the pyroelectric calorimeter, are plotted against $\overline{N}^{-1/3}$ in Figure 21.3.

Clearly, two different classes of silicon cluster could be identified by their size-dependent binding energies. One class of silicon clusters has binding energies proportional to $\overline{N}^{-1/3}$. This size dependence is expected for compact, more spherically shaped clusters, because the number of surface to volume atoms of these structures decreases with $\overline{N}^{-1/3}$ for increasing cluster size. Therefore, it is obvious to identify the silicon clusters with the enhanced binding energies, i. e. the more stable ones, as silicon cluster isomers having a more compact, spherical-like shape. The physical basis for this simple energetic argument is the so-called droplet model. With this model, it is possible to predict quantitatively the size dependence of the binding energies, taking the surface energy and the density of bulk silicon into account.

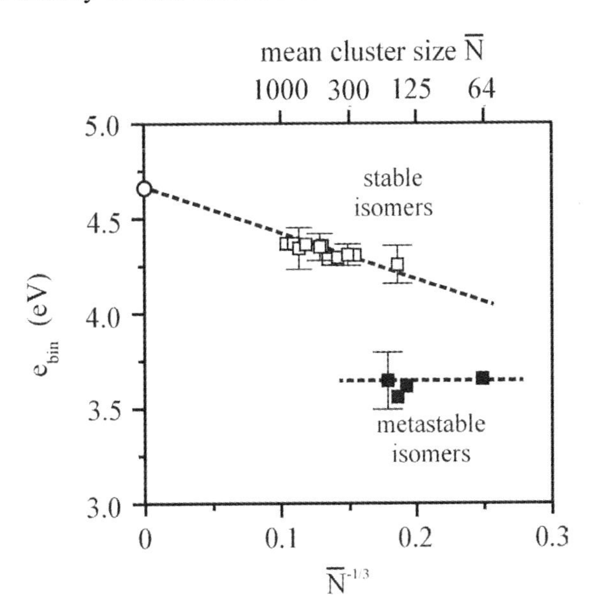

Figure 21.3. Binding energies per atom, e_{bin}, for neutral silicon cluster size distributions determined with the pyroelectric calorimeter. The half-width of the size distributions is about 50 atoms for smaller cluster sizes and about 250 atoms for larger cluster sizes. The open circle represents the bulk value of the cohesive energy of silicon.[18]

These more stable, larger silicon clusters have been found later in the cluster pulse.[13] If one investigates smaller clusters, which are present earlier in the cluster pulse, the size dependence of the binding energies changes dramatically. The binding energies of these silicon clusters are much smaller and do not depend on size within the accuracy of the measurements, as shown in Figure 21.3. Clearly, for $\overline{N}$ = 100–200 there is an overlap in cluster size between these two classes of silicon clusters. This indicates that, for a given cluster size, two different types of cluster isomers exist in the molecular beam. If the size dependence of the binding energies of this second class of isomers could be also explained in terms of the number of surface to volume atoms, one would expect this ratio to be size-independent for these silicon clusters. Therefore, these metastable clusters might have less compact structures compared to the more stable ones. One possibility to account for the size dependence of the measured binding energies might be elongated or cylindrically shaped clusters. However, other cluster structures might be equally likely to describe the measured binding energies. Since it is hardly possible to ascribe a definite structure or shape to the different cluster isomers from the binding energies, a second experiment is necessary, which allows one to obtain size-selected structural information on different cluster isomers in a more direct way.

21.4 Collision Cross-Sections and Shape of Neutral Silicon Cluster Isomers

Figure 21.2 (b) presents the intensities of Si_N clusters with $N = 50$ atoms measured with the TOF-MS with respect to the time of ionization for different nozzle tube temperatures. The measured intensity profiles for nozzle tube temperatures of $T = 300$ and 450 K have a gaussian-like shape. Therefore, the mean cluster velocity can be derived from the maximum of a gaussian function adjusted to the cluster intensity profile. A comparison of the widths at half-height of the cluster intensity profiles for $T = 300$ and 450 K with the shutter opening profile reveals that the effect of the cluster velocity distribution on the width at half-height is negligible. One can conclude that, under these source conditions, the molecular beam either contains only a single cluster isomer or the difference in the cross-sections between different isomers must be too small. However, this situation becomes different when the temperature of the nozzle tube is raised from 300 K to 350, 375 and 400 K. Then, the shape of the velocity distribution changes, becoming broader and asymmetric. This observation can be explained in terms of the existence of at least two structural isomers in the molecular beam, these being simultaneously present in the cluster beam at $T = 350$, 375 and 400 K. Different collision cross-sections of the cluster structural isomers lead to different final cluster velocities, which is manifested in the broadening of the cluster intensity profiles. At 300 K the faster and at 450 K the slower isomer predominates. Comparing these results with the measured binding energies, one can identify the faster clusters, which are predominately observed at 300 K, as the isomers having smaller binding energies, i.e. those that are less stable.

Within the hard-sphere model the slow-flying isomer must have a smaller collision cross-section and therefore a more compact structure than the fast-flying isomer. Hence, one gets an idea about the differences in shape if one assumes that the slow-flying isomers, which only appear in the mass spectrometer at higher nozzle-tube temperatures, have compact spherical structures, in agreement with the measurement of their binding energies. With this assumption one can derive from a simple acceleration model [14] the collision cross-section of the fast-flying isomers relative to the compact, spherically shaped cluster isomers. The enhanced final velocities of the second class of silicon cluster isomers can be explained in terms of an increase of the collision cross-section of these clusters as a result of their non-spherical cluster shape. If one assumes, for example, that these non-spherical isomers have a cylindrical shape with length *l* and diameter *d*, their geometry ratios *l*/*d* can be adjusted to the enlarged collision cross-sections. This is shown in Table 21.1 for Si_N clusters with N > 60. The *l*/*d* ratios of the assumed cylindrical cluster geometries increase with growing cluster size and show a significant deviation from a spherical cluster shape. In general, our adjusted *l*/*d* ratios are in good agreement with quantum chemical calculations [9] and drift mobility measurements on silicon cluster cations.[5, 7, 10] Therefore, the experimentally determined velocities for two different silicon cluster isomer classes support the idea that the non-spherical cluster isomers have a cylindrical shape, even if a more detailed interpretation of our results on the basis of the simple acceleration model and the assignment of a defined structure to the measured collision cross-section is not possible.

Table 21.1. Geometry ratio *l*/*d* of the fast-flying, metastable clusters with an assumed cylindrical shape calculated from the velocity difference of the two isomers compared with results obtained from quantum chemical calculations and drift mobility measurements.

cluster size N	geometry ratio (non-spherical isomer) *l*/*d*	
27	3	Jarrold [5]
50	3.5–4.5	Grossman, Mitas [9]
67	6.9 (+1.9/ -1.2)	our results [14]
84	7.5 (+2.0/ -1.3)	our results [14]
100	7–9	extrapolated from [9]
100	11	extrapolated from [5]
102	8.6 (+2.4/ –1.5)	our results [14]

21.5 Discussion of Growth Processes for Metastable, Elongated Clusters

It is still an important question as to how the metastable clusters are formed from the hot plasma of thousands of degrees as it is rapidly cooled down to about 300 K by the helium within microseconds. It is even un-clear as to whether these clusters grow by addition of single Si atoms, molecules or small clusters. They might also be generated by fragmentation from larger clusters. Although annealing of thin films of evaporated semiconductors at elevated temperatures reduces the number of voids to a certain extent, the voids cannot be removed entirely before the crystallization temperature is reached [19] and hence it is not surprising that unannealed metastable clusters are generated. What is surprising is that these unannealed species do not statistically occur in a broad variety of overall shapes but rather can be classified into two types that strongly differ in their shape and binding energies. The assumed cylindrical structure might result from a growth process that is initiated by a nucleus that is composed of two small spherical-like clusters. However, in many respects, this is a delicate process and we must discuss its likelihood under our experimental conditions. The clusters typically spend 1 ms in the cluster source before they enter the high-vacuum apparatus, and the number density in the cluster source can be estimated by deposition experiments in combination with mass distribution analysis to be of the order of 10^{14} clusters/cm^3.[20] This gives cluster–cluster collision rates of 10^4 s^{-1}, which means that about 10 cluster–cluster collisions take place in the source chamber. The most interesting result is that these cluster–cluster collisions do not generally lead to larger product clusters, meaning that there is a considerable activation barrier for the cluster growth. However, we observed the formation of larger clusters when we doped the clusters with metal atoms such as Co, V or Dy, indicating that the barrier can be overcome by generating through doping a kind of reactive site. This means that the barrier for cluster–cluster reactions will depend sensitively on the mutual positioning of the clusters. However, even if the positioning of the educt clusters is ideal and a cylindrical cluster is formed, one has to consider that the cluster is heated due to the new bonds that are formed. The clusters must then be efficiently cooled down by the cluster–helium collisions within a typical time scale small enough that the free enthalpy of activation $\Delta^{\ddagger}G$ for isomerization is not overcome. Then, a cylindrical cluster with an increased geometry ratio l/d is formed. The question then arises as to why this elongated cluster grows to larger l/d values. This may be due to reactive sites that are located at the tips of the rods, where efficient bonding of additional clusters is again possible thereby generating in a stepwise manner a kind of cluster chain consisting of small clusters. However, the mechanism for the formation of metastable cluster isomers is still undetermined.

21.6 Temperature Dependence of Collision Cross-Sections and Unimolecular Decay Model

21.6.1 Temperature-Dependent Collision Cross-Sections

From the observed broadening of the cluster intensity profiles in the velocity slip experiments one cannot determine how many isomers each of the assumed two isomer classes contains. However, it is possible to extract from the broadening of the cluster intensity profiles the ratio of all isomers having a more compact, spherical shape to all of the isomers having a less compact, elongated shape. From the areas of the two gaussians in Figure 21.2 (b) one obtains the temperature-dependent intensity ratio of the two silicon cluster isomers, which is the intensity of the spherical isomers divided by the intensity of the non-spherical isomers.

In order to determine the rate constant for the cluster isomerization reaction from the temperature-dependent intensity ratios, one needs a model for the mechanism of the transformation. The simplest model for the isomerization of more cylindrically shaped clusters Si_N^c to more spherically shaped clusters Si_N^s is based on the assumption that the cluster isomerization can be described by a unimolecular gas-phase reaction $Si_N^c \xrightarrow{k} Si_N^s$ with a single rate constant k.[21] Within the Eyring transition state theory for a unimolecular gas-phase reaction $Si_N^c \rightleftharpoons [Si_N]^\ddagger \rightarrow Si_N^s$, with the activated complex $[Si_N]^\ddagger$, the rate constant k is then related to the activation free enthalpy $\Delta^\ddagger G$ of the isomerization reaction.

21.6.2 Analysis of Isomerization by a Unimolecular Decay Model

The $\Delta^\ddagger G$ values for all investigated cluster sizes are between 50 and 110 kJ/mol.[21] These values are of the same order as the previously estimated activation energies obtained on the basis of RRK theory for the elongated to spherical transition of cationic silicon clusters.[12]

The activation enthalpy $\Delta^\ddagger H$ and the activation entropy $\Delta^\ddagger S$ can be derived from the temperature dependence of the $\Delta^\ddagger G$ values as a function of cluster size; for $N = 45$–60 values of $\Delta^\ddagger H = 5$ kJ/mol and $\Delta^\ddagger S = -180$ J/(K·mol) are typical.[21] This means that entropic effects play an important role in the cluster isomerization reaction, because the activation free enthalpies $\Delta^\ddagger G$ are an order of magnitude larger than the activation enthalpies $\Delta^{\ddagger}H$, which are only of the order of the thermal energy RT, with the ideal gas constant R.

The negative values for $\Delta^{\ddagger}S$ will lead, within the Eyring theory for unimolecular reactions, to frequency factors of about 10^2-10^4 s^{-1}. These frequency factors are extremely small, in contrast to investigations on gas-phase molecular isomerization of organic molecules, in which typical values for the frequency factors of the order of 10^{10}–10^{14} s^{-1} have been found.[22, 23] Although exceptions with values of 10^5 s^{-1} are known in the literature,[22] the rate-determining mechanism for the isomerization of silicon clusters seems to be different from those conventionally discussed for the typical molecular isomerizations of organic molecules. For a large number of organic molecules,[22, 23] the experimental $\Delta^{\ddagger}S$ values have been reasonably explained by a change in the vibrational frequency spectrum of the molecule in its initial and its transition state. In order to explain large values for $\Delta^{\ddagger}S$, a change in the rotational and vibrational partition function Q of the reactant and the activated complex $Q^{\ddagger}$ and a symmetry factor have been considered.[22, 23] However, it is hard to imagine that Q and $Q^{\ddagger}$ differ by some orders of magnitude for different silicon cluster isomers with $N \approx 50$. Even though it has been shown that the vibrational entropy can play a crucial role in determining the equilibrium structure of clusters,[24] the frequency factors derived from our thermal analysis of the rate constants are ca. 10^9 orders too small to be explained by this way.

However, for the isomerization of the silicon clusters, there might be an additional effect that could explain the observed $\Delta^{\ddagger}S$ values. Recent ab initio molecular dynamics simulations have demonstrated a process of bond breaking and formation on a picosecond time scale at 300 K,[11] i.e. a fluctuation between isomer structures separated by low energy barriers of the order of the thermal energy RT. In a kinetic picture, this means that there are energetically low-lying transition states, which allow different isomers to coexist at room temperature. These different isomers belong to one class, i.e. no overall change in shape takes place. Therefore, the situation is completely different compared to typical molecular isomerizations of organic molecules, where usually only one structural isomer has to be considered. The molecular dynamics simulations indicate that the investigated silicon clusters seem to have some weak chemical bonds. The weak chemical bonds allow the different silicon cluster isomers to rearrange without a global change in their external shape. In contrast to this, a global rearrangement of the cluster geometry from cylindrical to spherical shape needs more than an arbitrary breaking and reforming of some weak chemical bonds. The global isomerization process might need either the breaking of some strong chemical bonds or a transition state with several weak broken bonds located in certain critical positions. In the first case, one would expect $\Delta^{\ddagger}H$ values that are considerably larger than the observed experimental ones. In the second case one can indeed have low $\Delta^{\ddagger}H$ values in combination with low frequency factors as experimentally observed.

The possibility of having low $\Delta^{\ddagger}H$ values can, in principle, be explained by simple Hückel-like molecular orbital schemes describing the chemical bonding of the silicon clusters as has been discussed in ref. [21]. However, if the non-spherical, cylindrical clusters grow by addition of preformed small ring-like or polyhedral clusters, it has to be considered that these precursors need not be in their most stable configuration, i.e. that they are also energetically activated. This can store additional energy in the growing cylindrical clusters, thereby leading to energetically weakly stabilized cylindrical clusters with small $\Delta^{\ddagger}H$ barriers. However, these cylindrical isomers seem to be entropically stabilized through the negative $\Delta^{\ddagger}S$ found in the experiments. The resulting low frequency factors indicate that a large number of successive statistical fluctuations are necessary to find the critical combinations of broken bonds. In order to quantify this effect, one can estimate the corresponding frequency factor by assuming that a large number Ω of different configurations of cylindrically shaped isomers are accessible and thermally fluctuate into each other by breaking and forming a small number of bonds at 300 K. The mixing entropy S_{mix} due to the mixture of these configurations of the isomers is then $S_{mix} = k_B \ln \Omega$, where k_B is the Boltzmann constant. If one tries to explain the experimental results for $\Delta^{\ddagger}S$, the number of broken bonds has to be set to 5–7 to achieve $\Delta^{\ddagger}S = -152$ to -197 J/(K·mol).[21] This means that in our simple model the isomerization mechanism can be understood analogously to the isomerization of organic molecules as a one-step global rearrangement of the less compact, cylindrically shaped clusters, but at least 5–7 critical chemical bonds have to be broken in order to reach the transition state. The presence of the transition state bonding configuration is very improbable in comparison with all other configurations. At room temperature, it takes a random walk with a duration of 10^6 vibrations in order to find the transition state configuration with the small $\Delta^{\ddagger}H$ value, meaning that this reaction path follows a narrow valley on the energy surface between higher enthalpy.

21.7 Acknowledgements

We acknowledge the support of the Fonds der Chemischen Industrie.

21.8 References

[1] M. F. Jarrold, *Science* **1991**, *252*, 1085.
[2] H. Weller, *Angew. Chem. Int. Ed. Engl.* **1993**, *32*, 41.
[3] W. A. de Heer, *Rev. Mod. Phys.* **1993**, *65*, 611.
[4] J. A. Becker, *Angew. Chem. Int. Ed. Engl.* **1997**, *36*, 1391.
[5] M. F. Jarrold, V. A. Constant, *Phys. Rev. Lett.* **1991**, *67*, 2994.
[6] R. R. Hudgins, M. Imai, M. F. Jarrold, P. Dogourd, *J. Chem. Phys.* **1999**, *111*, 7865.
[7] A. A. Shvartsburg, B. Liu, M. F. Jarrold, K. M. Ho, *J. Chem. Phys.* **2000**, *112*, 4517.
[8] E. Kaxiras, K. Jackson, *Phys. Rev. Lett.* **1993**, *71*, 727.
[9] J. C. Grossman, L. Mitas, *Phys. Rev. B*, **1995**, *52*, 16735.
[10] K. M. Ho, A. A. Shvartsburg, B. Pan, Z. Y. Lu, C. Z. Wang, J. G. Wacker, J. L. Fye, M. F. Jarrold, *Nature* **1998**, *392*, 582.
[11] L. Mitas, J. C. Grossman, I. Stich, J. Tobik, *Phys. Rev. Lett.* **2000**, *84*, 1479.
[12] M. F. Jarrold, E. C. Honea, *J. Am. Chem. Soc.* **1992**, *114*, 459.
[13] T. Bachels, R. Schäfer, *Chem. Phys. Lett.* **2000**, *325*, 365.
[14] M. Rosemeyer, R. Schäfer, J. A. Becker, *Chem. Phys. Lett.* **2001**, *339*, 323.
[15] J. T. Stuckless, N. A. Frei, C. T. Campbell, *Rev. Sci. Instrum.* **1998**, *69*, 2427.
[16] T. Bachels, F. Tiefenbacher, R. Schäfer, *J. Chem. Phys.* **1999**, *110*, 10008.
[17] P. Milani, W. A. de Heer, *Rev. Sci. Instrum.* **1990**, *61*, 1835.
[18] I. Barin, *Thermochemical Data of Pure Substances*, VCH Verlag, Weinheim, 1993.
[19] S. R. Elliott, *Physics of Amorphous Materials*, Longman, London, 1983, p. 269 ff.
[20] C. Herwig, J. A. Becker, *Eur. Phys. J. D* **2001**, *16*, 51.
[21] M. Rosemeyer, J. A. Becker, R. Schäfer, *Z. Phys. Chem.* **2002**, *216*, 857.
[22] S. W. Benson, *The Foundations of Chemical Kinetics*, McGraw-Hill, New York, 1960.
[23] P. J. Robinson, K. A. Holbrook, *Unimolecular Reactions*, Wiley, London, 1971.
[24] J. P. K. Doye, F. Calvo, *Phys. Rev. Lett.* **2001**, *86*, 3570.

22 Investigation of the Influence of Oxidation and HF Attack on the Photoluminescence of Silicon Nanoparticles

F. Huisken*, G. Ledoux*‡, O. Guillois#, and C. Reynaud#

22.1 Introduction

Silicon is the most widely used material in the electronics industry. To develop silicon-based devices for optoelectronic applications, one would like to make silicon a photon-emitting material. Unfortunately, silicon is an indirect gap semiconductor and, thus, the efficiency of photon emission is extremely low since the radiative recombination of the electron–hole pair is not allowed without the assistance of a momentum-conserving phonon. Moreover, the existence of defects leads to an almost total quenching of this rather unlikely process.

In the early nineties, the studies of Canham [1] and Lehmann and Gösele [2] showed that a silicon wafer could be made to emit visible light when it was electrochemically etched in hydrofluoric acid thus producing a porous nanostructured surface. The observation of photoluminescence (PL) was explained on the basis of the quantum confinement leading to a widening of the band gap and a partial relaxation of the selection rules making silicon a somewhat more direct-gap material.[3, 4] An even more important reason for the enhanced efficiency is an effect termed spatial confinement which prevents the diffusion of the carriers to nonradiative recombination centers. Due to the reduced size, the probability of the carriers finding a defect in the core is drastically reduced. However, for this to occur it is important that the crystalline Si core is either isolated or surrounded by a higher band gap material, and that the silicon nanoparticle is perfectly crystalline and does not have any dangling bond.

Since the discovery of the intense red photoluminescence of porous silicon,[1,2] much work has been devoted to this particular nanostructured material [4, 5] and, in the meantime, also to silicon nanoparticles.[6, 7] An important issue in current studies is the influence of the passivation on the photoluminescence properties. It has already been noted that, in the quantum

* Address of the authors: Max-Planck-Institut für Strömungsforschung, Bunsenstraße 10, D–37073 Göttingen, Germany

*‡ Present address: Laboratoire de Physico Chimie des Materiaux Luminescents, Université Claude Bernard Lyon I, F-69622 Villeurbanne Cedex, France

CEA/DSM/DRECAM/SPAM, CE Saclay, F–91191 Gif-sur-Yvette Cedex, France

confinement model, it is essential that the surface is well passivated to avoid any dangling bond at the surface.[8] Being middle-gap defects, these dangling bonds will quench the photoluminescence. On the other hand, the surface itself may lead to surface states that can be the origin of another kind of photoluminescence.[9, 10]

During the last few years, we have studied silicon nanocrystals produced by CO_2 laser pyrolysis of silane and we have been able to show that, in these experiments, the PL characteristics can be unambiguously explained in terms of quantum confinement effects.[11–13] However, to observe the photoluminescence with the naked eye, we had to wait a few hours or even a few days. It appeared that the silicon nanocrystals were passivated by natural aerial oxidation and that, with time, the photoluminescence became more and more intense.

In this chapter, we present the most recent results of our investigations devoted to the initial steps of oxidation and its influence on the PL properties of Si nanoparticles. These studies have also given us new insight into the PL behavior of aged samples. We also demonstrate the effect of HF attack on the oxide shell of aged samples and the subsequent oxidation as far as the PL of these samples is concerned. A short account of this study has been given in a recently published Letter.[14] Using the etching technique with HF vapor and subsequent oxidation, it is possible to shift the size distribution of macroscopic samples of silicon nanoparticles collected in the exhaust line of a flow reactor to smaller sizes and to shift their PL from the near IR to the visible. All observations presented in this chapter can be explained in the frame of the quantum confinement model. Other origins of photoemission need not be invoked.

22.2 Experimental

Thin films of non-interacting silicon nanoparticles were produced by pulsed CO_2 laser pyrolysis of silane in a dedicated gas-flow reactor and molecular beam apparatus. This apparatus and the characterization techniques used have been described in detail elsewhere.[7, 11, 15, 16] Here, it is sufficient to concentrate on the most important aspects that will be discussed along with the schematic view of the experimental setup given in Figure 22.1. The radiation of a pulsed CO_2 laser is focused into the flow reactor to interact with the SiH_4 molecules, which will become dissociated. In this way, a saturated vapor of silicon atoms at elevated temperature (≥ 1300 K) is produced, giving rise to condensation and subsequent growth of silicon nanoparticles. The as-prepared Si clusters are extracted from the flow reactor through a conical nozzle of 0.2 mm diameter to form a pulsed molecular beam of non-interacting particles. Since their velocity is strongly correlated with their mass (the smaller nanoparticles are faster than the larger ones), they can be selected according to their size by using a simple molecular beam chopper.[7, 16]

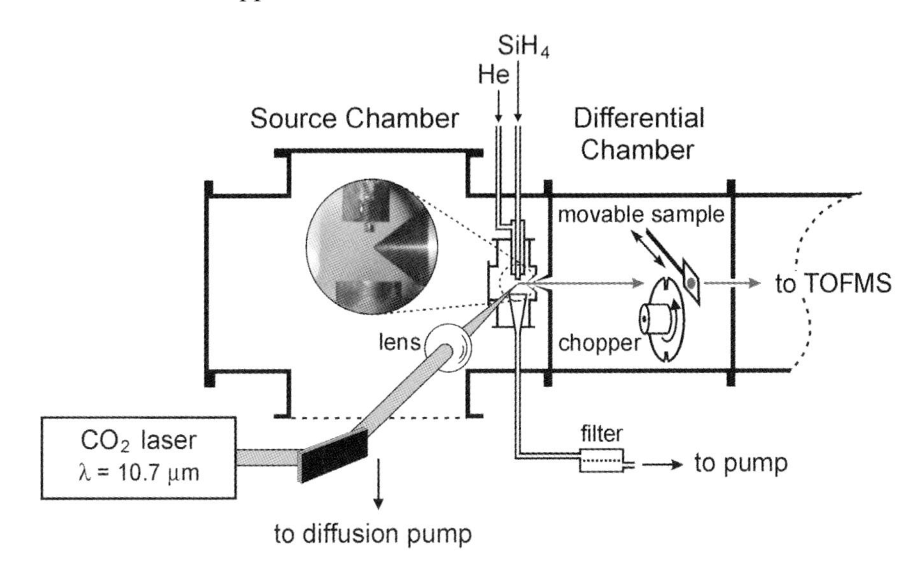

Figure 22.1: Schematic view of the experimental set-up.

Thin films of size-selected neutral silicon nanoparticles were then deposited at low energy on various substrates positioned behind the chopper. When the substrate is not in place the particles can proceed into the following chamber where they are analyzed by time-of-flight mass spectrometry (TOFMS) employing an ArF excimer laser to ionize them. The size distribution determined in the beam by TOFMS was further checked *ex situ* by atomic force microscopy (AFM) of a low coverage deposit. The two size distributions were found to be completely consistent in the small size regime (between 0 and 10 nm). As a result, one can conclude that TOFMS is very well suited to determine the size distribution of nanoparticles on the substrate if they are deposited at low energy.[11] The actual experimental parameters employed in the present studies are summarized here: flow of silane: 30 sccm; flow of helium: 1100 sccm; total pressure: 330 mbar; CO_2 laser line: 10μP28 (936.8 cm^{-1}); laser energy: 30 mJ; pulse length: 20 ns, repetition rate 20–30 Hz.

While only a minor part of the clusters and nanoparticles are extracted through the small conical nozzle into the evacuated source chamber, most particles produced during the CO_2 laser pulse are pumped away through the funnel and collected in a filter mounted outside of the apparatus. It is important to note that these particles are always larger than the ones extracted by the nozzle since their residence time in the laser-heated reaction zone is much longer. At the same time, their size distributions are also considerably broader. However, it is possible to collect many more particles in the filter than can be deposited from the beam.

Compared to the experiments reported previously,[7, 11, 16] the apparatus has been equipped with an additional vacuum chamber, mounted perpendicularly to the cluster beam axis. This chamber, not shown in Figure 22.1, can be separated from the differential chamber by means of a gate valve. After deposition, the samples were transferred under vacuum into this analysis chamber and their photoluminescence properties were studied by exciting them with the fourth harmonic of a pulsed Nd:YAG laser (λ = 266 nm) and monitoring the emitted light with a dedicated spectrometer. This PL spectrometer is very similar to the one used in Saclay and described previously.[11] The response of the S20 photocathode and the entire optical set-up has been carefully calibrated. It should be stressed that the sensitivity of the detection system is considerably lower in the infrared region than in the visible, leading to a lower signal-to-noise level for wavelengths above 750 nm.

To study the evolution of the PL as a function of the degree of oxidation, the samples were exposed to air under normal conditions for well-defined periods of time while keeping them in the analysis chamber. Before each PL measurement, the analysis chamber was evacuated. In this way, stable conditions during the data collection were ensured. In order to remove the oxide layer of already passivated samples, we exposed them for several minutes (up to 40 min) to the HF vapor simply by holding them over a Teflon crucible filled with a 37% solution of HF in water. Thereafter, the samples were returned to the vacuum chamber which was immediately pumped. For these HF-treated samples, the effect of oxidation on the PL behavior was studied as well.

Due to the larger size of the Si nanoparticles collected in the filter, most filter samples do not show any visible photoluminescence. However, as we could observe with an IR-sensitive video camera, they usually luminesce in the IR. For the following experiments, we selected a few samples that showed some weak PL below 700 nm that could be observed by eye. After HF treatment and subsequent oxidation, the samples clearly showed enhanced PL in the visible as a result of the reduced particle size. To accelerate the oxidation, the samples were exposed for 2–4 h to water vapor at ~150 °C in an oven.

22.3 Results and Discussion

In order to demonstrate the size-selection capabilities of our cluster beam apparatus, we present in Figure 22.2 a photo of a luminescent sample of silicon nanocrystals when it was exposed to the light of a laboratory UV lamp ($\lambda = 254$ nm). The sample was prepared with a clockwise rotating molecular beam chopper that distributed the transmitted nanoparticles from left to right according to their size. Details of the experiment are given in a recent publication.[17] Within the deposited film, the size of the Si nanoparticles varies from 2.5 nm (on the left side) to 8 nm (on the right side), as was determined with the TOF mass spectrometer. Accordingly, the color of the photoluminescence varies from yellow-orange to the IR. A selected set of PL curves measured along a horizontal line on the sample is plotted in the lower panel of Figure 22.2.

If we plot the peak positions of the PL bands in eV as a function of the nanocrystal size (d in nm),[17] the data points follow nicely the inverse power law

$$E_{PL}(d) = E_0 + 3.73/d^{1.39} \tag{1}$$

derived by Delerue and co-workers[18] on the basis of the quantum confinement model. The band gap of bulk silicon enters this formula as $E_0 = 1.17$ eV. In contrast to an earlier comparison,[11] the experimental data are much less scattered. This can be ascribed to the fact that all particle sizes are contained in one single sample that was prepared in a single run, thus avoiding any difference in the production conditions or the oxidation history after deposition. The conclusion that could be drawn from this experiment was that the photoluminescence of the Si nanocrystals, produced by CO_2 laser-assisted pyrolysis of silane and gently oxidized in air under normal conditions, can be perfectly explained on the basis of the quantum confinement model, that is, by the radiative recombination of electron–hole pairs confined in the nanocrystals.[17]

In the same study, we also determined the efficiency of the PL process by carefully measuring the ratio of the emitted and absorbed energies. A pronounced maximum was found for crystallite sizes around 3.5 nm. For this size, a PL yield of 30% was observed. On going to larger sizes, the PL yield decreases exponentially to reach a value of only 1% for 8-nm particles. On the other side, the PL yield of 2.5-nm particles is reduced to 10%.

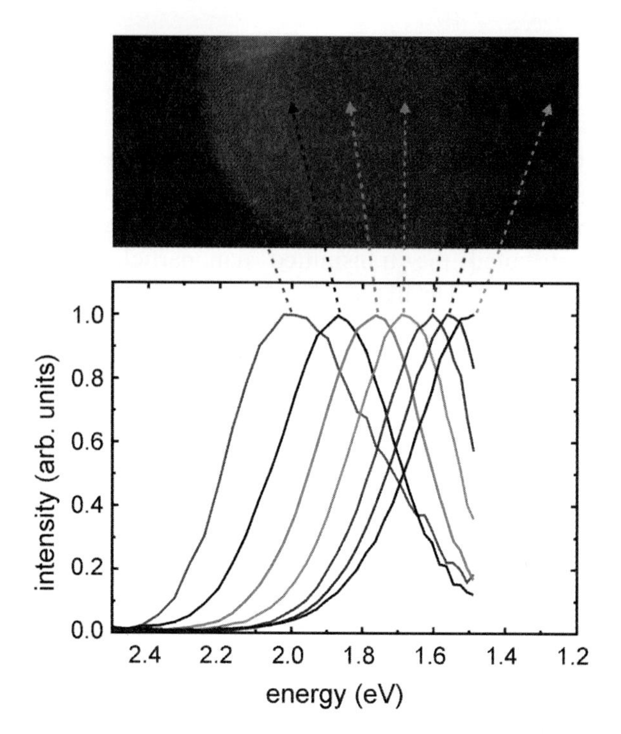

Figure 22.2. PL study of a sample produced by cluster beam deposition using a chopper for size separation. The upper panel shows a photo of the deposit when it was illuminated by a simple UV lamp. The lower panel reports the PL spectra recorded at the positions indicated by the arrows.

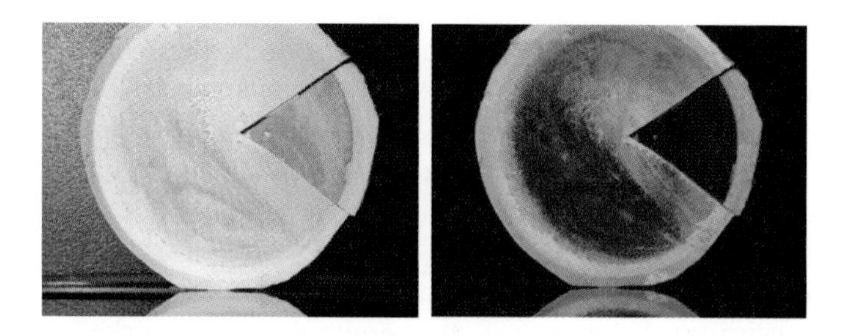

Figure 22.7. Photos of a filter with normal (left) and UV (right) light illumination. The quarter piece has been cut from the original sample. Its PL is very weak and mostly in the IR. The larger three-quarter piece was subjected to HF vapor and subsequently oxidized.

Now we want to discuss in some detail the effect of oxidation on the PL behavior of Si nanocrystals. For this purpose, we have carried out a set of dedicated experiments. We first prepared a size-selected sample with an average size of $\langle d \rangle$ = 3.6 nm and a size distribution of full-width at half-maximum (FWHM) of Δd = 0.6 nm. As described in the experimental section, the freshly prepared sample was transferred under vacuum into the analysis chamber where we tried to measure a PL spectrum. However, no PL could be detected unless we exposed the sample to air. Only after 20 min could some rather weak photoluminescence be observed. The maximum position of the corresponding PL curve was at 1.72 eV. This and the subsequent results are displayed in Figure 22.3. With increasing time of exposure to air, the PL became more intense and the PL peak position further shifted to the blue. After one day, the maximum position had already experienced a shift to 1.85 eV. At later times, the effect became smaller and seemed to go into saturation. Finally, after 25 days, the maximum position was found at 1.87 eV.

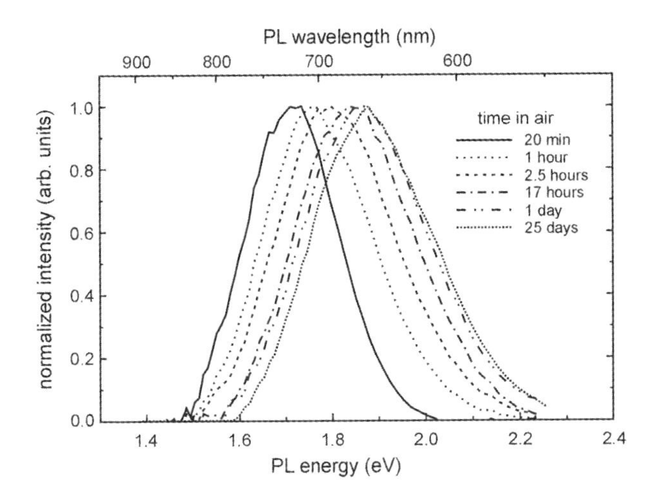

Figure 22.3. Evolution of the amplitude-normalized PL spectra for a size-selected sample exposed to air for different cumulative times given in the figure.

As the maximum of the PL band shifts to higher energies, its width also increases. Thus, for the same period, the FWHM varies from 0.23 eV for the first spectrum to 0.31 eV after 25 days. The evolution of the two parameters, position and width of the PL band, as a function of time is summarized in Figure 22.4. In this representation, it can also clearly be seen that the time dependence of these two parameters saturates after approximately one week (10^4 min).

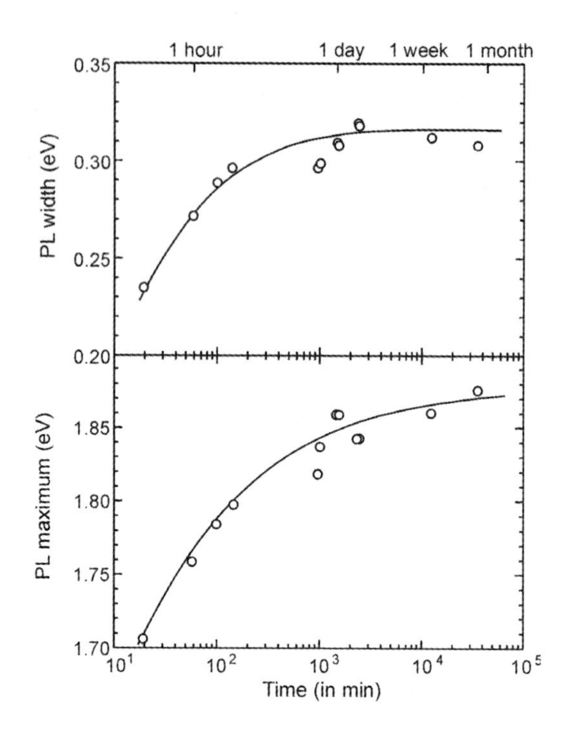

Figure 22.4. Evolution of the full-width at half-maximum (upper panel) and peak position (lower panel) of the PL bands as a function of the time for which the sample was exposed to air.

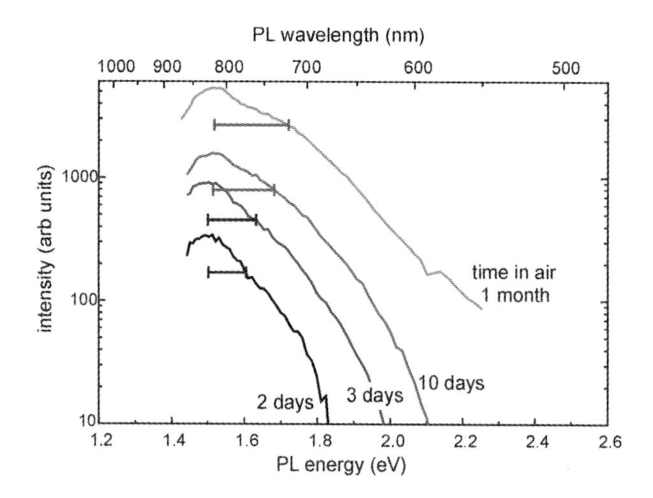

Figure 22.5. Relative PL spectra of a non-size-selected sample exposed to air for different cumulative times. The horizontal bars indicate the half-widths at half-maximum (HWHM) of the PL bands.

Figure 22.5 shows the time behavior of another sample that was prepared without size selection ($\langle d \rangle$ = 4.4 nm; Δd = 2 nm). During this study, we have paid particular attention to keeping the power of the exciting laser at a constant level. Therefore, the spectra can be readily compared as far as their PL efficiencies are concerned. As for the first sample, no PL could be observed immediately after preparation. In this case, however, the process of complete passivation seemed to take longer since no PL could be detected until 2 days had elapsed. Again, we observe a broadening of the PL band with time (from 0.2 to 0.4 eV). However, more important is the observation of a strong increase of the PL intensity. From 2 days to one month, it increases by a factor of 16. During this time, the maximum position of the PL band shifts only slightly from 1.49 to 1.52 eV.

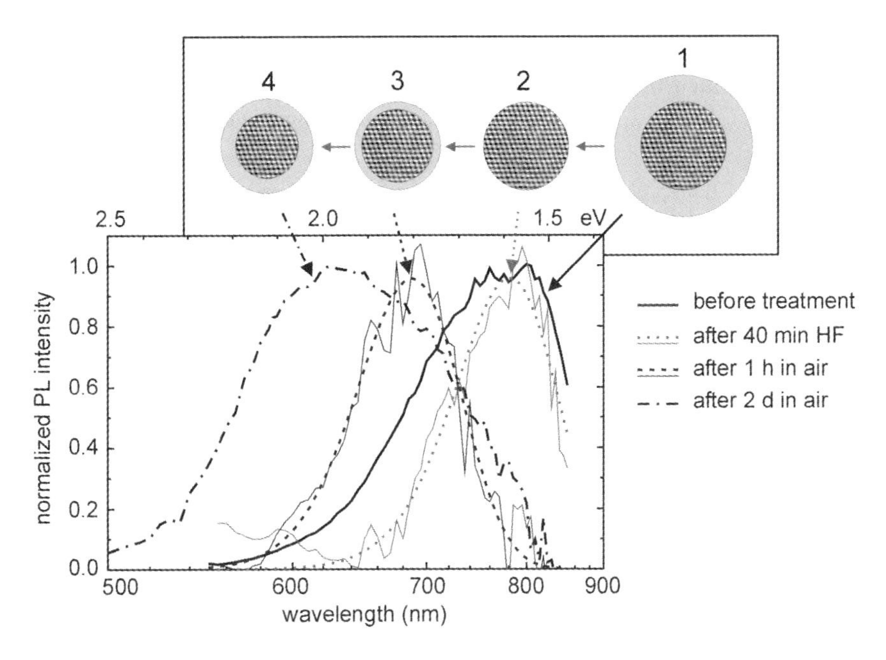

Figure 22.6. PL spectra of a non-size-selected sample after different successive treatments: after passivation in air for two months (thick solid curve), after exposure to HF vapor for 40 min (grey thin and dotted curves), after re-exposure to air for 1 h (black thin and dashed curves), and after continued exposure to air for 2 d (dash-dotted curve). The sketch in the upper part of the figure illustrates the effect of the various treatments on the core and oxide shell of the nanoparticles (schematic).

Figure 22.6 describes the results of another kind of experiment. In this study, we started from an already passivated sample. Its PL spectrum is given by the black solid curve that peaks at approximately 775 nm or 1.6 eV. The sample was then exposed to HF vapor for 40 min (see the experimental section for details). The spectrum recorded after this procedure is shown by the thin grey curve. Due to the low signal, the measured PL band is rather noisy. Therefore, we have fitted the experimental curve by a Gaussian, which is presented in the figure as the dotted grey curve. Then the sample was again exposed to air for definite periods of time. After 1 h, the spectrum represented by the thin black curve and the dashed fit was measured. Finally, the spectrum representing the latest stage of the evolution after 2 days in air has been plotted by the dash-dotted line. This rather broad curve peaks at approximately 2 eV.

The attack of HF results in a complete or almost complete removal of the oxide layers surrounding the Si nanoparticles. As can be seen in Figure 22.6, this treatment results in a substantial narrowing of the band (from 0.4 to 0.26 eV), while the position of the band maximum is not changed. On the other hand, the integrated intensity of the PL is considerably lower, having decreased by a factor of 70. This is the reason why the spectrum is rather noisy after HF treatment. Subsequent oxidation of the Si nanoparticles has the same effect as the oxidation of freshly prepared samples. The peak shifts to higher energies (from 1.59 through 1.81 to 1.97 eV), and the width increases from 0.26 through 0.28 to 0.55 eV. At the same time, the integrated PL intensity finally increases by a factor of 5.

In a previous study,[19] we showed by high-resolution transmission electron microscopy that aged silicon nanocrystals are surrounded by an oxide shell, the thickness of which corresponds to approximately 10% of the total particle diameter. It was found that, for a given particle size, the spacing of the {111} lattice plane fringes varies by ~2%. This variation, which can be explained by different degrees of oxidation and thus different stresses exerted on the crystalline lattice, results in an inhomogeneous "oxide-induced" PL bandwidth of 0.25 eV for a given particle size in an aged sample.[11] The final PL response of a given sample can be calculated by transforming the particle size distribution into a PL band (taking the correlation between particle size and band gap given by Delerue *et al.* [18]) and convoluting the resulting curve with a Gaussian line-shape function with a FWHM of 0.25 eV to account for the oxide-induced inhomogeneous broadening.[11] Since the oxide layer is clearly less pronounced for fresh samples, we expect much narrower PL bandwidths for these samples, provided that the size distribution is not too broad.

If we look at the results of the present study, we indeed find that, for relatively fresh samples, the spectra are quite narrow ranging in width from 0.2 to 0.23 eV (FWHM). These values are even smaller than the oxide-induced width of 0.25 eV of an ensemble of single-sized aged nanocrystals studied in ref. [11]. On the other hand, after one month of exposure to air, we end up with PL bandwidths of 0.31 to 0.4 eV. This is clearly due to an increase of the oxide-induced width. Another confirmation of this interpretation is given by the last experiment. Starting from an aged sample for which the PL peak is rather broad, the width is significantly reduced when the oxide layer is removed. This results in a relaxation of the stress exerted by the oxide layer. Conversely, when the sample is oxidized again, the width increases and even exceeds a value of 0.55 eV.

In the same conceptional framework, we can also understand the evolution of the peak position. As we have shown before, the position of the PL of aged samples is well correlated with the average size of the nanocrystallites in accordance with the theoretical law established by Delerue *et al.*[18] In the present experiment, we find that, with time, the PL peak gradually shifts to the blue. On the basis of the inverse power law [18] (see Eq. 1) and the experimental verification, one can derive that, for the first sample reported in Figure 22.3, the crystalline core varies from $d_i = 3.96$ nm at the beginning to $d_f = 3.33$ nm in the final state. This corresponds to a shrinking by a factor of 0.84.

Taking an average molecular weight of SiO_x of $M = 52$ (corresponding to $x = 1.5$) and assuming that the density of silicon oxide is rather close to that of silicon (2.33 g/cm^3),[11] it follows that the volume containing the same number of Si atoms is a factor of 1.86 larger for the oxide shell (SiO_x) than for crystalline silicon. Knowing, furthermore, that the thickness of the oxide layer of an aged Si nanoparticle is approximately 10% of the total diameter,[19] we can calculate that, after complete passivation, the size of the crystalline core is reduced by a factor of 0.87. This number compares rather favorably with the experimental value (0.84) derived in the previous paragraph.

For the second sample (Figure 22.5), the spectral shift observed between two days and one month corresponds to a change in size from 5.85 to 5.48 nm. Due to the poor detection efficiency in the IR, we could not observe the PL for the very fresh sample. Therefore, the determination of the oxide shell thickness is not straightforward.

The results obtained for the sample exposed to HF give new insight into the characteristics of the photoluminescence of Si nanocrystals. The initially measured PL curve (black curve in Figure 22.6) could be fitted rather well by assuming a nanocrystal size distribution with an average diameter $\langle d \rangle = 4.1$ nm and having a width of 2.2 nm (FWHM).[14] Corresponding to our earlier investigations,[11] the convolution was carried out with an oxide-induced width of $w = 250$ meV. The treatment with HF did not change the nanocrystal size distribution, as evidenced by the fact that the PL maximum did not shift. While keeping the original size distribution, we had to reduce the individual width w of a single nanocrystal size from 250 to less than 10 meV to obtain good agreement between simulation and measurement.[14] This gives us an upper limit for the intrinsic width of the PL response of a single Si nanocrystal of 10 meV. Upon oxidation, the crystalline cores of the nanoparticles are further reduced in size. An attempt to fit the experimental curve of the last measurement (green curve) with the same log-normal distribution shifted to smaller sizes and an oxide-induced width of w = 250 meV resulted in an average particle size of $\langle d \rangle$ = 2.8 nm. However, the agreement with the experimental data was not completely satisfactory, especially at higher energies.

Recently, Wolkin *et al.* [9] observed an upper limit of the PL emission energy of 2.1 eV in oxidized porous silicon even if the nanocrystal size became smaller than 2 nm. This behavior, which seems to contradict quantum confinement, was explained in terms of the formation of stabilized electronic states on Si=O bonds at the surface. For nanocrystals with diameters smaller than 2.8 nm, the widening of the band gap due to quantum confinement makes them appear as inner band gap states. Including the results of Wolkin *et al.* in our model calculations, we obtained nice agreement with the experimental data.[14]

The behavior of the PL band of silicon nanoparticles with diameters between 2 and 5 nm after exposure to air can be summarized as follows. The maximum position of the PL band shifts to higher energies while its intensity and bandwidth both increase with the time. All observations can be explained as resulting from a shrinking crystalline core and a growing oxide shell. From the fact that the time dependences go into saturation, it can be concluded that the oxidation of Si nanoparticles in this size regime is a self-limiting process. This is in agreement with our earlier electron microscopy study [19] and with previous studies on the progressive oxidation of silicon nanowires by Liu *et al.*[20] Finally, it should be stressed that the PL behavior of the Si nanoparticles investigated in the present study can be fully understood in terms of quantum confinement. Only if the nanocrystal size drops below approximately 2.8 nm does a new state within the band gap seem to evolve, limiting the maximum PL energy to 2.1 eV. This latter observation is in perfect agreement with the combined experimental and theoretical study of Wolkin *et al.*[9]

It has already been mentioned that the Si nanoparticles collected in the exhaust line of the flow reactor are larger due to their longer residence time in the reaction zone. As a result, the filters do not usually show visible photoluminescence if they are subjected to UV light; the PL is entirely in the IR. However, for the following experiment, we chose a filter whose PL was rather close to the visible, as manifested by its dark-red color under UV light. From this filter, we cut about one quarter to retain it for later comparison. The remaining filter was subjected to HF vapor for 40 min as described in the experimental section. Afterwards, the sample was passivated in a humid atmosphere in an oven at 150 °C. Figure 22.7 (displayed after Figure 22.2) shows photos of the two pieces of the filter under normal illumination (left) and under UV light (right). Whereas the original filter is still quite dark under UV light, the HF-treated filter exhibits rather intense PL in the red and even in the orange region of the spectrum. Furthermore, from the different colors of the luminescent sample, it can be seen that the laminar flow entering the small filter chamber results, to a certain extent, in a size separation. The original intention of the experiment was to apply the HF attack and the subsequent oxidation successively, in order to reduce the particle size at will and to shift the PL further to the blue. However, as can be seen particularly in the left photo, the density of Si nanoparticles on the filter was significantly diminished after the first cycle. This means that many silicon particles were already transformed to SiO_2 by the etching and oxidation procedure.

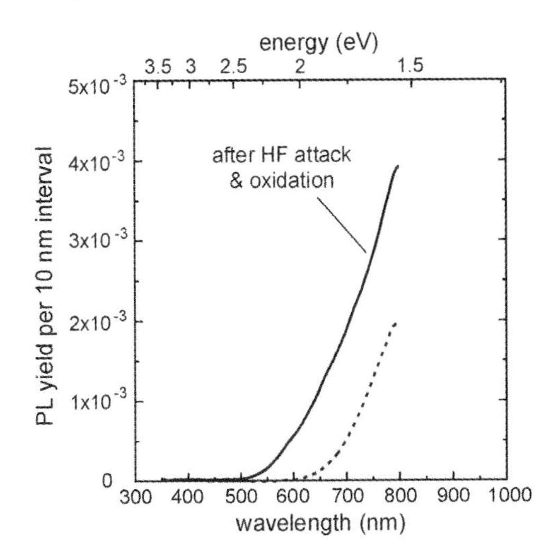

Figure 22.8. Averaged PL response measured from the original filter (dashed curve) and from the one treated with HF (solid curve).

We measured several series of PL spectra from both filter pieces. Because of the inhomogeneous size distribution of the particles on the filter, the PL curves were averaged, and the final result is displayed in Figure 22.8. Unfortunately, due to the limited IR response of the photocathode used, the curves are cut at 800 nm and do not extend into the IR. Nevertheless, the trend is clear. Whereas the dashed curve, which was measured from the original filter, shows only a very small contribution of visible PL, the solid curve shows that the HF attack and subsequent oxidation reduces the size of the silicon crystallites, thus shifting a substantial part of their photoluminescence into the visible. It should be emphasized that the maximum of this PL curve is still in the IR. Apparently, the procedure not only shifts, but also leads to a substantial broadening of the size distribution. This behavior seems to limit the possibility of applying subsequent cycles of HF attack and oxidation to reduce the size of

Si nanoparticles. However, to clarify this issue, it will be necessary to measure the complete PL spectra in the IR by employing a sensitive semiconductor detector.

22.4 Conclusions

CO_2 laser pyrolysis of silane in a gas flow reactor and the extraction of the resulting silicon nanoparticles into a cluster beam apparatus has been shown to offer an excellent means for the production of homogeneous films of size-separated quantum dots. Their photoluminescence varies with the size of the crystalline core. All observations are in perfect agreement with the quantum confinement model, that is, the photoluminescence is the result of the recombination of the electron–hole pair created by the absorption of a UV photon. Other mechanisms involving defects or surface states are not operative in our samples.

We have shown that, in order to exhibit intense PL, the Si nanocrystals must be perfectly passivated. A simple way to achieve this is through natural aerial oxidation. We have followed this process by measuring the photoluminescence as a function of time. Stable conditions are achieved after approximately 6 months. This indicates that the oxidation of Si nanoparticles is a self-limiting process.

The oxide shell of silicon nanoparticles can be etched away by exposing the samples to HF vapor. The subsequent oxidation reduces the size of the crystalline core and shifts the PL of the nanoparticles to shorter wavelengths. This technique can also be applied to reduce the size of Si nanoparticles collected in much larger amounts in the exhaust line of the flow reactor, and to shift their photoluminescence, which is normally in the IR, into the visible.

22.5 Acknowledgements

This work was supported by PROCOPE, a bilateral cooperation between France and Germany. G. L. thanks the Alexander-von-Humboldt Foundation for a fellowship. Finally, we are grateful to Jion Gong for preparing the size-selected deposit and to Henri Perez for his advice and help in the experiments devoted to the HF etching and passivation of the Si nanoparticles collected on the filter.

22.6 References

[1] L. T. Canham, *Appl. Phys. Lett.* **1990**, *57*, 1046.
[2] V. Lehmann, U. Gösele, *Appl. Phys. Lett.* **1991**, 58, 856.
[3] L. T. Canham, *Phys. Stat. Sol. (b)* **1995**, *190*, 9.
[4] A. G. Cullis, L. T. Canham, P. D. J. Calcott, *J. Appl. Phys.* **1997**, *82*, 909.
[5] P. M. Fauchet, *J. Luminescence* **1996**, *70*, 294.
[6] L. E. Brus, P. F. Szajowski, W. L. Wilson, T. D. Harris, S. Schuppler, P. H. Citrin, *J. Am. Chem. Soc.* **1995**, *117*, 2915.
[7] M. Ehbrecht, B. Kohn, F. Huisken, M. A. Laguna, V. Paillard, *Phys. Rev. B* **1997**, *56*, 6958.
[8] R. B. Wehrspohn, J.-N. Chazalviel, F. Ozanam, I. Solomon, *Eur. Phys. J. B* **1999**, *8*, 179.
[9] M. V. Wolkin, J. Jorne, P. M. Fauchet, G. Allan, C. Delerue, *Phys. Rev. Lett.* **1999**, *82*, 197.
[10] S. M. Prokes, *J. Mater. Res.* **1996**, *11*, 305.
[11] G. Ledoux, O. Guillois, D. Porterat, C. Reynaud, F. Huisken, B. Kohn, V. Paillard, *Phys. Rev. B* **2000**, *62*, 15942.
[12] G. Ledoux, O. Guillois, C. Reynaud, F. Huisken, B. Kohn, and V. Paillard, *Mater. Sci. Eng. B* **2000**, *69–70*, 350.
[13] G. Ledoux, O. Guillois, F. Huisken, B. Kohn, D. Porterat, C. Reynaud, *Astron. Astrophys.* **2001**, *377*, 707.
[14] G. Ledoux, J. Gong, F. Huisken, *Appl. Phys. Lett.* **2001**, *79*, 4028.
[15] M. Ehbrecht, H. Ferkel, V. V. Smirnov, O. M. Stelmakh, W. Zhang, F. Huisken, *Rev. Sci. Instrum.* **1995**, *66*, 3833.
[16] M. Ehbrecht and F. Huisken, *Phys. Rev. B* **1999**, *59*, 2975.
[17] G. Ledoux, J. Gong, F. Huisken, O. Guillois, C. Reynaud, *Appl. Phys. Lett.* **2002**, *80*, 4834.
[18] C. Delerue, G. Allan, M. Lannoo, *Phys. Rev. B* **1993**, *48*, 11024.
[19] H. Hofmeister, F. Huisken, B. Kohn, *Europ. Phys. J. D* **1999**, *9*, 137.
[20] H. I. Liu, D. K. Biegelsen, F. A. Ponce, N. M. Johnson, R. F. W. Pease, *Appl. Phys. Lett.* **1994**, *64*, 1383.

23. Localization Phenomena and Photoluminescence from Nano-structured Silicon and from Silicon/Silicon Dioxide Nanocomposites

S. Veprek* **and D. Azinovic**

23.1 Introduction

The first reports on visible photoluminescence (PL) from porous silicon[1, 2] triggered a large number of investigations in the hope that it will be possible to develop silicon-based optoelectronic devices. After more than a decade of intense research work on both PL and electroluminescence (EL), this goal is still far away and the available data suggest that it will be extremely difficult or impossible to achieve it. The reason is a low efficiency and short lifetime of such devices because the PL and EL – as far as they have been investigated for a given material in sufficient detail – originate from defect states within the layer that is used to passivate the silicon surface. Such a passivation is necessary because a single dangling bond at the Si surface is sufficient to cause a very fast, non-radiative electron–hole recombination. In this chapter, we concentrate on the PL from nanocrystalline silicon (nc-Si) passivated with thermally grown silicon dioxide, which also acts as a matrix separating the Si nanocrystals in order to ensure the localization of the photogenerated charge carriers.

The fundamental aspects of the physics of nano-sized semiconductors have been discussed in many excellent reviews to which we refer.[3, 4, 5, 6]. When crystallite size decreases to below about 10 nm, the number of eigenstates decreases to a few thousand or less. As a result, the Pauli repulsion decreases, the band gap increases,[7] and discrete states appear at the band edges. This has been shown for a variety of compound semiconductors that have a direct gap,[4] as well as for organosilicon molecules.[3, 8] In order to obtain an efficient PL, the surface of such nanocrystals has to be passivated by appropriate substituents [3] or encapsulated (overcoated) with another semiconductor having a large band gap.[9] By choosing the appropriate sizes of monodispersed nanocrystals, PL covering the whole range of colors of the visible spectrum can be obtained.[4, 7, 10]

Relaxation of the momentum conservation rule is another important change when the crystallite size decreases to 1–2 nm. This can be understood in terms of Heisenberg's uncertainty principle $\Delta p \cdot \Delta x \geq h$. In a large single crystal of

* Institute for Chemistry of Inorganic Materials, Technical University Munich, Lichtenberg Strasse 4, D-85747 Garching b. Munich, Germany.

$\Delta x \approx 1$ cm, the uncertainty of the momentum, Δp, is small. Thus, the momentum is a good quantum number and its conservation has to be obeyed in electronic transitions. In silicon, which has an indirect band gap, the recombination of electrons and holes requires a creation or annihilation of phonons and is therefore predominantly of non-radiative nature.[11] If, however, the size of an Si nanocrystal approches 1 nm, $\Delta x \rightarrow 1$ nm and Δp spans a significant part of the Brillouin zone. The momentum conservation is relaxed and electronic transitions (absorption of a photon with the formation of an electron–hole pair or their radiative recombination) become efficient.

In the following sections, we shall briefly discuss the preparation of nanocrystalline (nc-Si) and porous (PS) silicon and the post-treatment thereof needed to obtain efficient PL. Afterwards, the experimental evidence for the predicted increase of the band gap and electronic transition rates will be given, together with a brief summary of the properties of the PL obtained from nc-Si/SiO_2 nanocomposites. The last section will deal with selected examples of artefacts, such as impurities and defects, which cause a variety of different PL.

23.2 Preparation of Light Emitting Silicon

The majority of papers deal with the PL from porous silicon because of its relatively easy preparation by anodic dissolution.[1, 2] However, the nano-crystalline silicon prepared either by chemical transport in hydrogen plasma [12] or by plasma CVD from SiH_4 diluted with H_2 [13] offers a much better control of the particle size owing to the regular shape and separation of the nanocrystals. The as-deposited nc-Si shows a negligible PL because of the presence of dangling bonds on the surface of the nanocrystals and their small separation. Therefore, the grain boundaries of the deposited nc-Si have to be oxidized and the oxide post-treated in forming gas (FG, ca. 90 vol.% N_2 and 10 vol.% H_2). We refer to ref. [12] and papers quoted therein. This material will be discussed here. It should be emphasized that the PL attributed to nc-Si in many papers was actually due to impurities, such as silanol groups, organosilicon and metallic impurities, and others (see ref. [12] and papers quoted there). This is illustrated in Figures 23.1a and 23.1b. Fairly efficient "green" PL was also observed from the nc-Si/SiO_2 nanocomposites prepared by ultra-pure processing after their storage in air for several months.

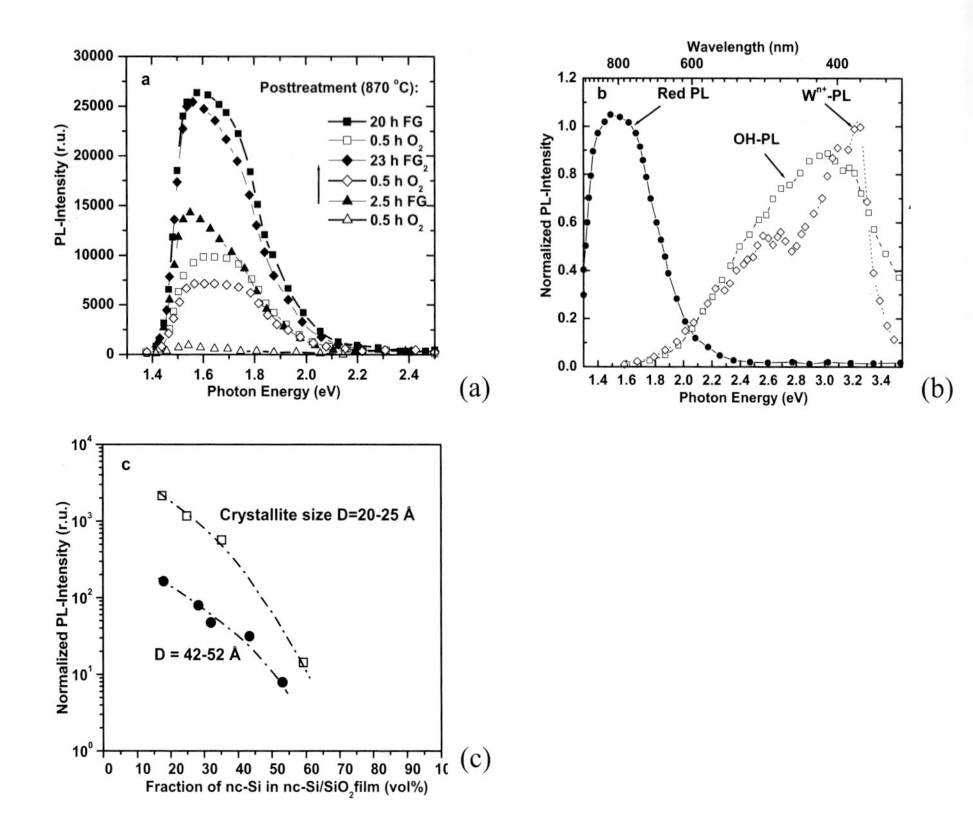

Figure 23.1. (a) Development of the PL from as-deposited nc-Si upon oxidation and annealing in forming gas; (b): "red" PL from nc-Si/SiO_2 prepared by the ultra-clean processing as in (a) and PL due to silanol and tungsten impurities (see ref. [12] and papers quoted there); (c): Effect of the separation of the Si nanocrystals on the "red" PL.[14]

23.3 Increase of the Band Gap of nc-Si with Decreasing Size and the Photoluminescense from nc-Si/SiO_2 Nanocomposites

As outlined above, the band gap of nanocrystalline semiconductors increases with decreasing crystallite size. Thus, direct gap semiconductors (GaAs, CdS, CdSe ...) with well passivated surfaces show optical absorption and PL that change color when the crystallite size is varied.[4, 9, 10] In the case of nc-Si/SiO_2 (and also porous Si), the situation is more complex because although the band gap increases with decreasing size as predicted by the theory, the spectral distribution of the PL remains unchanged as shown in Figure 23.2.

As can be seen in Figure 23.1a, the spectral distribution of the "red" PL remains almost unchanged, although during the subsequent oxidation and annealing in FG the crystallite size of nc-Si decreases, typically from about 10 to 1 nm.[14] Figure 23.1b shows, however, that with decreasing crystallite size the photon energy that is needed to excite this PL increases. This increase means that the band gap of the absorbing Si nanocrystals increases as shown in Figure 23.2a. This increase is in a very good agreement with the theoretical calculations of Delerue et al.[15] as well as with those of Delley and Steigmeier.[16] Obviously, the mechanism of the PL is complex and it has to involve the photogeneration of electron–hole pairs in the Si nanocrystals, energy transfer to some kind of radiation centers and the photon emission from there.

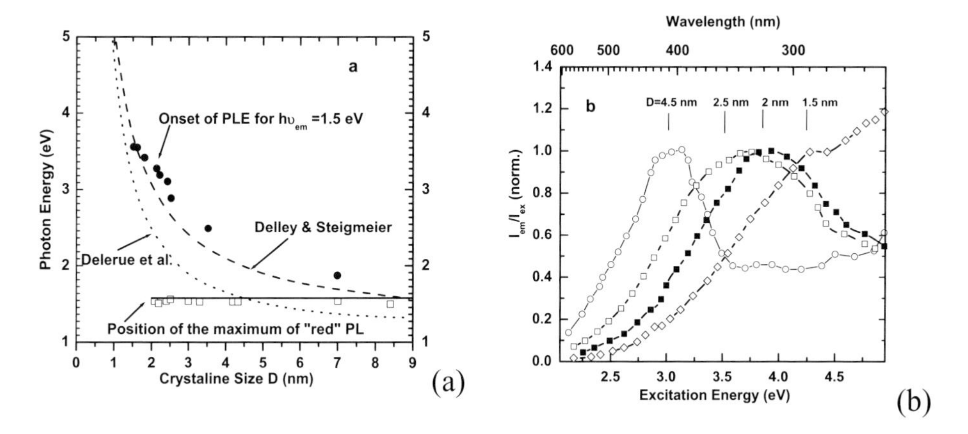

Figure 23.2. (a): Dependence of the band gap of nc-Si determined from the onset of the PL excitation spectra (shown in (b)) in comparison with theoretical calculations [15, 16] and the position of the maximum of the "red" PL from nc-Si/SiO_2.

The strong decrease of the PL with decreasing average separation between the Si nanocrystals shown in Figure 23.1c is due to an increasing delocalization of the electronic eigenstates of the crystals when the wave functions overlap. This conclusion has also been substantiated by microwave absorption measurements [13] as illustrated in Figure 23.3. The PL is observed only when there is no measurable microwave absorption, i.e. when the Si nanocrystals are sufficiently separated.

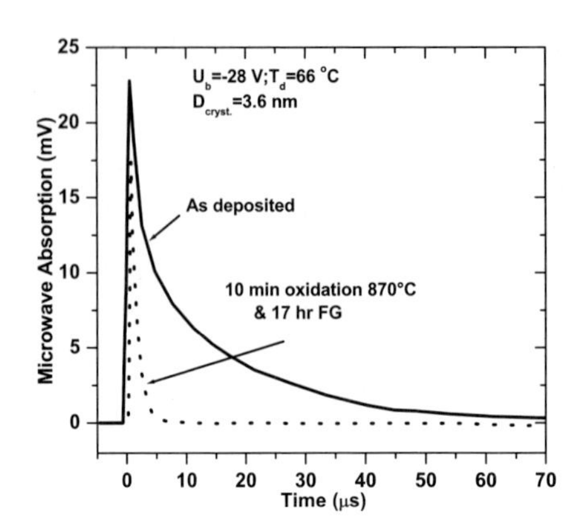

Figure 23.3. A strong microwave absorption is seen in as-deposited nc-Si when the electrons can pass through the grain boundaries between different Si crystals. It sharply decreases after oxidation of the grain boundaries and concomitant increase of the separation between the nanocrystals, when the PL appears.[13]

23.4 The Mechanism of the PL from nc-Si/SiO_2 Nanocomposites

The increase of the measured normalized intensity of the "red" PL (see Figure 23.4a) with decreasing crystallite size is in agreement with theoretical calculations of the transition rate for electron–hole recombination, which is similar to that of the electron–hole photogeneration due to the relaxation of the momentum conservation rule. The experimental observation that the decay time of the "red" PL remains unchanged (by analogy with the constancy of the decay time of the much faster PL from the silanol and W^+ impurity radiative centers, see Figure 23.4b), being too fast for samples with crystallite sizes larger than about 5 nm and too slow for smaller crystallites (Figure 23.4a), lends strong support to the above suggested mechanism, which is shown schematically in Figure 23.5.

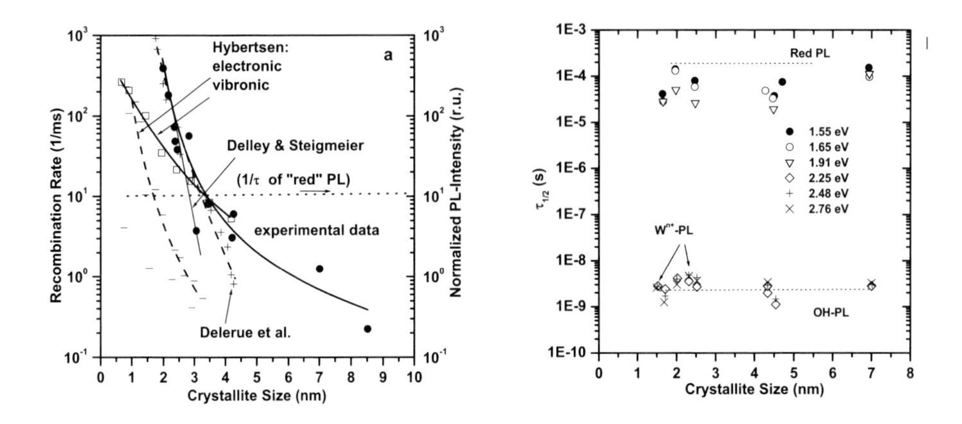

Figure 23.4. (a) Theoretical calculations of the electron–hole recombination rate and the measured intensity of the "red" PL (normalized throughout to the same volume fraction of silicon in the nc-Si/SiO_2 sample) vs. crystallite size. (b) Decay time of the "red" PL (upper line) and of the PL from the silanol groups and W^+-doped nc-Si/SiO_2.

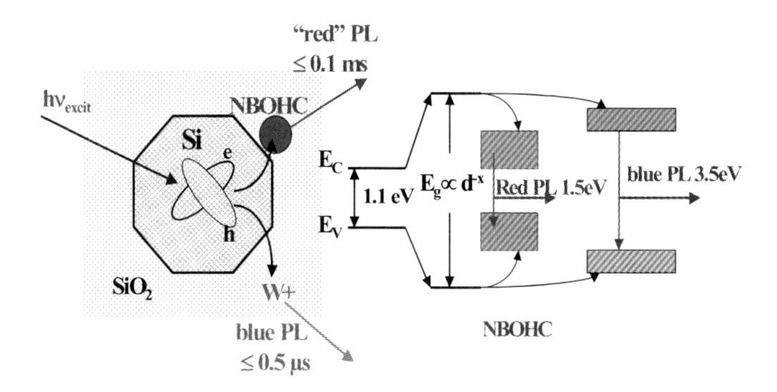

Figure 23.5. Mechanism of photoluminescence from nc-Si/SiO_2 nanocomposites: Left: pictorial illustration, right: energy levels involved in (1) the photogeneration of electron–hole pairs in the silicon nanocrystals, the band gap and transition probabilities of which increase with decreasing crystallite size, which is followed by an energy transfer to localized radiative centers, such as (2) a non-bridging oxygen hole center (NBOHC) in the SiO_2 layer or (2′) a metallic impurity.

Accordingly, the absorption of a photon with a sufficiently high energy, corresponding to the increasing band gap, results in the photogeneration of electron–hole pairs in the Si nanocrystals, in agreement with theoretical calculations of the quantum confinement. This step is followed by a fast energy transfer to the radiative centers, either at the Si/SiO_2 interface or within the SiO_2 matrix. In the case of metal-doped samples, such as with W^{n+}, these sites provide a fast PL the intensity of which reaches a maximum when the doping level corresponds to approximately one W^{n+} metallic ion per Si nanocrystal.[15] A similar mechanism was also found for so-called "spark-processed" light-emitting silicon.[16] Strongly supportive of this suggested mechanism is the absence of any polarization memory (Figure 23.6) for these PL's,[18, 19] whereas the PL from silanol (≡Si-O-H), alanol (=Al-O-H) and other hydrated metal oxides shows a strong memory.[17, 18, 19] (When one and the same dipole absorbs and emits a photon, the polarization of such PL correlates with that of the incoming exciting light.[20 21] The polarization memory is ,however, lost when an energy transfer occurs from the absorbing center to the emitting one.)

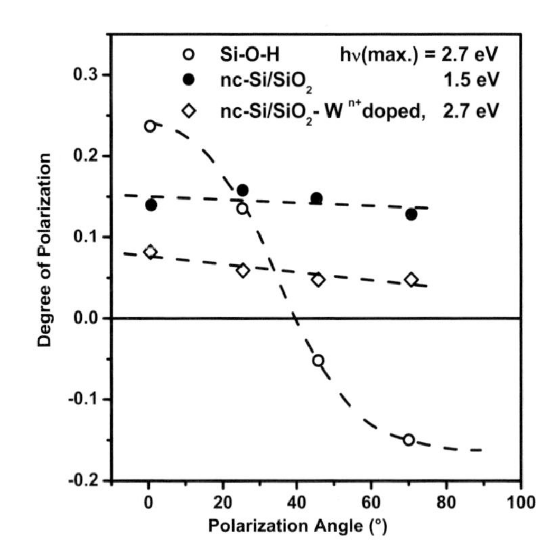

Figure 23.6. Polarization memory of the PL from silanol (≡Si-O-H) groups emitting around 2.7 eV (open circles) and its absence in the "red" PL from nc-Si/SiO_2 (filled circles) emitting around 1.5 eV and W^{n+}-doped nc-Si/SiO_2 (open diamonds) emitting at about 2.7 eV.

23.5 The Nature of the Emitting Centers

Because a variety of PL can be observed from various defects in SiO_2 as well as from metallic and organic impurities, detailed ESR studies have been carried out in order to elucidate the possible nature of the radiative centers that are responsible for the "red" PL from ultra-pure processed nc-Si/SiO_2. As a result, the non-bridging oxygen hole centers (NBOHC) were identified as those involved in the above described energy transfer and light emission. Figure 23.7 shows the correlation between the intensity of the PL and that of the ESR signal corresponding to the NBOHC.[22]

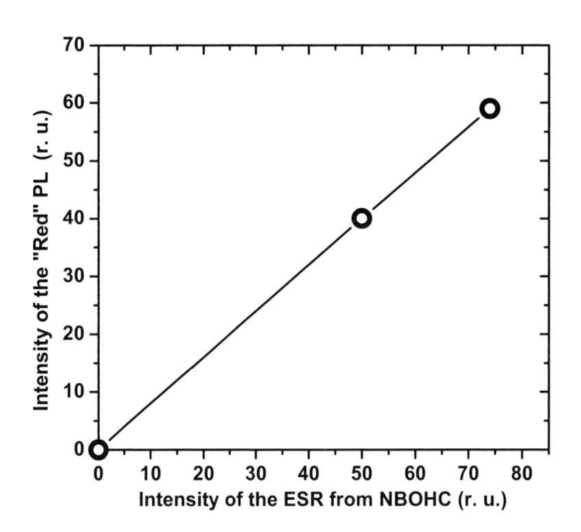

Figure 23.7. Correlation between the density of NBOHC radiative centers and the intensity of the "red" PL from ultra-pure processed nc-Si/SiO_2 nanocomposites. The PL was measured under identical conditions and normalized to the same amount of the material.

23.6 Origin of the Observed Small Changes of the PL Spectral Distribution

Occasionally, researchers have reported small changes of the PL from nc-Si passivated with an oxide layer and have discussed whether it could be attributed to phonon confinement phenomena. Figure 23.8a shows two examples reported by Schuppler et al.[23] and by Takagi et al.[24] These shifts are, however, much smaller than the corresponding change of the band gap (see Figure 23.2a and 23.8a) as well as the blue shift of the very efficient PL from organosilicon chains and ladders.[28] For the latter, the blue shift is observed when the chain or ladder length becomes comparable with the size of the exciton (or smaller), which extends over about 20 Si atoms.

Figures 23.8b and 23.8c show the changes of the spectral distribution and the corresponding shifts of the position of the maximum of the "red" PL from nc-Si/a-SiO_2 nanocomposites prepared by ultra-pure processing. Because in the course of the oxidation the crystallite size decreases whereas the PL shows a red shift, it is unambiguously clear that such minor changes as reported by Schuppler et al. and Tagaki et al. (Figure 23.8a) and as shown in Figures 23.8b and 23.8c cannot be attributed to quantum confinement. They are associated with small changes of the energy position of the (broadly distributed) NBOHC, as also observed by ESR.[23]

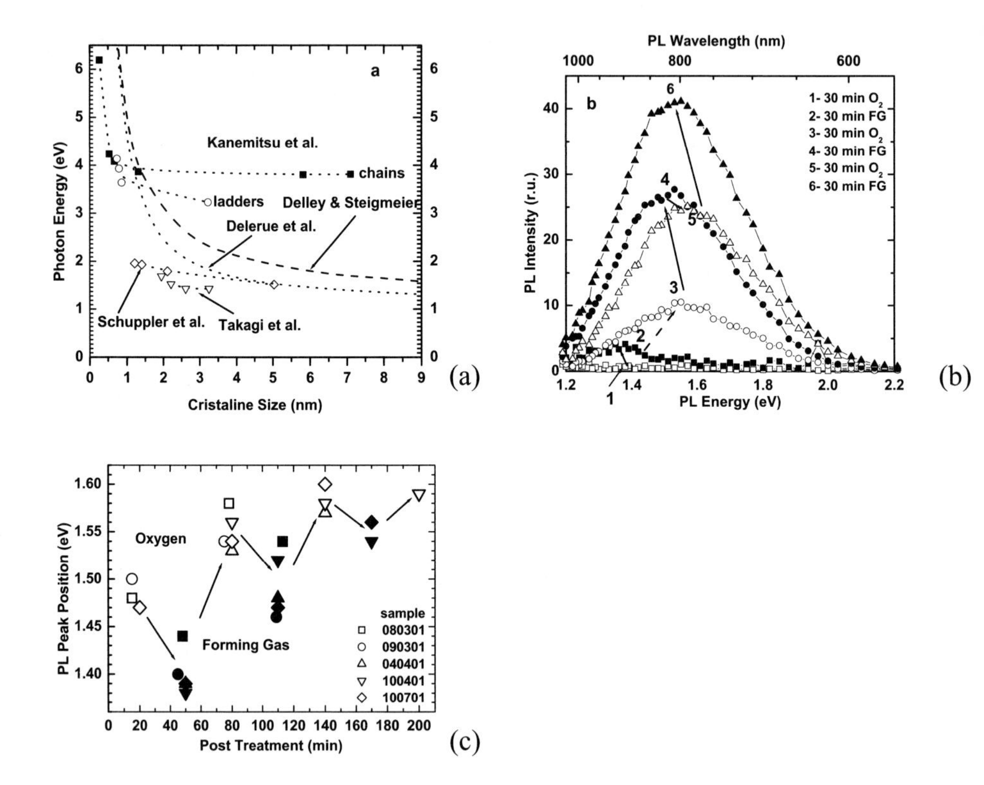

Figure 23.8. (a) Blue shift of the PL from nc-Si passivated with SiO_2 as reported by Schuppler et al.[26] and Tagaki et al.,[27] for organosilicon chains and ladders [28] and theoretical calculations [15, 16] of the change of the band gap; (b) changes of the spectral distribution of the "red" PL upon post-treatment (oxidation and FG annealing); (c) the corresponding shift of the position of the PL maximum.

The experimental results presented here clearly show that the "red" PL from nc-Si/SiO_2 nanocomposites prepared by ultra-pure processing originates from the non-bridging oxygen hole centers in the passivating SiO_2 matrix. The overall mechanism involves a photogeneration of electron–hole pairs within the silicon nanocrystals. Both the band gap and the transition rate increase with decreasing crystallite size, in very good agreement with the theoretical predictions. However, the recombination does not result in PL from the nanocrystals (which would have to show a similar energy dependence as found for the band gap of the nc-Si and for the PL of direct-gap semiconductor nanocrystals). Instead, an energy transfer to NBOHC radiative centers within the passivating SiO_2 matrix leads to the observed "red" PL with a nearly constant decay time and spectral distribution. The small variation of the PL spectral distribution can readily be attributed to small changes of the energies of the NBOHC sites. A variety of impurities, such as OH groups, metallic ions, and organic compounds, can be identified as the source of various PL reported in the literature and frequently attributed to quantum confinement phenomena.

23.7 Photodegradation of the PL from nc-Si

As already mentioned, electroluminescing devices based on nc-Si or PS-Si would have only a short lifetime. The relatively slow PL decay strongly limits the area of possible applications. Moreover, even nc-Si/SiO_2 nanocomposites show a photodegradation of the PL, as shown in Figure 23.9.

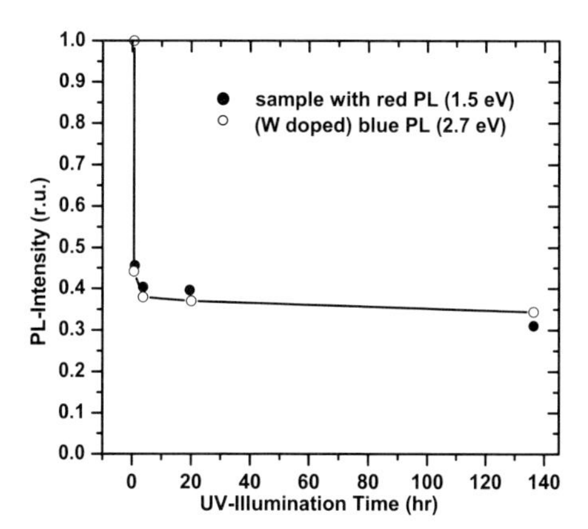

Figure 23.9. Photodegradation of the photoluminescence intensity from nc-Si/SiO_2 nanocomposites upon prolonged illumination with the excitation radiation for both the "red" PL and the fast PL from W^{n+}-doped nc-Si/SiO_2 nanocomposites. The photodegraded sample can be recovered by annealing in forming gas (FG) as described above, but this is not of any use for applications in devices.

These facts make the possible application of nanocrystalline silicon in optoelectronic devices unlikely. Also, the majority of organopolysilane compounds do not represent suitable candidates because the transition involved in the PL (e.g. Figure 23.8a) involves the $\sigma \leftrightarrow \sigma^*$ (HOMO $\leftrightarrow$ LUMO) transition which would also result in an irreversible photodegradation due to photolysis.

Possible and promising candidates that should not suffer from such problems are substituted silsesquioxanes $R_n(SiO_{1.5})_n$ because the observed intense PL from these compounds is due to the transition between the non-bonding states at oxygens and the LUMO states of the ligands. Silsesquioxanes do not show photodegradation, even when illuminated with a light from a strong mercury lamp for a period of several hours. Detailed studies of the absorption and PL spectra, together with a mechanism of the PL from silsesquioxanes of the type $R_8(SiO_{1-5})_8$, have been published in our recent papers.[25, 26]

23.8 Conclusions

The mechanism of the photoluminescence from nc-Si/SiO_2 nanocomposites includes a quantum confinement controlled photogeneration of electron–hole pairs within the silicon nanocrystals, followed by an energy transfer to non-bridging oxygen hole centers within the passivating SiO_2 matrix. The relatively inefficient and slow PL from such radiative centers is of little interest. A variety of other PL that have been reported in the literature and attributed to radiative recombination within the Si nanocrystals are most probably due to impurities. Several examples are shown to illustrate that extreme care has to be taken in order to avoid such artefacts.

The PL from substituted silsesquioxanes is efficient, fast and, because it is due to excitation from non-bonding states at the oxygen, does not show photodegradation, unlike that from the majority of polysilanes. This phenomenon makes them interesting for possible applications.

23. 9 Acknowledgment

We would like to thank the co-workers who were involved in this research in the course of several years of studies, and to Professor Y. Apeloig, Dr. D. Bravo-Zhivotovski and their co-workers for fruitful collaboration.

23.10 References

[1] L. T. Canham, *Appl. Phys. Lett.* **1990**, *57* 1046–1048.

[2] V. Lehmann, U. Gössele, *Appl. Phys. Lett.* **1991**, *58*, 856–858.

[3] L. Bruss, *J. Phys. Chem.* **1994**, *98*, 3575–3581.

[4] A. P. Alivisatos, *Science* **1996**, *271* 933–937.

[5] D. L. Feldheim, C. D. Keating, *Chem. Soc. Rev.* **1998**, *27*, 1–12.

[6] R. E. Hummel, P. Wißman (Eds.), *Handbook of Optical Properties*, Vol. II, Optics of Small Particles, Interfaces and Surfaces, CRC Press, Boca Raton, 1997.

[7] Remember that the energy states of an electron in a potential well increase with square of the width of that well.

[8] Y. Kanemitsu, K. Suzuki, Y. Nakayoshi, Y. Matsumoto, *Phys. Rev.* **1992**, *B 46,* 3916–3919.

[9] B. O. Dabbousi, J. Rodriguez-Viejo, F. V. Mikulec, J. R. Heine, H. Mattousi, R. Ober, K. F. Jensen, M. G. Bawendi, *J. Phys. Chem.* **1997**, *B 101,* 9436–9475.

[10] J. Lee, V. C. Sundar, J. R. Heine, M. G. Bawendi, K. F. Jensen, *Adv. Mater.* **2000**, *12,* 1102–1105.

[11] Therefore, direct band gap materials, such as GaAs, AlAs, $Ga_{1-x}Al_xAs$, GaN and others are used for light–emitting diodes and solid–state lasers because in indirect semiconductors the electron–hole recombination in a p/n junction biased in the forward direction produces only negligible light.

[12] S. Veprek, V. Marecek, *Solid State Electronics* **1968**, *11,* 683–684.

[13] For a review see S. Veprek, T. Wirschem in Ref. [6], p. 129–143 and references therein.

[14] This has been found for a large number of samples, but it is not shown here because of lack of space.

[15] S. Veprek, M. Rückschloß, Th. Wirschem, B. Landkammer, M. Fuss, X. Lin, *Appl. Phys. Lett.* **1995**, *67*, 2215–2217.

[16] S. Veprek, Th. Wirschem, J. Dian, S. Perna, R. Merica, M. G. J. Veprek-Heijman, R. Heinecke, V. Perina, *Thin Solid Films* **1997**, *297*, 171–175.

[17] S. Veprek, Th. Wirschem, M. Rückschloß, H. Tamura, J. Oswald, *Mater. Res. Soc. Symp. Proc.* **1995**, *358*, 99–110.

[18] H. Tamura, M. Rückschloß, Th. Wirschem, S. Veprek, *Thin Solid Films* **1995**, *255*, 92–95.

[19] M. Rückschloß, Th. Wirschem, H. Tamura, G. Ruhl, J. Oswald, S. Veprek, *J. Luminescence* **1995**, *63*, 279–287.

[20] A. C. Albrecht, *J. Molec. Spectr.* **1961**, *6,* 84.
[21] M. D. Galanin, *Luminescence of Molecules and Crystals*, Cambridge Int. Sci. Publ., Cambridge, U.K. 1996.
[22] S. M. Prokes, W. E. Carlos, S. Veprek, Ch. Ossadnik, *Phys. Rev.* **1998**, *58,* 15632–15635.
[23] S. Schuppler, S. L. Friedman, M. A. Marcus, D. L. Adler, Y.-H. Xie, E. M. Ross, T. D. Harris, W. L. Brown, Y. J. Chabal, L. E. Bruss, P. H. Citrin, *Phys. Rev. Lett.* **1994**, *72*, 2648–2651.
[24] H. Takagi, H. Ogawa, Y. Yamazaki, A. Ishizaki, T. Nakagiri, *Appl. Phys. Lett.* **1990**, *56*, 2379–2380.
[25] Ch. Ossadnik, S. Veprek, H. C. Marsmann, E. Rikowski, *Monatshefte f. Chemie (Chemical Monthly)* **1999**, *130*, 55–68.
[26] D. Azinovic, J. Cai, C. Eggs, H. König, H.C. Marsmann, S. Vepřek, *J. Luminescence* **2002**, *97*, 40–50.

Part III

Si-O Systems: From Molecular Building Blocks to Extended Networks

Networks based on Si-O-Si linkages are ubiquitous. All naturally occurring silicon-containing compounds (crystalline and amorphous silica, silicates, silica in biological systems) and many man-made materials (silicones, sol-gel materials) are constructed from tetrahedral units, with silicon as the central atom. The structural and compositional variety of such compounds originates from two facts. First, the average number of corners a tetrahedral unit shares with neighboring units, i.e. the average number of Si-O-Si bonds per silicon atom, can vary from two (as in silicones) to four (as in silica). The average degree of crosslinking is of key importance for the network structure and thus the macroscopic properties. The terminal groups, i.e. the groups not participating in the network formation (OH, O^-, OR, Cl, R, etc.), modify the properties of the siloxane network. Second, as the silicon-based tetrahedra only share corners, the structures resulting from the connected tetrahedra have many degrees of freedom and can adopt a variety of geometries.

Extended siloxane structures are in almost all cases formed by the stepwise condensation of silanols. The possibility of following the growth of siloxane structures from molecular units via oligomeric species to two- or three-dimensional extended systems is unprecedented in chemistry. This allows, inter alia, study of the influence of the precursor composition and the synthesis parameters on the structures and the properties of the extended systems.

Following this rationale, the first chapters of this Section deal with molecular precursors with Si-O bonds. The five- and six-coordinate compounds described in Chapter 24 by R. Tacke et al. give an outlook on what is possible beyond the "normal" four-coordinate silicon-oxygen compounds. It has been speculated that such higher-coordinate complexes may play a role in silicon biochemistry. The stability of silanols and hence the reactivity of Si-OH groups is a key issue in studying the formation of Si-O-Si bonds, and much can be learned from "stabilized" silanols (Chapters 25 and 26). U. Klingebiel et al. employ bulky organic substituents to prepare functionalized silanols, such as halo- or amino-substituted silanols, and use these compounds for the stepwise formation of linear or cyclic oligomers with Si-O-Si bonds. In the contribution of W. Malisch et al., silanols stable with respect to condensation reactions are obtained by bonding the silicon atom to a transition metal. The electron–releasing effect of the metal complex fragment reduces the acidity of the Si-OH proton and thus allows the isolation of transition metal substituted silanols, -diols, and -triols.

An important nexus between molecular compounds with Si-O bonds to three-dimensional silicate or siloxane networks are cage compounds. Such compounds can be constructed from silicon-based tetrahedral units alone, as in the silsequioxanes or sphero–silicates, or in combination with polyhedral units of other elements. Chapters 27 and 28 describe the use of silanols to assemble heteroatomic cage compounds. M. Veith et al. discuss the synthesis of the prototypical $(Ph_4Si_2O_5)_4Al_4(OH)_4$ from diphenylsilanediol and $^tBuOAlH_2$, while P. Jutzi et al. deal with the synthesis of polyhedral titanasiloxanes from silanetriols and titanium alkoxides. An important aspect of both contributions is how the initially obtained cage structures can be modified post-synthesis, a very important aspect with regard to the general question as to how extended oxide structures develop from small oligomeric units (nuclei).

Another strategy for synthesizing heteroatom-substituted silicate cages is to introduce the heteroatom after pre-assembly of the silicate fragment. Examples of this approach are given in the contribution of F. Edelmann (Chapter 29), which describes the preparation of various metallasilsesquioxanes by reacting incompletely condensed silsesquioxanes with metal salts.

The experimental finding that silsesquioxane units are isolators and do not allow considerable through-space and/or through-bond interactions across the cage is treated theoretically in Chapter 30 by W. W. Schoeller. It is shown that spin-delocalization is interrupted by the electronegative oxygen atoms linking the silicon centers.

While many methods are available for the characterization of both molecular and solid-state compounds, the reliable identification and characterization of oligomeric compounds is still a challenge to analytical chemistry. The contribution by R.-P. Krüger and G. Schulz (Chapter 31) gives an introduction to MALDI-TOF mass spectrometry and illustrates the great potential of this promising method with the characterization of several types of oligomeric and polymeric silicon-containing polymers.

There are two technically important methods by which extended Si/O systems can be formed from molecular precursors. The first is by reaction of chlorosilanes with oxygen at high temperatures, while the second is by hydrolysis and condensation reactions of chloro- or alkoxysilanes. Chapters 32 and 33 deal with the structural evolution of siloxane structures in such reactions from an experimental and theoretical viewpoint. M. Binnewies et al. compare the stepwise formation of Si-O networks from $SiCl_4$ for both the combustion and hydrolysis reactions. The stability and reactivity of intermediate chlorosiloxanes is an important issue in this work. Both the initial process in the reaction of $SiCl_4$ with O_2 and the growth of larger siloxane cages are investigated theoretically in the contribution of K. Jug.

The final four contributions in this Section deal with polymeric Si-O compounds. G. Kickelbick et al. (Chapter 34) discuss the formation and phase behavior of amphiphilic copolymers with defined siloxane blocks. It is shown that short–chain poly(dimethylsiloxane)-poly(ethylene oxide) diblock copolymers preferentially form lamellar phases. The topic of Chapters 35 and 36 is mesostructured silica materials with an ordered porosity. Such materials are prepared by templating with supramolecular assemblies of amphiphiles. N. Hüsing et al. give an overview on the synthesis of mesostructured films and describe different approaches for the functionalization of the walls of the mesopores. P. Behrens et al. describe the modification of mesostructured powders by either post-synthesis grafting of functional groups onto the pore walls or by co-condensation of differently substituted silanes during the formation of the network structures. The ultimate goal of any effort to synthesize hierarchically ordered structures is to reach the sophistication of natural structures. The final Chapter of this Section– the contribution of C. C. Perry (Chapter 37) on how Nature builds up complex siliceous structures – may therefore serve as a motivation for further efforts in this area and as a guide for where to go.

Peter Jutzi, Ulrich Schubert

24 Higher-Coordinate Silicon Compounds with *Si*O_5 and *Si*O_6 Skeletons

Reinhold Tacke, Oliver Seiler*

24.1 Introduction

The chemistry of silicon-oxygen compounds with *Si*O_4 skeletons has been studied in great depth. In contrast, compounds with *Si*O_5 and *Si*O_6 frameworks are significantly less well explored. This article is an account of contributions to the chemistry of molecular silicon compounds with *Si*O_5 and *Si*O_6 skeletons. Studies dealing with higher-coordinate silicon in crystalline silicon dioxide[1] and silicate phases[2,3] as well as in silicate glasses[4] will not be discussed.

24.2 Pentacoordinate Silicon Compounds with *Si*O_5 Skeletons

24.2.1 Anionic Species with *Si*O_5 Skeletons

The anion $[Si(OH)_5]^-$ can be regarded as the parent system for anionic silicon species with *Si*O_5 skeletons. Unlike $[SiF_5]^-$, the $[Si(OH)_5]^-$ anion has not yet been isolated and characterized. Recently, we succeeded in synthesizing some compounds that formally derive from $[Si(OH)_5]^-$ and its anhydride

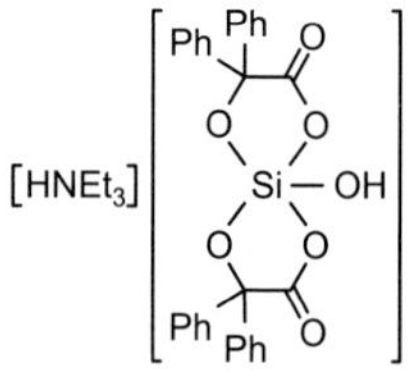

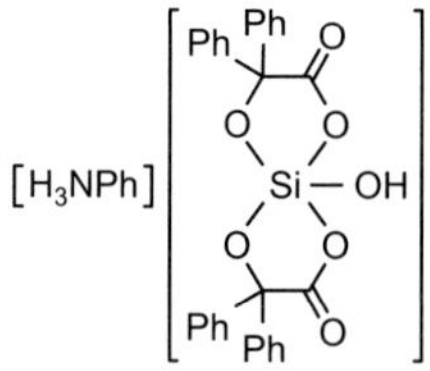

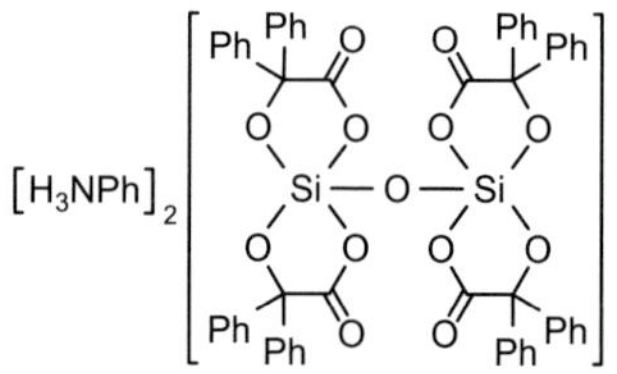

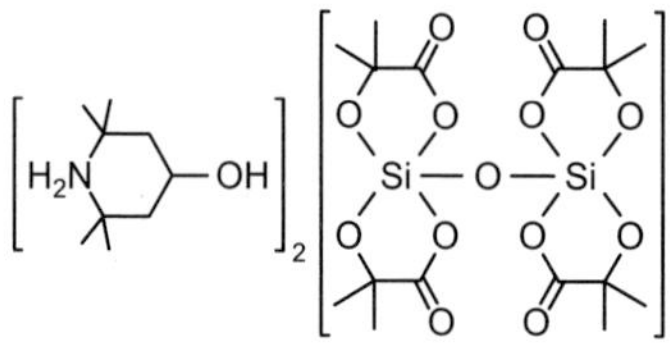

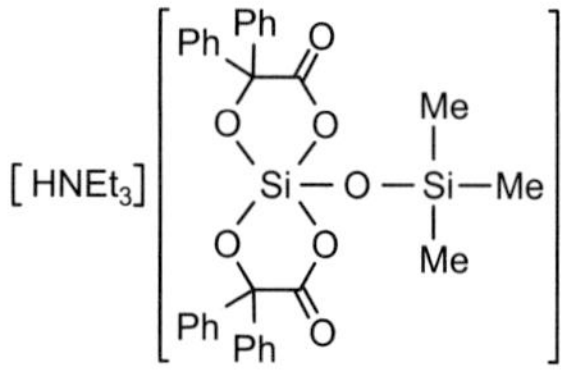

* Address of the authors: Institut für Anorganische Chemie, Universität Würzburg, Am Hubland, D-97074 Würzburg, Germany

$[(HO)_4Si–O–Si(OH)_4]^{2-}$, the λ^5*Si*-hydroxosilicates **1**[5] and **2**[5] as well as the λ^5*Si*,λ^5*Si′*-μ-oxo-disilicates **3**[5] and **4**[6] (synthesis of **2** and **3**: Scheme 24.1). The related compound **5** was also prepared.[6]

The racemic compounds **1**, **2**·THF (isolated as a conglomerate), **3**·$2CH_3CN$, and **5** as well as *meso*-**4** were crystallographically characterized. The Si coordination polyhedra of all the complexes are distorted trigonal bipyramids, with the carboxylate oxygen atoms of the two bidentate ligands in the axial positions. This is illustrated for the anions of **2**·THF and **3**·$2CH_3CN$ in Figure 24.1.

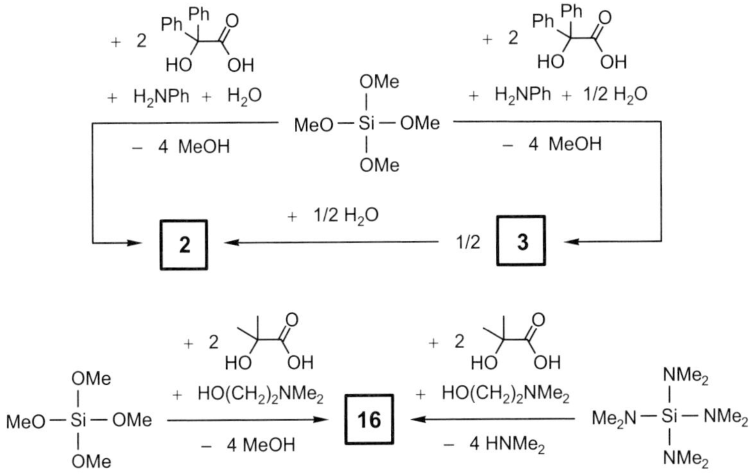

+ 2 $Ph_2C(OH)COOH$ + H_2NPh + H_2O, − 4 MeOH
$Si(OMe)_4$
+ 2 $Ph_2C(OH)COOH$ + H_2NPh + 1/2 H_2O, − 4 MeOH
2 ← (+ 1/2 H_2O) ← 1/2 **3**

$Si(OMe)_4$ → (+ 2 $Me_2C(OH)COOH$, + $HO(CH_2)_2NMe_2$, − 4 MeOH) → **16** ← (+ 2 $Me_2C(OH)COOH$, + $HO(CH_2)_2NMe_2$, − 4 $HNMe_2$) ← $Si(NMe_2)_4$

Scheme 24.1. Synthesis of compounds **2**, **3**, and **16**.

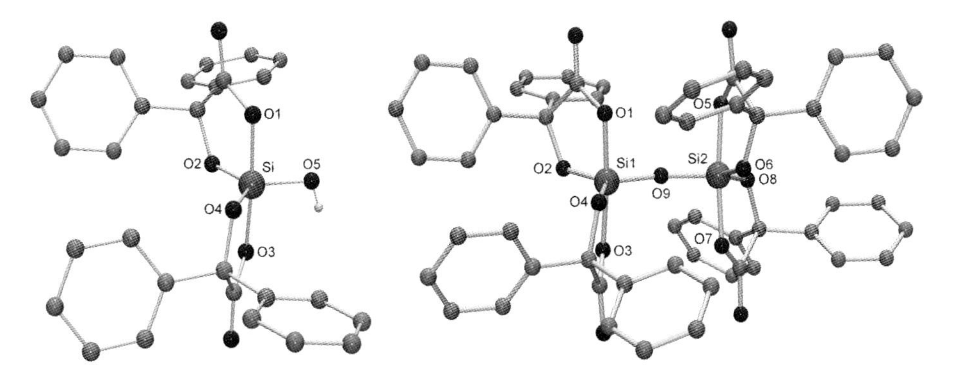

Figure 24.1. Structures of the anions in the crystals of *Λ*-**2**·THF (left, Si-O(ax) 1.798(3), Si-O(eq) 1.650(3)–1.660(2) Å) and **3**·$2CH_3CN$ (right, Si-O(ax) 1.805(2)–1.836(2), Si-O(eq) 1.615(1)–1.659(1) Å).

Distorted trigonal-bipyramidal Si coordination polyhedra have also been reported for the related λ^5*Si*-silicates **6**,[7] **7**,[8] **8**,[9] **10**,[8] **11**,[10] **12**,[10] **13**,[11] and **14**,[12] with the monodentate ligands in equatorial positions. In contrast, the Si-coordination polyhedron of **9**·MeOH[8] is best described as a distorted square pyramid, the four basal positions being occupied by the two bidentate ligands.

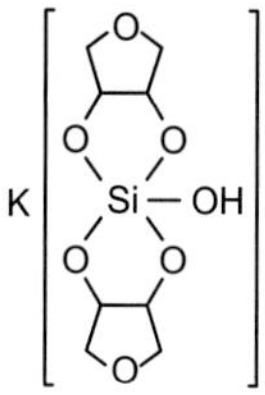

6

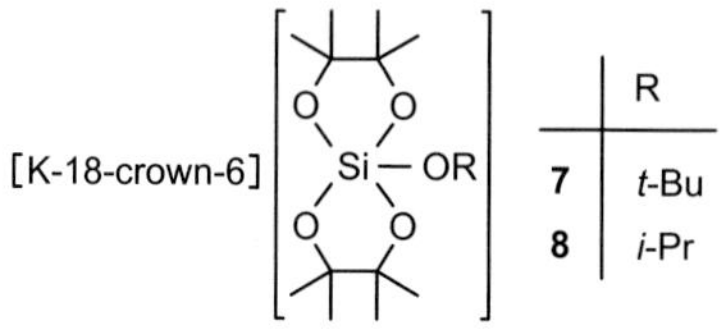

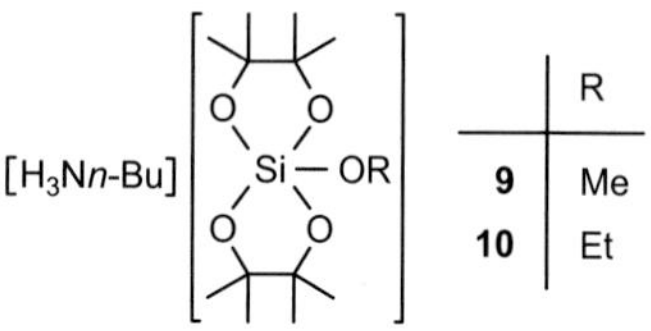

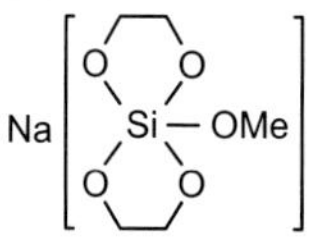

	R
7	*t*-Bu
8	*i*-Pr

	R
9	Me
10	Et

11

	M^+
12	Li^+
13	Na^+
14	K^+

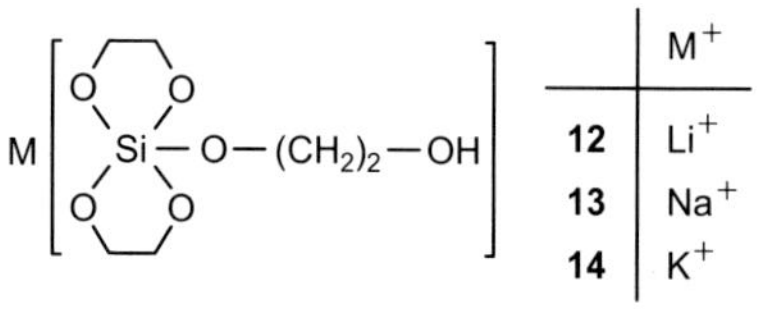

24.2.2 Neutral Species with *Si*O_5 Skeletons

In 1970, the synthesis of zwitterionic (neutral) λ^5*Si*-silicates with *Si*O_5 skeletons, such as **15**, has been claimed, based only on elemental analyses.[13] A few years ago, the existence of such compounds was established unequivocally. In the course of our systematic studies on zwitterionic λ^5*Si*-silicates,[14] we succeeded in synthesizing **16**,[15] **17**,[15] and **18**·DMF[16] (synthesis of **16**: Scheme 24.1) and characterized their structures by single-crystal X-ray diffraction.

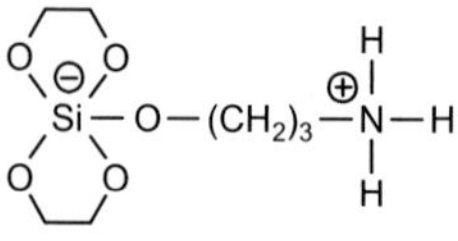

15

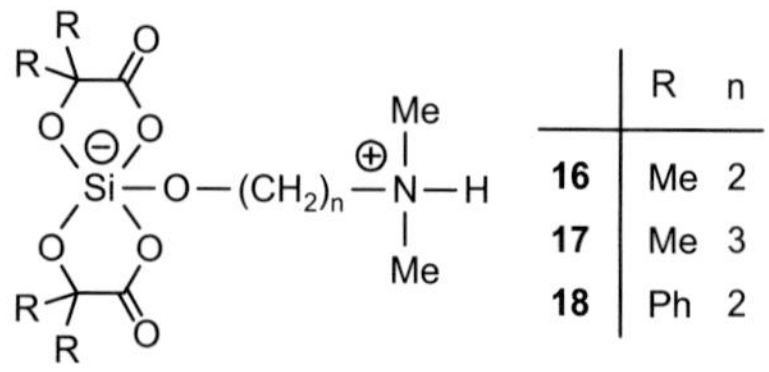

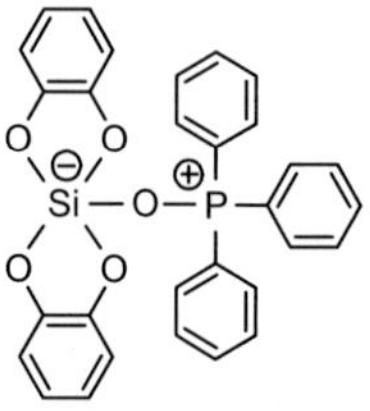

	R	n
16	Me	2
17	Me	3
18	Ph	2

19

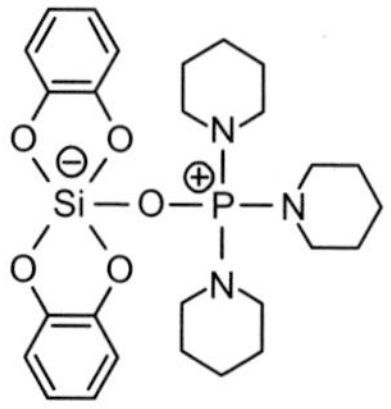

20

The Si coordination polyhedra of **16** (Figure 24.2), **17**, and **18**·DMF are distorted trigonal bipyramids, with the carboxylate oxygen atoms of the two bidentate ligands in the axial positions. A distorted trigonal-bipyramidal environment was also observed for the silicon atom of **19**, whereas the Si coordination polyhedron of **20**·CH_2Cl_2 is a square pyramid.[17]

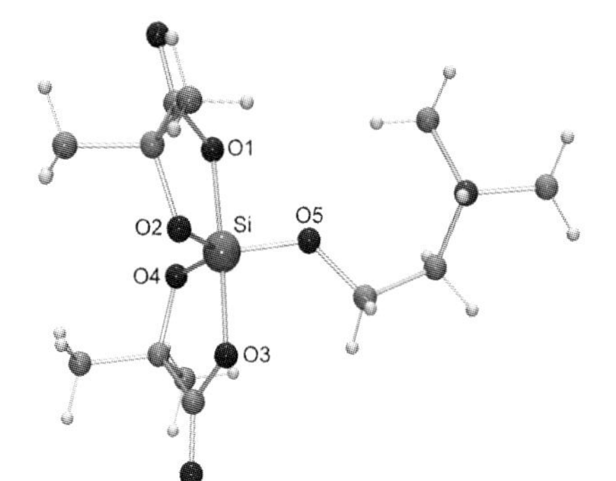

Figure 24.2. Structure of **16** in the crystal (Si-O(ax) 1.773(2) and 1.798(2), Si-O(eq) 1.643(2)–1.659(2) Å).

24.3 Hexacoordinate Silicon Compounds with *Si*O_6 Skeletons

24.3.1 Anionic Species with *Si*O_6 Skeletons

The dianion $[Si(OH)_6]^{2-}$ can be regarded as the parent system for dianionic silicon species with *Si*O_6 skeletons. Unlike $[SiF_6]^{2-}$, the $[Si(OH)_6]^{2-}$ dianion has not yet been isolated and characterized. However, the synthesis of the first derivative, the tris[benzene-1,2-diolato(2–)]silicate dianion, was reported as long ago as 1920. This silicon(IV) complex was obtained as the pyridinium salt **21**,[18] the identity of which was unequivocally established five decades later by single-crystal X-ray diffraction.[19] In the meantime, many derivatives of **21**, with different cations and/or substituted benzene-1,2-diolato(2–) ligands, were synthesized, and various aspects of their chemistry were studied (selected publications: refs.[20–32]. With the syntheses and crystal structure analyses of **22**·$2CH_3CN$ (Scheme 24.2, Figure 24.3) and **23**, we have also contributed to this chemistry.[33] The Si coordination polyhedra of **21**, **22**·$2CH_3CN$, **23**, and other tris[benzene-1,2-diolato(2–)]silicates[22,29,32] are distorted octahedra.

cat_2 [tris(benzene-1,2-diolato)silicate]

	cat^+
21	[HN(pyridinium)]$^+$
22	$[HO(CH)_2NMe_2H]^+$
23	$[HO(CH)_3NMe_2H]^+$

Ba[Si(OCH₂CH₂O)₃] **24**; cat₂[Si(C₂O₄)₃]

	cat⁺
25	$[NEt_4]^+$
26	$[NMe_4]^+$
27	$[NMe_3Ph]^+$
28	$[HO(CH_2)_2NH(C_4H_8)]^+$
29	$[HO(CH_2)_2NH(C_5H_{10})]^+$

In contrast to the well-established chemistry of the tris[benzene-1,2-diolato(2–)]silicate dianion (including derivatives with substituted ligands), the chemistry of the related tris[ethane-1,2-diolato(2–)]silicate dianion is significantly less well explored. To the best of our knowledge, **24**·3.25$HOCH_2CH_2OH$ (obtained by reaction of BaO and SiO_2 with an excess of ethane-1,2-diol) is the only example that has been structurally characterized.[34]

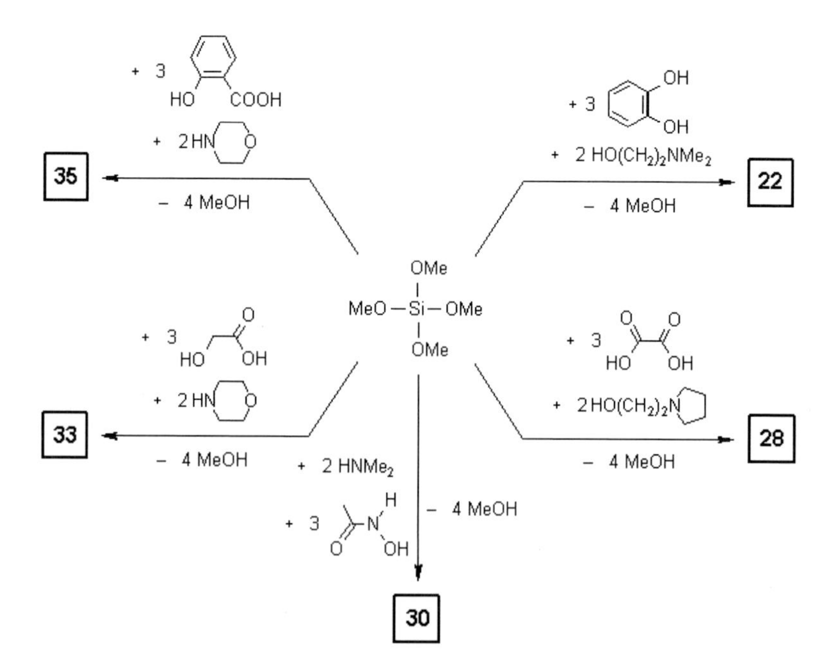

Scheme 24.2. Synthesis of compounds **22**, **28**, **30**, **33**, and **35**.

The tris[oxalato(2–)]silicate dianion represents another λ^6Si-silicate species with three symmetric bidentate ligands. The first example of this type of silicon(IV) complex, compound **25** (obtained by treatment of $SiCl_4$ with

[NEt_4]Br and silver oxalate (molar ratio 1:2:3)), was first described in 1969,[35] and ca. 25 years later the crystal structures of the derivatives **26** and **27** were reported.[36] In the meantime, various tris[oxalato(2–)]silicates with different cations were described.[36–40] As shown by crystal structure analyses, the Si coordination polyhedra of **26**,[36] **27**,[36] **28**[40] (synthesis: Scheme 24.2; structure: Figure 24.3), and **29**[40] are distorted octahedra.

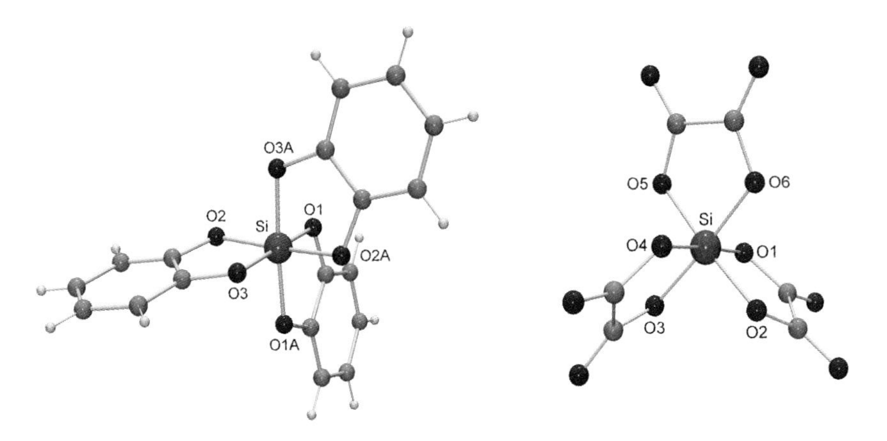

Figure 24.3. Structures of the dianions in the crystals of **22**·$2CH_3CN$ (left, Si-O 1.774(1)–1.798(1) Å) and **28** (right, Si-O 1.767(1)–1.788(1) Å).

Recently, we succeeded in developing syntheses of dianionic λ^6*Si*-silicates with *Si*O_6 skeletons containing three unsymmetric bidentate ligands. These ligands derive from (i) aceto- or benzohydroximic acid,[41] (ii) α-hydroxycarboxylic acids (such as glycolic acid and benzilic acid),[42] or (iii) ß-hydroxycarboxylic acids of the salicylic acid type.[43] In the case of such unsymmetric bidentate ligands, special stereochemical features have to be considered, i.e. *fac*/*mer*-isomerism and *Λ*/*Δ*-enantiomerism.

[H_2NMe_2]$_2$ [Si(ONC(R)O)$_3$]

	R
30	Me
31	Ph

[$H_3N(CH_2)_2NH_3$] [Si(ONC(Ph)O)$_3$]

32

Compounds **30**–**32** represent tris[hydroximato(2–)]silicate complexes.[41] They were prepared by using the same preparative approach (synthesis of **30**: Scheme 24.2) and isolated as the *fac*- (**30**, **31**·MeOH) or *mer*-isomer

(**32**·2MeOH). As shown by crystal structure analyses, the Si coordination polyhedra of these compounds are distorted octahedra.

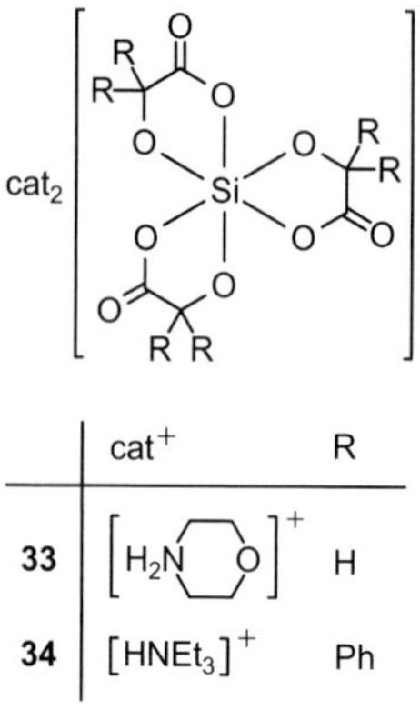

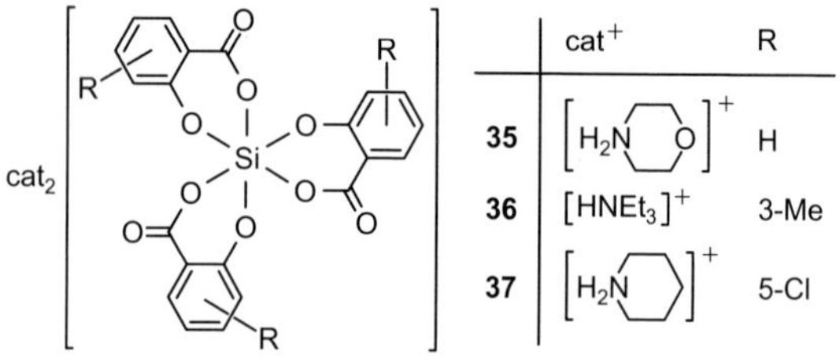

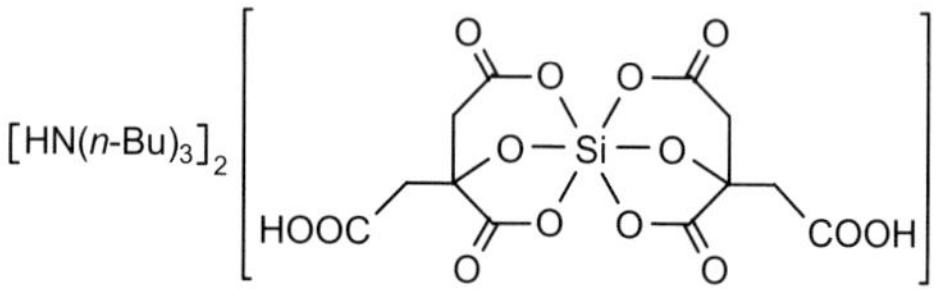

	cat$^+$	R
33	$[H_2N\ \ O]^+$	H
34	$[HNEt_3]^+$	Ph

	cat$^+$	R
35	$[H_2N\ \ O]^+$	H
36	$[HNEt_3]^+$	3-Me
37	$[H_2N]^+$	5-Cl

$[HN(n\text{-}Bu)_3]_2$ **38**

Compounds **33** and **34** are silicon(IV) complexes with three bidentate ligands derived from α-hydroxycarboxylic acids. They were synthesized according to Schemes 24.2 or 24.3 and were isolated as *mer*-**33** and *fac*-**34**·1/2$C_4H_8O_2$ ($C_4H_8O_2$ = 1,4-dioxane).[42]

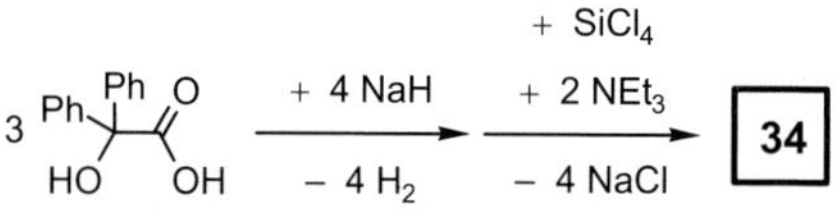

$$3\ Ph_2C(OH)COOH \xrightarrow[-\ 4\ H_2]{+\ 4\ NaH} \xrightarrow[-\ 4\ NaCl]{+\ SiCl_4,\ +\ 2\ NEt_3} \mathbf{34}$$

Scheme 24.3. Synthesis of compound **34**.

Compounds **35**–**37** are silicon(IV) complexes of the tris[salicylato(2–)]–silicate type.[43] They were prepared by using the same synthetic approach (synthesis of **35**: Scheme 24.2) and isolated as the *mer*-isomers (**37** was obtained as *mer*-**37**·2THF). In contrast to all the other aforementioned hexa-coordinate silicon(IV) complexes with three five-membered chelate rings, compounds **35**–**37** contain three six-membered chelate rings.

Recently, we succeeded in synthesizing the first silicon(IV) complex with two tridentate citrato(3–) ligands, compound **38** (Scheme 24.4, Figure 24.4).[44] As shown by single-crystal X-ray diffraction, the two tridentate ligands are each coordinated to the silicon atom through two carboxylato oxygen atoms and one alcoholato oxygen atom, forming two bicyclic moieties each with a five-, six-, and seven-membered ring. The Si coordination polyhedron is a distorted octahedron.

As demonstrated by crystal structure analyses, the Si coordination polyhedra of **33**–**38** are distorted octahedra, the Si-O(carboxylato) distances being significantly longer than the Si-O(alcoholato) distances.

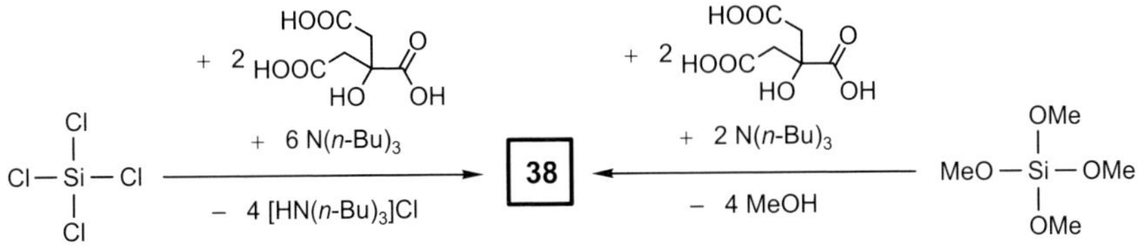

$SiCl_4$ + 2 HOOC–CH₂–C(OH)(CH₂COOH)–COOH + 6 N(*n*-Bu)₃ − 4 [HN(*n*-Bu)₃]Cl → **38** ← Si(OMe)₄ + 2 HOOC–CH₂–C(OH)(CH₂COOH)–COOH + 2 N(*n*-Bu)₃ − 4 MeOH

Scheme 24.4. Synthesis of compound **38**.

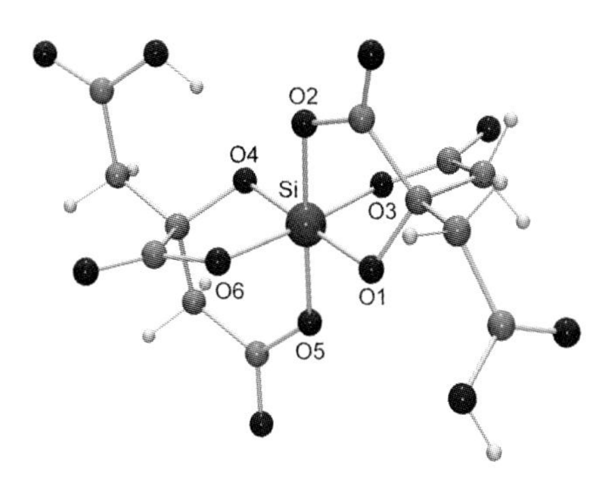

Figure 24.4. Structure of the dianion in the crystal of **38** (Si-O1 1.716(1), Si-O2 1.815(1), Si-O3 1.800(1), Si-O4 1.733(1), Si-O5 1.795(1), Si-O6 1.812(1) Å).

24.3.2 Cationic Species with SiO_6 Skeletons

To the best of our knowledge, cationic pentacoordinate silicon(IV) complexes with SiO_5 skeletons have not yet been described, whereas the chemistry of cationic hexacoordinate silicon(IV) species with SiO_6 skeletons is well established.

In 1964, the syntheses of the cationic tris[1-oxopyridin-2-olato(1–)]-silicon(IV) complexes **39**–**41** were claimed,[45] based on elemental analyses and ESCA data for **39**.[46] Very recently, we succeeded in synthesizing and characterizing the related complexes **42**–**44** (synthesis of **42**: Scheme 24.5).[47] Compound **42**·$C_5H_5NO_2$ ($C_5H_5NO_2$ = 1-hydroxy-2-pyridone or 2-hydroxy-pyridine *N*-oxide) was structurally characterized by single-crystal X-ray diffraction;[47] the Si coordination polyhedron was found to be a distorted octahedron.

In 1964, the first cationic tris[tropolonato(1–)]silicon(IV) complexes, **45** and **46**, were described (synthesis of **45**: Scheme 24.5).[48] However, their identity was established only by elemental analyses. In the meantime, further reports on cationic tris[tropolonato(1–)]silicon(IV) complexes (including derivatives with substituted ligands) have been published.[49–54] Very recently, the crystal structure of **47** was reported.[54] The Si coordination polyhedron of its cation is a distorted octahedron.

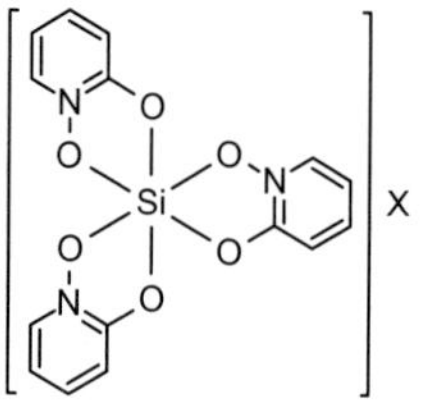

	X^-
39	Cl^-
40	$FeCl_4^-$
41	1/2 $SnCl_6^{2-}$
42	$CF_3SO_3^-$
43	$EtOSO_3^-$
44	*i*-$PrOSO_3^-$

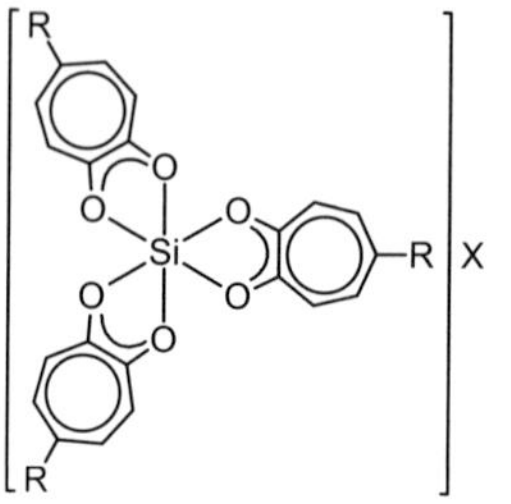

	X^-	R
45	Cl^-	H
46	PF_6^-	H
47	5-isopropyl-tropolonate$^-$	*i*-Pr

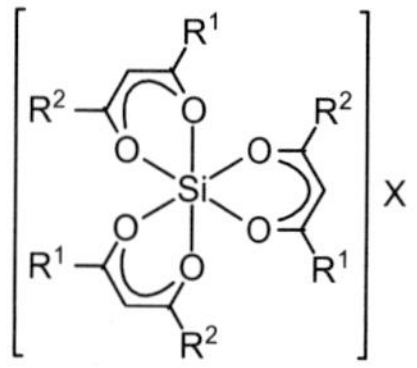

	X^-	R^1	R^2
48	HCl_2^-	Me	Me
49	$FeCl_4^-$	Me	Me
50	$AuCl_4^-$	Me	Me
51	$FeCl_4^-$	Me	Ph
52	$AuCl_4^-$	Me	Ph
53	$FeCl_4^-$	Ph	Ph
54	$AuCl_4^-$	Ph	Ph
55	$(TCNQ)_2^{\bullet -}$	Me	Me
56	HCl_2^-	Me	Ph
57	$AgCl_2^-$	Me	Me

The syntheses of the first cationic tris[ß-diketonato(1–)]silicon(IV) complexes, such as **48**–**54**, were described as long ago as 1903 (synthesis of **48**: Scheme 24.5).[55,56] Their identities were established by elemental analyses. Eight decades later, the crystal structure of **55** was reported.[57] In the meantime, further reports on cationic complexes of this type were published (selected publications: refs.[58–65]), including the crystal structure analyses of *mer*-**56**[64] and **57**·CH_2Cl_2.[65] The Si coordination polyhedra of **55**, *mer*-**56**, and **57**·CH_2Cl_2 are distorted octahedra.

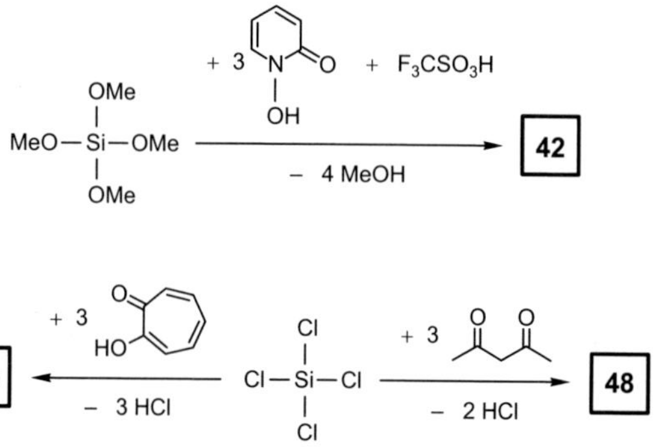

Scheme 24.5. Synthesis of compounds **42**, **45**, and **48**.

24.3.3 Neutral Species with SiO_6 Skeletons

In contrast to the well-established chemistry of anionic and cationic hexacoordinate silicon(IV) complexes with SiO_6 skeletons, neutral hexacoordinate silicon(IV) species with SiO_6 frameworks are significantly less well explored. Compounds **58**,[66–68] **59**,[69] and **60**[70] are examples of this type of compound.

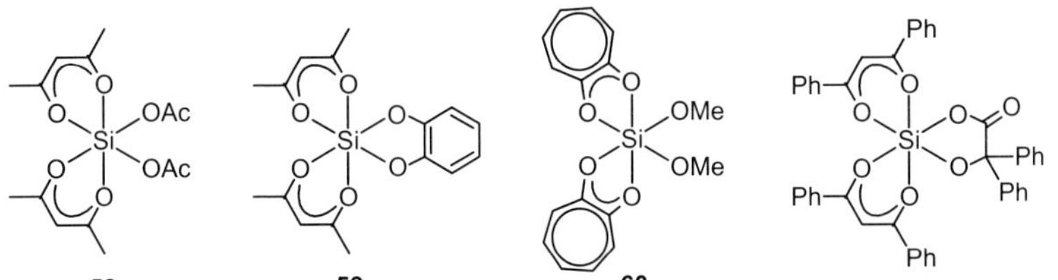

58 **59** **60** **61**

None of these complexes was characterized by crystal structure analysis. Very recently, we succeeded in synthesizing **61** (Scheme 24.6) and established its structure by single-crystal X-ray diffraction.[71] The Si coordination polyhedron of **61** is a distorted octahedron (Figure 24.5).

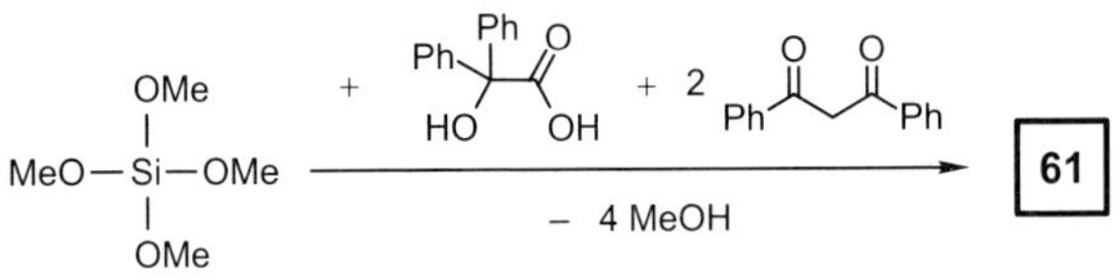

$Si(OMe)_4$ + $Ph_2C(OH)COOH$ + 2 $PhC(O)CH_2C(O)Ph$ → **61** (− 4 MeOH)

Scheme 24.6. Synthesis of compound **61**.

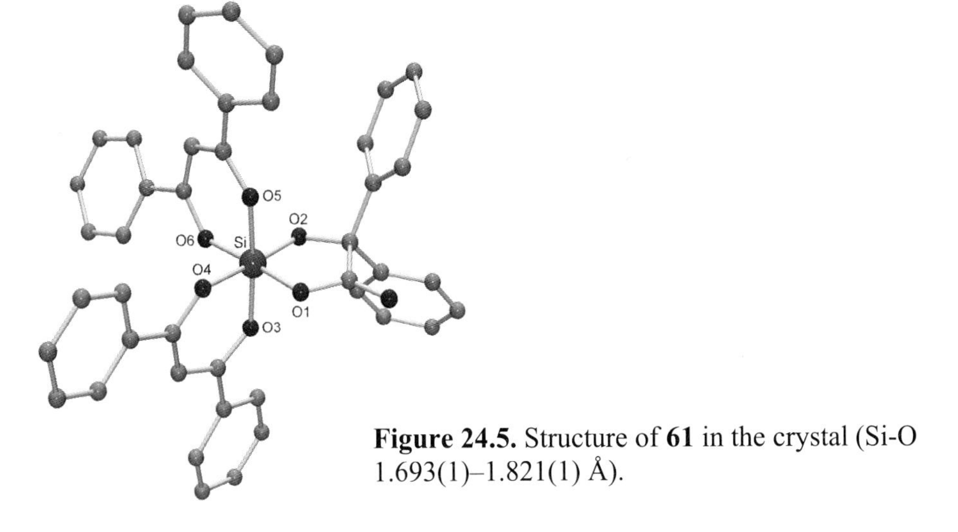

Figure 24.5. Structure of **61** in the crystal (Si-O 1.693(1)–1.821(1) Å).

24.4 Concluding Remarks

As is evident from sections 24.2 and 24.3, significant progress in the chemistry of molecular higher-coordinate silicon compounds (including anionic, cationic, and neutral species) with SiO_5 and SiO_6 skeletons has been made in recent years. Since the first reports on the syntheses of compounds with SiO_5 frameworks (1970; neutral species[13]) and SiO_6 skeletons (1903; cationic species[55,56]), a variety of anionic, cationic (only with SiO_6 frameworks), and neutral species of this type have been synthesized and structurally characterized. It should be mentioned that there is also increasing knowledge about their chemical reactivity and stereodynamics in solution (for further details, see the cited literature). The chemistry of higher-coordinate silicon species with SiO_5 and SiO_6 skeletons in aqueous solution is currently of special interest: it has been speculated that such higher-coordinate silicon(IV) complexes with ligands derived from organic hydroxy compounds (such as pyrocatechol derivatives, hydroxycarboxylic acids, and carbohydrates) may play a role in silicon biochemistry, controlling the transport of silicon, its concentration as a soluble silicon pool, and its deposition as silica (in this context, see ref.[44] and literature cited therein).

24.5 Acknowledgment

R. T. expresses his sincere thanks to his co-workers, without whose contributions this article could not have been written; their names are cited in the references. In addition, financial support by the Fonds der Chemischen Industrie is gratefully acknowledged.

24.6 References

[1] High-pressure silicon dioxide phase stishovite (rutile structure): W. Sinclair, A. E. Ringwood, *Nature* **1978**, *272*, 714.

[2] High-pressure $CaSi_2O_5$ phase: R. J. Angel, N. L. Ross, F. Seifert, T. F. Fliervoet, *Nature* **1996**, *384*, 441. Y. Kudoh, M. Kanzaki, *Phys. Chem. Min.* **1998**, *25*, 429. J. F. Stebbins, B. T. Poe, *Geophys. Res. Lett.* **1999**, *26*, 2521.

[3] High-pressure (Mg,Fe)SiO_3 perovskite phase (probably the most abundant mineral of the Earth): Q. Williams, E. Knittle, R. Jeanloz, in *Perovskite: A Structure of Great Interest to Geophysics and Material Science* (Eds.: A.

Navrotsky, D. J. Weidner), American Geophysical Union, Washington D. C., **1989**, pp. 1–12.
[4] High-pressure alkali silicate glasses: J. F. Stebbins, *Nature* **1991**, *351*, 638. X. Xue, J. F. Stebbins, M. Kanzaki, P. F. McMillan, B. Poe, *Am. Mineral.* **1991**, *76*, 8.
[5] R. Tacke, C. Burschka, I. Richter, B. Wagner, R. Willeke, *J. Am. Chem. Soc.* **2000**, *122*, 8480.
[6] R. Tacke, I. Richter, M. Penka, unpublished results.
[7] K. Benner, P. Klüfers, J. Schuhmacher, *Z. Anorg. Allg. Chem.* **1999**, *625*, 541.
[8] R. R. Holmes, R. O. Day, J. S. Payne, *Phosphorus, Sulfur, Silicon* **1989**, *42*, 1.
[9] K. C. Kumara Swamy, V. Chandrasekhar, J. J. Harland, J. M. Holmes, R. O. Day, R. R. Holmes, *J. Am. Chem. Soc.* **1990**, *112*, 2341.
[10] W. Donhärl, I. Elhofer, P. Wiede, U. Schubert, *J. Chem. Soc., Dalton Trans.* **1998**, 2445.
[11] G. J. Gainsford, T. Kemmitt, N. B. Milestone, *Acta Cryst. C* **1995**, *5*, 8.
[12] R. M. Laine, K. Y. Blohowiak, T. R. Robinson, M. L. Hoppe, P. Nardi, J. Kampf, J. Uhm, *Nature* **1991**, *353*, 642.
[13] C. L. Frye, *J. Am. Chem. Soc.* **1970**, *92*, 1205.
[14] R. Tacke, M. Pülm, B. Wagner, *Adv. Organomet. Chem.* **1999**, *44*, 221.
[15] R. Tacke, B. Pfrommer, M. Pülm, R. Bertermann, *Eur. J. Inorg. Chem.* **1999**, 807.
[16] R. Tacke, M. Mühleisen, *Inorg. Chem.* **1994**, *33*, 4191.
[17] E. Hey-Hawkins, U. Dettlaff-Weglikowska, D. Thiery, H. G. von Schnering, *Polyhedron* **1992**, *11*, 1789.
[18] A. Rosenheim, O. Sorge, *Ber. Dtsch. Chem. Ges.* **1920**, *53*, 932.
[19] J. F. Flynn, F. P. Boer, *J. Am. Chem. Soc.* **1969**, *91*, 5756.
[20] A. Rosenheim, B. Raibmann, G. Schendel, *Z. Anorg. Allg. Chem.* **1931**, *196*, 160.
[21] D. W. Barnum, *Inorg. Chem.* **1972**, *11*, 1424.
[22] D. Sackerer, G. Nagorsen, *Z. Anorg. Allg. Chem.* **1977**, *437*, 188.
[23] A. Boudin, G. Cerveau, C. Chuit, R. J. P. Corriu, C. Reye, *Angew. Chem.* **1986**, *98*, 473; *Angew. Chem. Int. Ed. Engl.* **1986**, *25*, 474.
[24] A. Boudin, G. Cerveau, C. Chuit, R. J. P. Corriu, C. Reye, *Organometallics* **1988**, *7*, 1165.
[25] A. Boudin, G. Cerveau, C. Chuit, R. J. P. Corriu, *J. Organomet. Chem.* **1989**, *362*, 265.
[26] G. Cerveau, C. Chuit, R. J. P. Corriu, L. Gerbier, C. Reyé, *Phosphorus, Sulfur, Silicon* **1989**, *42*, 115.
[27] D. F. Evans, J. Parr, E. N. Coker, *Polyhedron* **1990**, *9*, 813.

[28] L.-O. Öhman, A. Nordin, I. F. Sedeh, S. Sjöberg, *Acta Chem. Scand.* **1991**, *45*, 335.
[29] F. E. Hahn, M. Keck, K. N. Raymond, *Inorg. Chem.* **1995**, *34*, 1402.
[30] P. D. Lickiss, R. Lucas, *Polyhedron* **1996**, *15*, 1975.
[31] V. Chandrasekhar, S. Nagendran, Samiksha, G. T. Senthil Andavan, *Tetrahedron Lett.* **1998**, *39*, 8505.
[32] J. V. Kingston, B. Varghеese, M. N. S. Rao, *Main Group Chem.* **2000**, *3*, 79.
[33] R. Tacke, A. Stewart, J. Becht, C. Burschka, I. Richter, *Can. J. Chem.* **2000**, *78*, 1380.
[34] M. L. Hoppe, R. M. Laine, J. Kampf, M. S. Gordon, L. W. Burggraf, *Angew. Chem.* **1993**, *105*, 283; *Angew. Chem. Int. Ed. Engl.* **1993**, *32*, 287.
[35] P. A. W. Dean, D. F. Evans, R. F. Phillips, *J. Chem. Soc. A* **1969**, 363.
[36] K. J. Balkus, Jr., I. S. Gabrielova, S. G. Bott, *Inorg. Chem.* **1995**, *34*, 5776.
[37] G. Schott, D. Lange, *Z. Anorg. Allg. Chem.* **1972**, *391*, 27.
[38] K. Ueyama, G.-E. Matsubayashi, T. Tanaka, *Inorg. Chim. Acta* **1984**, *87*, 143.
[39] K. E. Bessler, V. M. Deflon, *Z. Anorg. Allg. Chem.* **1994**, *620*, 947.
[40] O. Seiler, C. Burschka, M. Penka, R. Tacke, *Z. Anorg. Allg. Chem.* **2002**, *628*, 2427.
[41] A. Biller, C. Burschka, M. Penka, R. Tacke, *Inorg. Chem.* **2002**, *41*, 3901.
[42] I. Richter, M. Penka, R. Tacke, *Inorg. Chem.* **2002**, *41*, 3950.
[43] O. Seiler, C. Burschka, M. Penka, R. Tacke, unpublished results.
[44] R. Tacke, M. Penka, F. Popp, I. Richter, *Eur. J. Inorg. Chem.* **2002**, 1025.
[45] A. Weiss, D. R. Harvey, *Angew. Chem.* **1964**, *76*, 818; *Angew. Chem. Int. Ed. Engl.* **1964**, *3*, 698.
[46] H. Meyer, G. Nagorsen, *Z. Naturforsch.* **1974**, *29b*, 72.
[47] R. Tacke, M. Willeke, M. Penka, *Z. Anorg. Allg. Chem.* **2001**, *627*, 1236.
[48] E. L. Muetterties, C. M. Wright, *J. Am. Chem. Soc.* **1964**, *86*, 5132.
[49] E. L. Muetterties, C. W. Alegranti, *J. Am. Chem. Soc.* **1969**, *91*, 4420.
[50] T. Ito, N. Tanaka, I. Hanazaki, S. Nagakura, *Inorg. Nucl. Chem. Lett.* **1969**, *5*, 781.
[51] T. Inoue, *Inorg. Chem.* **1983**, *22*, 2435.
[52] M. A. Al-Kadier, D. S. Urch, *J. Chem. Soc., Dalton Trans.* **1984**, 263.
[53] S. Azuma, M. Kojima, Y. Yoshikawa, *Inorg. Chim. Acta* **1998**, *271*, 24.
[54] M. Kira, L. C. Zhang, C. Kabuto, H. Sakurai, *Organometallics* **1998**, *17*, 887.
[55] W. Dilthey, *Ber. Dtsch. Chem. Ges.* **1903**, *36*, 923.
[56] W. Dilthey, *Ber. Dtsch. Chem. Ges.* **1903**, *36*, 1595.
[57] K. Ueyama, G.-E. Matsubayashi, I. Shimohara, T. Tanaka, K. Nakatsu, *J. Chem. Res. (S)* **1985**, 48; *J. Chem. Res. (M)* **1985**, 0801.
[58] W. Dilthey, *Liebigs Ann. Chem.* **1906**, *344*, 300.

[59] R. West, *J. Am. Chem. Soc.* **1958**, *80*, 3246.
[60] S. K. Dhar, V. Doron, S. Kirschner, *J. Am. Chem. Soc.* **1958**, *80*, 753.
[61] J. A. S. Smith, E. J. Wilkins, *J. Chem. Soc. A* **1966**, 1749.
[62] R. C. Fay, N. Serpone, *J. Am. Chem. Soc.* **1968**, *90*, 5701.
[63] T. Shimizutani, Y. Yoshikawa, *Inorg. Chem.* **1991**, *30*, 3236.
[64] U. Thewalt, U. Link, *Z. Naturforsch.* **1991**, *46b*, 293.
[65] D. Högerle, U. Link, U. Thewalt, *Z. Naturforsch.* **1993**, *48b*, 691.
[66] R. M. Pike, R. R. Luongo, *J. Am. Chem. Soc.* **1965**, *87*, 1403.
[67] C. E. Holloway, R. R. Luongo, R. M. Pike, *J. Am. Chem. Soc.* **1966**, *88*, 2060.
[68] D. T. Haworth, G.-Y. Lin, C. A. Wilkie, *Inorg. Chim. Acta* **1980**, *40*, 119.
[69] K. Yamaguchi, K. Ueno, H. Ogino, *Chem. Lett.* **1998**, 247.
[70] M. Kira, L. C. Zhang, C. Kabuto, H. Sakurai, *Chem. Lett.* **1995**, 659.
[71] O. Seiler, M. Penka, R. Tacke, unpublished results.

25 Functionalized Silanols and Silanolates

Susanne Kliem, Clemens Reiche, Uwe Klingebiel*

25.1 Introduction

The hydrolysis of chlorosilanes R_2SiCl_2 (R = alkyl, aryl) is a well known reaction for preparing silanols and silanediols.[1,2] With small alkyl groups the silanols and silanediols are unstable and condense spontaneously under formation of acyclic and cyclic siloxanes. Bulky organyl groups stabilize silanediols in such a manner that they do not condense, even under the influence of mineral acids or on heating in an autoclave.[3] Analogous reaction behavior is found for other H-acidic silicon compounds such as aminosilanes, R_3SiNH_2 and $R_2Si(NH_2)_2$,[3,4] or silylphosphanes, R_3SiPH_2, and $R_2Si(PH_2)_2$.[5]. This is in contrast to carbon chemistry, where compounds with two H-acidic groups bonded to the same carbon atom are unstable.

25.2 Halosilanols, Aminosilanols, Silanediols, and Siloxanediols

Bulky substituted difluorosilanes such as ${}^tBu_2SiF_2$ react with KOH in a 1:1 molar ratio to give stable fluorosilanols.[1,2,6] Chloro- and bromosilanols result from the reaction of silanediols with halides such as $PHal_5$.[2,7] In contrast to the fluorosilanol, the bromo- and chlorosilanols form aminosilanols upon reaction with NH_3 (Eq. 1).[2,8]

$$R_2Si(OH)_2 \xrightarrow[-\,POHal_3,\ -\,HHal]{+\,PHal_5} R_2Si(OH)Hal \xrightarrow[-\,NH_4Hal]{+\,2\,NH_3} R_2Si(OH)NH_2 \qquad Hal = Cl, Br \tag{1}$$

Di-*tert*-butylsilanediol and di-*tert*-butylaminosilanol crystallize in ladder-like chains via H-bridges.[2,9] The structure of the aminosilanol is a good example of a case where, in a compound with two basic atoms, the less electronegative element, here the nitrogen, forms stronger H-bridges (Figure 25.1).[10]

* Address of the authors: Institute of Inorganic Chemistry, University of Göttingen, Tammannstraße 4, D-37077 Goettingen, Germany.

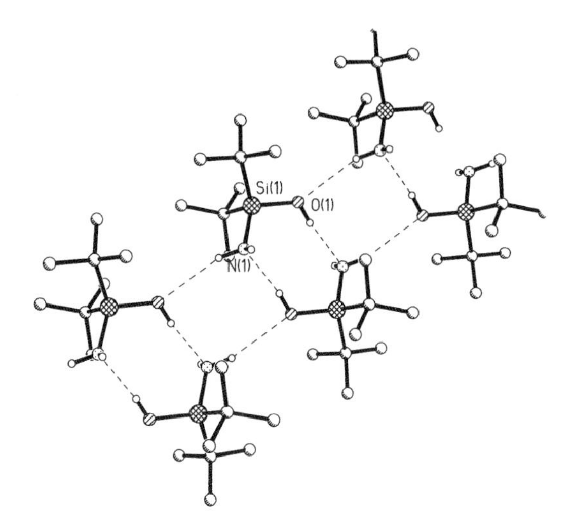

Figure 25.1. Crystal structure of $^tBu_2Si(OH)NH_2$ (N····HO = 285 pm, O····HN = 336 pm).

While trisiloxanediols like $R_2Si[OSi^tBu_2(OH)]_2$ (R = Me, ^{i}Pr) are dimers in the solid state,[2,11] the isoelectronic 1-amino-5-ol-trisiloxane is a monomer due to a hydrogen bridge between the lone pair of the nitrogen and the OH group.[2,11]

$(CMe_3)_2$ $(CMe_3)_2$ Me_2
Si Si Si
O O O O
277 pm
OH----O
Me_2Si H 282 pm $SiMe_2$ $(Me_3C)_2Si$ $Si(CMe_3)_2$
H
O---HO 286 pm
N---HO
O O
Si Si H H
$(CMe_3)_2$ $(CMe_3)_2$

25.3 Alkali Metal Silanolates and Siloxanolates

Alkali metal derivatives of stable functionalized silanols are very important for the stepwise formation of siloxane units of almost any size.[1,2] Detailed structural analyses are therefore important to understand the mechanism of their reactions.

Most information about mono-metallated compounds is available for lithium derivatives.[2,12,13] They crystallize as cubanes or dimers. The sodium and the potassium derivatives of $^tBu_2Si(OH)_2$ form hexagonal prisms instead. These structural elements are even stable in the gaseous state.[2,14]

The stepwise synthesis of Si-O chains and rings is feasible from alkali metal silanolates and fluorosilanes. The first heptasiloxane was prepared in this way (Eq. 2).[11]

$$F{-}Si{-}OLi + F_2Si< \xrightarrow{-\,LiF} F{-}Si{-}O{-}Si{-}F \xrightarrow[-\,KF]{+\,KOH} HO{-}Si{-}O{-}Si{-}F \xrightarrow[-\,n\text{-}C_4H_{10}]{+\,n\text{-}C_4H_9Li}$$

$$Me_2HC{-}Si(CHMe_2)(F){-}O{-}Si(CMe_3)_2{-}O{-}SiMe_2{-}O{-}Si(CMe_3)_2{-}O{-}SiMe_2{-}O{-}Si(CMe_3)_2{-}O{-}Si(CHMe_2)(F){-}CHMe_2 \qquad (2)$$

The alkali metal siloxanolates crystallize in different structures from polar or nonpolar solvents.[2,12,15] A dimer (Li-O four-membered ring) is formed in THF, and a trimer (Li-O six-membered ring) in *n*-hexane.

THF: $[\,{-}Si(OLi(THF)){-}O{-}Si(F){-}\,]_2$ *n*-hexane: $[\,Si{-}O{-}Si{-}F{-}Li{-}O\,]_3$

The hard Lewis acid Li^+ coordinates the hard Lewis base F^-. The sodium analogue reveals a surprisingly similar structure with three-coordinate sodium atoms.[15] The trimers might appear to be predisposed for forming four-membered Si-O-rings by intramolecular salt- (LiF or NaF) elimination. However, both the lithium and sodium salts are stable, even when melted, distilled or sublimed.

25.4 Six- and Eight-Membered Si-O Ring Systems

A simple way of synthesizing cyclotrisiloxanes is thermal LiF elimination from lithiated fluorosilanols (Eq. 3, for example).[2,6]

$$3\ (Me_3C)_2Si(OLi)F \xrightarrow[-\ 3\ LiF]{>300\ °C} [(Me_3C)_2Si-O]_3 \tag{3}$$

A better control of the reaction products is achieved by starting the synthesis from dilithiated disiloxanediols. Cyclotrisiloxanes are formed upon reaction with halides such as ${}^tBuSiF_3$ (Eq. 4).[2] Six-membered ring systems containing heteroatoms are also accessible by this method.

$$[(Me_3C)_2Si(OLi)]_2O + F_3SiCMe_3 \xrightarrow{-\ 2\ LiF} [(Me_3C)_2SiO]_2[Si(F)(CMe_3)O] \tag{4}$$

A third way of synthesizing six-membered rings is by intramolecular salt elimination from metallated halogen-substituted trisiloxanols. Lithium 1,1,5,5-tetra-*tert*-butyl-1-fluoro-5-hydroxy-3,3-dimethyl-trisiloxane is stable in *n*-hexane. Only by weakening the Li-O contact by using a donor solvent such as THF, can the elimination of LiF with concomitant formation of a six-membered ring be achieved (Eq. 5).[15]

$$Me_3C-Si(CMe_3)(OH)-O-SiMe_2-O-Si(CMe_3)(F)-CMe_3 \xrightarrow[-\ n\text{-}C_4H_{10}]{+\ n\text{-}C_4H_9Li}$$

$$n\text{-hexane:}\ \tfrac{1}{2}\,[Me_2Si(OSi(CMe_3)_2F)(OSi(CMe_3)_2OLi)]_2 \xrightarrow{THF} \quad THF,\ -\ LiF:\ Me_2Si[OSi(CMe_3)_2]_2O \tag{5}$$

1,5-dihydroxy-1,3,5-trisiloxanes crystallize as dimers via O···H bridges.[11] Thermal H_2O elimination from such siloxanes represents a fourth method of synthesizing six-membered Si-O rings (Eq. 6).

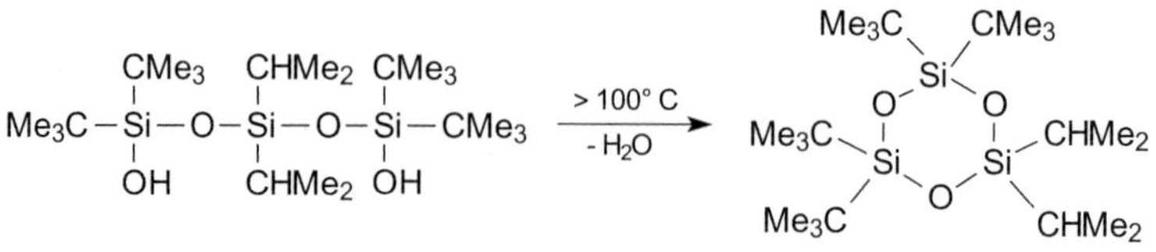

(6)

Cyclotetrasiloxanes are obtained in the reaction of dilithiated silanediols with di-, tri-, or tetrahalosilanes (Eq. 7).[2,10] Coupling reactions of the fluoro-functionalized rings with other silanols are possible.

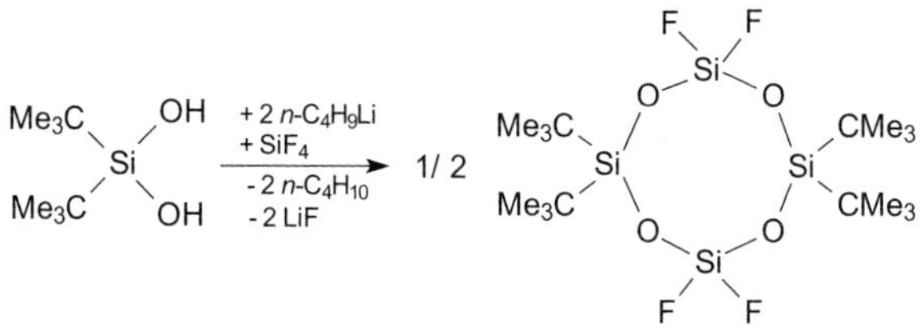

(7)

Another route for the preparation of cyclotetrasiloxanes is thermal LiF elimination from the lithium salts of 1-hydroxy-3-fluorodisiloxanes (Eq. 8):[10]

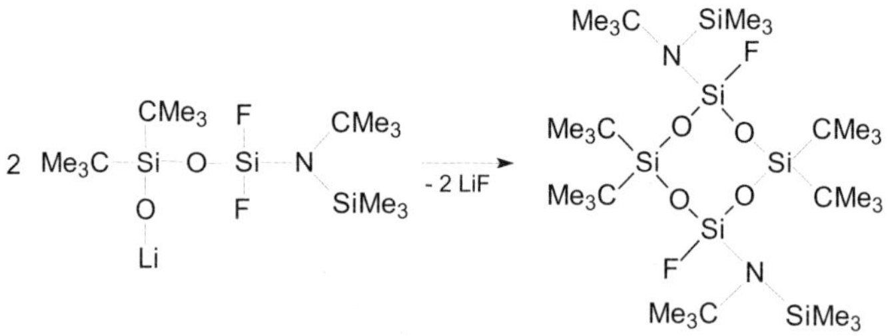

(8)

The most successful way of preparation of heteroatomic eight-membered rings utilizes bifunctional trisiloxanes as precursors. By reacting a dilithiated trisiloxanediol with dihalo-functional compounds, a variety of eight-membered ring systems can be obtained (Eq. 9).[2]

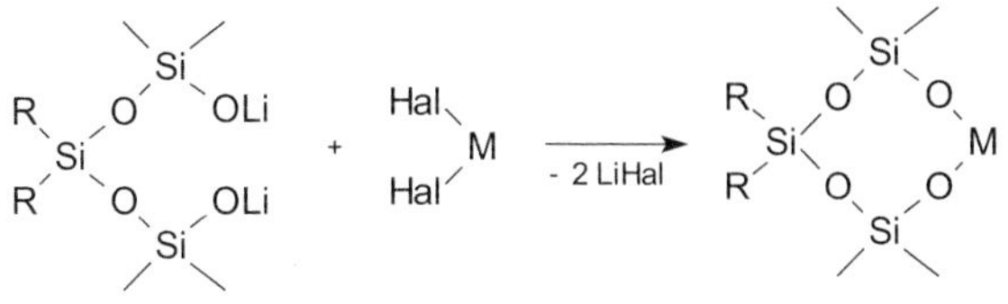

$$M = SiMe_2,\ SiF_2,\ BF,\ AlCl,\ GeCl_2,\ AsF,\ TiCl_2,\ PF \qquad (9)$$

25.5 Sodium and Potassium Aminosilanolates

In contrast to the lithium aminosilanolate, which forms an Li-O cubane structure in the solid state, the sodium- and potassium aminosilanolates form hexameric prisms (Eq. 10, Figure 25.2).[10]

$$(Me_3C)_2Si(OH)NH_2 + M \xrightarrow[-\frac{1}{2} H_2]{} (Me_3C)_2Si(OM)(NH_2) \qquad (10)$$

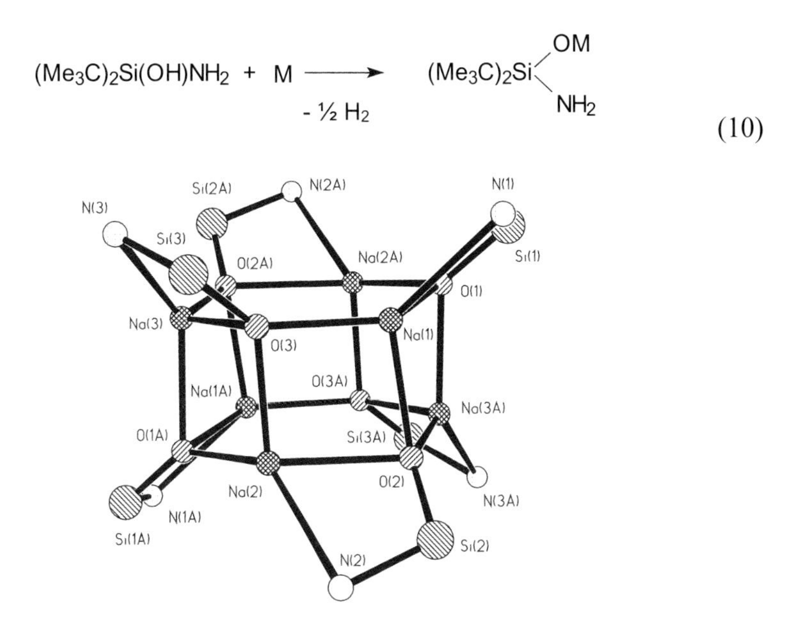

Figure 25.2. Crystal structure of $^tBu_2Si(NH_2)ONa$ (Si-O = 159.7(3), Si-N 176.4(4), Na(1)-O(1) 241.6(3), Na(1)-O(2) 229.0(3), Na(1)-N(1) 249.5(4), Na(2)-N(2) 252.9(4) pm).

The metal atoms are coordinated by three oxygen atoms and the nitrogen atom of the NH_2 group. The nitrogen thus brackets the Si-O-Na skeleton to form a planar four-membered ring of a monomeric unit.

25.6 From Aminosiloxanes to Six-Membered $(SiO)_2SiN$-Rings

1,5-Diamino-1,3,5-trisiloxanes are prepared by the reaction of the alkalimetal aminosilanolates with halosilanes.[2,10] NH_3 elimination leads to the formation of six-membered or spirocyclic $(SiO)_2SiN$-rings (Eq. 11 and Figure 25.3).[10]

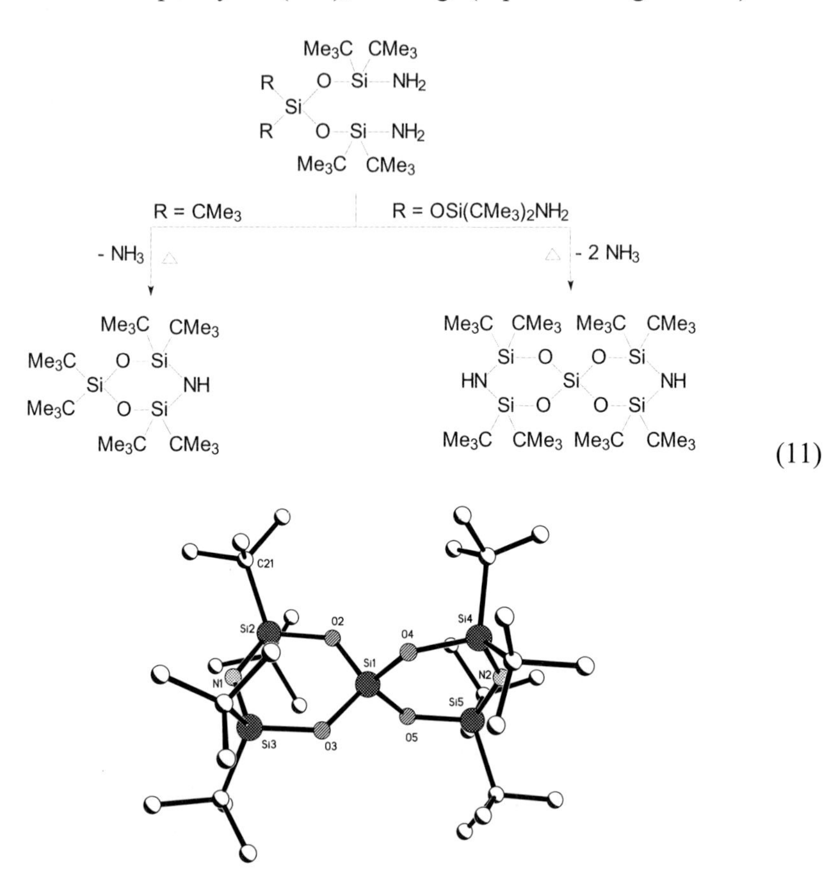

Figure 25.3. Crystal structure of $[HN(^tBu_2SiO)_2]_2Si$: (Si(1)-O(3) = 166.9(3), Si(3)-O(3) = 169.6(4), Si(3)-N(1) = 166.2(3) pm).

25.7 Four- and Eight-Membered Aza-oxa-cyclosiloxanes

The lithium derivatives of aminosiloxanes are usually isolated as 1,3-disilazan-1-olates (type A), i.e. a 1,3-(O-N) silyl group migration has occurred. This rearrangement can be prevented by bulky substituents, such that 1-lithium-amido-1,3-disiloxanes (type B) are formed.[10]

$(Me_3C)_2Si{-}O{-}Li$, N attached: $R_2Si(F)$, H (type A)

$(Me_3C)_2Si{-}O{-}SiR_2F$, N attached: H, Li (type B)

Eight-membered $(SiOSiN)_2$ rings are obtained in salt elimination reactions from halofunctional compounds of type A.[2,16–18] The eight-membered rings are often planar and invariably have large ring angles at the oxygen atoms (Si-O-Si = 178–180°) and smaller angles at the nitrogen atoms (Si-N-Si = 145–150°). Starting from di-*tert*-butyl-aminosiloxanols, eight-membered ring systems with one, two, and three NH groups are also available (Eq. 12).[2,19]

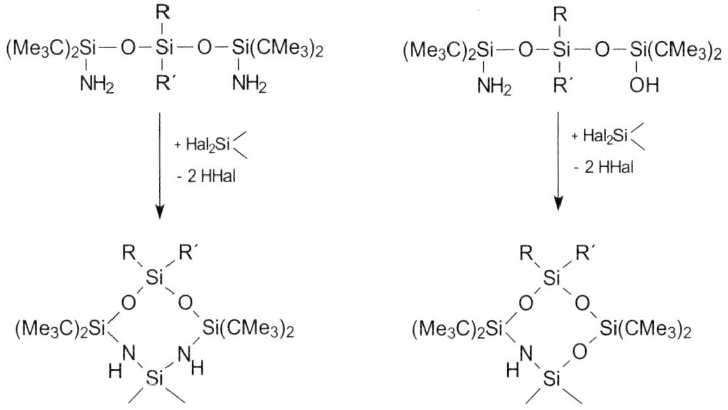

$$(Me_3C)_2Si(NH_2){-}O{-}SiRR'{-}O{-}Si(CMe_3)_2(NH_2) \xrightarrow[-\,2\,HHal]{+\,Hal_2Si<} \text{ring}$$

$$(Me_3C)_2Si(NH_2){-}O{-}SiRR'{-}O{-}Si(CMe_3)_2(OH) \xrightarrow[-\,2\,HHal]{+\,Hal_2Si<} \text{ring} \qquad (12)$$

Bulky substituents lead to the formation of small-ring compounds. For example, LiF elimination from 1-lithium-amido-3-fluoro-1,3-disiloxanes leads to four-membered (SiNSiO)-ring systems (Eq. 13).

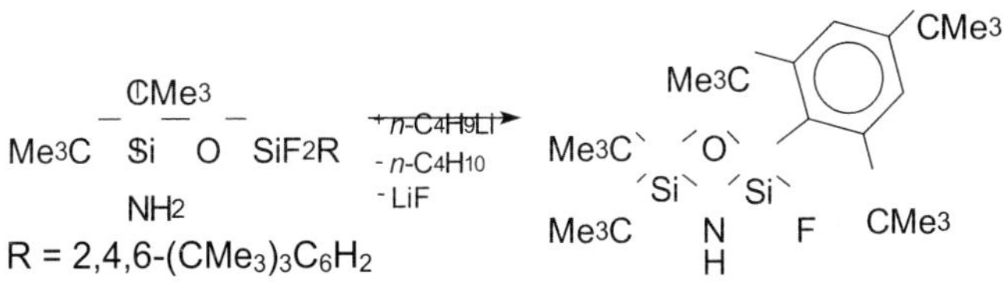

$$(Me_3C)_2Si(NH_2){-}O{-}SiF_2R \xrightarrow[-\,n\text{-}C_4H_{10},\ -\,LiF]{+\,n\text{-}C_4H_9Li} \text{(1-aza-3-oxa-2,4-disilacyclobutane)} \qquad (13)$$

$R = 2,4,6\text{-}(CMe_3)_3C_6H_2$

The atoms N, Si, O, and Si of this 1-aza-3-oxa-2,4-disilacyclobutane form a planar four-membered ring. The Si-O-Si (93.42(10)°) and Si-N-Si angles

(94.03(10)°) are slightly opened, which is accompanied by a contraction of the angles at the other ring atoms [N-Si-O = 87.15(9)°, 85.4(10)°]. The Si-O (167.2(2), 170.2(2) pm) bond distances are lengthened. The Si-N bonds (166.6(2), 169.1(2) pm) are short, even shorter than the Si-O bonds. The lithium salt of the four-membered ring crystallizes from *n*-hexane as a dimer (Figure 25.4). The four-membered $(Li\text{-}N)_2$ ring is folded by 10.7°. The dihedral angle between the Si-O-Si-N rings and the central $(LiN)_2$ ring is 86° and that between the two Si-O-Si-N rings 5.8°. The *tert*-butyl groups, and therefore the aryl group and the fluorine atom, are in *cis* positions. The Si-O distances are shorter than the Si-N distances. The ring angles at Si(1) (92.6(3)°) and Si(2) (90.9(3)°) are now larger than 90°, and the angles at O(1) (89.7(2)°) and N(1) (86.5(3)°) smaller than 90°. This puts the Si(1) and Si(2) atoms in closer contact to each other; the Si···Si nonbonding distance across the ring is only 237.2 pm. The transannular O···N distance is 245.1 pm.[20]

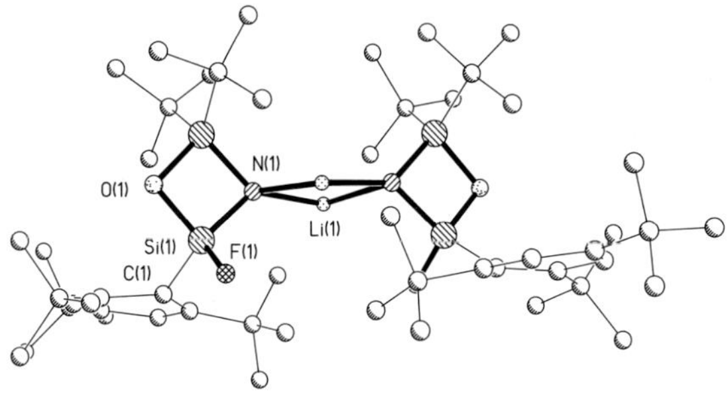

N(1)
O(1)
Li(1)
Si(1)
F(1)
C(1)

Figure 25.4. Crystal structure of $[RSiFONLiSi(CMe_3)_2]_2$, R = 2.4.6-$C_6H_2(CMe_3)_3$

The lithium atom of these four-membered rings can be replaced by other groups, for example $SiMe_3$ (Eq. 14).[20]

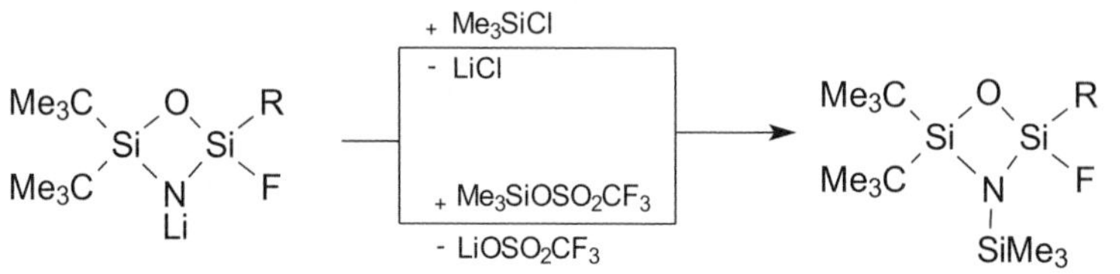

$$\underset{\text{Li}}{\overset{}{(Me_3C)_2Si(\mu\text{-}O)(\mu\text{-}N)Si(R)F}} \xrightarrow[\substack{-\,LiCl \\ +\,Me_3SiOSO_2CF_3 \\ -\,LiOSO_2CF_3}]{+\,Me_3SiCl} (Me_3C)_2Si(\mu\text{-}O)(\mu\text{-}N\text{-}SiMe_3)Si(R)F$$

R = 2,4,6-$(CMe_3)_3C_6H_2$ (14)

The (OSiNSi) four-membered ring in the silyl derivative is planar with ring angles at the Si atoms smaller than 90° and angles at the N and O atoms larger than 90°. The transannular Si···Si distance is 244.5 pm.

25.8 References

[1] P. D. Lickiss, *Adv. Inorg. Chem.* **1995**, *42*, 147.
[2] U. Klingebiel, S. Schütte, D. Schmidt-Bäse, in *The Chemistry of Inorganic Ring Systems*, Elsevier, Amsterdam, 1992, p. 75.
[3] L. H. Sommer, L. J. Tyler, *J. Am. Chem. Soc.* **1954**, *76*, 1030.
[4] R. Murugavel, M. Bhattacharfee, H. W. Roesky, *Appl. Organomet. Chem.* **1999**, *13*, 227.
[5] U. Klingebiel, N. Vater, *Angew. Chem.* **1982**, *94*, 870; *Angew. Chem. Int. Ed. Engl.* **1982**, *21*, 857.
[6] U. Klingebiel, *Angew. Chem.* **1981**, *93*, 696, *Angew. Chem. Int. Ed. Engl.* **1981**, *20*, 678.
[7] O. Graalmann, U. Klingebiel, *J. Organomet. Chem.* **1984**, *275*, C1.
[8] O. Graalmann, U. Klingebiel, W. Clegg, M. Haase, G. M. Sheldrick, *Angew. Chem.* **1984**, *96*, 904; *Angew. Chem. Int. Ed. Engl.* **1984**, *234*, 891.
[9] N. H. Buttrus, E. Eaborn, B. B. Hitchcock, A. K. Saxena, *J. Organomet. Chem.* **1985**, *284*, 291.
[10] C. Reiche, S. Kliem, U. Klingebiel, M. Noltemeyer, C. Voit, R. Herbst-Irmer, S. Schmatz, *J. Organomet. Chem.* submitted.
[11] S. Kliem, C. Reiche, U. Klingebiel, *Organosilicon Chemistry 5*, N. Auner, J. Weis (Eds.), Wiley-VCH, Weinheim, 2002, in press.
[12] D. Schmidt-Bäse, U. Klingebiel, *Chem. Ber.* **1990**, *123*, 449.
[13] S. Schütte, U. Pieper, U. Klingebiel, D. Stalke, *J. Organomet. Chem.* **1993**, *446*, 45.
[14] S. Schütte, U. Klingebiel, D. Schmidt-Bäse, *Z. Naturforsch.* **1993**, *48b*, 263.
[15] D. Schmidt-Bäse, U. Klingebiel, *Chem. Ber.* **1989**, *122*, 815.
[16] S. Kliem, U. Klingebiel in *Silicon for the Chemical Industry VI,* H.A. Øye, H.M. Rong, L. Nygaard, G. Schüssler, J. Tuset (Eds.), Norwegian University of Science and Technology, Trondheim, Norway, 2002, p. 139.
[17] S. Walter, U. Klingebiel, *Coord. Chem. Rev.* **1994**, *130*, 481.
[18] J. Hemme, U. Klingebiel, *Adv. Organomet. Chem.* **1996**, *39*, 159.
[19] D. Schmidt-Bäse, U. Klingebiel, *J. Organomet. Chem.* **1989**, *363*, 313.
[20] U. Klingebiel, M. Noltemeyer, *Eur. J. Inorg. Chem.* **2001**, 1889.

26 Transition Metal Fragment Substituted Silanols of Iron and Tungsten – Synthesis, Structure, and Condensation Reactions[#]

Wolfgang Malisch,[*] Marco Hofmann, Matthias Vögler, Dirk Schumacher, Andreas Sohns, Holger Bera, Heinrich Jehle[†]

26.1 Introduction

Due to their important role as intermediates in the technical synthesis of silicones, organosilanols of the general type $R_{4-n}Si(OH)_n$ (R = alkyl, aryl; n = 1–3) have been the subject of extensive studies concerning intermolecular condensation to give siloxanes.[1] This tendency rises with temperature and the number of OH groups, as well as with decreasing steric demand of the organic substituents. Therefore, the isolation especially of silanediols or silanetriols has been achieved preferentially with bulky organic ligands and by using special reaction conditions.[2]

In the context of our studies on the reactivity of functionalized silicon transition metal complexes,[3] we have established a new type of silanol with a transition metal fragment directly bonded to silicon. These metallo-silanols of the general type $L_mM{-}SiR_{3-n}(OH)_n$ (n = 1–3) are characterized by a remarkably high stability towards self-condensation due to the electron-releasing effect of the metal fragment reducing dramatically the acidity of the Si-OH proton. This property allows the isolation of a number of transition metal substituted silanols,[4] including examples with stereogenic metal and silicon atoms [5] and even of metallo-silanediols [6] and -triols.[7]

26.2 Metallo-silanols

26.2.1 Preparative Procedures

Two routes proved to be most efficient for the generation of metallo-silanols: the hydrolysis of metallo-chlorosilanes in the presence of an auxiliary base and

---

[#] Part 26 of the series "Metallo-Silanols and Metallo-Siloxanes". In addition, Part 54 of the series "Synthesis and reactivity of Silicon Transition Metal Complexes". Part 25/53 see ref. [25].

[*] Corresponding author

[†] Address of the authors: Institut für Anorganische Chemie der Universität Würzburg, Am Hubland, D-97074 Würzburg, Germany.

the reaction of metallo-hydridosilanes with the oxygen transfer agent dimethyldioxirane (Eq. 1).[8]

$$L_nM-Si(R)_{3-x}Cl_x \xrightarrow[- n\ [Et_3NH]Cl]{+ n\ H_2O,\ + n\ Et_3N} L_nM-Si(R)_{3-x}(OH)_x \xleftarrow[- n\ \text{acetone}]{+ n\ \text{dimethyldioxirane}} L_nM-Si(R)_{3-x}H_x \quad (1)$$

$x = 1\text{-}3$ $\quad L_nM = C_5R'_5(OC)_2Fe/Ru$ (R' = H, Me); $Cp(OC)(Ph_3P)Fe$; $C_5R'_5(OC)_2(Me_3P)Mo/W$ (R' = H, Me) $\quad$ R = Me, Ph, *o*-Tol

Electrophilic oxygen insertion is especially productive for metallo-silanes with electron-rich Si-H bonds provided by metal fragments of high donor capacity. This is the case for phosphane-substituted metal fragments such as $Cp(OC)(RPh_2P)Fe$ [R = Ph, N(H)C*(Me)(Ph)H]. Even the diastereomerically pure ferrio-silanol $Cp(OC)(Ph_3P)Fe–Si(Me)(Ph)OH$ [5] and the enantiomerically pure ferrio-silanediol $Cp(OC)[H(Me)(Ph)C^*N(Me)Ph_2P]Fe–Si(Ph)(OH)_2$ [9] were isolated. The hydrolysis of the analogous metallo-chlorosilanes fails for these systems due to insufficient electrophilicity of silicon. In an extension of the oxygenation procedure, a catalytic method was developed using urea/hydrogen peroxide adduct in the presence of $MeReO_3$.[10] For example, $C_5Me_5(OC)_2(Me_3P)W\text{-}Si(OH)_3$ was obtained in better yields from $C_5Me_5(OC)_2(Me_3P)W–SiH_3$ by this method than by using dimethyldioxirane.[11]

Other procedures involve the selective, hydrolytic cleavage of the Co-Si unit in heterodinuclear complexes of the type $(OC)_4Co–SiR_2–Fe(CO)_2Cp$ obtained from $Cp(OC)_2Fe–SiR_2H$ and $Co_2(CO)_8$. For R = Me, the H/OH exchange at silicon can even be performed with catalytic amounts of $Co_2(CO)_8$.[12]

26.2.2 Condensation Reactions to Metallo-siloxanes and -heterosiloxanes

The metallo-silanols with an M-Si bond are characterized by a high stability towards self-condensation. For example, the phosphane-substituted ferrio-silanetriol $Cp(OC)(Ph_3P)Fe–Si(OH)_3$ is recovered unchanged after heating to 60 °C for 5 d in THF. In the case of $Cp(OC)_2Fe–SiMe_2OH$, a ß-hydrogen elimination of the Si-OH proton is observed, leading to the formation of the metal hydride complex $Cp(OC)_2Fe–H$ and, formally, dimethyl silanone "$Me_2Si=O$", which is converted into hexamethylcyclotrisiloxane.

However, the facile isolation of metallo-silanols allows controlled condensation reactions with organochlorosilanes in the presence of an auxiliary base leading to transition metal substituted oligosiloxanes which can be considered as model compounds for silica-immobilized catalytic systems (Eq. 2a).

Especially those metallo-siloxanes bearing an Si-H function at the γ silicon have a high synthetic potential, since oxidative addition to electronically unsaturated metal centers gives easy access to Si-O-Si bridged binuclear metal complexes, as illustrated in Eq. 2b.[13]

2	a	b
R	Me	p-Tol

3	a	b
R	Me	p-Tol

(2)

The ferrio-silanols $Cp(OC)_2Fe–Si(Me)(R')OH$ (R' = Me, Ph), although characterized by a lowered acidity compared to organosilanols, react with trialkylalanes, -gallanes, and -indanes to yield the transition metal fragment substituted heterosiloxanes **4a–d** under elimination of alkane. Molecular weight determination of **4a,b** shows the presence of dimers in solution, deriving from an intermolecular interaction of the Lewis acidic group 13 element and the Lewis basic oxygen atom of the Si-O unit. In the case of the derivatives **4c,d** with stereogenic silicon atoms, the aggregation to dimers can easily be determined by NMR spectroscopy due to the formation of diastereomers. In addition, the dimeric structure of **4a,b,d** in the solid state is confirmed by X-ray structure analyses.[14] These results clearly indicate that an increased O-E π-interaction (E = In, Ga), induced by the $Cp(OC)_2Fe$ fragment at the silicon, is not sufficient to suppress dimerization.

4	a	b	c	d	e	f
E	In	Ga	Al	Ga	Ga	In
R	Me	Me	Ph	Ph	OH	OH

5

The reactions of the ferrio-silanediol $Cp(OC)_2Fe–Si(Me)(OH)_2$ with trimethylgallane and -indane, respectively, result in the formation of the heterosiloxanols **4e,f**. Only one OH function of the ferrio-silanediol is transformed, irrespective of the amount of trimethylgallane or -indane used. The same is found for the reaction of the ferrio-silanetriol $Cp(OC)_2Fe–Si(OH)_3$ with trimethylgallane. In *n*-hexane at room temperature, $Cp(OC)_2Fe–Si(OH)_2–OGaMe_2$ is initially formed. However, in the presence of two equivalents of THF at 65 °C, the drum-shaped heterosiloxane **5** with iron-substituted silicon atoms is produced.[15] In this reaction, THF acts as a donor to stabilize the formed intermediates and to guarantee a controlled condensation.[16]

26.3 Bis(metallo)silanols and -siloxanes

The "transition metal effect", operative for most of the metallo-silanols, can be doubled by attaching two transition metal fragments to the same silicon atom. These bis(metallo)silanols are characterized by a very high stability towards self-condensation which even provides rare examples of Si-H and Si-Cl functionalized silanols $R_2Si(X)OH$ (X = H, Cl). These are accessible from the same starting compound $[Cp(OC)_2Fe]_2Si(H)Cl$ (**6**) by making use of the two complementary synthetic routes, hydrolysis and oxygenation. Et_3N-assisted hydrolysis of **6** yields the corresponding SiH-functional bis(ferrio)silanol **7a**, while the bis(metallo)chlorosilanol **7b** can be prepared by treating **6** with dimethyldioxirane.[17] However, in spite of the two iron substituents, **7b** rapidly decomposes at room temperature.

The bis(metallo)chlorosilanol **7b** is especially interesting with regard to the generation of stable silanones,[18] which could be derived from **7b** by intramolecular HCl elimination.

In analogy to the synthesis of **7b**, the bis(ferrio)silanediol **8** is obtained in good yield from $[Cp(OC)_2Fe]_2SiH_2$ [19] under standard conditions in acetone at –78 °C. The oxofunctionalization of an Si-H bond using dimethyldioxirane also proved to be successful for the synthesis of the first hetero-bismetalated silanol $[Cp(OC)_2Fe][Cp(OC)_2(Me_3P)W]Si(Me)OH$.[17]

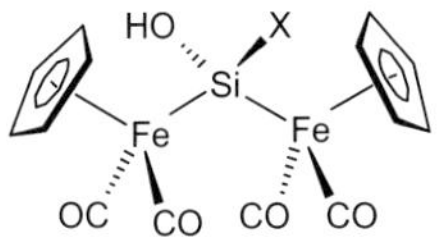

7	a	b
X	H	Cl

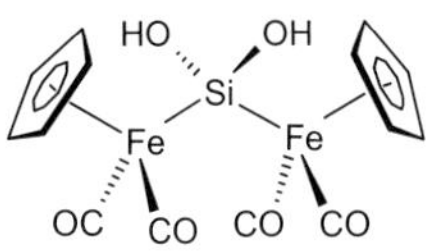

8

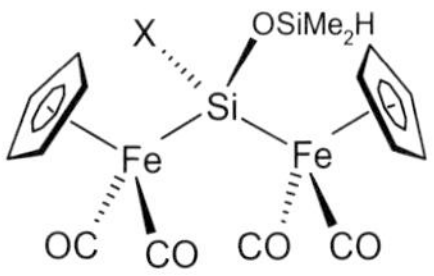

9	a	b
X	H	$OSiMe_2H$

Despite their greatly reduced reactivity with respect to self-condensation, the Si(OH) units in the stable bis(metallo)silanols **7a** and **8** can be used to carry out the typical condensation reactions of silanols with chlorodimethylsilane to yield the bis(metallo)siloxanes **9a,b**, which are useful for further modifications involving the Si-H function.

26.4 The Transition Metal Effect: Regiospecific Hydroxylation

Metallodisilanes $L_nM\text{-}SiX_2\text{–}SiX_3$ (X = H, Cl) represent interesting model compounds for the study of the "transition metal effect", because these species contain both a transition metal substituted- as well as a metal-free Si-X unit in the same molecule. Et_3N-assisted hydrolysis of the ferrio-pentachlorodisilane $Cp(OC)_2Fe\text{–}Si_2Cl_5$ yields exclusively the ferriodichloro-trihydroxy-disilane **10**, demonstrating the reduced susceptibility to nucleophilic attack at the silicon in the α-position to the transition metal atom.[20]

On the other hand, the oxygenation of $C_5R_5(OC)_2Fe\text{–}Si_2H_5$ (R = H, Me) with dimethyldioxirane results in the formation of the ferrio-dihydroxydisilanes **11a,b**, which reveals the strong activation of the α-SiH_2 unit towards electrophilic oxygenation.[20]

These regiospecific hydroxylation reactions allow access to new types of Si-H and Si-Cl functional metallo-disilanols by simple reaction steps due to the influence of the transition metal fragment.

An analogous regiospecific hydroxylation with dimethyldioxirane results in the formation of the bis(ferrio)disiloxanol **12** from $[Cp(OC)_2Fe]_2Si(H)OSiMe_2H$ (**9a**). As in the case of **11a,b**, the exclusive formation of **12** shows that electrophilic attack of dimethyldioxirane at the metal-activated α-Si-H bond of **9a** is strongly favored.[17]

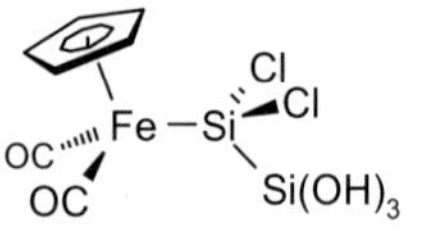

10

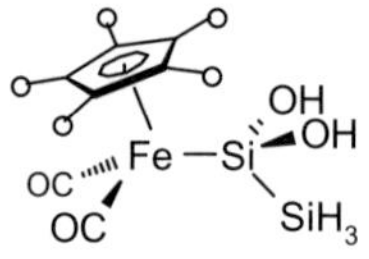

11	a	b
-o	H	Me

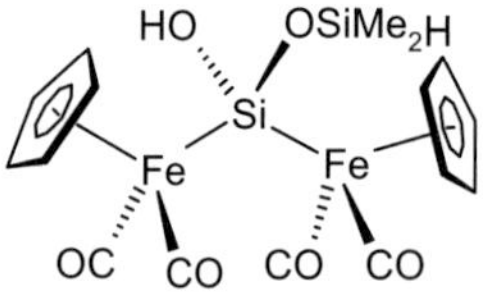

12

26.5 Metallo-silanols with the Silicon and Transition Metal Separated by a Spacer Group

It is not possible to obtain metal fragment substituted polysiloxanes by controlled self-condensation reactions with an M-Si bond due to their great condensation stability, a consequence of the direct influence of the transition metal fragment on the silanol unit. Therefore, it seemed reasonable to decrease this stabilizing effect by separation of the metal and the silanol group. In this context, we have used in a first approach a methylene group or a η^5-cyclopentadienyl unit as suitable spacer groups. We envisage access to polynuclear metal fragment substituted oligo- and polysiloxanes from these systems.

26.5.1 Ferriomethyl- and Tungstenmethyl-diorganosilanols

The ferriomethyl-silanols **13a–c** are produced by Et_3N-assisted hydrolysis of $C_5R_5(OC)_2Fe–CH_2–Si(Me)(R')Cl$ (R = H, Me; R' = Me; R = H, R' = Ph) in good to moderate yields.[21] The dioxirane route is also applied for **13a** by oxygenation of the Si-H functional ferriomethyl-silane $Cp(OC)_2Fe–CH_2–SiMe_2H$.[22] However, in this case the oxygenation method is inferior to the hydrolysis as it results in lower yields of the silanol **13a**.

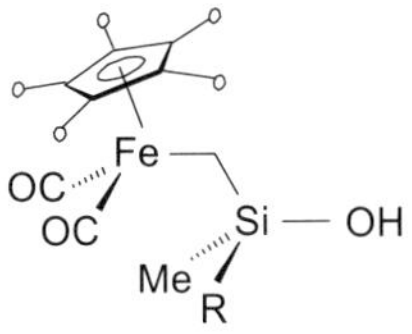

13	a	b	c
-o	H	H	Me
R	Me	Ph	Me

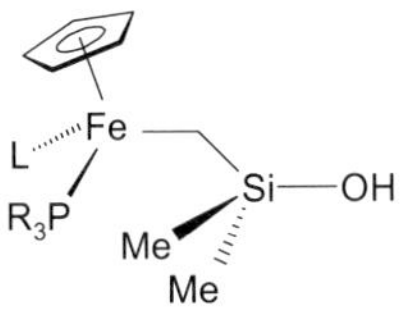

14	a	b
L	CO	Me_3P
R	Ph	Me

In contrast to ferrio-silanols with a direct metal-silicon bond, a modification of the ligand sphere around the metal center of the ferriomethyl-silanol **13a** is possible without decomposition. Photo induced CO/PR_3 (R = Me, Ph) exchange leads to the mono- and bis-phosphane-substituted ferriomethyl-silanols **14a,b**, which can be isolated in moderate yields. In addition, **14a,b** can be synthesized by Et_3N-assisted hydrolysis of the corresponding ferriomethyl-chlorosilanes.[23]

X-ray structure analyses of **13a–c** reveal the formation of hydrogen-bonded superstructures leading to OH···O-bridged tetramers in the case of **13a,b**, which are stacked on top of each other to build up a tube-like arrangement with additional weak CH···OC interactions between a Cp and a CO ligand of neighboring molecules (Figure 26.1, left). These tubes can be considered as hydrophilic channels inside a lipophilic structure composed of the remaining ferriomethyl fragments. In the case of **13c**, an infinite linear chain cover is produced, the result of a disordered Si-OH proton which guarantees the linkage of the molecules into two directions by OH···O bridges (Figure 26.1, right).[21]

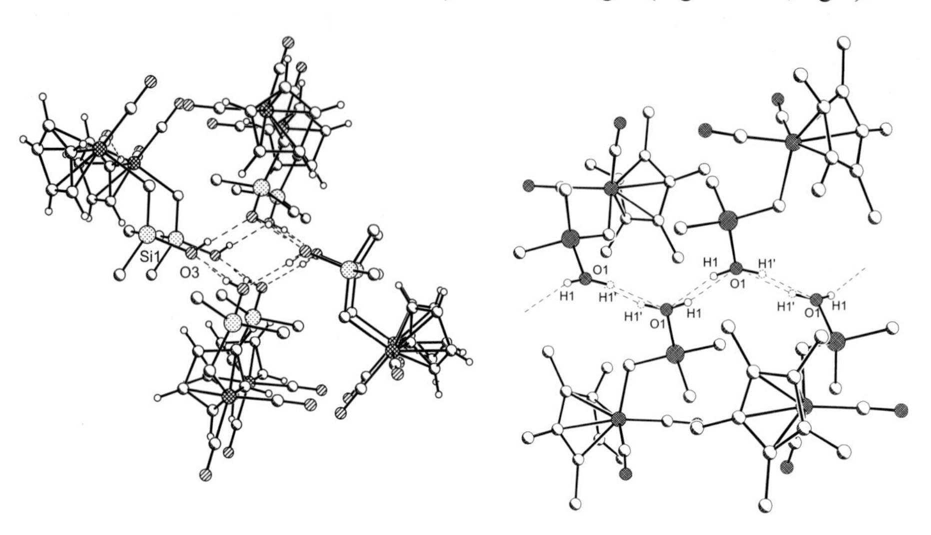

Figure 26.1. Left: Tube-like arrangement of tetrameric units of **13a**; O···O 2.74 Å. Right: Chain structure of **13c**; O···O 2.77 Å.

The methods used for the preparation of ferriomethyl-diorganosilanols can be extended to the analogous tungsten derivatives $Cp(OC)_2(R_3P)W–CH_2–SiMe_2OH$ (R=Me, Ph).[24]

26.5.2 Ferriomethyl-disiloxanes and -heterosiloxanes

The ferriomethyl-silanol $Cp(OC)_2Fe–CH_2–SiMe_2OH$ (**13a**), which can be considered as a derivative of trimethylsilanol with a methyl hydrogen substituted by the $Cp(OC)_2Fe$ fragment, exhibits an enhanced tendency towards self-condensation compared to the analogous Fe-Si complex. Thus, **13a** undergoes slow self-condensation at room temperature leading to the 1,3-(bisferriomethyl)-disiloxane **15a**. $C_5R_5(OC)_2Fe–CH_2–Si(Me)(R')OH$ [R = H, R' = Ph (**13b**); R = R' = Me (**13c**)] are stable with respect to this process due to steric (**13b**) or electronic (**13c**) factors.

The disiloxane **15a** can also be obtained by the reaction of the silanol **13a** with $Cp(OC)_2Fe–CH_2–SiMe_2Cl$ in the presence of Et_3N. The disiloxanes **15b,c** with only one iron fragment are analogously available using the organochlorosilanes Me(R)Si(H)Cl (R = Me, *p*-Tol).

The Si-OH function of ferriomethyl-diorganosilanols is also suitable for condensation reactions with group 4 metal chlorides, as proved for

$Cp(OC)_2Fe–CH_2–SiMe_2OH$ (**13a**). The Et_3N-assisted reaction of **13a** with titanocene or zirconocene dichloride offers an easy access to the bismetalated heterosiloxanes **16a,b**. Further condensation of **16b** with **13a** to give the trinuclear species **16c** proceeds almost quantitatively.[25]

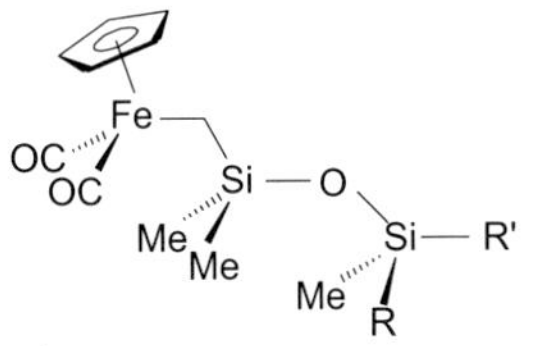

15	a	b	c
R	Me	Me	*p*-Tol
R'	$CH_2Fe(CO)_2Cp$	H	H

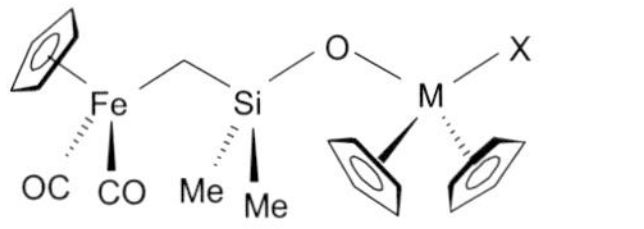

16	a	b	c
M	Ti	Zr	Zr
X	Cl	Cl	$OSiMe_2CH_2Fe(CO)_2Cp$

26.5.3 Ferriomethyl-silanediols and -silanetriols

The synthesis of the silanediol **17a** and silanetriol **18** can be accomplished either by the oxygenation or the hydrolysis method starting from $Cp(OC)_2Fe–CH_2–Si(R)H_2$ (R = Me, H) [26] or the chlorofunctional ferriomethyl-silanes $Cp(OC)_2Fe–CH_2–Si(R)Cl_2$ (R = Me, Cl).[27] Additional access to **18** is provided by the hydrolysis of $Cp(OC)_2Fe–CH_2–Si(OMe)_3$ with an excess of water in the presence of acetic acid.

Both the silanediol **17a** and the silanetriol **18** can be isolated in pure form and can be stored for several weeks at –20 °C without significant change. However, at room temperature, the reactivity of **17a** and **18** in solution is significantly enhanced giving rise to the formation of complex product mixtures, presumably due to self-condensation. A greater stability towards self-condensation is observed for the phosphane-substituted ferriomethyl-silanediol **17b**, which can be obtained by Et_3N-assisted hydrolysis of the corresponding ferriomethyl-dichlorosilane.

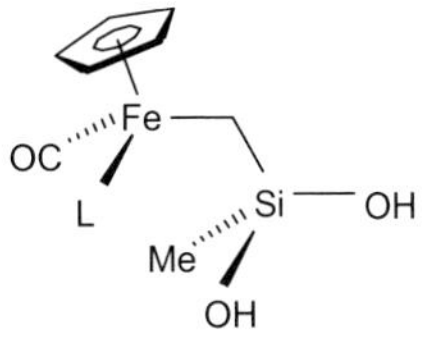

17	a	b
L	CO	PPh_3

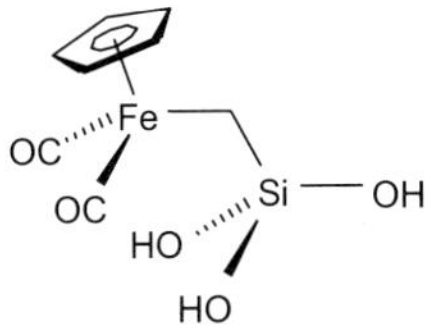

18

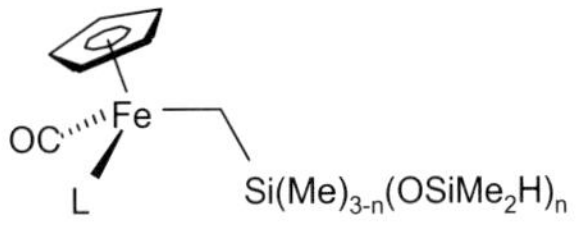

19	a	b	c
L	CO	PPh_3	CO
n	2	2	3

The condensation reactions of **17a,b** and **18** with dimethylchlorosilane in the presence of triethylamine proceed in a controlled manner yielding the ferriomethyl-tri- and -tetrasiloxanes **19a–c,** respectively, in good yields as orange-brown to dark brown oils.[26]

26.5.4 Photochemical Reactions of Ferriomethyl-siloxanes

The Si-H functions of the ferriomethyl-siloxanes **19a–c** offer the possibility for further functionalization reactions. For example, photochemical treatment induces elimination of CO and subsequent reaction steps leading to rearrangement of the siloxane backbone. In the case of the ferriomethyl-tri- and -tetrasiloxanes **19a,c**, UV irradiation in benzene results in the formation of the six-membered cyclo(ferra)trisiloxanes **20a,b** (Scheme 26.1).[27]

20a, which bears two methyl groups at the γ-silicon atom, is isolated as a single compound, whereas **20b** is obtained as mixture of two isomers in a ratio of ca. 60:40, originating from the relative orientation of the substituents at the iron atom and the γ silicon atom.

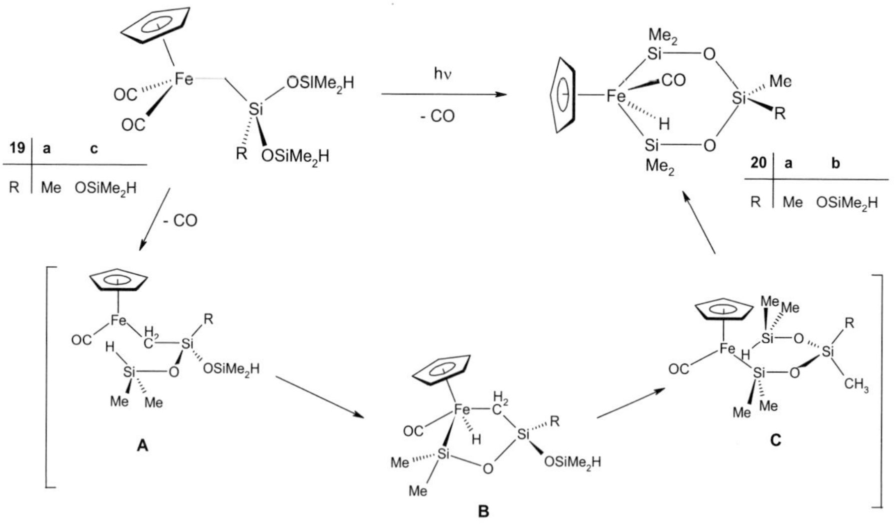

Scheme 26.1. Rearrangement of **19a,c** to **20a,b** induced by UV irradiation.

Mechanistically, the reaction shown in Scheme 26.1 involves an initial CO elimination from **19a,c** upon UV irradiation leading to the 16-electron species **A**, which is stabilized by an intramolecular oxidative addition reaction of one of the terminal Si-H units to the iron center. The resulting five-membered ring **B** is transformed via a reductive elimination of the iron-bound hydrogen and methylene unit into **C**, giving rise to the formation of a new Si-Me group. Finally, **C** is again stabilized by an oxidative addition reaction of the remaining Si-H function to the iron center, leading to the formation of **20a,b**.

26.5.5 Metallo-silanols with η^5-C_5H_5 Spacer Units

The synthesis of this special type of metallo-silanol starts most efficiently with an appropriate silyl metal complex, such as $Cp(OC)_2Fe$–$SiMe_2R$ (R = H, OMe). An anionic shift of the silyl group from the iron to the cyclopentadienyl unit can be induced with lithium diisopropylamide, leading to the metallates **21a,b**. Methylation with methyl iodide produces the neutral methyl iron complexes $(C_5H_4SiMe_2R)(OC)_2Fe$–Me (R = H, OMe), which can be converted either, for R = H, by the $Co_2(CO)_8$ method, or, for R = OMe, by hydrolysis in the presence of acetic acid, into the corresponding silanol **22**.

21	a	b
R	H	OMe

22

Silanols with the silicon directly linked to the cyclopentadienyl group are stable towards self-condensation at room temperature. However, if the silanol unit is attached to C_5H_4 via a propylidene chain, self-condensation to the corresponding siloxanes is observed.

26.6 Conclusions

The chemistry of metallo-silanols is strongly influenced by the electron-releasing transition metal fragment, which is especially demonstrated in the decreased tendency for self-condensation. This property has been used to synthesize a number of transition metal substituted silanols and to study their structural features as well as their reactivity, especially the controlled condensation with organochlorosilanes leading to transition metal fragment-substituted oligosiloxanes. The transition metal effect is impressively manifested in some regiospecific hydroxylation reactions of metallo-disilanes and bis(metallo)disiloxanes having metal-bound as well as "metal-free" Si-X functions (X = H, Cl). Extension of these studies to derivatives with a methylene spacer group between metal fragment and silanol unit demonstrates the weakening of the transition metal effect, expressed by the increased tendency for self-condensation of the ferriomethyl-silanols $Cp(OC)_2Fe–CH_2–Si(Me)_{3–n}(OH)_n$ (n = 1–3).

26.7 References

[1] W. Noll, *Chemie und Technologie der Silicone*, Verlag Chemie, Weinheim, 1968.

[2] P. D. Lickiss, *Adv. Inorg. Chem.* **1995**, *42*, 147. R. Murugavel, V. Chandrasekhar, H. W. Roesky, *Acc. Chem. Res.* **1996**, *29*, 183.

[3] S. Schmitzer, U. Weis, H. Käb, W. Buchner, W. Malisch, T. Polzer, U. Posset, W. Kiefer, *Inorg. Chem.* **1993**, *32*, 303. W. Malisch, R. Lankat, S. Schmitzer, R. Pikl, U. Posset, W. Kiefer, *Organometallics* **1995**, *14*, 5622.

[4] L. S. Chang, M. P. Johnson, M. J. Fink, *Organometallics* **1991**, *10*, 1219. W. Adam, U. Azzena, F. Prechtl, K. Hindahl, W. Malisch, *Chem. Ber.* **1992**, *125*, 1409. R. Goikhman, M. Aizenberg, H.-B. Kraatz, D. Milstein, *J. Am. Chem. Soc.* **1995**, *117*, 5865; S. H. A. Petri, D. Eikenberg, B. Neumann, H.-G. Stammler, P. Jutzi, *Organometallics* **1999**, *18*, 2615.

[5] W. Malisch, M. Neumayer, O. Fey, W. Adam, R. Schuhmann, *Chem. Ber.* **1995**, *128*, 1257.

[6] W. Malisch, R. Lankat, O. Fey, J. Reising, S. Schmitzer, *J. Chem. Soc., Chem. Commun.* **1995**, 1917.

[7] C. E. F. Rickard, W. R. Roper, D. M. Salter, L. J. Wright, *J. Am. Chem. Soc.* **1992**, *114*, 9682. W. Malisch, R. Lankat, S. Schmitzer, J. Reising, *Inorg. Chem.* **1995**, *34*, 5701.

[8] W. Adam, R. Mello, R. Curci, *Angew. Chem.* **1990**, *102*, 916. *Angew. Chem. Int. Ed. Engl.* **1990**, *29*, 890.

[9] W. Malisch, M. Neumayer, K. Perneker, N. Gunzelmann, K. Roschmann, in N. Auner, J. Weis (Eds.), *Organosilicon Chemistry III: From Molecules to Materials*, VCH, Weinheim, 1998, p. 407.

[10] W. A. Herrmann, J. G. Kuchler, J. K. Felixberger, E. Herdtweck, W. Wagner, *Angew. Chem.* **1988**, *100*, 420. W. A. Herrmann, F. E. Kühn, *Acc. Chem. Res.* **1997**, *30*, 169.

[11] W. Malisch, H. Jehle, C. Mitchel, W. Adam, *J. Organomet. Chem.* **1998**, *566*, 259.

[12] W. Malisch, M. Vögler, in N. Auner, J. Weis (Eds.), *Organosilicon Chemistry IV: From Molecules to Materials*, VCH, Weinheim, 2000, p. 442.

[13] W. Malisch, M. Hofmann, G. Kaupp, H. Käb, J. Reising, *Eur. J. Inorg. Chem.* **2002**, 3235.

[14] W. Malisch, B. Schmiedeskamp, D. Schumacher, H. Jehle, M. Nieger, *Eur. J. Inorg. Chem.*, in press.
[15] W. Malisch, D. Schumacher, M. Nieger, unpublished.
[16] R. Murugavel, A. Voigt, M. G. Walawalkar, H. W. Roesky, *Chem. Rev.* **1996**, *96*, 2205.
[17] W. Malisch, M. Vögler, D. Schumacher, M. Nieger, *Organometallics* **2002**, *21*, 2891.
[18] G. Raabe, J. Michl, *Chem. Rev.* **1985**, *85*, 419.
[19] W. Malisch, M. Vögler, H. Käb, H.-U. Wekel, *Organometallics* **2002**, *21*, 2830.
[20] W. Malisch, H. Jehle, S. Möller, C. Saha-Möller, W. Adam, *Eur. J. Inorg. Chem.* **1998**, 1585.
[21] W. Malisch, M. Hofmann, M. Nieger, W. W. Schöller, A. Sundermann, *Eur. J. Inorg. Chem.* **2002**, 3242.
[22] K. H. Pannell, *J. Organomet. Chem.* **1970**, *21*, P17. J. E. Bulkowski, N. D. Miro, D. Sepelak, C. H. Van Dyke, *J. Organomet. Chem.* **1975**, *101*, 267.
[23] D. Schumacher, M. Hofmann, W. Malisch, M. Nieger, *J. Chem. Soc., Dalton Trans.*, submitted.
[24] M. Hofmann, W. Malisch, H. Hupfer, M. Nieger, *Z. Naturforsch. B* **2003**, *58*, 36.
[25] M. Hofmann, W. Malisch, D. Schumacher, M. Lager, M. Nieger, *Organometallics* **2002**, *21*, 3485.
[26] W. Malisch, M. Hofmann, M. Nieger, in N. Auner, J. Weis (Eds.), *Organosilicon Chemistry IV: From Molecules to Materials*, VCH, Weinheim, 2000, p. 446.
[27] M. Hofmann, W. Malisch, M. Nieger, *Organometallics*, submitted.

27 Rational Syntheses of Cyclosiloxanes and Molecular Alumo- and Gallosiloxanes

Michael Veith, Andreas Rammo*

27.1 Selective and Stepwise Synthesis of Cyclosiloxanes with Mixed and "Inorganic" Substituents

Cyclic siloxanes with organic substituents on the silicon atoms have often been described in the literature,[1] whereas reports about Si-O ring systems with halogen, sulfur or nitrogen as ligands at the silicon atoms are rare.[2–4] Cyclosiloxanes with only amino substituents were unknown until recently. Moreover, cyclosiloxanes with halogen substituents are not only difficult to prepare but are only obtained in moderate yields.[1,5]

We have examined in more detail new and modified syntheses of cyclotri- and cyclotetrasiloxanes with inorganic substituents using two novel general procedures.[6–9]

The first route (useful for the synthesis of a variety of amino-substituted cyclotrisiloxanes) proceeds in a stepwise fashion. [6] The initial reaction is the controlled hydrolysis of diaminodichlorosilanes in the presence of the HCl scavenger triethylamine to furnish diaminochlorosilanols. The diaminochlorosilanol is then transformed into the silanolate by treatment with ^{n}BuLi in n-hexane (Eq. 1). Both intermediates, the chlorosilanol as well as the lithium silanolate, are remarkably stable and can be isolated by sublimation without decomposition (neither HCl elimination nor LiCl abstraction is observed).

$$Me_2Si(NR)_2SiCl_2 \xrightarrow[-\,Et_3N\,*HCl]{H_2O\,/\,Et_3N} Me_2Si(NR)_2Si(Cl)OH$$

$$\xrightarrow[-\,BuH]{+\,BuLi} Me_2Si(NR)_2Si(Cl)OLi \qquad (1)$$

The silanol and the lithium derivative in Eq. 1 are crystalline and form tetramers in the solid state, as determined by X-ray structure determinations. Whereas O-H-O bridges between the silanol groups are responsible for the agglomeration to four molecules,[6, 8, 10–12] the mutual interactions of the Li-O

* Address of the authors: Institut für Anorganische Chemie der Universität des Saarlandes, D-66440 Saarbrücken

units in the lithium salt lead to an Li_4O_4 cube as the central structural polyhedron.[6, 8, 13–15]

Although the lithium silanolate can be recrystallized from hot toluene without LiCl elimination, a change to the polar solvent THF allows the selective formation of six-membered siloxane rings under LiCl elimination (Eq. 2).

$$3\ Me_2Si(NR)_2Si(Cl)OLi \xrightarrow[-3\ LiCl]{\Delta} [Me_2Si(NR)_2SiO]_3 \quad (2)$$

The preparative potential of the second route is even more versatile. In this method, the auxiliary reagent hexaorganodistannoxane is used as the oxygen source for the Si-O-Si cycles. The starting material in this procedure is a dichlorosilane with additional organic or "inorganic" substituents R, which can be transformed into a di(stannoxo)silane by the use of this reagent.[16] Depending on the steric demand of the substituents at the silicon atom, a substitution of only one chlorine atom by an R'_3SnO group is sometimes observed at room temperature; the reaction solution then has to be heated to complete the double substitution.[7, 8]

The intermediate $R_2Si(OSnR_3)_2$ is a key product in the procedure, as either six- or eight-membered rings can be obtained depending on the reaction partners. Six-membered Si_3O_3 rings can be synthesized by treating $R_2Si(OSnR_3)_2$ with 1,3-dihalogenodisiloxanes,[7, 8] eight-membered rings by the reaction of $R_2Si(OSnR_3)_2$ with dichlorosilanes in an equimolar ratio (Eq. 3).[9] Condensations to the Si_3O_3 or Si_4O_4 rings occur even at room temperature in diethyl ether solution, with concomitant cleavage of triorganotin chloride. A selection of new rings obtained from these routes is shown in Scheme 27.1.

$$2\ R_2SiCl_2 \xrightarrow[-4\ R'_3SnCl]{+4\ (R'_3Sn)_2O} 2\ R_2Si(OSnR'_3)_2$$

Left branch: $+ 2\ R''_2SiX_2 / - 4\ R'_3SnX$ → eight-membered ring $[R_2SiO]_2[R''_2SiO]_2$ (Si–O ring with Si(R)(R) and Si(R'')(R'') alternating)

Right branch: $+ 2\ (ClSiR''_2)_2O / -4\ R'_3SnCl$ → 2 six-membered rings $R_2Si[OSiR''_2]_2O$

(3)

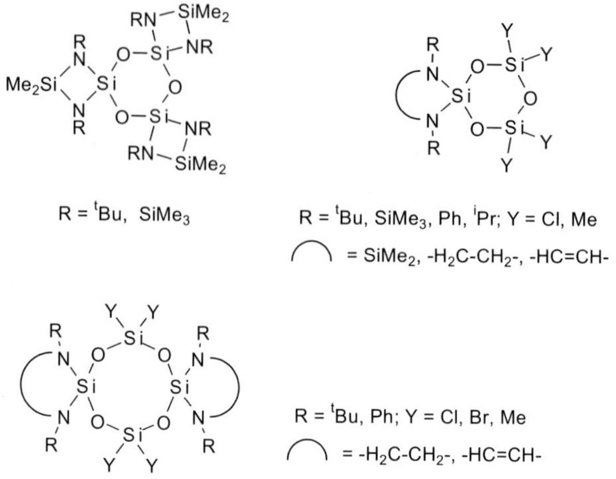

R = ^tBu, $SiMe_3$

R = ^tBu, $SiMe_3$, Ph, ^iPr; Y = Cl, Me

= $SiMe_2$, $-H_2C-CH_2-$, -HC=CH-

R = ^tBu, Ph; Y = Cl, Br, Me

= $-H_2C-CH_2-$, -HC=CH-

Scheme 27.1. Selected six- and eight-membered cyclosiloxanes with different substituents on silicon.

The cyclosiloxanes substituted by amino groups or halides may be used for further reactions. In some cases, we were able to selectively cleave the Si-N-bonds with a solution of HCl in THF without destroying the Si_4O_4 ring framework (Eq. 4). Compounds such as cyclo-$(Cl_2SiO)_2(Me_2SiO)_2$ (Eq. 4) may of course be used for further modification reactions or may serve as starting compounds for silicon polymers.

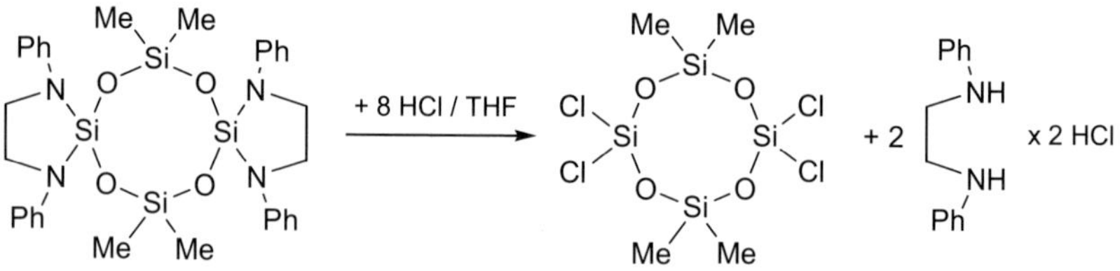

+ 8 HCl / THF → + 2 x 2 HCl (4)

27.2 Molecular Alumo-Siloxanes

The formal systematic replacement of SiO_2 units in quartz or other polymorphs of silicon dioxide by AlO(OH) or $M^+AlO_2^-$ (M = monovalent metal) is well known to lead either to sheet silicates or to tectosilicates, the zeolites being the most remarkable representatives.[17–19] All these solid-state compounds have as a common feature interstitial holes, which are occupied by easily extractable cations, or have three-dimensional frameworks which can be modified and used for many applications (ion exchange, catalysts, specific coordination sites, molecular sieves, etc.). Along these lines, it seems interesting to ask whether modifications of silicone chains or polyhedral silsesquioxanes can also be made by replacing some of the silicon atoms by aluminum [20–26] or gallium.[27,28] These modifications should influence the structures and may induce properties similar to those of the above mentioned solid-state compounds.

We used the reaction of diphenylsilanediol, $Ph_2Si(OH)_2$, with *tert*-butoxyalane, $^tBuOAlH_2$, to form alumosiloxanes.[29] As can be seen from Eq. 5, the reaction between these two molecules seems to proceed by initial condensation of $Ph_2Si(OH)_2$ to $(HO)SiPh_2–O–SiPh_2(OH)$ and subsequent reaction with $^tBuOAlH_2$, whereby all substituents on aluminum are eliminated as H_2 or *tert*-butanol.

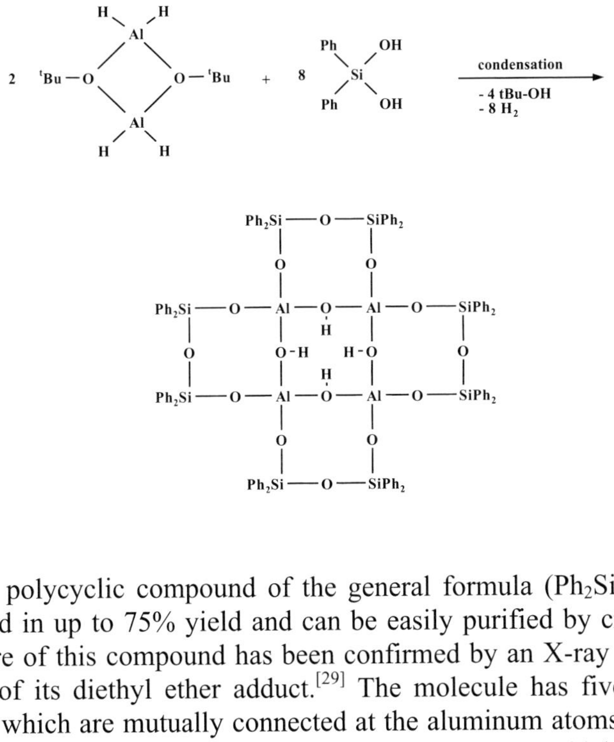

(5)

The polycyclic compound of the general formula $(Ph_2SiO)_8[Al(O)OH]_4$ is obtained in up to 75% yield and can be easily purified by crystallization. The structure of this compound has been confirmed by an X-ray structure determination of its diethyl ether adduct.[29] The molecule has five eight-membered cycles, which are mutually connected at the aluminum atoms: the center of the molecule thus consists of a pure $Al_4(OH)_4$ ring, to which four O-$SiPh_2$-O-$SiPh_2$-O chains are fused. As the aluminum atoms are in an AlO_4 environment, the $Si_2O_4Al_2$ rings are either directed up or down with respect to the central $Al_4(OH)_4$ ring. The whole molecule has almost S_4-symmetry, which means that the Al_4O_4 ring does not have a crown conformation but rather a conformation similar to S_4N_4. The four hydroxy groups are alternately pointing in opposite directions. Phenyl groups are located on the outer sphere of the molecule, wrapping up the interior inorganic skeleton.

From a simple ball-and-stick model of $(Ph_2SiO)_8[Al(O)OH]_4$ (Figure 27.1), it becomes clear that chemical access to the inner skeleton of the molecule is hindered by the steric crowding of the phenyl groups.[8] We have established discrete different reactivities depending on the nature of the reactant, as explained in the following sections.

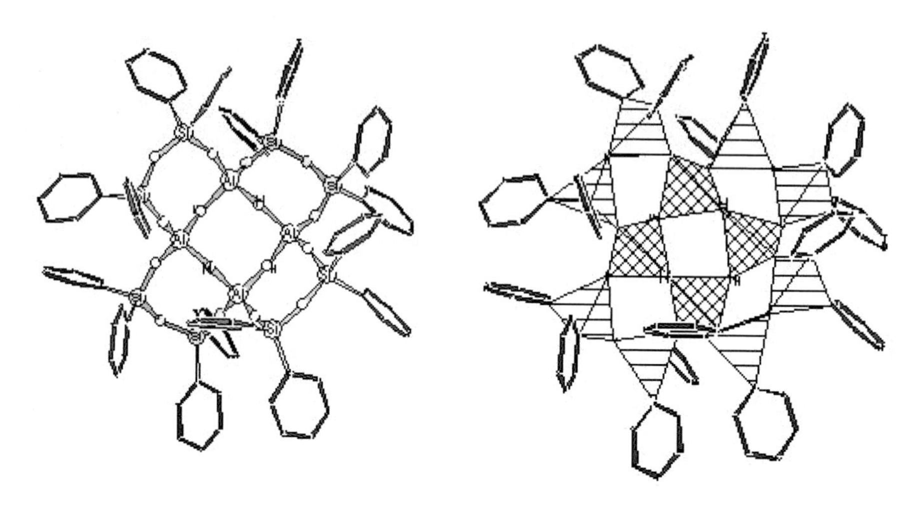

Figure 27.1. Ball and stick model (left) and polyhedra representation around Al (cross hatching) and Si (simple hatching) (right) for $(Ph_2SiO)_8[Al(O)OH]_4$.

27.2.1 Reactions of $(Ph_2SiO)_8[Al(O)OH]_4$ with Organic Nitrogen and Oxygen Bases

Organic nitrogen and oxygen bases such as triethylamine, diethyl ether, or pyridine have different steric demands, the shielding of the basic atom being most effective in triethylamine. While $(Ph_2SiO)_8[Al(O)OH]_4$ forms a 1:2 adduct with NEt_3, it is 1:3 for OEt_2 and 1:4 for pyridine; all adducts have hydrogen bridges between the OH groups of the $Al_4(OH)_4$ cycle and the corresponding nitrogen and oxygen atoms of the bases.[29,30] Although NEt_3 is the most powerful base in this series (which can also be seen from an effective attraction of the acidic hydrogen of the OH group to nitrogen, almost forming a $HNEt_3^+$ ion), it can only bind to two of the OH groups because of the steric requirement, the two other OH groups being shielded by the conformational change in the polycycle and the cut-off through the phenyl groups. The situation is less dramatic for diethyl ether, as this base can coordinate three OH groups of $(Ph_2SiO)_8$ $[Al(O)OH]_4$, and even less for pyridine, which coordinates all four OH groups. In the latter case, the flat pyridine molecules are almost "sandwiched" by the phenyl groups at the silicon atoms. The

$(Ph_2SiO)_8[Al(O)OH]_4$ polycycle thus functions as a size selective, hydrogen-mediated base selector.

27.2.2 Reactions of $(Ph_2SiO)_8[Al(O)OH]_4$ with Ammonia and Water

Apart from the acidic hydrogen atoms of the OH groups in $(Ph_2SiO)_8[Al(O)OH]_4$, the aluminum atoms themselves may also act as Lewis acids. However, as yet we have not found any attack of organic bases on the aluminum atoms (probably because of the steric congestion around these atoms). This situation becomes different when the bases become smaller in size. Water and ammonia can easily access the aluminum atoms of $(Ph_2SiO)_8[Al(O)OH]_4$ and, due to rearrangements within the Lewis acid/base adducts, dramatically change the composition and structure of the alumosiloxane. The reactions proceed differently depending on the molar ratios of reactants and the nature of the solvent. We were able to characterize $NH_4^+\{[(Ph_2Si)_2O_3]_6[Al_5O_2(OH)_3{\cdot}Et_2O][Al(NH_3)_2]\}^-$ and $(NH_4^+)_2$ $\{[(Ph_2Si)_2O_3]_6[Al_8O_4(OH)_6]{\cdot}2Et_2O\}^{2-}$ as crystalline compounds using ammonia as the base, and $[(Ph_2Si)_2O_3]_6[Al_6(OH)_3][Al(OH)_6]{\cdot}3Et_2O$ using water. The structures of the two products from the ammonolysis reactions are shown in Figure 27.2.

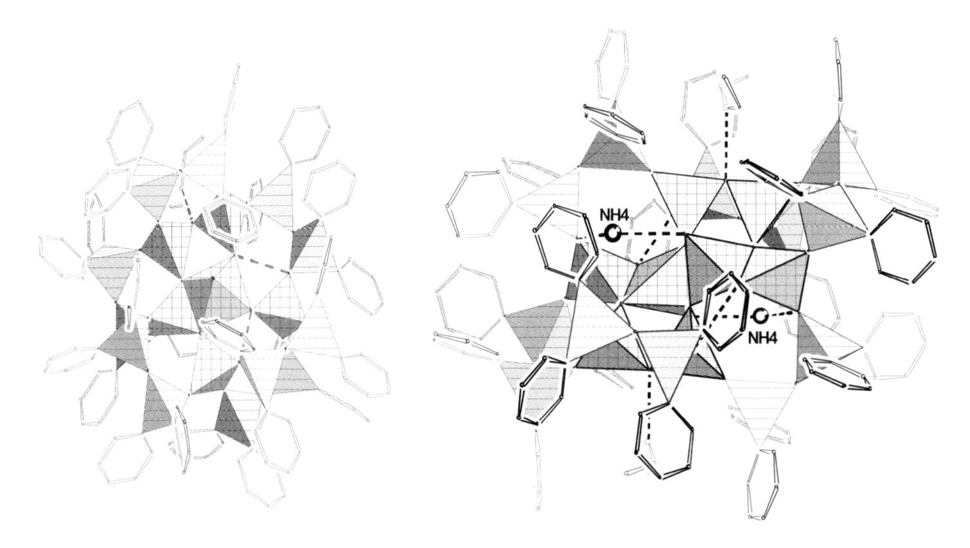

Figure 27.2. Polyhedra around Si (simple hatching) and Al (cross hatching) in the anion $\{[(Ph_2Si)_2O_3]_6[Al_5O_2(OH)_3{\cdot}OEt_2][Al(NH_3)_2]\}^-$ (left) and in $(NH_4^+)_2$ $\{[(Ph_2Si)_2O_3]_6[Al_8O_4(OH)_6]{\cdot}2OEt_2\}^{2-}$ (right). The dashed lines represent the hydrogen bridges.

The most remarkable structural feature in the compound $NH_4^+\{[(Ph_2Si)_2O_3]_6[Al_5O_2(OH)_3{\cdot}OEt_2][Al(NH_3)_2]\}^-$ is an octahedron around one of the aluminum atoms with two oxo, two hydroxo, and two NH_3 ligands. Further, the octahedron is fused through a common edge with an AlO_4 tetrahedron and through corners with the four other aluminum-centered tetrahedra. Altogether there are six polyhedra with aluminum in the center and twelve tetrahedra around the silicon atoms, maintaining the silicon/aluminum ratio at 2:1 as in the starting compound $(Ph_2SiO)_8[Al(O)OH]_4$. While the anionic part of the ion pair, $\{[(Ph_2Si)_2O_3]_6[Al_5O_2(OH)_3{\cdot}OEt_2][Al(NH_3)_2]\}^-$, can easily be distinguished in the crystal, the ammonium cation appears to be distributed over several places (split positions).

As regards $(NH_4^+)_2\{[(Ph_2Si)_2O_3]_6[Al_8O_4(OH)_6]{\cdot}2OEt_2\}^{2-}$, several differences are apparent compared to the other ammonolysis products: the Si/Al ratio has changed to 3:2, besides aluminum-centered tetrahedra AlO_5 trigonal bipyramids are also present, and the ammonium cations are incorporated in the polycyclic Al-O-Si-O framework, occupying holes. This has some resemblance to ammonium aluminosilicates. In the middle of the centrosymmetric molecule, four aluminum-centered polyhedra are linked through common edges, while the remaining four AlO_4 tetrahedra are attached through corners to the central ribbon. As in the other reaction products of $(Ph_2SiO)_8[Al(O)OH]_4$, the central aluminum-oxygen framework is wrapped by the diphenylsiloxane units.

When $(Ph_2SiO)_8[Al(O)OH]_4$ is allowed to react with limited amounts of water, the product that crystallizes from diethyl ether solutions is $[(Ph_2Si)_2O_3]_6[Al_6(OH)_3][Al(OH)_6]{\cdot}3OEt_2$.[31] The silicon/aluminum ratio is again altered (2:1.167), as is the O/(OH) ratio (from 3:1 to 2:1) compared to the starting compound. From the X-ray structure analysis it becomes clear that a central $Al(OH)_6$ octahedron is linked to six AlO_4 tetrahedra through hydroxo corners, which are connected in pairs through other hydroxo corners. The aluminum-centered polyhedra are again interconnected in the center of the molecule and are efficiently wrapped by peripheral diphenylsiloxane units. A sketch of the connected polyhedra in $[(Ph_2Si)_2O_3]_6[Al_6(OH)_3][Al(OH)_6]{\cdot}3OEt_2$, from which the phenyl groups have been omitted, is shown in Figure 27.3; it also illustrates the high point symmetry of D_3 (32) in the crystal.

27.2.3 Formation of "Molecular Alumosilicates"

As the hydrogen atoms in $(Ph_2SiO)_8[Al(O)OH]_4$ are quite acidic, an obvious step was to replace them by electropositive elements such as lithium, potassium or divalent tin and thus to create molecular alumosilicates. The reactions can easily be performed with reactants like butyllithium, phenyllithium, lithium amide, potassium *tert*-butoxide, tin di(*tert*-butoxide), bis(hexamethyldisilazyl)-

tin, etc.[8,9,30] The molarity of the reactant with respect to $(Ph_2SiO)_8[Al(O)OH]_4$, as well as the reaction conditions, such as solvent and temperature, are crucial for obtaining reproducible products. A list of isolated and structurally characterized compounds is given in Table 27.1.

Table 27.1. Selection of "molecular alumosilicates" obtained by hydrogen–substitution from $(Ph_2SiO)_8[Al(O)OH]_4$ [9]

(a) Original inner skeleton maintained:
$(Ph_2SiO)_8(AlO_2)_4(Li\cdot Et_2O)_4$
$(Ph_2SiO)_8(AlO_2)_4(Li\cdot Et_2O)_2(Li\cdot NH_3)(Li\cdot OEt_2,NH_3)$
$(Ph_2SiO)_8(AlO_2)_2\,[AlO(OH)]_2Sn$
$(Ph_2SiO)_8(AlO_2)_4Sn_2$

(b) Modified inner skeleton:
$(Ph_2SiO)_8[AlO_{1.5}(OH)]_4Li_4\cdot 2\ OEt_2$ *2 THF
$(Ph_2SiO)_8[AlO(OH)]_2[AlO_{1.5}(OH)]_4(Li\cdot OEt_2)_4$
$(Ph_2SiO)_8(AlO_2)_2(Li\cdot NH_3)[Li(OH_2)_2(THF)]$
$(Ph_2SiO)_{12}(AlO_2)_2[AlO_{1.5}(OH)]_4Li_4(Li\cdot THF)_2$
$(Ph_2SiO)_{12}(AlO_2)_2[AlO_{1.5}(OH)]_4Li_5[Li\cdot OEt_2]$
$(Ph_2SiO)_{12}(AlO_2)_4[AlO(OH)]_4K_4\cdot 2\ OEt_2$

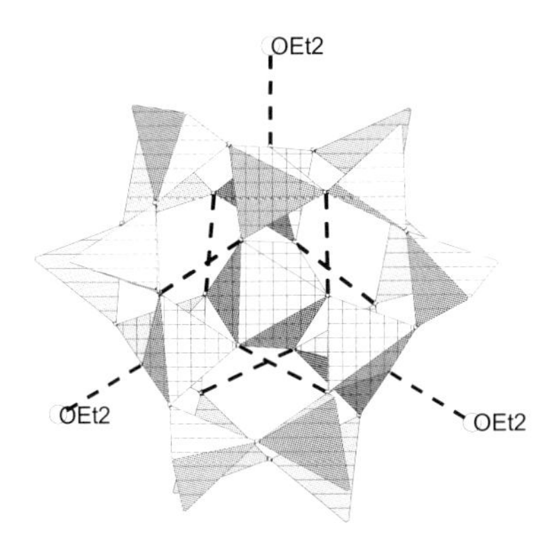

Figure 27.3. Polyhedra around Si and Al in $[(Ph_2Si)_2O_3]_6[Al_6(OH)_3]$-$[Al(OH)_6]\cdot 3Et_2O$. The phenyl groups have been omitted for clarity; the dashed lines show hydrogen bridges.

Two sorts of reaction products can be distinguished. In the first case (Table 27.1a), the compounds are structurally very similar to the starting material $(Ph_2SiO)_8[Al(O)OH]_4$, with the exception that the hydrogen atoms are replaced by metallic elements. In the second case (Table 27.1b), the skeleton is considerably modified, creating oxygen coordination holes which are occupied by the cations such as Li^+ and K^+. Of the structures assembled in Table 27.1, two are depicted in more detail (Figure 27.4). The structure of the inner skeleton of $(Ph_2SiO)_8(AlO_2)_4Sn_2$ is shown on the left-hand side of Figure 27.4, illustrating the bridging ability of the two-valent tin atoms. In $(Ph_2SiO)_8(AlO_2)_4Sn_2$, the metallic atoms form part of the polycyclic system and can be considered as covalently bonded. In $Li_5[Li{\cdot}OEt_2](Ph_2SiO)_{12}(AlO_2)_2[AlO_{1.5}(OH)]_4$ (Figure 27.4, right), the situation is different as the lithium cations either occupy holes in the crystal (one of them) or in the polycycles (the other five).

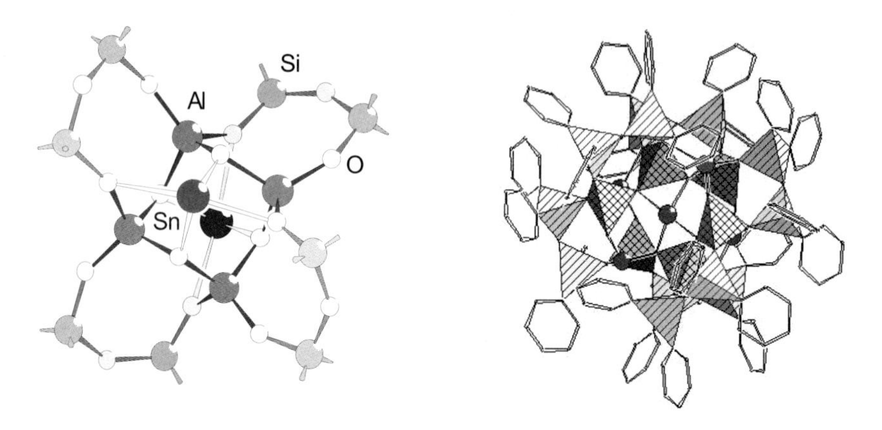

Figure 27.4. Left: Si-O-Al-O-Sn skeleton in $(Ph_2SiO)_8$-$(AlO_2)_4Sn_2$ with almost S_4 symmetry. Right: Polyhedral representation of $Li_5(Ph_2SiO)_{12}(AlO_2)_2$ $[AlO_{1.5}(OH)]_4$ with the lithium cations in the holes of the polycyclic structure.

27.3 Molecular Gallo-Siloxanes

We were surprised to find that the reaction of *tert*-butoxygallane with diphenyl-silanediol, analogous to that of *tert*-butoxyalane (see Section 27.2), did not give the corresponding products. Instead, we found that $(^tBuO\text{-}GaH_2)_2$ reacted with silanols such as $(HO)Ph_2Si–O–SiPh_2(OH)$ and $(HO)Ph_2Si–O–SiPh_2–O–SiPh_2(OH)$ to give different polycycles, mostly with only partial loss of the substituents at gallium.[32] To illustrate this peculiar behavior we have selected the reaction of $(^tBuO–GaH_2)_2$ with $(HO)Ph_2Si–O–SiPh_2–O–SiPh_2(OH)$, which leads to the parallel formation of two distinct products: $\{[(O)Ph_2Si–O–SiPh_2–O–SiPh_2(O)]GaH\}_2$ and $\{[(O)Ph_2Si–O–SiPh_2–O–SiPh_2(O)]Ga–O^tBu\}_2$ (see Eq. 6). In the first product, the gallium atom has retained one of the hydride ligands, whereas in the second the *tert*-butoxy group is retained. As the ratio of the products is almost 1:1, this demonstrates that the hydride and the alkoxy groups behave similarly towards siloxy groups, and that the reactivity of these groups at gallium is reduced in comparison to those at aluminum.

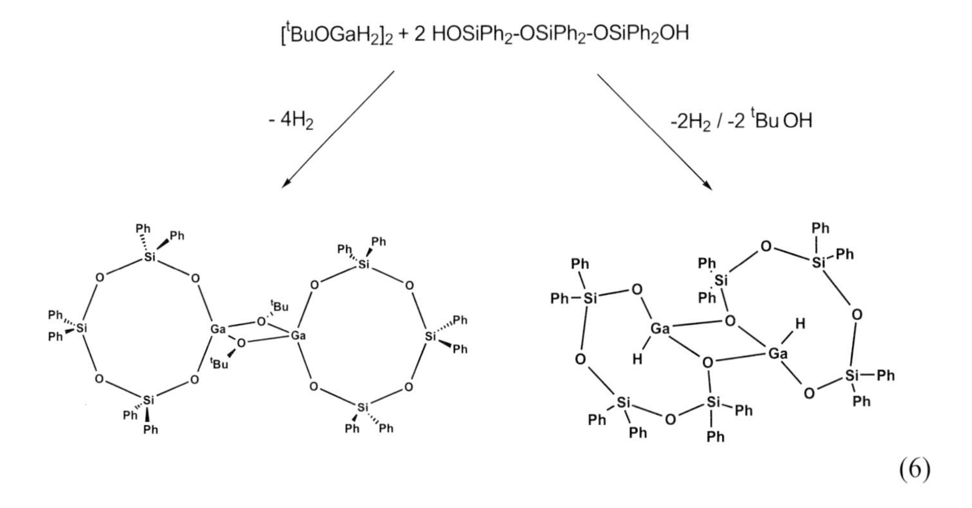

(6)

27.4 Acknowledgements

The Fonds der Chemischen Industrie is greatly acknowledged for financial support. We would also like to thank the following scientists for their interest, support, and collaboration during the period of the "Schwerpunktprogramm": Kira Fries, Sascha Wieczorek, Maria Jarczyk, Matthias Gasthauer, Oliver Schütt, Heidi Vogelgesang, Volker Huch (X-ray), and Michael Zimmer (NMR).

27.5 References

[1] J. Haiduc, *The Chemistry of Inorganic Ring Systems*, part 1, Wiley, London, 1990.

[2] U. Wannagat, G. Bogedain, A. Schervan, H. C. Marsmann, D. J. Brauer, H. Bürger, F. Dörrenbach, G. Pawelke, C. Krüger, K.-H. Claus, *Z. Naturforsch.* **1991**, *46b*, 931. A. Kornick, M. Binnewies, *Z. Anorg. Allg. Chem.* **1990**, *587*, 157.

[3] A. A. Zhadanov, A. B. Zachernyuk, G. V. Solomatin, Y. T. Struchkov, V. E. Shklover *Dokl. Akad. Nauk. SSSR* **1980**, *251,* 121.

[4] Yu. E. Ovchinnikov, V. E. Shklover, Y. T. Struchkov, A. B. Zachernyuk, A. A. Zhadanov, V. B. Isaev, *Metalloorg. Khim.* **1988**, *1*, 1121.

[5] D. Seyferth, C. Prud'Homme, G. H. Wiseman, *Inorg. Chem,* **1983**, *22*, 2163.

[6] M. Veith, A. Rammo, *Phosphorus, Sulfur, Silicon,* **1997**, *123*, 75.

[7] M. Veith, A. Rammo, M. Gießelmann, *Z. Anorg. Allg. Chem.* **1998**, *624*, 419.
[8] M. Veith, A. Rammo, M. Jarczyk, V. Huch, *Monatsh. Chem.* **1999**, *130,* 15.
[9] M. Veith et al., unpublished.
[10] W. Clegg, *Acta Cryst.* **1983**, *C39,* 901.
[11] P. D. Lickiss, A. D. Redhouse, R. J. Thompson, W. A. Stanczyk, K. Rozga, *J. Organomet. Chem.* **1993,** *435*, 53.
[12] D. Seebach, R. Armstutz, T. Laube, W. B. Scheizer, J. D. Dunitz, *J. Am. Chem. Soc.* **1985**, *107,* 5403.
[13] D. Schmidt-Base, U. Klingebiel, *Chem. Ber.* **1989**, *122,* 815.
[14] H. J. Gais, U. Dingerdissen, C. Kruger, K. Angermund, *J. Am. Chem. Soc.* **1987**, *109*, 3775.
[15] K. Thiele, H. Görls, W. Seidel, *Z. Anorg. Allg. Chem.* **1998**, *624,* 1391.
[16] D. A. Armitage, A. Tarassoli, *Inorg. Nucl. Chem. Lett.* **1973**, *9,* 1225.
[17] A. F. Holleman, E. Wiberg, Lehrbuch der Anorganischen Chemie, 101. de Gruyter, Berlin/NewYork, 1995, p. 930.
[18] L. Puppe, *Chemie i. u. Zeit* **1986**, *20*, 117.
[19] W. Hölderich, M. Hesse, F. Naumann, *Angew. Chem.* **1988**, *100*, 232. *Angew. Chem. Int. Ed. Engl.* **1988**, *27*, 226.
[20] Y. K. Gun`ke, R. Reilly, V. G. Kessler, *New J. Chem.* **2001**, *25*, 528.
[21] C. N. McMahon, S. G. Bott, L. B. Alemany, H. W. Roesky, A. R. Barron, *Organometallics* **1999**, *18*, 5395.
[22] R. Duchateau, R. A. van Santen, G. P. A. Yap, *Organometallics* **2000**, *19*, 809.
[23] A. Klemp, H. Hatop, H. W. Roesky, H.-G. Schmidt, M. Noltemeyer, *Inorg. Chem.* **1999**, *38*, 5832.
[24] F. J. Feher, R. Terroba, J. W. Ziller, *Chem. Commun.* **1999**, 2153.
[25] F. T. Edelmann, G. Giessmann, A. Fischer, *J. Organomet. Chem.* **2001**, *620*, 80.
[26] M. D. Skowronska-Ptasinska, R. Duchateau, R. A. van Santen, *J. Inorg. Chem.* **2001**, 133.
[27] F. J. Feher, T. A. Budzuichowski, J. W. Ziller, *Inorg. Chem.* **1997**, *36*, 4082.
[28] A. Voigt, R. Murugavel, E. Parisini, H. W. Roesky, *Angew. Chem.* **1996**, *108*, 823. *Angew. Chem. Int. Ed. Engl.* **1996**, *35*, 748.
[29] M. Veith, M. Jarczyk, V. Huch, *Angew. Chem.* **1997**, *109*, 140; *Angew. Chem. Int. Ed.* **1997**, *36*, 117.

[30] M. Veith, M. Jarczyk, V. Huch, *Angew.Chem.* **1998**, *110*, 109; *Angew. Chem. Int. Ed.* **1998**, *37*, 105.
[31] M. Veith, M. Jarczyk, V. Huch, *Phosphorus, Sulfur, Silicon* **1997**, *124&125*, 213.
[32] M. Veith, H. Vogelgesang, V. Huch, *Organometallics* **2002**, *21*, 380.

28 Synthesis, Structure, and Reactivity of Novel Oligomeric Titanasiloxanes

Peter Jutzi*, Hans Martin Lindemann, Jörn-Oliver Nolte, Manuela Schneider

28.1 Introduction and Background

Amorphous and crystalline Ti-containing silicas represent an important family of modern materials which find industrial application in many areas. Amorphous titania-silicas are used extensively as heterogeneous catalysts and as supports for a wide variety of reactions, especially as oxidation catalysts.[1] In addition, they can be utilized as protective coatings, as antireflex coatings, and as glass materials with low thermal coefficients and high refractive indices.[1] The most widely used method for their preparation is sol-gel processing. Here, microdomain formation of anatase-like TiO_2 due to the differences in the hydrolysis and condensation rates of Ti- and Si-alkoxides is an unwanted reaction and the major problem. Numerous different structures are feasible in the amorphous gels, depending on the reactivity of the precursors and on the reaction conditions, and a remarkable restructuring may occur in the subsequent aging, drying, and calcination steps.[2] Furthermore, only materials with a low TiO_2 content can be obtained. At higher Ti contents, TiO_2 crystallites tend to be formed as a separate phase.

Micro- and mesoporous crystalline titanium silicas, such as Ti silicalite TS-1 and the Ti-containing MCM-41 zeolite, represent another important family of modern materials and find versatile application as catalysts for oxidation reactions and in absorption technologies.[3] These materials are prepared under hydrothermal conditions using tetraalkylammonium hydroxide, titanium tetraalkoxide, and tetraethyl orthosilicate as precursors. Only limited amounts of Ti can enter the SiO_2 framework and the actual bonding situation at Ti is not yet fully understood. Several data suggest that Ti substitution for Si in the lattice positions retaining the tetrahedral coordination is important for catalysis.

---

* Address of the authors: Fakultät für Chemie, Universität Bielefeld, Universitätsstraße 25, 33615 Bielefeld, Germany.

28.2 Strategy

Our own work deals with the synthesis and chemistry of oligomeric titanasiloxanes. Such compounds are regarded as model systems for the first condensation steps in the sol-gel processing of TiO_2-/SiO_2 solids and as suitable subjects for the investigation of catalytic effects. They can be prepared following a known procedure[4] by condensation reactions of organosilanetriols with titanium tetraalkoxides. It was intended to investigate in more detail the structure directing influence of steric effects of the organic substituents at silicon and of the alkoxy substituents at titanium on the process of framework formation. Furthermore, it was intended to investigate whether the substituents at the titanium and silicon centers in the preformed polyhedral structures can be exchanged under preservation of the respective TiOSi core. The introduction of functionalized substituents would allow the preparation of novel inorganic/organic hybrid materials[5] in a controlled way with polyhedral titanasiloxane cores as molecular building blocks. Furthermore, it would be possible to synthesize novel dendrimers possessing titanasiloxane cores. A necessary prerequisite for these strategies is the leaving group character of the substituents at silicon and titanium; therefore, we have chosen different kinds of cyclopentadienyl substituents (Cp^R)[6] at silicon and alkoxy or acetyl-acetonato groups at titanium in the condensation reactions. Finally, it was intended to investigate whether novel homogeneous Ti-O-Si materials can be prepared by condensation reactions between polyhedral titanasiloxanes under preservation of the respective core structures. Such reactions might occur by proton transfer from a TiOH group to the Cp^R substituent at silicon and by subsequent Cp^RH elimination under concomitant formation of a novel TiOSi bond. The formation of a TiOH unit by thermal alkene elimination from a TiOR group is documented in the literature.[7]

28.3 Synthesis and Structures of Polyhedral and Cyclic Titana-siloxanes

Condensation reactions of Cp^Rsilanetriols with titanium alkoxides have led to the formation of novel oligomeric titanasiloxanes. The chosen substituents at silicon (Cp^R) and at titanium (OR) are presented in Figure 28.1.

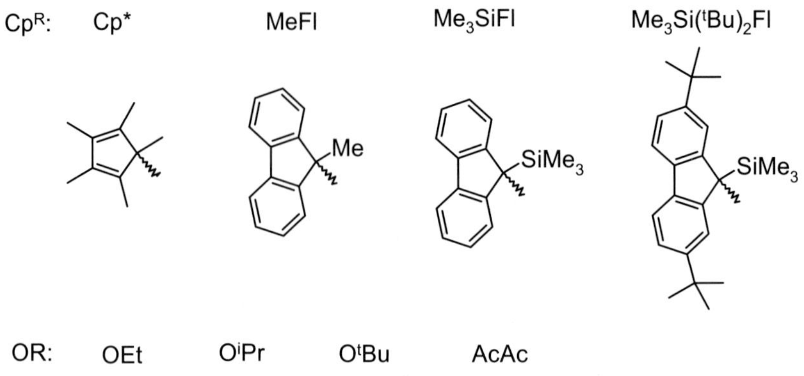

Cp^R: Cp* MeFl Me_3SiFl $Me_3Si(^tBu)_2Fl$

Me SiMe$_3$ SiMe$_3$

OR: OEt O^iPr O^tBu AcAc

Figure 28.1. Summary of substituents at silicon (Cp^R) and at titanium (OR).

Reactants	Conditions	Products
4 $(Me_3SiFl)Si(OH)_3$ + 4 $Ti(O^tBu)_4$	rt, hexane	$[(Me_3SiFl)Si]_4O_{12}[TiO^tBu]_4$ (**1**) + 12 ^tBuOH
4 $Cp^*Si(OH)_3$ + 4 $Ti(O^tBu)_4$	rt, hexane	$[Cp^*Si]_4O_{12}[TiO^tBu]_4$ (**2**) + 12 ^tBuOH
4 $(Me_3SiFl)Si(OH)_3$ + 4 $Ti(O^iPr)_4$	hexane,-78°C, THF, rt	$[(Me_3SiFl)Si]_4O_{12}[TiO^iPr]_4$ (**3**) + 12 ^iPrOH
4 $(Me_3SiFl)Si(OH)_3$ +6 $Ti(O^iPr)_4$ + 2 H_2O	rt, hexane	$[(Me_3SiFl)Si]_4O_{12}[TiO^iPr]_6[\mu_2\text{-}OiPr]_2[\mu_3\text{-} O]_2$ (**4**) + 16 ^iPrOH
2 $(MeFl)Si(OH)_3$ + 6 $Ti(O^iPr)_4$ + 4 H_2O + 2 $PhNH_2$	rt, hexane	$[(MeFl)Si]_2O_8[TiO^iPr]_6[\mu_2\text{-}OiPr]_4[\mu_3\text{-} O]_2[PhNH_2]_2$ (**5**) + 16 ^iPrOH
2 $(Me_3SiFl)Si(OH)_3$ +4 $Ti(OEt)_4$	rt, hexane	$[(Me_3SiFl)Si]_2O_5[TiOEt]_4[\mu_2\text{-}OEt]_6[\mu_4\text{-} O]$ (**6**) + 6 EtOH
6 $(Me_3Si(^tBu)_2Fl)Si(OH)_3$ +10 $Ti(O^iPr)_4$ + 5 H_2O	rt, hexane	$([(Me_3Si(^tBu)_2Fl)Si]_3O_{10}[TiO^iPr]_4[\mu_2\text{-}OiPr]_2[\mu_3\text{-}O]Ti)_2O$ (**7**) + 28 ^iPrOH
2 $(MeFl)Si(OH)_3$ +2 $Ti(O^iPr)_2(acac)_2$	rt, hexane	$[(MeFl)Si(OiPr)]_2O_4[Ti(acac)_2]_2$ (**8**) + 2 ^iPrOH

Scheme 28.1. Synthesis of oligomeric titanasiloxanes.

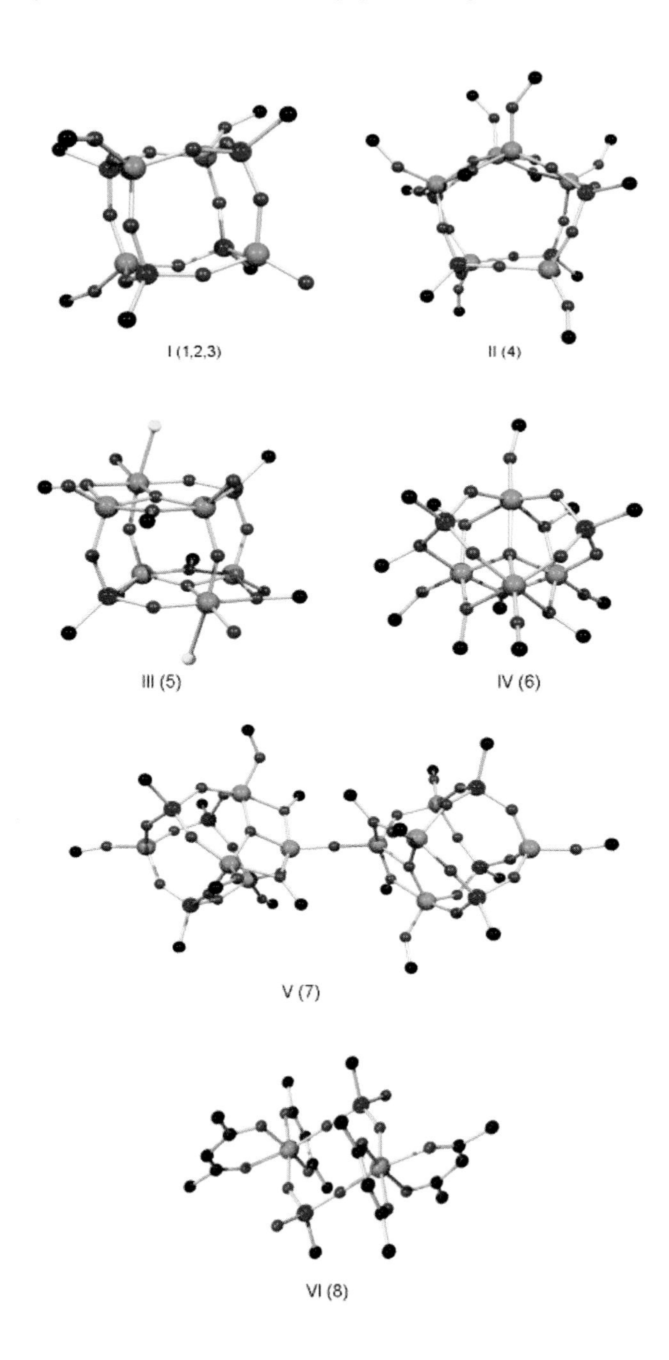

Figure 28.2. Titanasiloxane frameworks:[8] for clarity, the organic substituents are represented only by their ipso-C atoms; the aniline molecules are represented by the N atoms only.

With a given substrate ratio of 1:1, the expected compounds with a cubic core structure $[Cp^RSi]_4O_{12}[TiOR]_4$, possessing terminal Cp^R and OR and bridging Si-μ_2O-Ti units, were isolated in high yields only in a few cases. Quite often, a mixture of products was obtained, from which only the crystalline species, which were formed in rather low yield, could be unambiguously characterized. Interestingly, those compounds were obtained in nearly quantitative yield when the appropriate substrate ratio was used, including the presence of further substrates such as water or aniline. Furthermore, product formation has been shown to depend on the reaction temperature and even on the solvent. The equations with the appropriate stoichiometry of substrates for the synthesis of the oligomeric titanasiloxanes **1**–**8** are given in Scheme 28.1; in Figure 28.2, the different titanasiloxane frameworks I–VI are presented in the form of ball-and-stick models.

The more complex polyhedral structures II–V show, in addition to the structural features of the cubic compounds I, (μ^3-O)Ti_3, (μ^4-O)Ti_4, Ti(μ^2-OR)Ti, and Si(μ^2-OR)Ti units. Silicon atoms are always four-coordinate, whereas titanium atoms are four-, five-, or six-coordinate. The presence of TiOTi units indicates competing TiOR or TiOH condensation reactions. Intermediate TiOH species are formed by hydrolysis of TiOR units; the necessary water molecules arise from SiOH condensation reactions or from the reaction of SiOH groups with liberated ROH molecules. Another high yield product from a 1:1 substrate ratio is the eight-membered ring system of type VI (Figure 28.2) containing six-coordinate titanium centers; the acetylacetonato substituents are not involved in exchange reactions and thus prevent the formation of a polyhedral structure. The condensation reactions performed with $(Me_3SiFl)Si(OH)_3$ and $Ti(O^iPr)_4$ demonstrate that the product formation also depends on the solvent and temperature (see Scheme 28.1).

Further condensation reactions, which lead to oligomeric titanasiloxane species, are presented in Scheme 28.2. As already mentioned, the first compound with a cubic core structure (**9**) was prepared from the reaction by the aminosilanetriol $[(2,6\text{-}^iPr_2H_3C_6)N(SiMe_3)]Si(OH)_3$ with $Ti(OEt)_4$.[4] Reaction of the same aminosilanetriol with the titanium species $(Cp^*TiMe(\mu^3\text{-O}))_3$ led to the formation of the titanasiloxane **10** possessing an adamantanoid core structure with three six-coordinate Ti centers.[4] Finally, the reaction of the silanetriol $(MeFl)Si(OH)_3$ with Cp^*TiMe_3 afforded the cyclic oligotitanasiloxane **11** possessing one four- and four six-coordinate Ti centers. The structure of **10** and **11** are presented in Figure 28.3.

$4\ [(2,6\text{-}{}^{i}Pr_2H_3C_6)N(SiMe_3)]Si(OH)_3 + 4\ Ti(OEt)_4 \xrightarrow{rt} [(2,6\text{-}{}^{i}Pr_2H_3C_6)N(SiMe_3)Si]_4O_{12}[TiOEt]_4$ (**9**) $+ 12\ EtOH$

$[(2,6\text{-}iPr_2H_3C_6)N(SiMe_3)]Si(OH)_3 + (Cp^*TiMe(\mu_3\text{-}O))_3 \xrightarrow{rt} [(2,6\text{-}iPr_2C_6H_3)N(SiMe_3)Si]O_3[Cp^*TiO]_3$ (**10**) $+ 3\ MeH$

$(MeFl)Si(OH)_3 + Cp^*TiMe_3 \xrightarrow{-78°C} [(MeFl)Si]_4O_{10}[Cp^*Ti]_4[TiO_4]$ (**11**) + oligomers

Scheme 28.2. Synthesis of further oligomeric titanasiloxanes.

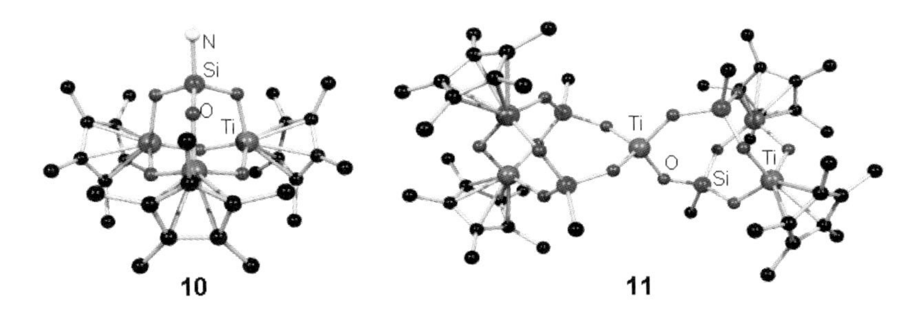

Figure 28.3. Titanasiloxane frameworks: for clarity, the peripheral carbon atoms of the substituents at silicon are omitted.

It is evident from the information given in this chapter that small changes in the steric demand of substituents at silicon and at titanium drastically influence the pathway in condensation reactions. It is also evident that the chosen reaction conditions also influence the product formation. Thus, the pathway in the condensation reaction of $(Me_3SiFl)Si(OH)_3$ with $Ti(O^iPr)_4$ is shown to depend on the solvent and on the temperature.

28.4 Core Functionalization

Following our strategy, it was an important target to investigate substitution reactions at the silicon and titanium atoms of the polyhedral frameworks. Earlier experiments in titanasiloxane chemistry have already shown that TiOSi units are very sensitive towards acid- and base-catalyzed oligomerizations.[9] Thus, the choice of reactants for selective functionalization under conservation of the core structure was very limited. We have investigated in more detail the cubic titanasiloxanes **1**, **2**, and **3** and the polyhedral titanasiloxane **4**. We have found selective reactions at the titanium centers; substitution chemistry at the silicon centers turned out to be much more difficult.

The first indication of selective substitution reactions came from experiments with **1** and **2** as catalysts for alkene epoxidation.[10] NMR experiments have shown that both compounds catalyze the epoxidation of cyclohexene with *t*-butyl hydroperoxide (TBHP). The catalytic activity is comparable to that of the model compound $Hex_7Si_7O_{12}Ti(O^iPr)$.[11] The measured turn over numbers indicate that all four Ti centers are involved in the catalytic process. The catalysts could be recovered quantitatively, a proof of core-functionalization and for the core stability during many catalytic cycles. A more detailed catalytic study has recently been performed with the cubic titanasiloxane $[(2,6\text{-}^iPr_2C_6H_3)\ (Me_3Si)NSi]_4O_{12}[TiO^tBu]_4$ (**12**).[12] This compound was prepared by the reaction of **9** with *t*-butanol and catalyzes the epoxidation of cyclohexene with TBHP. The titanium ^{t}butylperoxo intermediate could be isolated after a stoichiometric reaction with TBHP. This intermediate then reacted with cyclohexene to produce cyclohexene oxide. A schematic representation of the catalytic process is given in Figure 28.4.

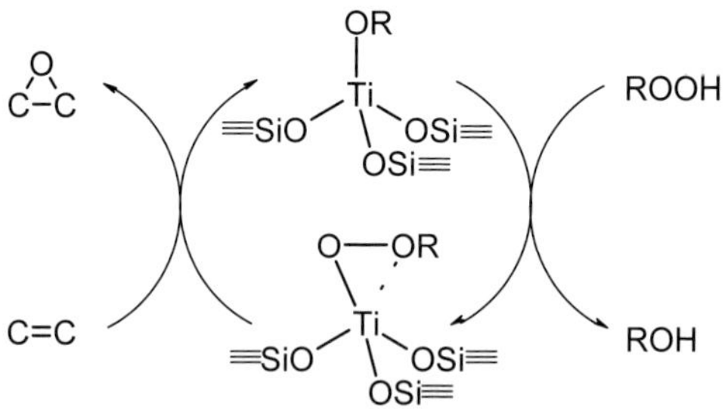

Figure 28.5. Catalytic cycle in the epoxidation of alkenes by cubic titanasilsesquioxanes (represented by $(\equiv SiO)_3TiOR$).

Further substitution reactions at the titanium atoms in polyhedral titanasiloxanes are collected in Scheme 28.3. Core degradation was observed in the reaction of the titanasiloxanes **1** or **3** with methanol or ethanol; cage compounds bearing methoxy or ethoxy substituents could not be detected. We conclude from the above observations that it is impossible to introduce a less bulky alkoxy group without core degradation. Reaction of **1** and **3** with four equivalents of triphenylsilanol proceeded readily at room temperature to afford the triphenylsiloxy-substituted derivative **13** almost quantitatively.[10] Treatment of **1** and **2** with four equivalents of dimethylhydroxylamine afforded the substitution products **14** and **15**, respectively, with bidentate dimethylhydroxylamido substituents. The five-coordinate titanium atoms were confirmed by X-ray crystallography.[10] Reaction of **2** with four equivalents of acetylacetone afforded the substitution product **16** bearing bidentate acetylacetonato substituents. Single crystals obtained from diethyl ether/dioxane contain four additional dioxane molecules coordinated to the four titanium atoms, as shown by X-ray crystallography. This arrangement leads to six-coordinate titanium centers.[10] Reaction of the polyhedral titanasiloxane **4** with six equivalents of acetylacetone led to the four-fold substitution product **17**. The core structure corresponds to that of the starting material (resembling a cube-like polyhedron with an additional "roof" created by insertion of a Ti-(μ_3-O)$_2$-Ti fragment into one of the six faces of the cube structure). Only the isopropoxy groups of the four-coordinate titanium atoms are substituted by acetylacetonato groups, whereas the two five-coordinate titanium atoms still bear isopropoxy substituents. The presence of five-coordinate (trigonal bipyramidal) and six-coordinate (octahedral) titanium atoms was proven by X-ray crystallography.[10]

9 + 4 ^{t}BuOH ⟶ [(2,6-^{i}Pr$_2$H$_3$C$_6$)N(SiMe$_3$)Si]$_4$O$_{12}$[TiO^{t}Bu]$_4$ (**12**) + 4 EtOH

1 + 4 Ph$_3$SiOH ⟶ [(Me$_3$SiFI)Si]$_4$O$_{12}$[TiOSiPh$_3$]$_4$ (**13**) + 4 ^{t}BuOH

3 + 4 Ph$_3$SiOH ⟶ [(Me$_3$SiFI)Si]$_4$O$_{12}$[TiOSiPh$_3$]$_4$ (**13**) + 4 ^{i}PrOH

1 + 4 Me$_2$NOH ⟶ [(Me$_3$SiFI)Si]$_4$O$_{12}$[TiONMe$_2$]$_4$ (**14**) + 4 ^{t}BuOH

2 + 4 Me$_2$NOH ⟶ [Cp*Si]$_4$O$_{12}$[TiONMe$_2$]$_4$ (**15**) + 4 ^{t}BuOH

2 + 4 AcacH ⟶ [Cp*Si]$_4$O$_{12}$[Tiacac]$_4$ (**16**) + 4 ^{t}BuOH

4 + 4 AcacH ⟶ [(Me$_3$SiFI)Si]$_4$O$_{12}$[TiO^{i}Pr]$_2$[Tiacac]$_4$[μ_2-iPrO]$_2$[μ_3-O]$_2$ (**17**) + 4 ^{i}PrOH

Scheme 28.3. Substitution reactions at the Ti-atoms in polyhedral titanasiloxanes.

With the reactions in Scheme 28.3 it has been demonstrated that polyhedral titanasiloxanes can be functionalized selectively at their four-coordinate titanium centers. The choice of the reagents turns out to be crucial to avoid core degradation. Most suitable are those reagents which lead to an increase of the coordination number (chelating ligands).

Substitution at the silicon atoms has been investigated with the cubic titanasilsesquioxanes **1** and **3** as representative examples. If Cp^R substitution takes place, then in most cases it is with concomitant degradation of the core structure. In compound **3**, the Cp*-Si bonds are easily cleaved by treatment with stoichiometric amounts of ethanol, water, or hydrogen chloride to afford pentamethylcyclopentadiene (Cp*H) and gel-like materials that form an insoluble solid after removal of the solvents. Interestingly, the formation of trimethylsilylfluorene (Me_3SiFlH) was not observed in comparable reactions with compound **1**; the slow formation of insoluble solids indicates catalytic cage opening followed by oligomerization reactions (attack of the electrophile at the SiOTi unit).

28.5 Core Condensation to Homogeneous TiOSi Materials

Preliminary results show that, in accord with our strategy, Cp^R-substituted titanasilsesquioxanes are versatile precursors for homogeneous TiOSi materials with high titanium content. Thermal treatment of compounds **1** and **2** at 300 °C leads to the elimination of isobutene and to the liberation of the protonated Cp^R substituents ($(Me_3Si)FlH$ and Cp*H, respectively. These reactions might occur under formation of a TiOH unit by thermal alkene elimination from a TiO^tBu group, followed by proton transfer to the Cp^R substituent at silicon and by subsequent Cp^RH elimination. In the case of compound **2**, the additional formation of significant amounts of tetramethylfulvene (TMF) is observed. The formation of TMF and Cp*H is a consequence of the disproportionation of Cp* radicals,[13] a well-known process in the gas-phase deposition of Cp*-silanes.[14] The homolytic cleavage of the Si-Cp* bonds allows further condensation processes to take place. After thermolysis, **1** and **2** show significant carbon contents due to the incorporation of organic material. The carbon contents can be reduced to values lower than 5 % by calcination in a flow of dry oxygen.

The TiOSi materials prepared by thermolysis of **1** and **2** and subsequent calcination have been analyzed by EDX, PXRD, IR-, Raman- and UV/vis spectroscopy. As expected, the samples possess a Ti/Si ratio close to one. PXRD measurements indicate amorphous, porous structures without greater TiO_2 domains. The UV/vis spectra suggest the presence of titanium centers possessing octahedral as well as tetrahedral symmetry. Octahedral titanium sites may arise from the coordination of H_2O molecules or the presence of TiO_2-nanodomains (< 5 nm; not observed by XRD). The IR spectra show significant amounts of Ti-O-Si hetero-linkages involving tetrahedrally coordinated titanium atoms.

Further studies, including DTA-MS and PXRD measurements and the determination of surface areas and porosities, are in progress.[15]

28.6 Conclusions

We have shown that novel polyhedral titanasiloxanes bearing Cp^R substituents at silicon and alkoxy substituents at titanium can be selectively prepared in high yields by the co-condensation of silanetriols and titanium alkoxides. The core-structure and thus the Ti/Si ratio of the titanasiloxanes can be determined by the choice of substituents at silicon and at titanium and even by variation of the solvent and temperature. The alkoxy groups at the titanium atoms can be selectively functionalized by suitable substituents without core degradation. The leaving group character of the substituents at silicon and at titanium can be used for the synthesis of homogeneous TiOSi materials by thermal or photochemical treatment of the compounds. Core condensation of selectively preformed polyhedral titanasiloxanes under formation of materials with Ti/Si ratios corresponding to those of the precursor molecules represents a novel strategy in TiOSi chemistry.

28.7 References

[1] X. Gao, J. E. Wachs, *Catal. Today* **1999**, *51*, 233–254

[2] C. Beck, T. Mallat, T. Burgi, A. Baiker, *J. Catal.* **2001**, *204*, 428–439

[3] T. Armaroli, F. Mitella, B. Notari, R. J. Willey, and G. Busca, *Topics in Catalysis* **2001**, *15*, 63–71; A. Corma. M. T. Navarro, J. Pérez Pariente, *J. Chem. Soc., Chem. Commun.* **1994**, 147–148.

[4] A. Voigt, R. Murugavel, V. Chandrasekhar, N. Winkhofer, H. W. Roesky, H. G. Schmidt, I. Uson, *Organometallics* **1996**, *15*, 1610. R. Murugavel. M. Bhattacharjee, H. W. Roesky, *Appl. Organomet. Chem.* **1999**, *13*, 227–243; R. Murugavel, A, Voigt, M. G. Walawalkar, H. W. Roesky, *Chem. Rev.* **1996**, 2205–2236; Reviews on metal compounds with silanolate, silanediolate, and silanetriolate substituents: H. Schmidbaur, *Angew. Chem.* **1965**, *77*, 206; L. King, A. C. Sullivan, *Coord. Chem. Rev.* **1999**, *189*, 19. V. Lorenz, A. Fischer, S. Gießmann, J. W. Gilje, Y. Gun'ko, K. Jakob, F. T. Edelmann, Coord. *Chem. Rev.* **2000**, *206–207*, 321.

[5] U. Schubert, *Chem. Mater.* **2001**, *13*, 3487. C. Sanchez, G. J. de A. A. Soler-Illia, F. Ribot, T. Lalot, C. R. Mayer, V. Cabuil, *Chem. Mater.* **2001**, *13*, 3061. M. D. Skowronska-Ptasinska, M. L. W. Vorstenbosch, R. A. van Santen, H. C. L. Abbenhuis, *Angew. Chem.* **2002**, *114*, 659; *Angew. Chem. Int. Ed. Engl.* **2002**, *41*, 637–639.

[6] P. Jutzi, *Comments Inorg. Chem.* **1987**, *6*, 123. P. Jutzi, *J. Organomet. Chem.* **1990**, 400. P. Jutzi, G. Reumann, *J. Chem. Soc., Dalton Trans.* **2000**, 2237–2244. M. Schneider, B. Neumann, H. G. Stammler, P. Jutzi, *Organosilicon Chemistry - From Molecules to Materials IV,* (Eds.: N. Auner, J. Weis), VCH, Weinheim, **2000**.

[7] M. P. Coles, C. G. Lugmair, K. W. Terry, T. D. Tilley, *Chem. Mater.* **2000**, *12*, 122.

[8] Color-code for the ball-and-stick models: Titanium, green; Silicon, red; Oxygen, blue; Nitrogen, yellow; Carbon, black.

[9] D. Hoebbel, M. Nacken, H. Schmidt, V. Huch, M. Veith, *J. Mater. Chem.* **1998**, *8 (1)*, 171.

[10] M. Schneider, Thesis, Univ. Bielefeld, **1999**. H. M. Lindemann, M. Schneider, B. Neumann, H. G. Stammler, A. Stammler, P. Jutzi, *Organometallics* **2002**, *21*, 3009. J. O. Nolte, M. Schneider, B. Neumann, H. G. Stammler, P. Jutzi, *Organometallics*, in press.

[11] T. Maschmeyer, M. C. Klunduk, C M. Martin, D. S. Shephard, J. M. Thomas, B. F. G. Johnson, *J. Chem. Soc., Chem. Commun.* **1997**, 1847. S. Krijnen, H. C. L. Abbenhuis, R.W. J. M. Hansen, J. H. C. van Hooff, R. A. van Santen, *Angew. Chem.* **1998**, *110*, 374; *Angew. Chem. Int. Ed. Engl.* **1998**, *37*, 356–358.

[12] M. Fujiwara, H. Wessel, P. Hyung-Suh, H. W. Roesky, *Tetrahedron* **2002**, *58*, 239.

[13] A. G. Davies, J. Lusztyk, *J. Chem. Soc., Dalton Trans.* **1981**, 692.

[14] J. Dahlhaus, P. Jutzi, H.-J. Frenck, W. Kulisch, *Adv. Mater.* **1993**, *5*, 377.

[15] H. M. Lindemann, P. Jutzi, M. Fröba, in preparation.

29 Metallasilsesquioxanes – Synthetic and Structural Studies

Frank T. Edelmann*

29.1 Introduction

Polyhedral silsesquioxanes of the general formula $(RSiO_{1.5})_n$ form an interesting class of organosilicon compounds which currently have a tremendous impact on catalysis research [1, 2] and materials science [3] alike. Due to their chemical composition they can be viewed as hybrids between silica, (SiO_2), and the silicones, $(R_2SiO)_n$. In accordance with several unique properties, the polyhedral silsesquioxanes have been termed the "smallest particles of silica possible" [4] or "small soluble chunks of silica".[2] The fact that silsesquioxane molecules like **1** contain covalently bonded reactive functionalities also makes them promising monomers for the development of novel hybrid materials which offer a variety of useful properties.[3, 4] With respect to catalysis the chemistry of metallasilsesquioxanes also receives considerable current interest.[1, 2, 5, 6] Incompletely condensed silsesquioxanes such as $Cy_7Si_7O_9(OH)_3$ (**1**, Cy = *c*-C_6H_{11}, Scheme 29.1) share structural similarities with β-cristobalite and β-tridymite and are thus quite realistic models for the silanol sites on silica surfaces.[7–11] Metal complexes derived from **1** are therefore regarded as "realistic" models for metal catalysts immobilized on silica surfaces.[2, 8]

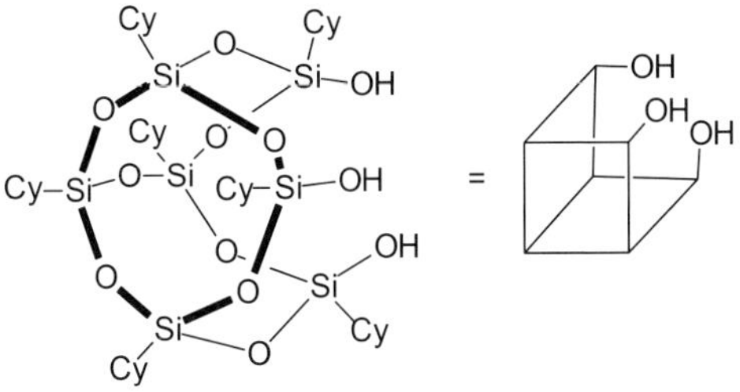

1: Cy = *c*-C_6H_{11}

Scheme 29.1. Schematic representation of silsesquioxane **1**.

* Address of the author: Prof. Frank T. Edelmann, Chemisches Institut der Otto-von-Guericke-Universität Magdeburg, Universitätsplatz 2, D-39106 Magdeburg (Germany), Fax.: +49-391-6712933;
e-mail: frank.edelmann@vst.uni-magdeburg.de

29.2 Group 1 and Group 2 Metal Derivatives

Various synthetic routes are available for the introduction of metal atoms into the cage system of **1**. The most straightforward route involves protonation of metal alkoxides, amides or organometallic compounds by the free trisilanol **1**.[1, 2, 8, 10] The reactivity of **1** can be modified by silylation of the Si-OH functions using Me_3SiCl/NEt_3 in different stoichiometric amounts.[6, 12] In this way, the silylated derivatives $Cy_7Si_7O_9(OH)_2(OSiMe_3)$ (**3**), $Cy_7Si_7O_9(OH)(OSiMe_3)_2$ (**4**), and $Cy_7Si_7O_9(OSiMe_3)_3$ (**5**) can be selectively prepared. Another possible means of modifying the reactivity of **1**, **3**, and **4** is to prepare the corresponding thallium silanolates by treatment of the free silanols with TlOEt.[13] Early attempts to fully deprotonate **1** were frustrated by a report by Feher et al., who found that treatment of **1** with strong bases such as NaO*t*Bu caused the silsesquioxane cage to collapse.[8] More recently, we[1,14] and Aspinall et al.[15] independently discovered that fully metallated derivatives of **1** are readily available by using *n*BuLi or $MN(SiMe_3)_2$ (M = Li, K) as deprotonating agents. With the use of *n*-BuLi, Aspinall et al. succeeded in preparing unsolvated $Cy_7Si_7O_9(OLi)_3$ (**6**) as an air-stable (!) white solid.[15] Crystalline materials suitable for X-ray diffraction can be prepared by treatment of **1** with $MN(SiMe_3)_2$ (M = Li, K) in the presence of coordinating solvents such as acetone, Et_2O, THF or DME. Thus far, the alkali metal silsesquioxane derivatives $[Cy_7Si_7O_9(OLi)_3]_2 \cdot 3\ Me_2CO$ (**7**), $[Cy_7Si_7O_9(OLi)_3]_2 \cdot 3\ Et_2O$ (**8**), $[Cy_7Si_7O_9(OLi)_3]_2 \cdot 3$ THF (**9**), and $[Cy_7Si_7O_9(OK)_3]_2 \cdot 4$ DME (**10**) have been fully characterized. [14] X-ray diffraction studies revealed that they consist of dimeric molecules in which two silsesquioxane cages are bridged by a box-shaped M_6O_6 polyhedron (M = Li, K). As a typical example, the molecular structure of $[Cy_7Si_7O_9(OLi)_3]_2 \cdot 3\ Me_2CO$ (**7**) [1, 14] is depicted in Figure 29.1.

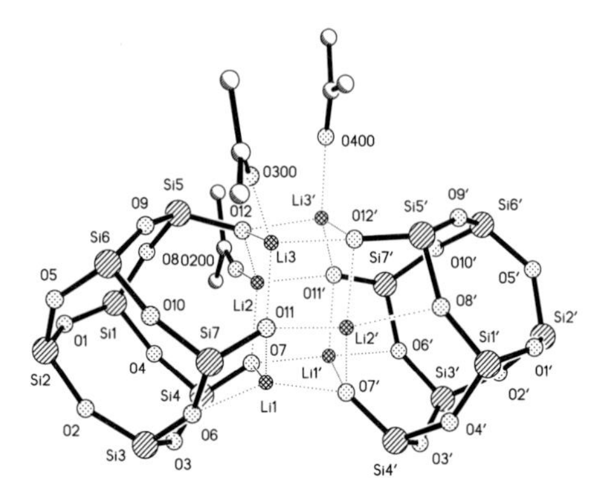

Figure 29.1. Molecular structure of $[Cy_7Si_7O_9(OLi)_3]_2 \cdot 3\ Me_2CO$ (**7**) (cyclohexyl substituents omitted for clarity)

Silsesquioxane derivatives of the alkaline earth metals are very little explored, although it was shown by Liu that the heterobimetallic complex $[Cy_7Si_7O_{12}MgTiCl_3]_n$ (**11**) displays a high catalytic activity for ethylene polymerization in the presence of triethylaluminum.[16a] An unprecedented tetranuclear Mg complex was isolated by Abbenhuis et al. from the reaction of **1** with MeMgCl.[16b] The use of **6** as starting material enabled us to achieve the synthesis and structural characterization of the first beryllasilsesquioxane.[17] Compound **6** was generated *in situ* by treatment of **1** with three equivalents of $LiN(SiMe_3)_2$ in THF. Subsequent reaction with anhydrous $BeCl_2$ according to Scheme 29.2 afforded the novel beryllasilsesquioxane $[Cy_7Si_7O_{12}BeLi]_2 \cdot 2$ THF (**12**) in 81% isolated yield. A single-crystal X-ray diffraction study established the presence of dimeric molecules of the composition $[Cy_7Si_7O_{12}BeLi]_2 \cdot 2$ THF in the solid state. Self-assembly under formation of a dimer occurs in a rather unsymmetrical manner. Two Li and two Be ions are bridged by the deprotonated silsesquioxane ligands in such a way that a different coordination environment results for each metal ion.

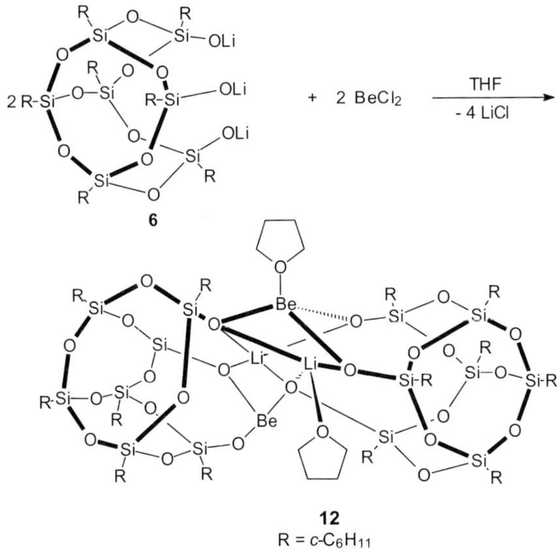

2 **6** + 2 $BeCl_2$ $\xrightarrow[\text{- 4 LiCl}]{\text{THF}}$ **12**

R = c-C_6H_{11}

Scheme 29.2. Preparation of the beryllasilsesquioxane **12**.

29.3 Early Transition Metal Derivatives: Titanium, Zirconium, and Tantalum

Apparently, the most thoroughly investigated class of complexes in this area are Ti complexes because of their promising catalytic applications.[1, 8] The first

Ti derivatives were made by Feher et al. and included Ti(III)[18,19] and Ti(IV) silsesquioxanes.[20, 21] The former were prepared by reacting **1** with either $Ti[N(SiMe_3)_2]_3$ or $TiCl_3(NMe_3)_2$. Blue dimeric $[Cy_7Si_7O_{12}Ti]_2$ (**13**) is the initial product in these reactions. Treatment of **13** with pyridine affords the bis-adduct $[Cy_7Si_7O_{12}Ti(py)]_2$ (**14**).[18, 19] Corner-capping of the silsesquioxane framework with a TiCp unit is also readily achieved yielding monomeric $Cy_7Si_7O_{12}TiCp$ (**15**).[20, 21] Other titanasilsesquioxanes have been designed as model compounds for titanosilicates, which are industrially important as oxidation catalysts.[22] Examples of such realistic models include $(MeC_5H_4)_4Ti_4(Si\textit{t}Bu)_4O_{12}$ (**16**),[23] $[2,6\text{-}\textit{i}Pr_2C_6H_3NH_3][(RSiO_3)_3Ti_4Cl_4(\mu_3\text{-}O)]$ [**17**, R = $2,6\text{-}\textit{i}Pr_2C_6H_3N(SiMe_3)$],[24] $[Cy_7Si_7O_{12}MgTiCl_3]_n$ (**11**, n = 1,2),[16a] $[(c\text{-}C_5H_9)_7Si_7O_{12}(SiMe_2R)]_2Ti$ (**18a**: R = vinyl; **18b**: R = allyl), and $[R_7Si_7O_{12}TiO\textit{i}Pr]_n$ (**19a**: R = $c\text{-}C_5H_9$; **19b**: $c\text{-}C_6H_{11}$; n = 1,2).[25] It has now become widely accepted that various titanasilsesquioxanes are veritable catalysts themselves, e.g. in olefin polymerizations and epoxidation reactions.[2] Treatment of $Cy_7Si_7O_{12}Ti(O^iPr)$ (**19b**) with MeOH affords the six-coordinate titanasilsesquioxane dimer $[Cy_7Si_7O_{12}Ti(\mu\text{-}OMe)(MeOH)]_2$ (**20**) which has been structurally characterized.[25] Related monomeric complexes have also been isolated from reactions of **1** with $Ti(CH_2Ph)_4$, $Ti(NMe_2)_4$ or $Ti(OSiMe_3)_4$ yielding $Cy_7Si_7O_{12}TiCH_2Ph$ (**21**), $Cy_7Si_7O_{12}TiNMe_2$ (**22**), and $Cy_7Si_7O_{12}TiOSiMe_3$ (**23**), respectively.[26] We started our work in this field by investigating reactions of **1** with titanium tetraalkoxides. While this work was in progress, several closely related titanasilsesquioxane alkoxides, including $[R_7Si_7O_{12}TiO\textit{i}Pr]_n$ (**19a**: R = $c\text{-}C_5H_9$; **19b**: $c\text{-}C_6H_{11}$; n = 1,2) and $[Cy_7Si_7O_{12}Ti(\mu\text{-}OMe)(MeOH)]_2$ (**20**), were published by Crocker and co-workers.[26] In our laboratory, the hitherto unknown ethoxide derivative **24** (Scheme 29.3) was prepared in an analogous manner.

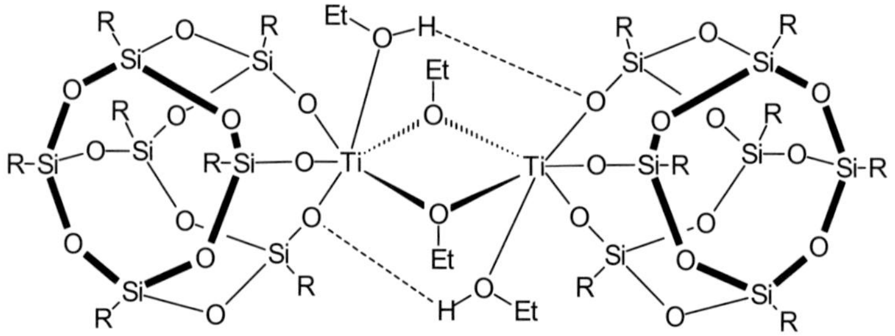

24 (R = cyclohexyl)

Scheme 29. 3. Schematic representation of $[Cy_7Si_7O_{12}Ti(\mu\text{-}OEt)(EtOH)]_2$ (**24**).

It was shown by Crocker et al. that treatment of $Cy_7Si_7O_9(OH)_2(OSiMe_3)$ (**3**) with one equivalent of $Ti(O\textit{i}Pr)_4$ produces monomeric $[Cy_7Si_7O_{11}(OSiMe_3)]Ti(O\textit{i}Pr)_2$ (**25**) as the sole product. A homoleptic bis(silsesquioxane) titanium complex, $[Cy_7Si_7O_{11}(OSiMe_3)]_2Ti$ (**26**), can be prepared by reacting **3** with tetrabenzyltitanium, $Ti(CH_2Ph)_4$.[26b] Compound **26** is also readily accessible by reacting $Ti(OEt)_4$ or $Ti(O\textit{i}Pr)_4$ with **3**. Both reactions afford white crystalline **26** in almost quantitative yield (96–97%) employing commercially available reagents. An X-ray diffraction study published by Crocker et al. revealed that the central Ti atom in **26** is tetrahedrally coordinated by the two silsesquioxane cages (Scheme 29.4).[26b]

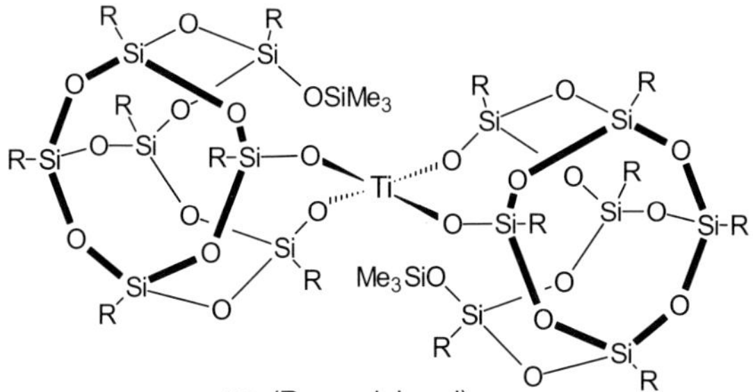

26 (R = cyclohexyl)

Scheme 29.4. Schematic representation of $[Cy_7Si_7O_{11}(OSiMe_3)]_2Ti$ (**26**).

Several titanocene derivatives containing silsesquioxane ligands have also been prepared and characterized.[27] A typical example is the reaction of $Cy_7Si_7O_9(OH)_2(OSiMe_3)$ (**3**) with Cp_2TiCl_2 in toluene solution in the presence of triethylamine. In this case, the μ-oxo dititanium complex $(\mu\text{-}O)[Cy_7Si_7O_{11}(OSiMe_3)TiCp]_2$ (**27**) was isolated in ca. 70% yield in the form of orange crystals (Scheme 29.5).

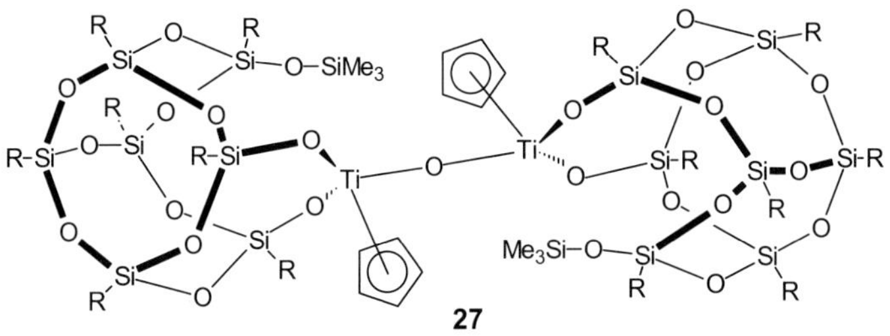

27

R = cyclohexyl

Scheme 29.5. Schematic representation of $(\mu\text{-}O)[Cy_7Si_7O_{11}(OSiMe_3)TiCp]_2$ (**27**).

When $Cy_7Si_7O_9(OH)_2(OSiMe_3)$ (**3**) was reacted with the corresponding pentamethylcyclopentadienyl complex $Cp^*_2TiCl_2$ in the presence of triethylamine,[27] it was possible to isolate and fully characterize the red crystalline trinuclear 1,3,5-trititana-2,4,6-trioxane derivative $Cp^*_2Ti_3O_3[Cy_7Si_7O_{11}(OSiMe_3)]_2$ (**28**) (Scheme 29.6).

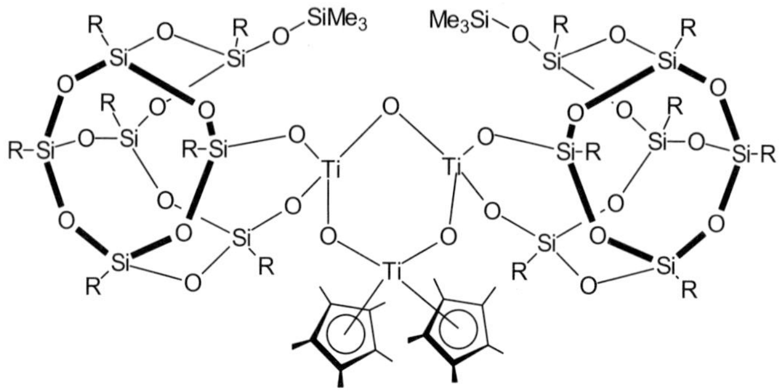

28 (R = cyclohexyl)

Scheme 29.6. Schematic representation of $Cp^*_2Ti_3O_3[Cy_7Si_7O_{11}(OSiMe_3)]_2$ (**20**).

The molecular structure of **28** has been elucidated by an X-ray structure analysis. The central structural motif of **28** is an unsymmetrically substituted six-membered Ti_3O_3 ring. Two pentamethylcyclopentadienyl ligands are coordinated to one titanium atom, while the other two are free of Cp*. They are both part of eight-membered $TiSi_3O_4$ ring systems within the silsesquioxane frameworks.

In a more straightforward manner, a bis(pentamethylcyclopentadienyl)-titanium(III) silsesquioxane complex became available according to Scheme 29.7.[27, 28] This synthesis involves addition of silsesquioxane precursors across the Ti-C bond of the "tucked-in" fulvene titanium complex $Cp^*Ti(C_5Me_4CH_2)$.[29] The main advantage of this "fulvene route" is that it is a salt-free method by which bis(pentamethylcyclopentadienyl)titanium complexes can be obtained without the need of separating any by-products. The method was first successfully employed by Teuben et al., who prepared various new Cp^*_2Ti derivatives by reacting $Cp^*Ti(C_5Me_4CH_2)$ with protic reagents such as alcohols, thiols, etc.[29b,c] It has since been found that the fulvene complex $Cp^*Ti(C_5Me_4CH_2)$ is also the reagent of choice to make novel titanium silsesquioxanes. For example, treatment of $Cp^*Ti(C_5Me_4CH_2)$ with one equivalent of the disilylated silsesquioxane precursor **4** resulted in clean formation of the Ti(III) silsesquioxane complex **29**.

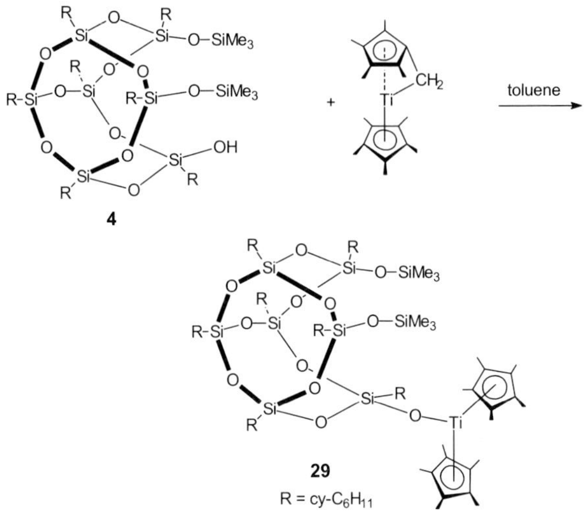

Scheme 29.7. Synthesis of **29** *via* the fulvene route.

Simple crystallization from the concentrated reaction mixture afforded **29** in the form of air-sensitive, dark-green crystals. An X-ray diffraction study of **29** revealed that a Cp^*_2Ti unit had been generated upon protonation of the coordinated tetramethylfulvene ligand. The Ti-O bond length in **29** is 192.7(2) pm. The "fulvene route" has also been successfully employed in the preparation of a compound which can be regarded as one of the most advanced molecular models for a catalytically active Ti center on a silica surface.[27,28] When $Cp^*Ti(C_5Me_4CH_2)$ was reacted with **3** in refluxing toluene, a bright-yellow material was isolated which was subsequently shown to be the novel *mono*(pentamethylcyclopentadienyl)titanium(IV) silsesquioxane complex **30** (69% yield, Scheme 29.8).

The surprising outcome of this reaction is the exclusive formation of a Cp^*Ti^{IV} complex in which two silsesquioxanes are bonded in different ways to a single Ti center. A mono(pentamethyl-cyclo-pentadienyl)titanium unit resides on a "model silica surface" formed by one chelating and one monodentate silsesquioxane ligand. A unique feature of **30,** which makes this compound a particularly "realistic" model system, is a silanol function in close proximity to the titanium center. A very weak hydrogen bonding interaction of this silanol group with a cage oxygen atom apparently prevents intramolecular protonation of the remaining Cp* ligand, thus "taming" the reactivity of the Si-OH function.

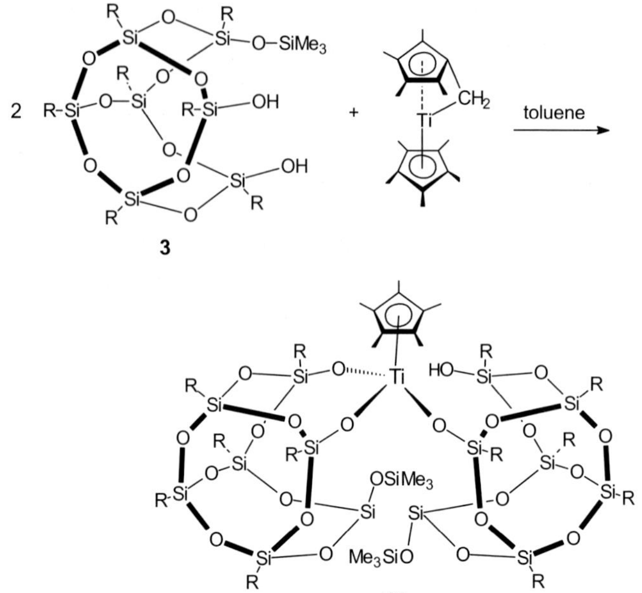

30

R = cyclohexyl

Scheme 29.8. Preparation of **30** *via* the fulvene route.

Known zirconium silsesquioxane complexes include the species $Cy_7Si_7O_{12}ZrCp^*$ (**31**),[30] $[(c\text{-}C_5H_9)_7Si_7O_{11}(OSiMe_3)]_2Zr(THF)$ (**32**),[31] and $[\{(c\text{-}C_5H_9)_7Si_7O_{12}\}Zr(CH_2Ph)]_2$ (**33**).[31] The zirconocene $[(c\text{-}C_5H_9)_7Si_7O_{11}(OSiMePh_2)]ZrCp_2$ (**34**) was obtained by treatment of Cp_2ZrMe_2 with an aluminosilsesquioxane.[32] The availability of permetallated $Cy_7Si_7O_9(OLi)_3$ (**6**) enabled us to prepare the first heterobimetallic zirconium silsesquioxane complex $(Cy_7Si_7O_{12})_2Zr[Li(O{=}CMe_2)]_2$ (**35**) according to Scheme 29.9.[14]

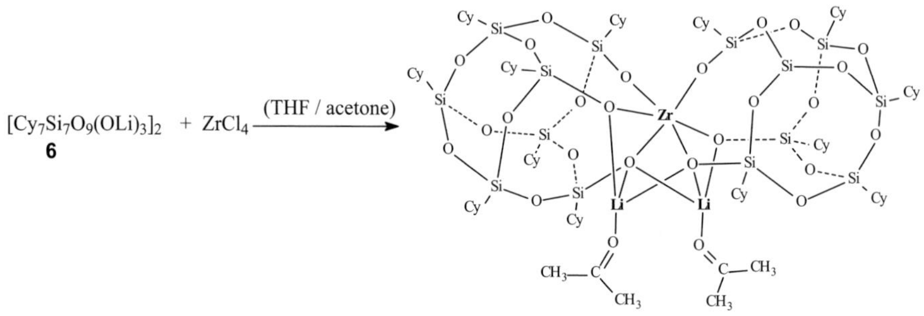

35

Scheme 29.9. Synthesis of $(Cy_7Si_7O_{12})_2Zr[Li(O{=}CMe_2)]_2$ (**35**).

Significantly less is known about metallasilsesquioxanes incorporating Group 5 metals. The vanadyl silsesquioxane $Cy_7Si_7O_{12}V{=}O$ (**36**) was reported more than ten years ago and has been shown to be a single-site catalyst for olefin polymerization.[33] During the course of our investigations, we found that various tantalum silsesquioxanes are readily accessible via a versatile amide route.[34] Protonation and liberation of 3 equivalents of dimethylamine took place upon reaction of $Ta(NMe_2)_5$ with **1** in a 1:1 molar ratio (Scheme 29.10). The "half-sandwich" complex $Cy_7Si_7O_{12}Ta(NMe_2)_2$ (**37**) was isolated in 92% yield. Similar treatment of $Ta(NMe_2)_5$ with **1** in a 1:2 molar ratio did not lead to the formation of the neutral tantalum(V) species $Cy_7Si_7O_{12}Ta[Cy_7Si_7O_{11}(OH)]$ with one Si-OH function remaining intact. Instead, the anionic bis(silsesquioxane) "sandwich" complex $[Ta(Cy_7Si_7O_{12})_2]^-$ was obtained in the form of its dimethylammonium salt **38** (89% yield) (Scheme 29.10). By adapting the amide route to more complex systems, the amido tantalacarborane species $(C_2B_9H_{11})Ta(NMe_2)_3$ was found to react cleanly with **1** to afford the novel "mixed-sandwich" complex $Cy_7Si_7O_{12}Ta(C_2B_9H_{11})$ (**41**) as an orange crystalline solid (82% yield). Compound **41** is the first representative of a novel class of inorganic cage compounds in which a carborane and a silsesquioxane framework are linked through a single metal center.

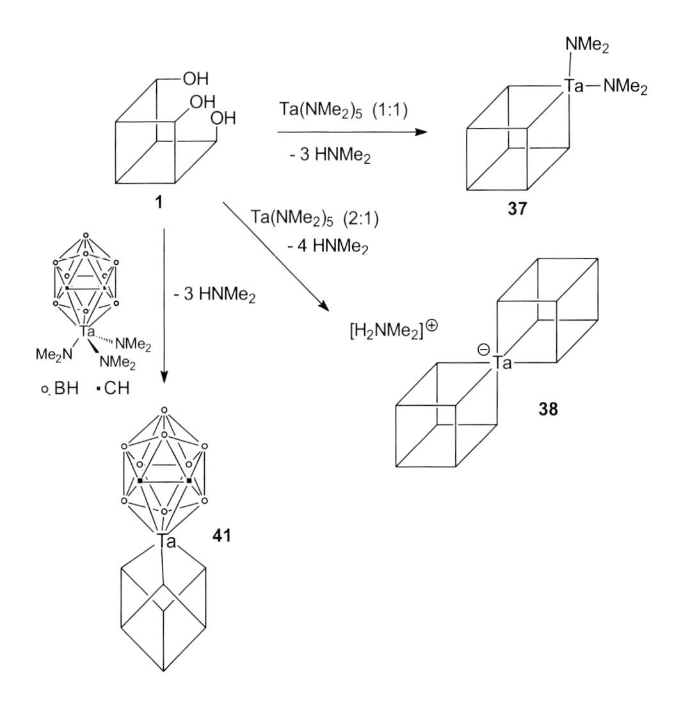

Scheme 29.10. Preparation of tantalasilsesquioxanes.

29.4 Lanthanide Derivatives: Cerium and Ytterbium

The first metallasilsesquioxanes incorporating lanthanides were described in 1994 by Herrmann et al.[35] More recently, several Ln silsesquioxanes resulting from reactions of **1** with lanthanide silylamides and aryloxides have been reported.[36] Our own efforts in this field resulted in the isolation of a novel Ce(IV) silsesquioxane complex [37] as well as the structural characterization of a bimetallic Yb/Li derivative.[14] Treatment of $[Ce\{N(SiMe_3)_2\}_3]$ or anhydrous $CeCl_3$ with two equivalents of $Cy_8Si_8O_{11}(OH)_2$ (**42**) in diethyl ether in the presence of an excess of pyridine exclusively afforded the diamagnetic complex $(Cy_8Si_8O_{13})_2Ce(py)_3$ (**43**, Scheme 29.11). Quite surprisingly, in both cases cerium was oxidized to the tetravalent oxidation state.

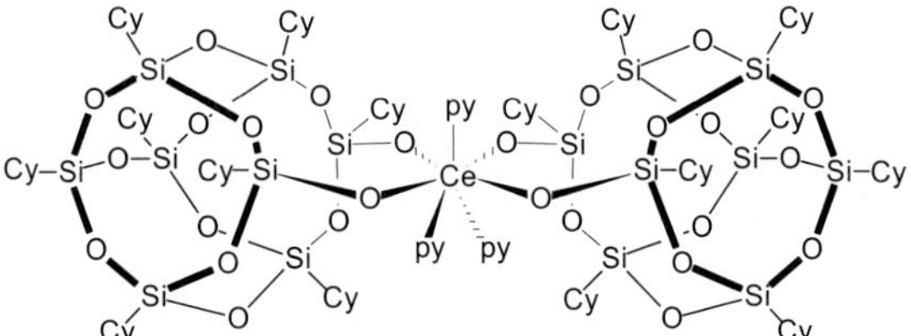

43

Scheme 29.11. Schematic representation of the cerium(IV) silsesquioxane **43**.

An unprecedented functional Yb silsesquioxane was obtained as outlined in Scheme 29.12. Trisilanol **1** was lithiated *in situ* using an excess of $LiN(SiMe_3)_2$, followed by treatment with $YbCl_3$. In bimetallic **44**, a reactive ytterbium bis(trimethylsilyl)amide unit resides on a model silica surface formed by two lithium-linked silsesquioxane cages.

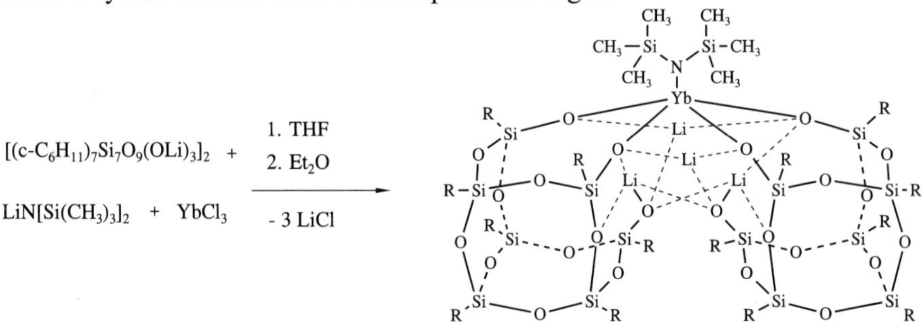

$[(c\text{-}C_6H_{11})_7Si_7O_9(OLi)_3]_2$ + $LiN[Si(CH_3)_3]_2$ + $YbCl_3$ —(1. THF; 2. Et_2O; - 3 LiCl)→ **44**

Scheme 29.12. Preparation of the amidoytterbium silsesquioxane **44**.

29.5 Acknowledgements

Special thanks are due to my co-workers, whose names appear in the list of references, and to the Otto-von-Guericke-Universität and the Fonds der Chemischen Industrie for financial support.

29.6 References

[1] V. Lorenz, A. Fischer, S. Gießmann, J. W. Gilje, Yu. K. Gun'ko, K. Jacob, F. T. Edelmann, *Coord. Chem. Rev.* **2000**, *206–207*, 321.

[2] H. C. L. Abbenhuis, *Chem. Eur. J.* **2000**, *6*, 25.

[3] P. G. Harrison, *J. Organomet. Chem.* **1997**, *542*, 141.

[4] J. D. Lichtenhan, *Comments Inorg. Chem.* **1995**, *17*, 115.

[5] M. G. Voronkov, V. L. Lavrentyev, *Top. Curr. Chem.* **1982**, *102*, 199.

[6] F. J. Feher, D. A. Newman, J. F. Walzer, *J. Am. Chem. Soc.* **1989**, *111*, 1741.

[7] F. J. Feher, T. A. Budzichowski, R. L. Blanski, K. J. Keller, J. W. Ziller, *Organometallics* **1991**, *10*, 2526.

[8] F. J. Feher, T. A. Budzichowski, *Polyhedron* **1995**, *14*, 3239.

[9] P. G. Harrison, *J. Organomet.Chem.* **1997**, *542*, 141.

[10] R. Murugavel, A. Voigt, M. G. Walawalkar, H. W. Roesky, *Chem. Rev.* **1996**, *96*, 2205.

[11] T. W. Hambley, T. Maschmeyer, A. F. Masters, *Appl. Organomet. Chem.* **1992**, *6*, 253.

[12] F. J. Feher, D. A. Newman, *J. Am. Chem. Soc.* **1990**, *112*, 1931.

[13] F. J. Feher, K. Rahimian, T. A. Budzichowski, J. W. Ziller, *Organometallics* **1995**, *14*, 3920.

[14] V. Lorenz, A. Fischer, F. T. Edelmann, *manuscript in preparation.*

[15] J. Annand, H. C. Aspinall, A. Steiner, *Inorg. Chem.* **1999**, *38*, 3941.

[16] a) J.-C. Liu, *Chem. Commun.* **1996**, 1109; b) R. W. J. M. Hanssen, A. Meetsma, R. A. van Santen, H. C. L. Abbenhuis, *Inorg. Chem.* **2001**, *40*, 4049.

[17] V. Lorenz, A. Fischer, F. T. Edelmann, *Inorg. Chem. Commun.* **2000**, *3*, 292.

[18] F. J. Feher, S. L. Gonzales, J. W. Ziller, *Inorg. Chem.* **1988**, *27*, 3440.

[19] F. J. Feher, J. F. Walzer, *Inorg. Chem.* **1990**, *29*, 1604.

[20] F. J. Feher, T. A. Budzichowski, K. Rahimian, J. W. Ziller, *J. Am. Chem. Soc.* **1992**, *114*, 3859.

[21] L. D. Field, C. M. Lindall, T. Maschmeyer, A. F. Masters, *Aust. J. Chem.* **1994**, *47*, 1127.

[22] R. Murugavel, H. W. Roesky, *Angew. Chem.* **1997**, *109*, 491; *Angew. Chem., Int. Ed. Engl.* **1997**, *36*, 476.

[23] N. Winkhofer, A. Voigt, H. Dorn, H. W. Roesky, A. Steiner, D. Stalke, A. Reller, *Angew. Chem.* **1994**, *106*, 1414; *Angew. Chem. Int. Ed. Engl.* **1994**, *33*, 1352.

[24] A. Voigt, R. Murugavel, M. L. Montero, H. Wessel, F.-Q. Liu, H. W. Roesky, I. Usón, T. Albers, E. Parisini, *Angew. Chem.* **1997**, *109*, 1020; *Angew. Chem. Int. Ed. Engl.* **1997**, *36*, 1001.

[25] (a) T. Maschmeyer, M. C. Klunduk, C. M. Martin, D. S. Shephard, J. M. Thomas, B. F. G. Johnson, *Chem. Commun.* **1997**, 1847; (b) K. Wada, K. Yamada, D. Izuhara, T. Kondo, T. Mitsudo, *Chem. Lett.* **2000**, 1332.

[26] (a) M. Crocker, R. H. M. Herold, A. G. Orpen, *Chem. Commun.* **1997**, 2411; (b) M. Crocker, R. H. M. Herold, A. G. Orpen, M. T. A. Overgaag, *J. Chem. Soc., Dalton Trans.* **1999**, 3791.

[27] F. T. Edelmann, S. Gießmann, A. Fischer, *J. Organomet. Chem.* **2001**, *620*, 80.

[28] S. Gießmann, A. Fischer, F. T. Edelmann, *Chem. Commun.* **2000**, 2153.

[29] (a) J. E. Bercaw, *J. Am. Chem. Soc.* **1974**, *96*, 5087; (b) J. W. Pattiasina, C. E. Hissink, J. L. de Boer, A. Meetsma, J. H. Teuben, *J. Am. Chem. Soc.* **1985**, *107*, 7785; (c) J. W. Pattiasina, *Ph. D. Thesis*, Rijksuniversiteit Groningen, **1988**.

[30] F. J. Feher, *J. Am. Chem. Soc.* **1986**, *108*, 3850.

[31] R. Duchateau, H. C. L. Abbenhuis, R. A. van Santen, A. Meetsma, S. K.-H. Thiele, M. F. H. van Tol, *Organometallics* **1998**, *17*, 5663.

[32] M. D. Skowronska-Ptasinska, R. Duchateau, R. van Santen, G. P. A. Yap, *Organometallics* **2001**, *20*, 3519.

[33] (a) F. J. Feher, J. F. Walzer, *Inorg. Chem.* **1991**, *30*, 1689; (b) F. J. Feher, J. F. Walzer, R. L. Blanski, *J. Am. Chem. Soc.* **1991**, *113*, 3618.

[34] Z. Fei, S. Busse, F. T. Edelmann, *Chem. Commun.* **2002**, 2587.

[35] W. A. Herrmann, R. Anwander, V. Dufaud, W. Scherer, *Angew. Chem.* **1994**, *106*, 1338; *Angew. Chem. Int. Ed. Engl.* **1994**, *33*, 1285.

[36] (a) J. Annand, H. C. Aspinall, *J. Chem. Soc., Dalton Trans.* **2000**, 1867; (b) J. Annand, H. C. Aspinall, A. Steiner, *Inorg. Chem.* **1999**, *38*, 3941.

[37] Yu. K. Gun'ko, R. Reilly, F. T. Edelmann, D. Stalke, *Angew. Chem.* **2001**, *113*, 1319; *Angew. Chem. Int. Ed.* **2001**, *40*, 1279.

30 Spin-Spin Interactions in Silsesquioxanes and Transition Metal Substitution

Wolfgang W. Schoeller*, Dirk Eisner

30.1 Introduction

Silsesquioxanes of the general formula $(RSiO_{3/2})_n$, n = 4, 6, 8 .... etc., are probably some of the best studied members of the large class of Si-O compounds. A variety of synthetic procedures has been reported during the last years, as summarized lucidly in a recent article.[1] For n = 8, they possess a cube-like structure, as shown in **1**, in which each silicon atom is surrounded by three oxygen atoms.

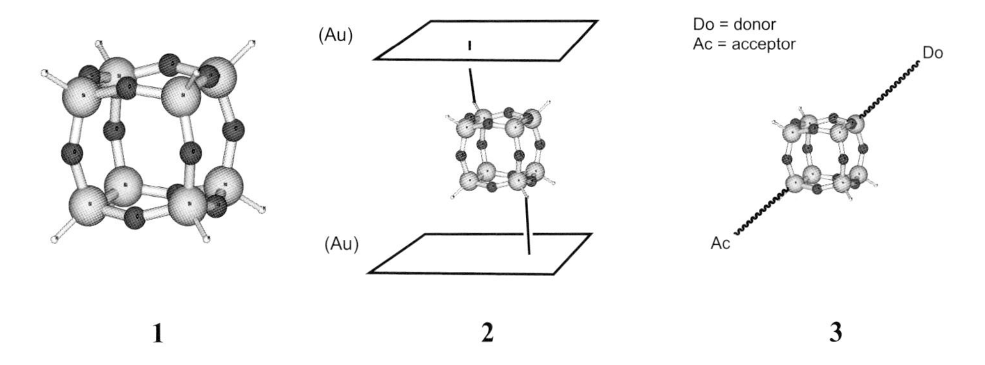

1 **2** **3**

The silsesquioxane $H_8Si_8O_{12}$ is an ideal compound for further experimental investigations. Besides the fact that its crystal structure has been analyzed in much detail,[2, 3] its derivatization by transition metal fragments has been reported.[4, 5] In addition, it reveals interesting material properties, e.g. as paints between metal surfaces[6], as schematically indicated in **2**. A current topic of debate is whether or not silsesquioxanes can mediate electron transfer, such as across the donor acceptor linkage **3**. The experimental investigations indicate that the silsesquioxanes mainly behave as isolators and that they do not allow considerable through-space and/or through-bond interactions through the cage.

* Address of the authors: Fakultät für Chemie der Universität Bielefeld, Postfach 10 01 31, 33503 Bielefeld, Germany, Fax: 0521 106 6467. E-mail: wolfgang.schoeller@uni-bielefeld.de

30.2 Results and Discussion

This facet of the Si-O linked system is somewhat surprising since it is known that polysilanes themselves reveal low band gaps.[7] In this respect, it is of great interest to investigate the bonding properties of the corresponding radical and biradical structures which emerge from the silsesquioxanes, when the ligands at the silicon centers are omitted. The biradicals can be viewed schematically as follows: the radicals can either reside in neighboring positions, as in the SiO structure, or in distal positions, as in the 1.3- and 1.4-centered biradicals.

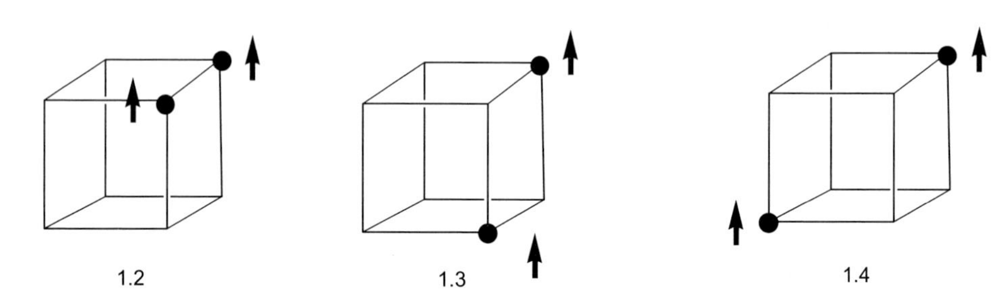

A biradical is usually defined as a chemical structure in which the electrons cannot be properly distributed by two-electron two-center bonds, i.e. they are defined as *non-Kekulé* compounds. Two of the archetypal biradicals are trimethylene **4**,[8] and the cyclobutanediyl-1.3 **5**.[9] The former is an intermediate in cyclopropane isomerization and the latter the transition state for bicyclobutane inversion. These structures are biradicals in the context of organic chemistry since they possess fairly small singlet-triplet (S-T) energy separations, of the order of a few kcal/mol.[10] Consequently, the methylene groups between the two radical positions function as “chemical isolators”, although there are some through-space and/or through-bond interactions.[11]

There is an essential difference with regard to the analogous silicon compounds. It is known that polysilylenes tend to fragment into silylene units.[12] As a consequence of *orbital non-hybridization*[13] silylene possesses a singlet rather than a triplet ground state, as known for methylene.[14]

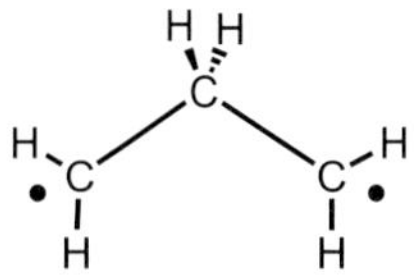

4 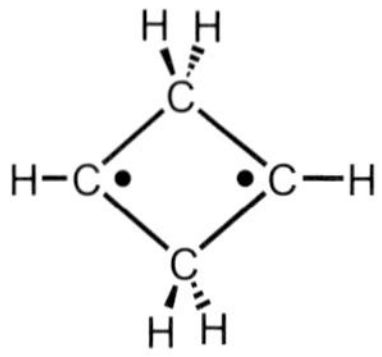 **5**

This gives rise to the view of trisilylene, which is the simplest representative of a polysilylene chain, as a donor-acceptor coupled ensemble of three singlet silylenes, **6b**, rather than as a biradical, **6a**.

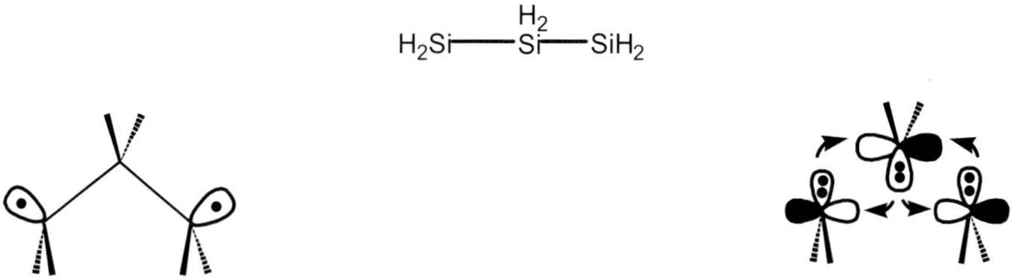

H_2Si—SiH_2—SiH_2

a **b**

6

This is supported by quantum chemical calculations, as well as by chemical experiments, which reveal for the tetrasilabicyclobutane structure a rather long (2.412 Å) Si-Si central bond.[15] Intermediate cases with a strong preference for a singlet ground state are exemplified by the boron systems **7**, [16] in which a singlet fragment (PR_2^+) is flanked by two radical positions (BR′).

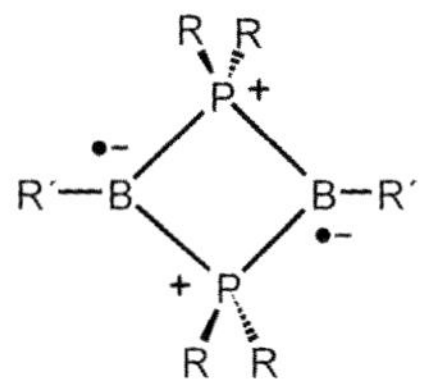

R, R′ = alkyl, aryl
7

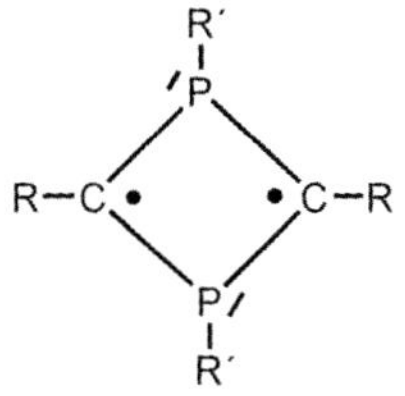

R, R′ = alkyl, aryl
8

In **8**, the intervening units are PR′ fragments which act as donors.[17] These findings[16, 17] extend our understanding of biradicals. Such systems range from species with small S-T energy separations (e.g. trimethylene), in which the two radical positions are coupled through a triplet fragment, to biradical structures with larger S-T separations, in which the radicals are coupled through singlet fragments or (weak) donors, such as the phosphino group. In this regard, the intervening units are preferentially a) an acceptor (SiH_2, PH_2^+) or b) a donor (PR, NR) unit. For the former, the resulting highest occupied molecular orbital (HOMO) is symmetric, i.e. it results in *transannular π-bonding,* while for the latter it is antisymmetric, thus leading to *transannular π-antibonding*. On this basis, the former structures can easily accommodate ring-closure to the corresponding bicyclobutane structures. while for the latter this process is symmetry forbidden on the basis of the Woodward-Hoffmann rules. This is consistent with the hitherto reported experimental investigations. We have termed the former biradical structures as *type I* and the latter as *type II* biradicals.[18]

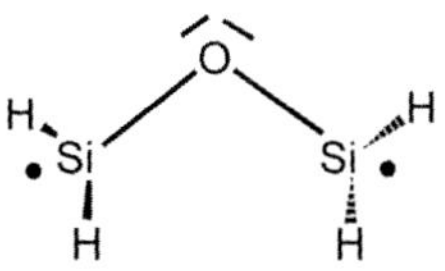

9

In the silsesquioxanes, it is oxygen that couples the two radical functions, as shown schematically in the model compound **9**. Quantum chemical calculations on **4** and **9** substantiate the qualitative assertions (Table 30.1). The singlet state of the C_2 conformation is substantially lower in energy (117.7 kJ/mol) than the planar structure (C_{2v} geometry), in agreement with the known strong pyramidalization forces in silyl radicals.

Table 30.1. MR-MP2 (MCSCF) calculations on **4** and **9**, S-T energy separations in kJ/mol.

Structure	symmetry	-E (S-T)[a]	c (HOMO)[b]	c (LUMO)[b]
4	C_{2v}	22.2	0.738	–0.675
9	C_{2v}	36.9	0.838	–0.545
	C_2	28.1	0.837	–0.547

a) CAS(2,2)/6-31g(d) optimized, plus MR-MP2 for dynamical electron correlation correction.
b) Expansion coefficients of the MCSCF wavefunction.

The trisilylene is a closed-shell structure, which emphasizes the nature of three silylenes coupled by mutual donor-acceptor interaction. Most notably, the biradical structure **9**, the motif for bonding in the silsesquioxanes, reveals a larger S-T separation than trimethylene. Pyramidalization at the silicon atoms (C_2 symmetry) reduces the S-T separation. In other words, the coupling of the two unpaired electrons depends strongly on the *orientation* of the nonbonding orbitals at the terminal silicon centers.

These aspects are also indicated for the various $H_6Si_8O_{12}$ biradicals (Table 30.2). For the calculations, the octahedral structure of the cube was imposed and the corresponding hydrogen atoms were removed. They thus refer to a "vertical" formation of the corresponding biradical structures from the cube.

Table 30.2. S-T energy separations (negative, in kJ/m), at a) TCSCF/ECP-31g(d) (ECP from Stevens, Basch, and Krauss) and b) UDFT (B3LYP/6-31g(d)) level on the various biradicals in silsesquioxanes.

Structure	TCSCF[a]	UDFT[b]
1.2-Biradical	3.0	6.7
1.3-Biradical	0.6	2.5
1.4-Biradical	0.1	2.5

a) Two-configuration SCF, b) Values after spin-contamination correction in the energy lowest singlet.

The $<S^2>$ values (≈ 1) reflect the biradical character in the silsesquioxanes (the lowest singlet and triplet energies strongly mixed in the unrestricted wavefunction), in agreement with the fairly small S-T energy separations. In fact, the S-T separations are smaller than in trimethylene (see Table 30.1). This indicates that these species, exemplified here by the various cube structures, behave as "electronic isolators", as witnessed in the experimental investigations.[6] At least this is strongly dictated for the 1.4-biradical structure-type.

There are further data which confirm these findings. The calculated UV spectra (TD/6-31g(d)) predict absorptions for the three lowest-energy UV transitions at 110.6 ($f = 0.0$), 106.3 ($f = 0.0$), 106.3 ($f = 0.0$) nm, in agreement with the colorless nature of these compounds. Furthermore, the structure of the silsesquioxane is well described quantum chemically even at the RHF level of sophistication (Table 30.3).

Table 30.3. Bonding parameters for $H_8Si_8O_{12}$ **1** (bond lengths in Å, bond angles in degrees).

Method	Si-O	< Si-O-Si	Si-H	< HSi-O
RHF/6-31g(d)	1.626	149.4	1.452	110.0
B3LYP/6-31g(d)	1.643	148.6	1.464	109.3
B3LYP/6-311g(d)	1.624	148.6	1.456	109.7
Experiment	1.62	148.4	1.48	109.5

On this basis, the theoretically calculated vibrations from the quantum chemical treatment are fairly well predicted at the RHF level of sophistication (Figure 30.1).

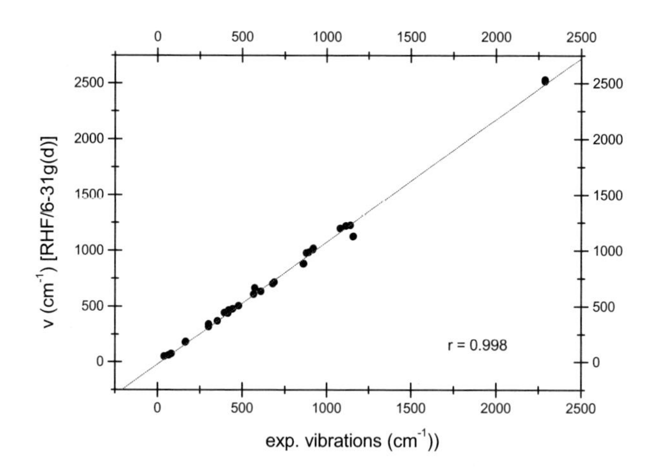

Figure 30.1. Experimental versus calculated (RHF/6-31g(d)) vibrations.

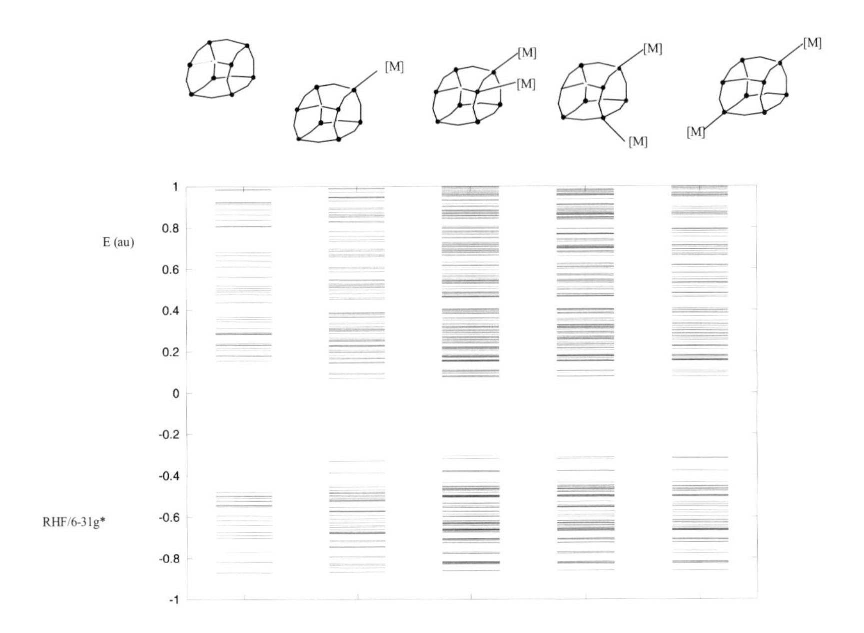

Figure 30.2. Plots of molecular orbitals of the parent cube and TM-substituted derivatives (RHF/6-31g(d)), [M] = $Co(CO)_4$.

They agree perfectly with the skillful analysis presented by Calzaferri et al.[19] The analysis of the molecular orbitals in the cube reveals the highest occupied molecular orbital to be a_{2g} (within O_h symmetry) and constituted from pure p orbitals at the oxygen atoms. In agreement with the previous considerations, the energy splitting between occupied and unoccupied molecular orbitals is considerable (Figure 30.2) and is not significantly affected by transition metal substitution ([M] = $Co(CO)_4$).

The structural parameters for the transition metal substitution, shown here for the TM fragment $Co(CO)_4$, are significantly affected by the electronegativity of the substituent attached to the silicon center. This is reproduced by quantum chemical calculations on various substituted model compounds of the type $R_3SiCo(CO)_4$ (Table 30.4).

With increasing electronegativity of the substituent R, the transition metal is more strongly bound to the silicon center. Concomitantly, the positive charge at the silicon is increased. The TM substitution has no significant effect on the HOMO-LUMO energy separation in the cube (see Figure 30.2). On this basis, the biradical character should be maintained, even if the TM fragments are oxidized (not presented in detail here).

Table 30.4. Transition metal substitution at an SiR_3 center (B3LYP/ECP-31g(d), ECP from Stevens, Basch, and Krauss), bond lengths in Å.

R	Si-Co	$Co\text{-}(CO)_{ax}$	BO^a (Si-Co)	q_{Co}^b	q_{Si}^b
H	2.380	1.814	0.785	–1.578	0.306
F	2.277	1.825	0.884	–1.551	1.118
OH	2.339	1.819	0.818	–1.552	0.931
SH	2.337	1.819	0.808	–1.558	0.345
SeH	2.359	1.819	0.831	–1.583	0.456
TeH	2.392	1.817	0.760	–1.595	0.321

a) Löwdin bond orders; [b]partial charges according to Löwdin.

Interestingly, the quantum chemical calculations on the TM-substituted silsesquioxanes, and on the various investigated substituted silyl model compounds indicate that the Si-[M] vibrations are easily identified from the other vibrations of the compounds. This is shown for the silyl-[M] structure (Figure 30.3).

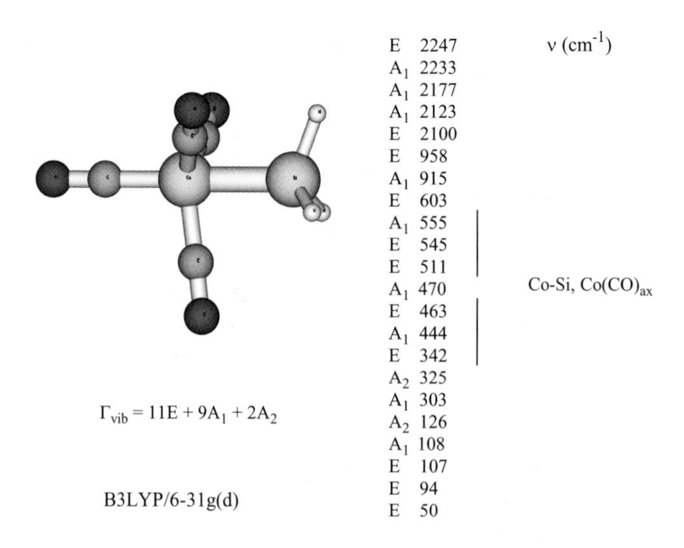

Figure 30.3. Calculated (harmonic) vibrations for H_3Si-[M], [M] = $Co(CO)_4$ (B3LYP/6-31g(d)).

The symmetric stretching vibrations (a_1 within C_{3v}) are strongly coupled with the CO ligand vibrations. The lowest-energy vibrations relate to wagging motions of CO ligands. Overall, the whole variety of vibrations relate to the representations $\Gamma = 11E + 9\,A_1 + 2\,A_2$.

A similar shortening of the Si-[M] bond is also obtained for the transition metal fragments which contain the cyclopentadienyl ligand (Figure 30.4). There, detailed relevant experimental data are provided in the work of Malisch et al.[20] Further population analysis of the substituted silanes shows that the polarization at the silicon as a function of the distance of the silicon center attached to the transition metal is fairly small. In other words, a change is only obtained in the charge for the neighboring silicon atom.

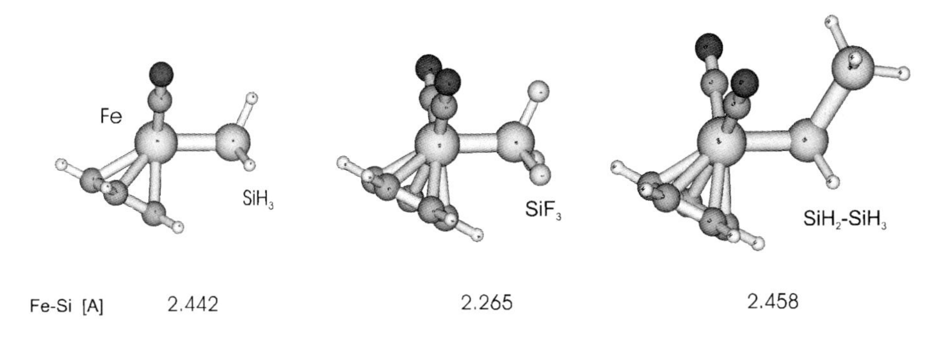

Figure 30.4. $Fe(cp)(CO)_2$ ligand substitution at silicon.

The transition metal substitution does not import a further strengthening of the spin-spin interactions in the silsesquioxanes. This is shown here for the dicoordinate structure (Figure 30.5).

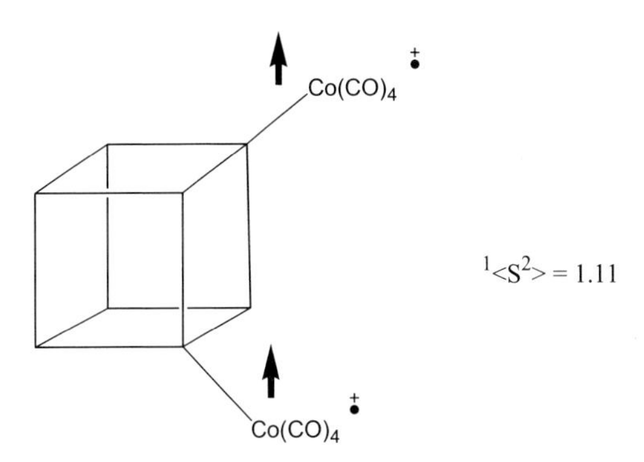

Figure 30.5. $\langle S^2 \rangle$ value for the dicoordinated cube, $[M] = Co(CO)_4$.

The <S^2> value is consistant with the previous considerations on the 1.2- to 1.4-biradical structures, with fairly small S-T energy separations. It indicates that in the TM-substituted silsesquioxanes the transition metal fragment can be readily oxidized at the metal center, but a delocalization over the cube does not take place. More generally, spin-delocalization is interrupted by the electronegative atoms linking the silicon centers. A similar phenomenon is observed for the spin-spin interactions in the resulting cubes with phosphaneiminato ligands.[21]

According to our calculations, the 1.4-biradical and to a lesser extent the 1.3-biradical structure, seem to be fairly good "isolators" for electronic communication for the unpaired spins, while our investigations also indicate a weak electronic coupling for the 1.2-type biradical.

30.3 References

[1] G. Calzaferri, in *Tailor-made Silicon-Oxygen Compounds, From Molecules to Materials*, R. Corriu and P. Jutzi, (Eds.), Vieweg Braunschweig 1995, pp. 149.

[2] T. P. Auf der Heyde, H.-B. Bürgi, H. Bürgy, K. W. Törnroos, *Chimia* **1991**, *45*, 38.

[3] K. W. Törnroos, *Acta Cryst.* **1994**, *C50*, 1646.

[4] G. Calzaferri, R. Imhof, K.W. Törnroos, *J. Chem. Soc., Dalton Trans.* **1993**, 3741.

[5] M. Rattay, D. Fenske, P. Jutzi, *Organometallics* **1998**, *17*, 2930.

[6] K. Nicholson, K. Zhang, M.M. Banaszak Holl, F. R. McFeely, G. Calzaferri, U. C. Pernisz, *Langmuir* **2001**, *17*, 7879.

[7] T. Gahimer, W. J. Welsh, *Polymer* **1996**, *37*, 1815.

[8] W. T. Borden, *J. Chem. Soc., Chem. Commun.* **1998**, 1919; and references cited therein.

[9] a) K. A. Nguyen, M.S. Gordon, J.A. Boatz, *J. Am. Chem. Soc.* **1994**, *116*, 9241.
b) K. A. Nguyen, M.S. Gordon, *J. Am. Chem. Soc.* **1995**, *117*, 3835.

[10] W.T. Borden, *Diradicals*, Wiley, New York 1982.

[11] a) R. Hoffmann, *J. Am. Chem. Soc.* **1969**, *90*, 175. b) R. Hoffmann, *Acc. Chem. Res.* **1971**, *4*, 1.

[12] H. Sakurai, M. Yoshida, K. Sakamoto, *J. Organomet. Chem.* **1996**, *521*, 287; and further references therein.

[13] (a) W. W. Schoeller, T. Dabisch, *Inorg. Chem.* **1987**, *26*, 1081.
(b) W. W. Schoeller, T. Dabisch, T. Busch, *Inorg. Chem.* **1987**, *26*, 4383.

[14] M. Driess, H. Grützmacher, *Angew. Chem.* **1996**, *108*, 900; *Angew. Chem. Int. Ed. Engl.* **1996**, *35*, 829; and further references therein.
[15] T. Iwamoto, D. Yin, C. Kabuto, M. Kira, *J. Am. Chem. Soc.* **2001**, *123*, 12730.
[16] D. Scheschkewitz, H. Amii, H. Gornitzka, W. W. Schoeller, D. Bourissou, G. Bertrand, *Science* **2002**, 1880.
[17] W. W. Schoeller, C. Begemann, E. Niecke, D. Gudat, *J. Phys. Chem. A* **2001**, *105*, 10731.
[18] W. W. Schoeller, A. B. Rozhenko, D. Bourissou, G. Bertrand, manuscript in preparation.
[19] C. Marcolli, P. Lainé, R. Bühler, G. Calzaferri, J. Tomkinson, *J. Phys. Chem. B* **1997**, *101*, 1171.
[20] a) W. Malisch, M. Hofmann, M. Nieger, W. W. Schoeller, A. Sundermann, *Eur. J. Inorg. Chem.,* in press.
b) W. Malisch, M. Hofmann, W. W. Schoeller, D. Eisner, manuscript in preparation.
[21] A. Sundermann, W. W. Schoeller, *J. Am. Chem. Soc.* **2000**, *122*, 4729.

31 Characterization of Silicon-Containing Polymers by Coupling of HPLC-Separation Methods with MALDI-TOF Mass Spectrometry

Jana Falkenhagen, Ralph-Peter Krüger*, Günter Schulz

31.1 Introduction

The worldwide investigation of new silicon-containing polymers as innovative high-tech materials requires the development of modern analytical techniques. The methods should allow the characterization of the starting compounds as well as a comprehensive analysis of the main and by-products with regard to their molar masses and molar mass distribution. Precise information on the distribution of the chemical constituents and functional groups is also needed in order to understand polymer formation processes, degradation behavior, and structure-property relationships.

The first step in the characterization of polymers is to fractionate the unknown sample, and then to determine the structures of the separated fractions. Separation is best performed by modern liquid chromatographic methods. Depending on the kind of heterogeneity it is necessary to select the most suitable chromatographic method, i.e. either Size-Exclusion Chromatography (SEC) or Liquid Adsorption Chromatography (LAC). Furthermore, the meanwhile well-established Liquid Adsorption Chromatography at Critical Conditions (LACCC) is also used. A separation system that operates near critical conditions sometimes has to be applied.

Whereas SEC, well-known for the determination of molar mass distribution, has been used since 1959,[1] LACCC was first mentioned by Belenkii et al. in 1976 [2] and then by Entelis et al. in 1986.[3] The theory and experimental details for the determination of critical conditions have been described by Pasch.[4] The separation of polymers under critical conditions of adsorption allows, for example, the elution of homopolymers with the same repeat unit irrespective of their molar mass, according to the number and nature of functional groups.

To operate in these HPLC modes, more and more highly sensitive methods for the detection and identification of separated species are required. Matrix-Assisted Laser Desorption/Ionization Time-Of-Flight Mass Spectrometry

---

* Address of the authors: Federal Institute for Materials Research and Testing (BAM), Unter den Eichen 87, D-12205 Berlin, Germany.

(MALDI-TOF-MS) has the potential to identify the fractions separated by SEC, LACCC or LAC. This method was developed by Karas et al.[5] and by Tanaka et al.[6] in 1988. The method was first applied for investigations of silicon-containing polymers in 1995, for example by Lorenz [7] and by Tang.[8] MALDI-TOF-MS represents a powerful tool not only for the determination of chemical structure but also for SEC calibration. SEC calibration for specific polymers is still problematic, since there are no commercially available standards. Therefore a calibration by the coupling of different modes of liquid chromatography with MALDI-TOF-MS is preferred.

In this article, some results concerning the characterization of siloxanes and silsesquioxanes are reported. It should be emphasized that only the described combination of separation and identification methods is capable of determining the polymer heterogeneity in a comprehensive manner.

31.2 Methods

31.2.1 HPLC

The separation of macromolecules in ideal size-exclusion chromatography is determined by an entropy-controlled process, i.e., by a change of conformation without interaction with the surface of the separation material. It depends on the hydrodynamic volume of the macromolecule. Large molecules elute earlier, because they cannot enter small pores. In contrast, in ideal adsorption chromatography, all repeat units can interact with the surface, resulting in an enthalpy change. Long molecules elute later. As can be seen in Figure 31.1, the separation efficiency in different molecular weight ranges is quite variable in the different LC modes (ΔV_1, ΔV_2). Depending on the column material, on the thermodynamic quality of the solvent mixture and on the temperature, one of the two processes will dominate. At a certain solvent/non-solvent composition at a given temperature, critical conditions are reached where the enthalpic and entropic contributions compensate each other and all molecules with the same chemical composition (repeat units, end groups) but different molar masses will elute at the same elution volume. Experimentally, the critical solvent composition is determined with polymers of the same type and different molar masses. The elution behavior in the three different modes is schematically demonstrated in Figure 31.1.

The LACCC and LAC experiments were conducted on a Hewlett-Packard (HP1090) HPLC system using a diode array UV Detector and an evaporative light-scattering detector. SEC measurements were performed on a modular system consisting of a Rheos 4000 pump and UV and refractive index detectors.

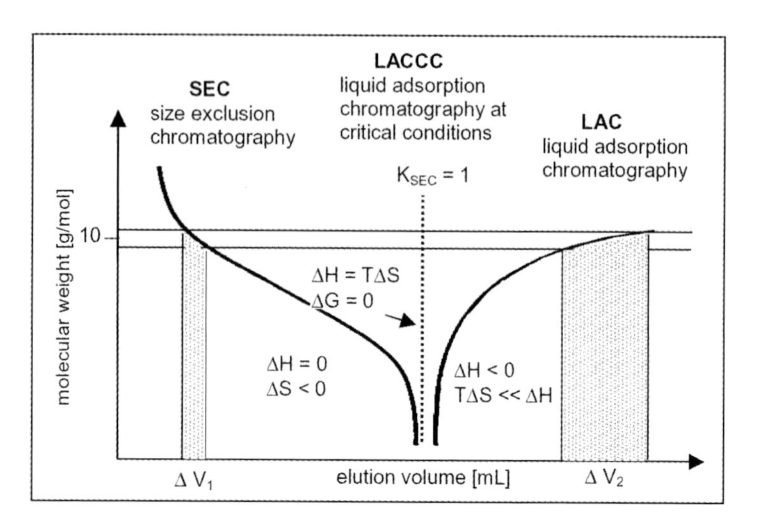

Figure 31.1. Elution behavior of macromolecules in the different modes of HPLC of polymers.

31.2.2 MALDI-TOF Mass Spectrometry

In the last few years MALDI-TOF-MS has become established as a powerful technique for the determination of the masses of large biomolecules and synthetic polymers. The technique involves mixing an appropriate matrix with the dissolved sample. The matrix preferentially strongly absorbs at the wavelength of the laser. The sample/matrix mixture is placed on a target, and is then subjected to laser desorption and ionization. The individual ions are accelerated to a fixed energy in an electrostatic field and directed into a field-free flight tube. The ions then impact onto an ion detector, and the time intervals between the pulse of the laser light and impact on the detector are measured (TOF).

By MALDI-MS, molecular weight and molecular weight distribution information can be obtained for polymers of narrow polydispersity. Monomer and end group masses can be deduced from the accurate measurement of the mass of individual oligomers. The ability to ionize a broad range of synthetic polymers, fast sample preparation, high sensitivity, broad mass range, and the absence of fragmentation are characteristics of the technique.

MALDI-TOF mass spectrometry was carried out on Kratos Kompact MALDI III and Bruker Reflex III spectrometers. 0.5 μL of a solution of the sample (1 mg/mL in THF) and 0.5 μL of the matrix solution (20 mg/mL of 2-nitrophenyl octyl ether and 10 mg/mL silver trifluoroacetate) were mixed on a stainless steel sample slide. The solvent was evaporated in a stream of air at ambient temperature. Bovine insulin or polyethylene oxide was used for calibration. Conditions for the measurements: polarity positive, flight path reflection, 20 kV acceleration voltage, nitrogen laser (λ = 337 nm). For the creation of 3D MALDI spectra, the sprayed target was analyzed continuously. The result is a three-dimensional plot giving the variation of peak intensity with shots and masses. Shot 1 corresponds to the start and shot 300 to the end of the sprayed trace, respectively.

31.2.3 Coupling of HPLC and MALDI-TOF-MS

A commercially available interface LC 500 (LabConnections) was used for semi-on-line coupling of liquid chromatography and mass spectrometry (Figure 31.2). The matrix was added continuously through a mixing T-fitting to the eluent after the separation column outlet by means of a secondary pump. To prevent damage of the pump, the matrix solution was filled in a loop (> 1 mL) which was placed between the pump and T-fitting. The eluate, containing the polymer, the solvent, and the matrix, was sprayed onto the target, and the solvent was simultaneously evaporated in a nitrogen gas stream at 120 °C. This allows an optimal embedding of the sample within the matrix and the use of a great variety of matrices.

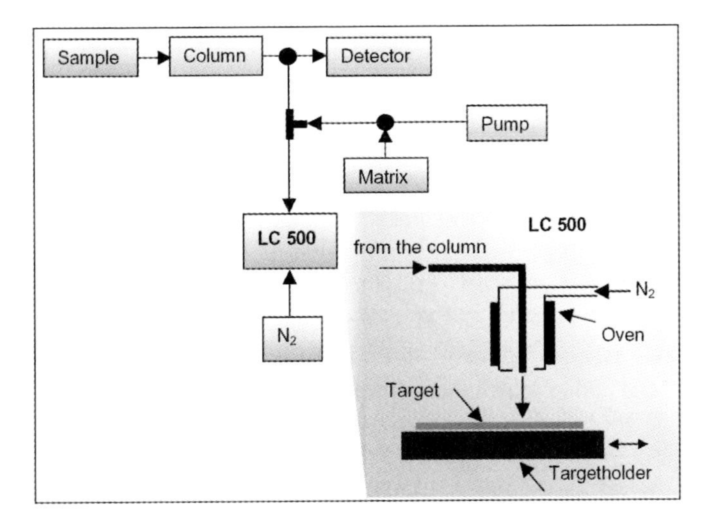

Figure 31.2. Semi-on-line coupling of HPLC and MALDI-TOF-MS.

31.3 Results

31.3.1 LAC / MALDI-TOF-MS Coupling

The liquid chromatogram of a polydimethylsiloxane (PDMS) sample with OH end groups is shown in Figure 31.3a. The separation was achieved by using a Nucleosil silica column (pore size 120 Å, 200 × 2.1 mm ID). The mobile phase was *n*-hexane/ethyl acetate, the flow rate 0.4 mL/min.[9]

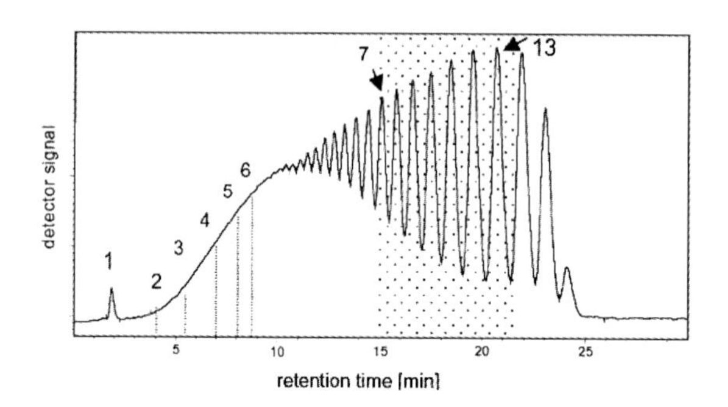

Figure 31.3a. LAC of PDMS with OH end groups.

According to the elution order it appears to be a chromatogram in SEC mode, but it is actually a case of adsorption chromatography. The separation takes place according to the intensity of interaction of OH end groups with the silica gel column material. The intensity of interaction depends on the number of OH groups in proportion to the molecular weight. The content of OH groups is higher in lower molecular mass species, which results in longer retention times.

Peak 1 in Figure 31.3a has no OH end groups, but only CH_3 end groups, as verified by MALDI-TOF-MS. A determination of the molecular weight distribution of this sample is in principle possible by the application of SEC software. For this purpose a calibration of the system is necessary. This was achieved by coupling with MALDI-TOF-MS. The shaded part of the chromatogram in Figure 31.3a (peaks 7–13) was deposited on the target for MALDI-MS. The associated monodisperse MALDI-TOF mass spectra (Figure 31.3b, left) demonstrate the exact chromatographic separation – there is only a single MALDI peak for each chromatographic fraction. Fractions taken from the higher molecular weight range (positions 2–6) and the associated MALDI spectra (Figure 31.3b, right) allow the construction of a calibration curve for the whole elution range (Figure 31.3c). It is thus possible to carry out a substance-specific calibration for unknown PDMS samples.

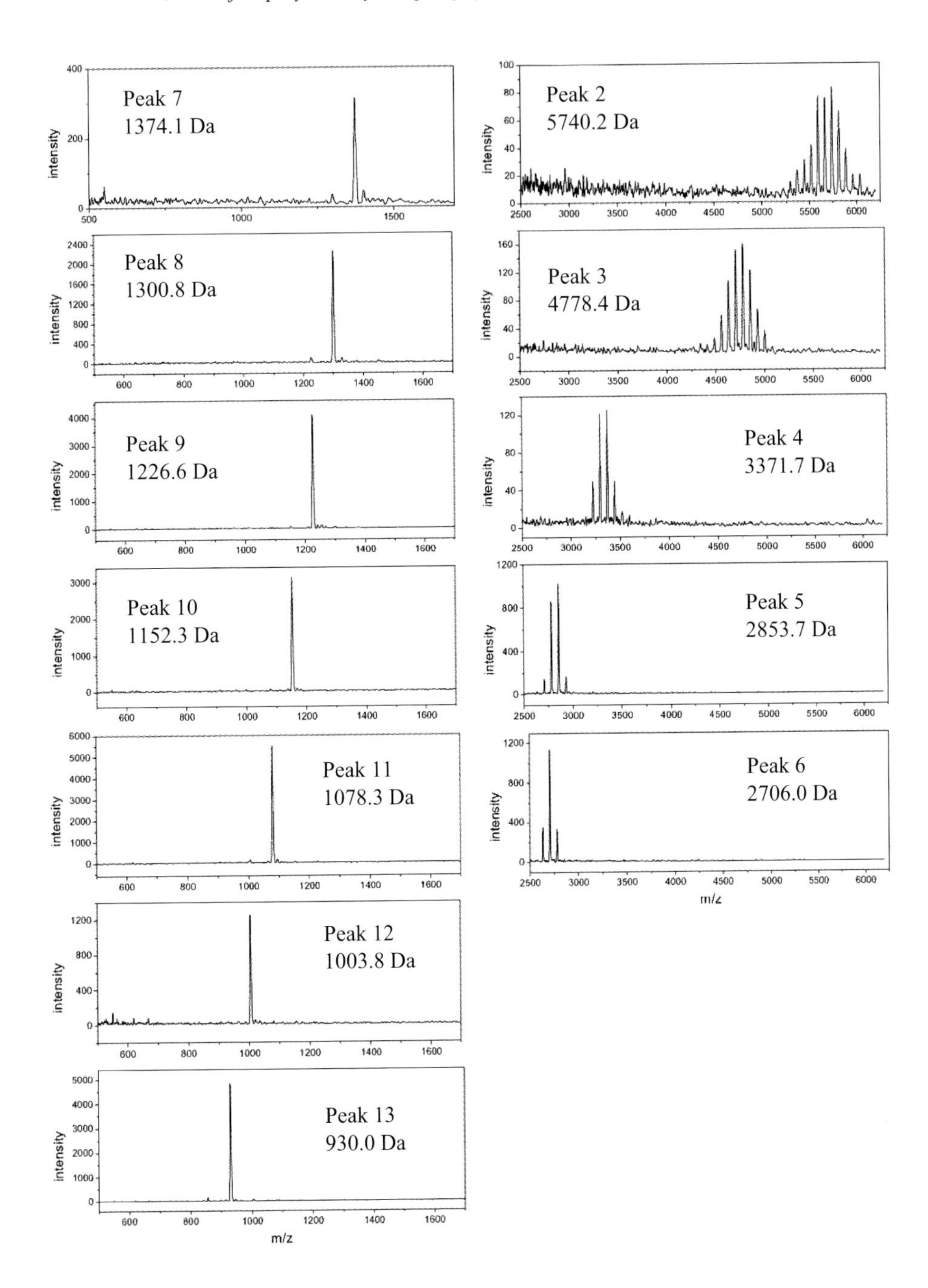

Figure 31.3b. MALDI-TOF mass spectra of different positions in the chromatogram.

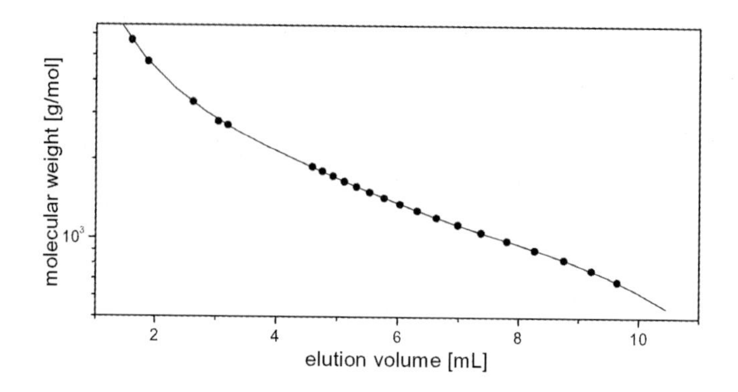

Figure 31.3c. Resulting calibration curve for determination of the molecular weight distribution.

31.3.2 LACCC / MALDI-TOF-MS Coupling

The following investigations demonstrate the ability of LACCC to separate a polymer sample with respect to different architectures.

The critical solvent composition (csc) for PDMS standards was determined using a silica column (pore size 100 Å, average particle size 5 μm, 200 × 4 mm ID) and toluene/*iso*-octane mixtures. When plotting the retention factor k versus solvent/nonsolvent composition, the csc is indicated by the intersection point demonstrated in Figure 31.4a. Under these conditions, PDMS samples of different molecular weights elute at the same retention time. Any separation of an unknown sample into different peaks can be ascribed to a heterogeneity other than molecular weight distribution.

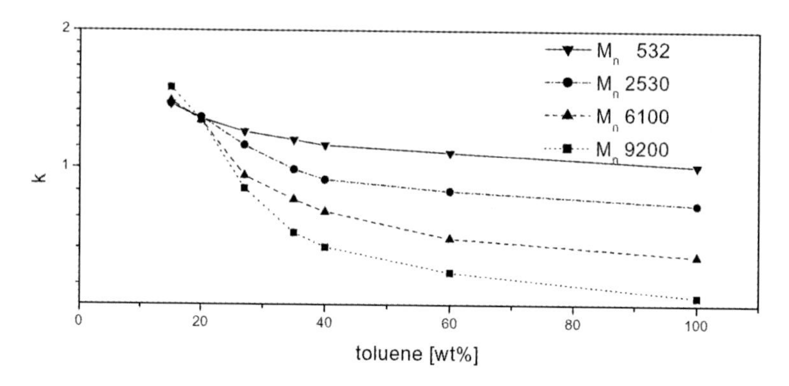

Figure 31.4a. Determination of critical conditions of adsorption of PDMS.

The LACCC elution curve of an unknown PDMS sample is shown in Figure 31.4b. MALDI-TOF-MS is used for characterization of the two peaks. The spectra are shown in Figure 31.4c. The first peak could be identified as being due to a cyclic structure of PDMS, whereas the second peak is due to linear species. Investigations on similar systems have been described previously.[10]

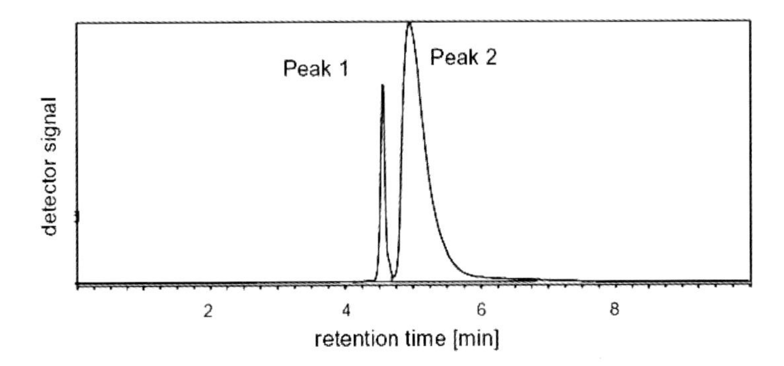

Figure 31.4b. LACCC of a PDMS sample.

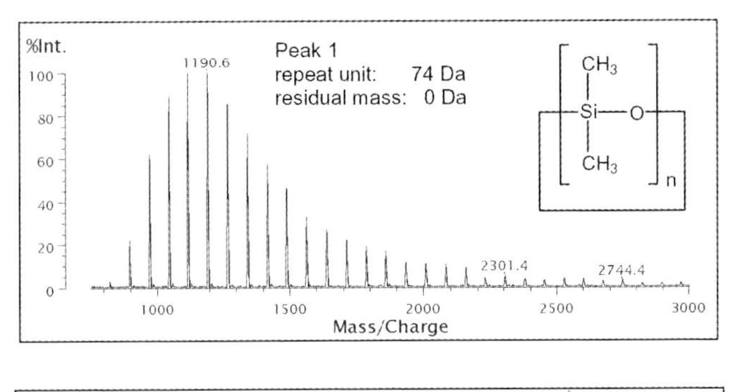

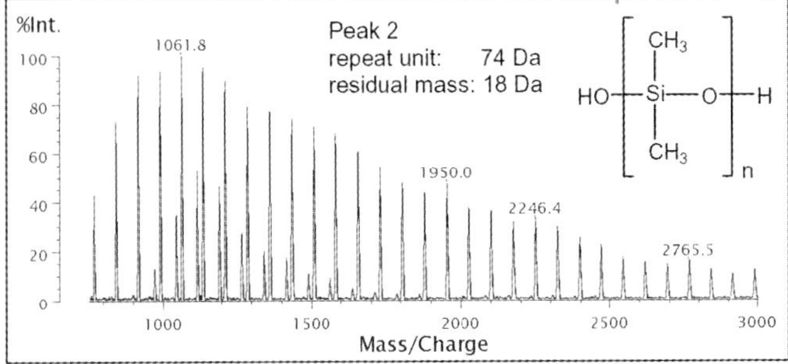

Figure 31.4c. MALDI-TOF mass spectra of the two peaks in Figure 31.4b (cationization with Ag).

31.3.3 SEC / MALDI-TOF-MS Coupling (Heterogeneity)

The coupling of SEC and MALDI-TOF mass spectrometry can also be a powerful tool for a more complete determination of polymer heterogeneities. In the following example, the SEC elution curve of the sample containing $Oct_8Si_8O_{12}$ (Oct = *n*-octyl) is shown (Figure 31.5a). A small portion of higher molecular weight species is visible. The MALDI-TOF mass spectra of the two peaks are reproduced in Figures 31.5b and 31.5c. The mass of 1430.7 Da represents $[OctSiO_{1.5}]_8$, the mass of 2542.3 Da for the first peak corresponds to two silsesquioxane polyhedrons linked by an Si-O-Si bond, i.e. $[Oct_7Si_8O_{12}]_2O$.

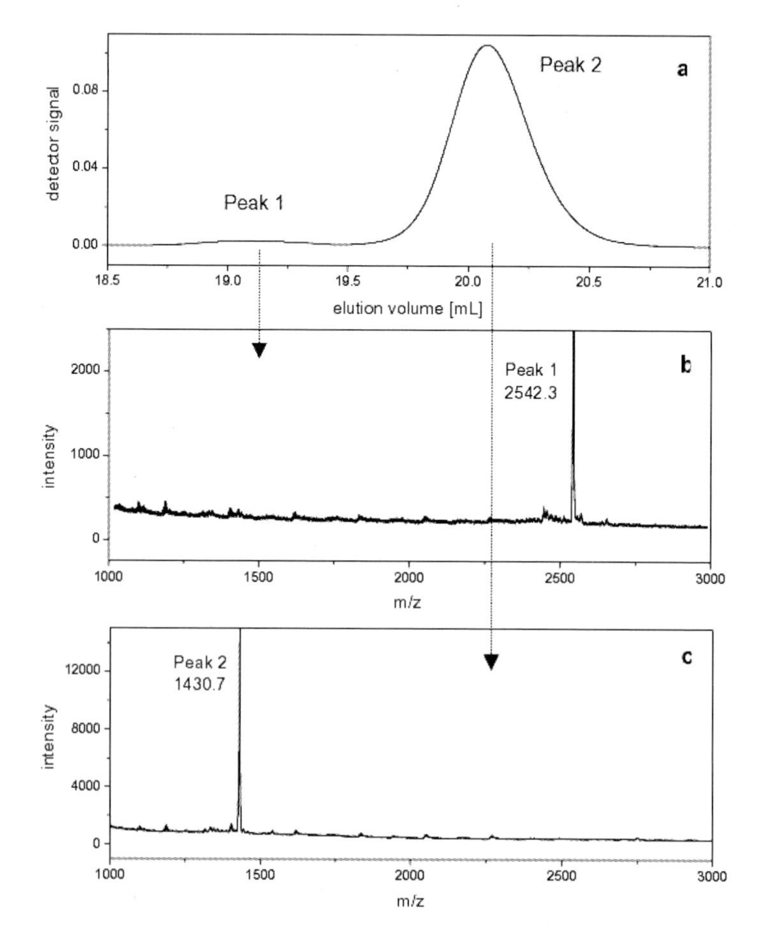

Figure 31.5. SEC elugram (a) and MALDI spectra (b,c) of $[OctSiO_{1.5}]_8$ containing a substituted species.

A three-dimensional MALDI-TOF-MS representation of the silsesquioxane shows even more results (Figure 31.6a). An additional peak (see circle) at 3652 Da was found. This peak represents the trimer $[Oct_7Si_8O_{12}O]_2[Oct_6Si_8O_{12}]$.

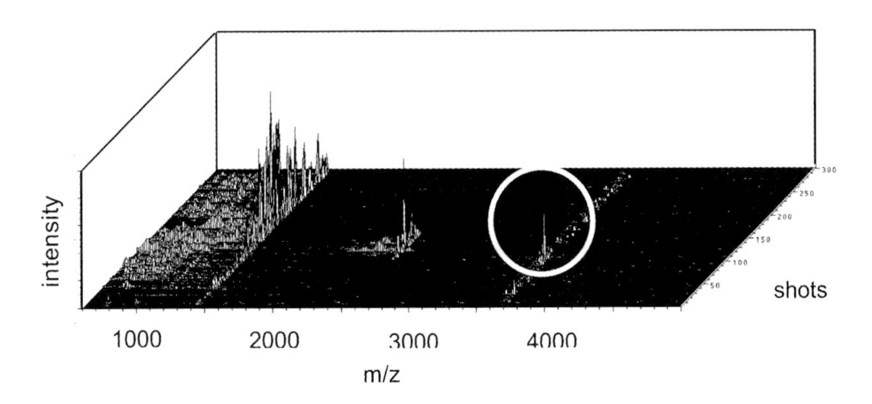

Figure 31.6a. MALDI-TOF-MS 3D-plot indicating single, dimeric, and trimeric polyhedra.

A single shot-mass spectrum (Figure 31.6b) shows all three masses that make up the three silsesquioxane structures.

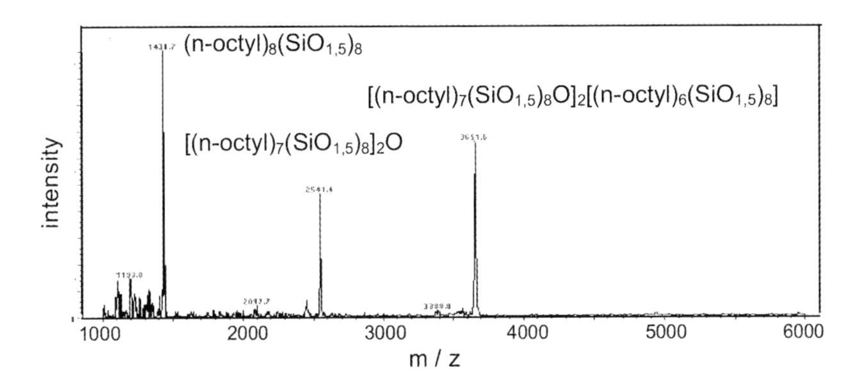

Figure 31.6b. MALDI-MS; single shot with different polyhedra.

Cage rearrangement is an established procedure to prepare polyhedral silsesquioxanes $(RSiO_{1.5})_n$ with $n > 8$.[11] Normally, the portion of compounds with $n > 12$ is very small and difficult to characterize. As shown previously,[12] SEC in combination with MALDI-TOF-MS is a very powerful tool to identify these products. In the case of the rearrangement of $[OctSiO_{1.5}]_8$, sizes up to $n = 40$ were found.

In the following, a new example of the cage rearrangement is demonstrated. In Figure 31.7a, the SEC elution curve of the sample containing $Hep_8Si_8O_{12}$ (Hep = *n*-heptyl) is shown.

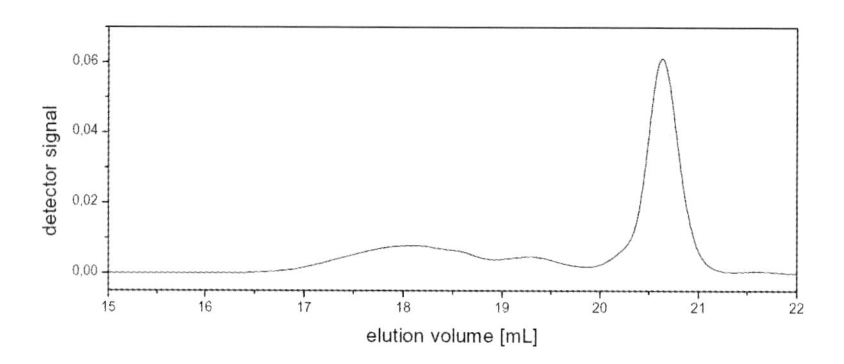

Figure 31.7a. SEC of a silsesquioxane sample containing $Hep_8Si_8O_{12}$ and higher oligomers.

The presence of higher molecular weight species is remarkable. By coupling of SEC/MALDI-TOF-MS, sizes of $n = 8$ up to $n = 28$ can be found. The 3D image of the MALDI-TOF mass spectra in Figure 31.7b proves the presence of higher polyhedra without vacancies.

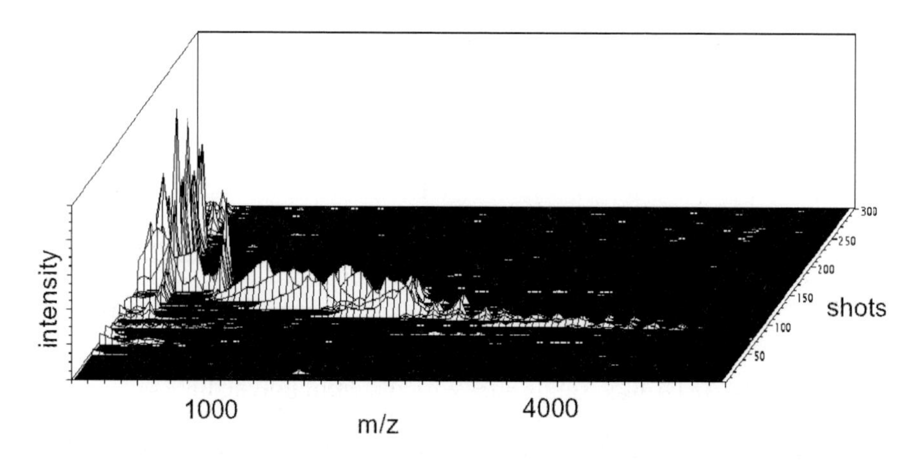

Figure 31.7b. MALDI-TOF-MS 3D-plot a silsesquioxane sample containing $Hep_8Si_8O_{12}$ and higher oligomers.

31.4 Conclusions

The research on modern silicon polymers by different chromatographic methods and their combination with structure-specific and molar mass-sensitive detectors has shown that the characterization of such Si-containing polymers is possible. The choice of coupled methods for use in such investigations depends on the properties of the polymer sample and has to be established in each individual case. Because of poor solubility, it is better to work with oligomers in the low molecular weight range. Moreover, it is easier to establish critical conditions of adsorption for a specific polymer in this range and to apply MALDI-TOF-MS. The developed semi-on-line coupling of liquid chromatography with MALDI-TOF-MS allows the characterization of the whole sample continuously at each point of the chromatogram.

It is reasonable to use the potential of the presented combinations of methods for future research on copolymers, blends, and silicon polymers of different architectures. The direct linking of separation, identification, and characterization is a great advantage.

31.5 Acknowledgements

The syntheses were performed in the groups of E. Rikowski (formerly University Paderborn, Germany) and R. Elsässer (formerly University Saarbrücken, Germany). For other experimental work, we thank R. Laging and B. Manger (BAM).

31.6 References

[1] J. Porath, P. Flodin, *Nature* **1959**, *183*, 1657.
[2] B. G. Belenkii, E. S. Gankina, M. B. Tennikov, L. Z. Vilenchik, *Dokl. Acad. Nauk USSR* **1976**, *231*, 1147.
[3] S. G. Entelis, V. V. Evreinov, A. V. Gorshkov, *Adv. Polym. Sci.* **1986**, *76*, 129.
[4] H. Pasch, B. Trathnigg, *HPLC of Polymers,* Springer, Heidelberg, 1997.
[5] M. Karas, F. Hillenkamp, *Anal. Chem.* **1988**, *60*, 2299.
[6] K. Tanaka, H. Waki, Y. Ido, S. Akita, Y. Yoshida, T. Yoshida, *Rapid Commun. Mass Spectrom.* **1988**, *2*, 151.
[7] K. Lorenz, R. Mülhaupt, H. Frey, *Macromolecules* **1995**, *28*, 6657.

[8] X. Tang, P. A. Dreifuss, A. Vertes, *Rapid Commun. Mass Spectrom.* **1995**, *9*, 1141.

[9] P. Montag, J. Falkenhagen, R.-P. Krüger, *Chem. Labor Biotech.* **1999**, *7*, 253.

[10] U. Just, R.-P. Krüger in *Organosilicon Chemistry II*, (Eds.: N. Auner, J. Weis), Wiley-VCH, Weinheim, 1996, p. 625.

[11] R.-P. Krüger, H. Much, G. Schulz, E. Rikowski in *Organosilicon Chemistry IV*, (Eds.: N. Auner, J. Weis), Wiley-VCH, Weinheim, 2000, p. 545.

[12] R.- P. Krüger, H. Much, G. Schulz, E. Rikowski, *Monatsh. Chem.* **1999**, *130*, 163.

32 The Stepwise Formation of Si-O Networks

Michael Binnewies*, **Nicola Söger**†
With the cooperation of M. Jerzembeck, A. Kornick, H. Quellhorst, C. Sack and A. Wilkening

32. 1 Introduction

For about 30 years scientists have been interested not only in structures, and the physical and chemical properties of solids, but increasingly in their formation pathways as well. Whereas cluster compounds were considered as rare exceptions in the 1960s,[1] more and more cluster compounds have been detected in the interranging years and are now discussed as connecting links between molecules and solids. The principle of condensed clusters [2] shows their role as links between molecules and solids. Colloids – giant clusters with 10^3 to 10^9 atoms – are also included in the discussion.[3] In most cases, cluster structures represent small sections of the solids' structure. From the present point of view, the traditional separation between molecules, complexes, and solids does not seem to be wholly adequate. As the structural borderline between molecules and solids becomes less well defined, the physical properties of compounds in this transition range become increasingly interesting.

This article focusses on the stepwise formation of solid SiO_2 from several molecular precursors.

The Gibbs energy ΔG_R^0 of the formation of oxidcs from chlorides is negative in most cases. Therefore, reaction equilibria are close to the oxide, e.g. for the reaction given in Eq. 1, ΔG_R^0 (298) = –233 kJ/mol:[4, 5]

$$SiCl_4(g) + O_2(g) \longrightarrow SiO_2(s) + 2\,Cl_2(g) \qquad (1)$$

In contrast to the thermodynamically expected equilibrium position, $SiCl_4$ does not react with oxygen at temperatures below 700–800 °C. In the temperature range between 800 and 1000°C, the reaction rate is enhanced, but instead of SiO_2 a multitude of molecular compounds, chlorosiloxanes $Si_xO_yCl_z$, are formed. Higher temperatures lead to the formation of particles with compositions close to SiO_2.

* Corresponding author
† Address of the authors: Institut für Anorganische Chemie, Universität Hannover, Callinstr. 9, D-30167 Hannover, Germany.

Chlorosiloxanes are also formed in the course of the reaction in Eq. 2 (ΔG_R^0 (298) = –137.7 kJ/mol [4, 5]).

$$SiCl_4(l) + 2\ H_2O(l) \longrightarrow SiO_2(s) + 4\ HCl(g) \quad (2)$$

The chlorosiloxanes can be considered as intermediates formed during both reactions (Eqs. 1 and 2). Their investigation allows an insight into the formation pathway of solid SiO_2 in a stepwise manner. The characterization of intermediate chlorosiloxanes becomes possible if reaction temperatures of about 900 °C (Eq. 1) or very small amounts of water (partial hydrolysis) (Eq. 2) are used.

32.2 Chlorosiloxanes

32.2.1 Composition

The literature on perchlorosiloxanes has been previously reviewed.[6] Investigations have shown that especially reaction (1) leads to the formation of numerous high-molecular chlorosiloxanes with molar weights up to 7000 D and up to 70 silicon atoms. The formulas of about 250 different chlorosiloxanes with molar masses of 5000 D were reported from mass spectrometric investigations.[7] These chlorosiloxanes can be classified as belonging to a homologous series of the general formula $Si_nO_{n+x}Cl_{2(n-x)}$ (x = –1–24). It has been observed that oxygen-rich chlorosiloxanes (large x) are particularly favored in large molecules (large n). This is to be expected, because x cannot exceed n. For x = n, the limit of Si_nO_{2n}, i. e. SiO_2, is obtained.

32.2.2 Structure

The structures of a substantial number of chlorosiloxanes are known, particularly from ^{29}Si NMR studies, but also from crystal structure analyses.[8–16] These investigations have demonstrated the relationship between the structure of the molecules and their composition. The chlorosiloxanes $Si_nO_{n+x}Cl_{2(n-x)}$ with x = –1 form chains, where unbranched and branched catena structures appear.[6, 8, 12] For x = 0, monocyclic Si-O rings are observed, often with side groups.[16] It should be mentioned that there are no four-membered rings of the Si_2O_2 type among the prepared chlorosiloxanes. This is in agreement with the structural chemistry of SiO_2 and the silicates. Quantum chemical calculations predict the existence of this small ring, but its stability is lower than that of the $Si_3O_3Cl_6$ and $Si_4O_4Cl_8$ rings.[17] Three-dimensional polycyclic Si-O structures arise when $x \geq 1$. The largest chlorosilsesquioxane with an experimentally determined structure is $Si_8O_{12}Cl_8$,[15] which has a cubane-like cage structure. Larger molecules have eluded experimental structure determination so far, but the results of quantum chemical calculations are available on the basis of which cage-like structures are expected for siloxanes of up to 40 Si atoms.[18, 19] These structures usually contain six-, eight-, ten-, and twelve-membered rings, which are connected in different ways.

32.2.3 Stability

Heats of formation and entropies have not been experimentally determined for any of the chlorosiloxanes described herein. A discussion on their thermodynamic stability therefore has to rely on quantum chemical calculations, estimates,[27] and experience.

The fact that these compounds are formed at temperatures of slightly less than 1000 °C suggests that they are thermodynamically stable. However, there is also literature data about the thermal decomposition of a chlorosiloxane into $SiCl_4$ and a variety of oxygen-rich chlorosiloxanes.[7, 20, 22] The number of silicon atoms in the resultant product(s) from the thermolysis of a chlorosiloxane can be either smaller (a), the same (b), or larger and smaller (c).

(a) A gaseous chlorosiloxane decomposes under cleavage of $SiCl_4$:

$$Si_nO_{n+x}Cl_{2(n-x)} \longrightarrow SiCl_4 + Si_{n-1}O_{n+x}Cl_{2(n-x-2)} \quad (3)$$

$$\text{e.g., } Si_4O_3Cl_{10} \longrightarrow SiCl_4 + Si_3O_3Cl_6 \ (n = 4, x = -1) \quad (4)$$

(b) A chlorosiloxane with n Si atoms decomposes under formation of an oxygen-richer and an oxygen-poorer chlorosiloxane with n Si atoms in each:

$$2\ Si_nO_{n+x}Cl_{2(n-x)} \longrightarrow Si_nO_{n+x+1}Cl_{2(n-x-1)} + Si_nO_{n+x-1}Cl_{2(n-x+1)} \quad (5)$$

$$\text{e.g., } 2\ Si_4O_4Cl_8 \longrightarrow Si_4O_5Cl_6 + Si_4O_3Cl_{10} \ (n = , x = 0) \quad (6)$$

(c) A chlorosiloxane with n Si atoms decomposes into a larger [(n + 1) Si atoms] and a smaller [(n - 1) Si atoms] siloxane:

$$2\ Si_nO_{n+x}Cl_{2(n-x)} \longrightarrow Si_{n+1}O_{n+x+1}Cl_{2(n-x+1)} + Si_{n-1}O_{n+x-1}Cl_{2(n-x-1)} \quad (7)$$

$$2\ Si_4O_4Cl_8 \longrightarrow Si_5O_5Cl_{10} + Si_3O_3Cl_6 \ (n = , x = 0) \quad (8)$$

In each case, the total number of Si-O and Si-Cl bonds remains unchanged during the reaction. Since one can assume that the binding energies in such chemically similar systems have similar values, heats of reaction close to zero are to be expected. The entropy terms of the reactions of types (a)–(c), however, do have to be considered: If a chlorosiloxane decomposes into several smaller molecules (type (a)), such a reaction will invariably proceed with an increase of the mole number, and therefore entropy is gained. This reaction entropy should be close to 120–160 J/(K·mol) for a change of mole number $\Delta\nu$ by +1. If a chlorosiloxane decomposes into two comparatively equal siloxanes, which are distinguished only by their oxygen content (type b), such a reaction occurs without a change of the mole number and its reaction entropy is therefore close to zero. The same holds for a reaction of type (c). Of course, mixed forms of reactions are conceivable. For case (a), a temperature-independent ($\Delta_R H^0 = 0$) equilibrium constant of $e^{120/R} - e^{160/R} \approx 10^6 - 10^8$, i.e. an almost complete decomposition, is to be expected from the thermodynamic point of view. For cases (b) and (c), equilibrium constants of about $e^0 = 1$ are expected. The decomposition products should have similar concentrations as the reactant. For each of the resulting decomposition products (except $SiCl_4$), analogous considerations can be made.

The most favored decomposition reaction is the cleavage of $SiCl_4$, which, after numerous repetitions, should lead to SiO_2 and $SiCl_4$. Considering the discussed decomposition reactions from a molecular view point (kinetic aspect) also leads to a clear preference for decomposition reaction (a): For reaction (a) to occur, only the necessary activation energy from a collision with an energy-rich molecule or the (hot) wall of the reaction vessel has to be supplied. For reactions (b) and (c) to become operative, however, a collision of two molecules of equal composition is necessary. This is rather improbable in the complex reaction mixture. Unimolecular reactions which lead to products under (b) or (c) are not conceivable.

The results of these considerations clearly show that from a thermodynamic and kinetic point of view all chlorosiloxanes must be unstable. They decompose into $SiCl_4$ with concomitant formation of oxygen-richer siloxanes. This agrees with all experimental evidence.[7, 20, 22]

32.3 Build-up and Decomposition Reactions

In this section, the formation of Si-O networks during the high–temperature combustion and the low-temperature hydrolysis of $SiCl_4$ will be compared.

32.3.1 Combustion of $SiCl_4$ with O_2

In the course of this reaction, an impressive number of chlorosiloxanes of quite different compositions with molar weights up to 7000 D is generated.[7] We suspect that further chlorosiloxanes with even much higher molar weights are formed, which are undetected.[24] The appearance of such a wealth of reaction products in this seemingly simple reaction leads to the conjecture that a highly reactive molecule is formed intermediately, which reacts non-specifically with numerous reaction partners, thereby generating the broad spectrum of products.

It was shown by matrix-isolation spectroscopy that $O{=}SiCl_2$ is formed in a primary step in both the reaction of $SiCl_4$ with O_2 and the thermolysis of Si_2OCl_6 at about 1000 °C.[23] Thermodynamic calculations and experimental work have shown that $O{=}SiCl_2$ (silaphosgene) appears in the temperature range 1100–1700 K, with a variable concentration that reaches a maximum at 1400 K.[23] An appreciable concentration of $O{=}SiCl_2$ first appears at a temperature which is identical to the minimum temperature required for the reaction of $SiCl_4$ with O_2 forming chlorosiloxanes. This would suggest that $O{=}SiCl_2$ is involved in the formation of chlorosiloxanes. The thermolysis of Si_2OCl_6 also leads to silaphosgene in a primary step of the reaction,[23] as well as to the build-up of higher chlorosiloxanes.[7, 20, 21] Here, the decisive role of silaphosgene in the build-up of catena-siloxanes could be shown.[23] The chemical properties of $O{=}SiCl_2$ are not known, but it is expected to be a highly reactive molecule. Although the value of its standard heat of formation is highly negative (–440 kJ/mol [23]), a decomposition into solid SiO_2 and $SiCl_4$ is expected by reference to other literature data on the stability of silicon compounds.[4, 5] In the case of the reaction of $SiCl_4$ and O_2, a great excess of $SiCl_4$ is available as a potential reaction partner. Silaphosgene could insert into its Si-Cl bond under formation of Si_2OCl_6 (Eq. 9).

$$\text{Si} + \text{O{=}Si} \rightleftharpoons \text{Si{-}O{-}Si} \qquad (9)$$

In fact, Si_2OCl_6 is the main product of the reaction of $SiCl_4$ with O_2, if the reaction is conducted in the temperature range 700–950 °C. As soon as the Si_2OCl_6 is formed, it is also available as a reaction partner for silaphosgene.

$$\text{Si{-}O{-}Si} + \text{O{=}Si} = \text{Si{-}O{-}Si{-}O{-}Si} \qquad (10)$$

For catena-siloxanes with four or more silicon atoms, several isomers appear, which can be separated by gas chromatography.[24] This growth of unbranched and branched chains comes to an end for chain lengths of about ten Si atoms.[24] Larger catena-siloxanes have not been observed up to the present.

This mechanism gives an understanding of the growth of chains. However, it does not provide an explanation for the formation of mono- and oligocyclic siloxanes (see ref.[6]). With each insertion of $O{=}SiCl_2$ during a growth reaction, silicon and oxygen atoms are introduced in the ratio 1:1 into newly formed siloxanes. In this way, no chlorosiloxanes with more oxygen than silicon atoms can be formed. Since these are undoubtedly formed, the insertion of $O{=}SiCl_2$ cannot be the only reaction responsible for the build-up of the Si-O framework in the oxygen-rich siloxanes and eventually in solid SiO_2.

It was discussed above that all chlorosiloxanes must decompose for entropy reasons, preferably according to Eq. 3. Thus, the stepwise built-up chlorosiloxane is decomposed, but the stoichiometry differs from that of the formation step. Whereas Si and O atoms are always introduced in equal numbers in the growth reaction (Eq. 9 and 10), only Si and Cl atoms are cleaved in the decomposition reaction (Eq. 3). All oxygen atoms remain in the rest of the decomposing molecule. In this way, the molecules become smaller in each decomposition step, but at the same time oxygen-richer.

Figure 32.1 presents the interplay between growth and decomposition. The ratio (n+x)/n for chlorosiloxanes $Si_nO_{n+x}Cl_{2(n-x)}$ is plotted versus the number of Si atoms per molecule (n) for a series of x values.

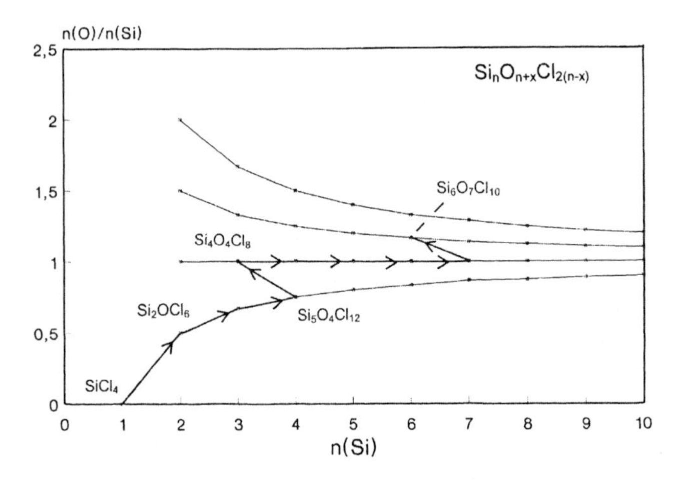

Figure 32.1. Growth and decomposition of chlorosiloxanes leading to the formation of SiO_2.

The generated sequence of points can be connected by hyperbola-like curves. Each of these curves corresponds to a homologous series of chlorosiloxanes; the lowest to the catena-siloxanes (x = –1), the next to the cyclosiloxanes (x = 0), etc.. Growth and decomposition steps are symbolized as bold lines with arrows in Figure 32.1. Starting from $SiCl_4$, Si_2OCl_6 is formed first, then the chain length increases, for example up to $Si_5O_4Cl_{12}$. If $Si_5O_4Cl_{12}$ decomposes under formation of $SiCl_4$, $Si_4O_4Cl_8$ is formed which can grow, for example, up to $Si_7O_7Cl_{14}$, before this compound decomposes into the bicyclic $Si_6O_7Cl_{10}$. In this way, the stepwise transformation of $SiCl_4$ to SiO_2 can be followed almost continuously.

For the thermolysis of Si_2OCl_6 in a closed vessel, it could be shown [22] that the rate of the growth reaction exceeds that of the decomposition reaction. Therefore, more and more chlorosiloxanes are formed at constant temperature. If the influence of higher temperature on the formation of chlorosiloxanes is studied, it is found that the number of oxygen-rich siloxanes increases.[20, 22] This means that the rate of the decomposition reaction, which leads to the formation of oxygen-rich compounds, has a higher temperature coefficient than that of the growth reaction. Therefore, with increasing temperature a point must be reached at which the overall decomposition reactions proceed more quickly than those of the growth processes. This temperature T_s is at first unknown. Also unknown is to what (average) molar weight the molecules have grown at this point. It is certain that the reaction process discussed on the basis of experimental results takes place below this temperature.

In summary, $O{=}SiCl_2$ is formed in a first step of the reaction between $SiCl_4$ and O_2. This highly reactive molecule reacts with all reaction partners with a Si-Cl bond under formation of an Si-O-Si group. This growth reaction leads to a variety of catenasiloxanes starting from $SiCl_4$. However, siloxanes with a ratio n(O)/n(Si) larger than one are formed from chlorosiloxanes by entropy-driven cleavage of $SiCl_4$. In this way, molecules become smaller again, but oxygen-richer. Their composition eventually approaches that of the final product SiO_2.

The reaction could be followed experimentally up to about 70 Si atoms and a molar weight of 7000 D. It is unlikely that there will be any change in the basic reaction pattern thereafter. If we extrapolate the described process, two alternative possibilities arise depending on the reaction conditions:

The reaction temperature is always below T_s. As long as the formed siloxanes are at the reaction temperature and $O{=}SiCl_2$ is available as a reaction partner, the molecules will keep growing and become more and more oxygen-rich. This growth cannot proceed indefinitely, because the volatility of the siloxanes decreases with increasing molar weight. This means that condensed phases must occur. This is most likely to happen if the siloxanes are retained in the reaction zone for a long time during which they can grow. As an alternative to the condensation of high-molecular chlorosiloxanes, the formation of aerosol particles with diameters 10^{-7}–10^{-3} cm appears likely. The formation of aerosol particles, which do not show a tendency for condensation, is not unusual.

The reaction temperature exceeds T_s. The decomposition reactions (in particular $SiCl_4$ cleavage) proceed more quickly than the formation reactions. However, such molecules cannot completely decompose in this way. If $SiCl_4$ is cleaved, Si and Cl atoms are lost in the ratio 1:4. If all Cl atoms are eliminated, Eq. 11 can be formulated.

$$Si_nO_{n+x}Cl_{2(n-x)} \longrightarrow 0.5\,(n - x)\,SiCl_4 + Si_{(n+x)/2}O_{n+x} \quad (11)$$

A "molecule" $(SiO_2)_{(n+x)/2}$ would have to be formed, i.e. an SiO_2 cluster with a molar weight not much below that of the chlorosiloxane from which it originated.

Since a fundamental change of the whole Si-O framework is unlikely in this process, a structural relationship between the formed SiO_2 and the high-molecular chlorosiloxanes can be expected. Quantum chemical calculations on the structure and stability of SiO_2 clusters seem quite interesting in this context.

It is found from experiment that the conversion of $SiCl_4$ with oxygen leads to larger and larger amounts of more and more oxygen-rich chlorosiloxanes with increasing temperature. Depending on the experimental conditions, a point just below 1000 °C is reached, at which the reaction behavior abruptly changes within a temperature interval of a few degrees.[24] This sudden change in the reaction behavior could be connected with the marked intersection point T_s. No more chlorosiloxanes with typically molecular properties (relatively high fugacity, solubility in nonpolar organic solvents) are formed, but rather a white, X-ray amorphous solid which mainly consists of "SiO_2" with variable chlorine content (< 5%). The globular particles have diameters of about 500 nm and are reminiscent of the industrial product "Aerosil" (Degussa).

The final question to be addressed is why the SiO_2 formed is X-ray amorphous. It is obvious that SiO_2 formed in fractions of a second in an oxyhydrogen flame cannot be well ordered and crystalline. An equally amorphous product is formed if an oxygen stream loaded with $SiCl_4$ passes a heating section of about 40 cm [24] with a relatively small flow velocity. The thermolysis of Si_2OCl_6 in a closed system for several weeks also leads to amorphous products.[22] This leads to the conclusion that the reaction time is not very important. In the discussion of the structures of the initially formed chlorosiloxanes, it was pointed out that rings of different size appear in the monocyclic and oligocyclic siloxanes. In the course of the reaction, these rings are linked to increasingly complicated three-dimensional structures. Quantum chemical calculations have been performed on the stabilities of isomeric siloxanes, particularly for the silsesquioxanes. These show that starting from about 48 silicon atoms, structures are favored where the skeleton resembles that of bulk, crystalline SiO_2.[6, 18] In smaller molecules containing less than 48 Si atoms, fullerene-like cage structures with 12 ten-membered and 14 twelve-membered rings are more stable.[19] The growth of fullerene-like structures can occur in two steps in the presence of water, whereby HCl is eliminated. The insertion mechanism has been described previously.[19] If the intermediately formed chlorosiloxanes were to show bulk-like structures in the initial stages of their growth, we would expect crystalline SiO_2. However, this is not the case, as was previously made clear. If the growth of Si-O frameworks starts in a way different from the formation and condensation of Si-O twelve-membered rings (as found in crystalline SiO_2) and molecules of the fullerene type are initially formed, it is difficult to conceive that a rearrangement to a bulk-like, albeit more stable structure occurs. It seems therefore that the initial growth of chlorosiloxane molecules with a few dozen Si atoms is responsible for the non-occurrence of crystalline bulk SiO_2.

32.3.2 Hydrolysis of Chlorosilanes and -siloxanes

In contrast to combustion, the hydrolysis of $SiCl_4$ (or other Si-Cl compounds) proceeds at ambient or even lower temperatures without significant inhibition. Because this reaction also leads to numerous siloxanes as intermediates, the formation of a highly reactive molecule in a primary step is expected. Mass spectrometric investigations of the hydrolysis of both $SiCl_4$ and Si_2OCl_6 showed the formation of the gaseous silanols $SiCl_3OH$ and Si_2OCl_5OH.[25, 26] Obviously, an analogous formation of silanols in the liquid phase can be assumed. The chlorosilanols are very reactive because of their strong tendency to undergo condensation reactions with the concomitant formation of Si-O-Si bonds.

The reaction course and product distribution are significantly influenced by the composition and reactivity of the initially formed silanols. In contrast to the previously described high-temperature reaction, the composition of the reactive molecule can be determined by selecting a specific precursor. For example, hydrolysis of Si_2OCl_6 leads to the formation of Si_2OCl_5OH. The course of this reaction is shown in Figure 32.2. The use of other chlorosiloxanes as precursors ($Si_3O_2Cl_8$, $Si_4O_4Cl_8$) leads to the formation of siloxanols in which the siloxane skeleton is retained. Consequently, the number of different products is considerably reduced. Products requiring cleavage of a precursor's Si-O bond do not arise. Only siloxanes containing odd numbers of Si atoms are built during partial hydrolysis of Si_2OCl_6. Hydrolysis of $Si_3O_2Cl_8$ leads to products with a number of Si atoms divisible by three.

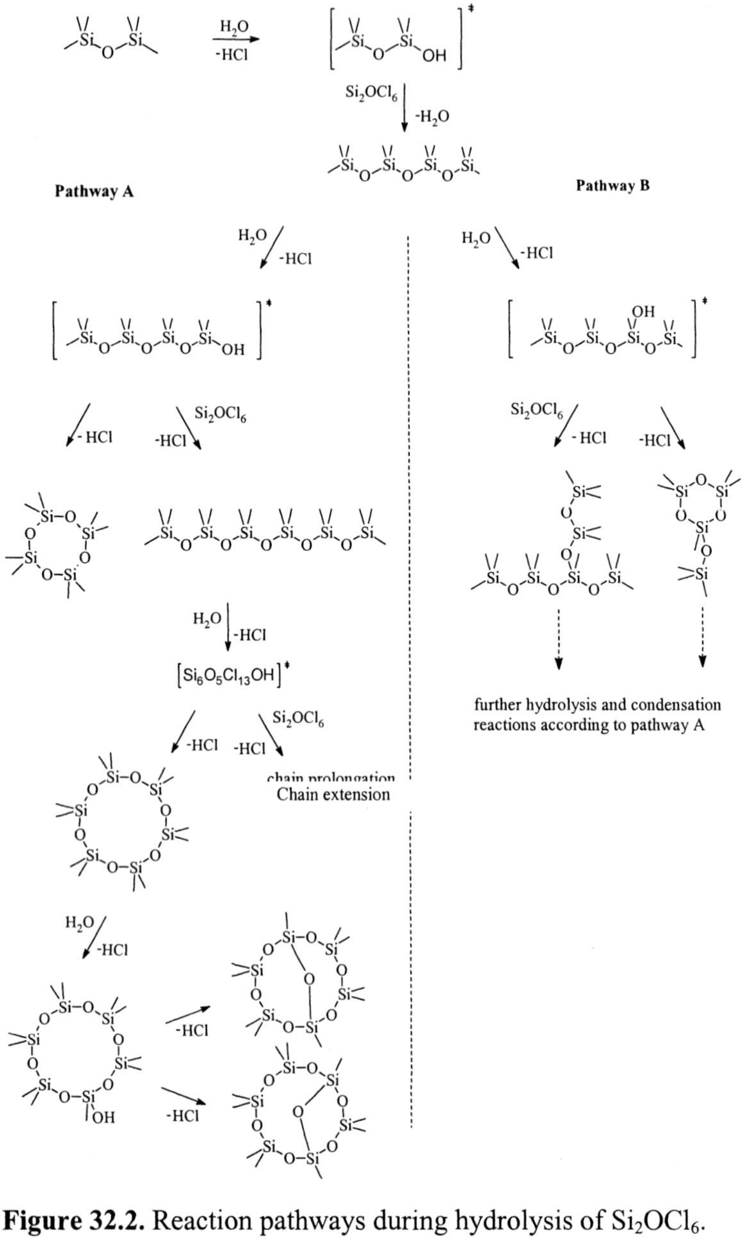

Figure 32.2. Reaction pathways during hydrolysis of Si_2OCl_6.

In the same way, we succeeded in preparing higher-condensed cyclic siloxanes by hydrolysis of the monocyclic $Si_4O_4Cl_8$. In each of the detected products $Si_8O_9Cl_{14}$, $Si_8O_{10}Cl_{12}$, $Si_{12}O_{14}Cl_{20}$ and $Si_{12}O_{15}Cl_{18}$, eight-membered Si-O rings are retained as basic structural components from the precursor.

Previous investigations [12, 27–31] on the hydrolysis of $SiCl_4$ exclusively identified siloxanes of the catena type. By means of GC-MS methods, we were able to prove that mono- and bicyclic siloxanes are also formed during hydrolysis reactions. The formation of these molecules through cyclization processes represents the essential step of the growth mechanism from molecules to the three-dimensionally extended solid SiO_2.

During the previously described high-temperature reaction, molecules with a higher oxygen content are formed by cleavage of $SiCl_4$. In contrast, intramolecular condensation is responsible for the formation of such molecules in the hydrolysis of Si-Cl compounds. Condensation is the essential step for increasing the siloxane's oxygen content (see Figure 32.2). Thus, the formation of SiO_2 (or $SiO_2{\cdot}xH_2O$, respectively) by hydrolysis of $SiCl_4$ or other Si-Cl compounds is based on two types of reaction step, i.e. formation of a silanol or siloxanol by hydrolysis of an Si-Cl bond and condensation of this silanol/siloxanol with another Si-Cl function. The second reaction step may occur intermolecularly (condensation leads to the association of two molecules through a new Si-O-Si bond) or intramolecularly (condensation leads to an additional Si-O-Si bond within the same molecule and thus to the stepwise formation of three-dimensional structures). The formation of polymeric silicic acids has to occur by a combination of both types of condensation reaction.

32.4 References

[1] H. Schäfer, H. G. v. Schnering, *Angew. Chem.* **1964**, *76*, 833.

[2] A. Simon, *Angew. Chem.* **1981**, *93*, 23; *Angew. Chem. Int. Ed. Engl.* **1981**, *20*, 1.

[3] G. Schmid, *Clusters and Colloids*, VCH, Weinheim, 1994.

[4] M. Binnewies, E. Milke, *Thermochemical Data of Elements and Compounds*, Wiley-VCH, Weinheim, 1999.

[5] *JANAF Thermochemical Tables* 3rd Ed. ACS, NBSB, New York, 1985.

[6] M. Binnewies, K. Jug, *Eur. J. Inorg. Chem.* **2000**, 1127.

[7] M. Binnewies, M. Jerzembeck, A. Wilkening, *Z. Anorg. Allg. Chem.* **1997**, *623*, 1875.

[8] H. C. Marsmann, E. Meyer, M. Vongehr, E. F. Weber, *Makromol. Chem.* **1983**, *184*, 1817.

[9] E. Bertling, H. C. Marsmann, *Phys. Chem. Minerals* **1988**, *16*, 295.
[10] E. Bertling, H. C. Marsmann, *Z. Anorg. Allg. Chem.* **1989**, *578*, 166.
[11] H. C. Marsmann, E. Meyer, *Makromol. Chem.* **1987**, *188*, 887.
[12] H. C. Marsmann, E. Meyer, *Z. Anorg. Allg. Chem.* **1987**, *548,* 193.
[13] H. Borrmann, M. Binnewies, *Z. Krist.* **2002**, *217*, 324..
[14] J. Magull, M. Binnewies, *Z. Krist.* **2002**, *217*, 325.
[15] K. W. Törnroos, G. Calzaferri, R. Imhof, *Acta Cryst. C* **1995**, *51, 1732.*
[16] U. Wannagat, D. Burgdorf, H. Bürger, G. Pawelke, *Z. Naturforsch. B* **1991**, *46*, 1039.
[17] K. Jug, D. Wichmann, *J. Mol. Struct. (Theochem.)* **1994**, *313*, 155.
[18] K. Jug, D. Wichmann, *J. Mol. Struct. (Theochem.)* **1997**, *398–399,* 365.
[19] D. Wichmann, K. Jug, *J. Phys. Chem.* **1999**, *103*, 10087.
[20] Yu. Borisova, G. S. Gol'din, L. S. Baturina, V. P. Voloshenko, S. V. Sin'ko, *Russ. J. Gen. Chem.* **1993**, *63*, 269.
[21] D. Wichmann, K. Jug, *J. Chem. Phys.* **1998**, *236,* 87.
[22] A. Wilkening, M. Binnewies, *Z. Naturforsch. B.* **2000**, *55 b,* 21.
[23] M. Junker, A. Wilkening, M. Binnewies, H. Schnöckel, *Eur. J. Inorg. Chem.* **1999**, 1531.
[24] H. Quellhorst, A. Wilkening, M. Binnewies, *Z. Anorg. Allg. Chem.* **1997**, *623*, 1871.
[25] H. Quellhorst, A. Wilkening, N. Söger, M. Binnewies, *Z. Naturforsch. B* **1999**, *54*, 577.
[26] A. Kornick, M. Binnewies, *Z. Anorg. Allg. Chem.* **1990**, *587,* 157.
[27] M. Jerzembeck, Dissertation, Universität Hannover, **1993**.
[28] W. C. Schumb, A. J. Stevens, *J. Am. Chem. Soc.* **1947**, *69*, 726.
[29] J. Goubeau, R. Warncke, *Z. Anorg. Allg. Chem.* **1949**, *259*, 109.
[30] A. Schervan, Dissertation, Technische Universität Braunschweig, **1988**.
[31] H. Bürgy, G. Calzaferri, I. Kamber, *Mikrochim. Acta* **1988**, *1*, 401.

33 Mechanism of Ring and Cage Formation in Siloxanes

Karl Jug*

33. 1 Introduction

The transition between molecules and solids has become a fascinating area of research. Such compounds have been termed clusters, and their size ranges from a few atoms up to thousands of atoms. With growing cluster size, the properties of these compounds gradually change towards solid-state properties. It is therefore essential to understand the growth mechanism of clusters in order to design compounds with particular useful properties.

This article deals with the essential steps in the formation of large siloxanes. Such compounds are formed by reaction of chlorosilanes with oxygen in the gas phase [1] or by hydrolysis of substituted silanes in solution.[2]

The reaction of $SiCl_4$ with O_2 is of technological importance for the chemical vapor deposition (CVD) of solid SiO_2 films.[3–5] SiO_2 and Cl_2 are formed above 1000 °C. However, in the temperature range 800–1000 °C, the product mixture consists of a great variety chlorosiloxanes of the general formula $Si_nO_mCl_l$.[1,6] These siloxanes can be considered as intermediates in the formation of solid SiO_2. They form chains, rings or cages. The mixtures were first separated by gas chromatography and then characterized by mass spectrometry.[6] It is difficult to determine the structures experimentally and even more the processes by which the siloxanes are formed.

This gap is bridged by means of quantum chemical calculations. In Section 33.2 of this article, the initial process is illuminated starting from $SiCl_4$ and O_2, and is followed up to the formation of the smallest ring $Si_2O_2Cl_4$. The formation of such a ring of the composition $Si_2O_2R_2R'_2$ from disilenes was recently reviewed,[7] but no information on the detailed steps starting from $SiCl_4$ is available. In Section 33.3, the growth of larger siloxane cages is described in relation to the synthesis of mixtures of higher silsesquioxanes by base-catalyzed cage rearrangements of octasilsesquioxanes [8] and subsequent characterization by MALDI-TOF mass spectrometry.[9]

---

* Address of the author: Theoretische Chemie, Universität Hannover, Am Kleinen Felde 30, D-30167 Hannover, Germany

33.2 Siloxane Ring Formation from $SiCl_4$ and O_2

Our early work was concerned with the structure and stability of chlorosiloxanes,[10] the growth pattern of chlorosiloxanes,[11] and the decomposition of perchlorodisiloxane.[12] However, it was not clear as to how the formation of siloxanes is initiated. In particular, the role of the elusive $O{=}SiCl_2$ [13] remained unclear. In recent work,[14] we have tried to elucidate the mechanism of siloxane ring formation from $SiCl_4$ and O_2. We investigated three pathways (A–C). Pathway A involves the formation of the Cl_2SiO_2 ring. The energy profile along this pathway is given in Figure 33.1. The highest barrier of 76 kcal/mol involves the transition state of the reaction of Cl_2SiO_2 with $SiCl_4$. The structures of the reactants, intermediates, transition states, and products (**1–12**) along this pathway are shown in Figure 33.2. They were optimized by density functional calculations at the B3LYP/6-311G* level.

An alternative pathway (B) involves the dimer $(Cl_3SiO)_2$ bound through an O-O single bond. The highest energy barrier results from the breaking of this O-O single bond, which is 72.7 kcal/mol higher than the energy of the reactants. A third possibility (pathway C) starts with the abstraction of a chlorine atom from $SiCl_4$. This leads to a barrier of 95.7 kcal/mol and is therefore not competitive with the other two pathways. Pathway C is also excluded by experimental evidence, i.e. the non-occurrence of $SiCl_3$ radicals. The investigation shows a rather complicated initiation of the siloxane formation.

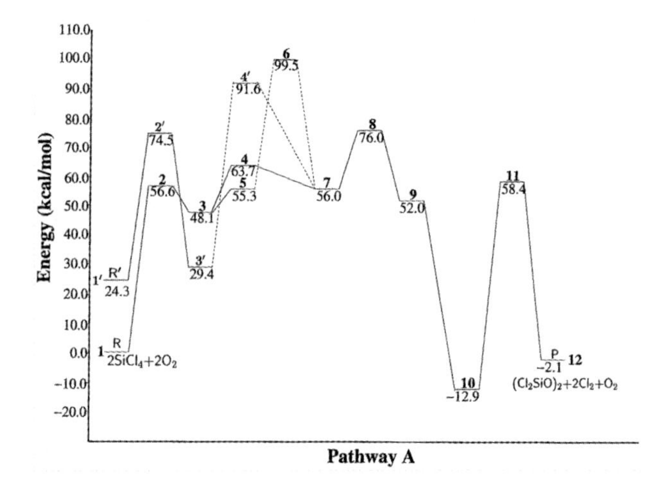

Figure 33.1. Energy profile of the reaction $SiCl_4 + O_2$ along pathway A; bold numbers refer to the structures in Figure 33.2.

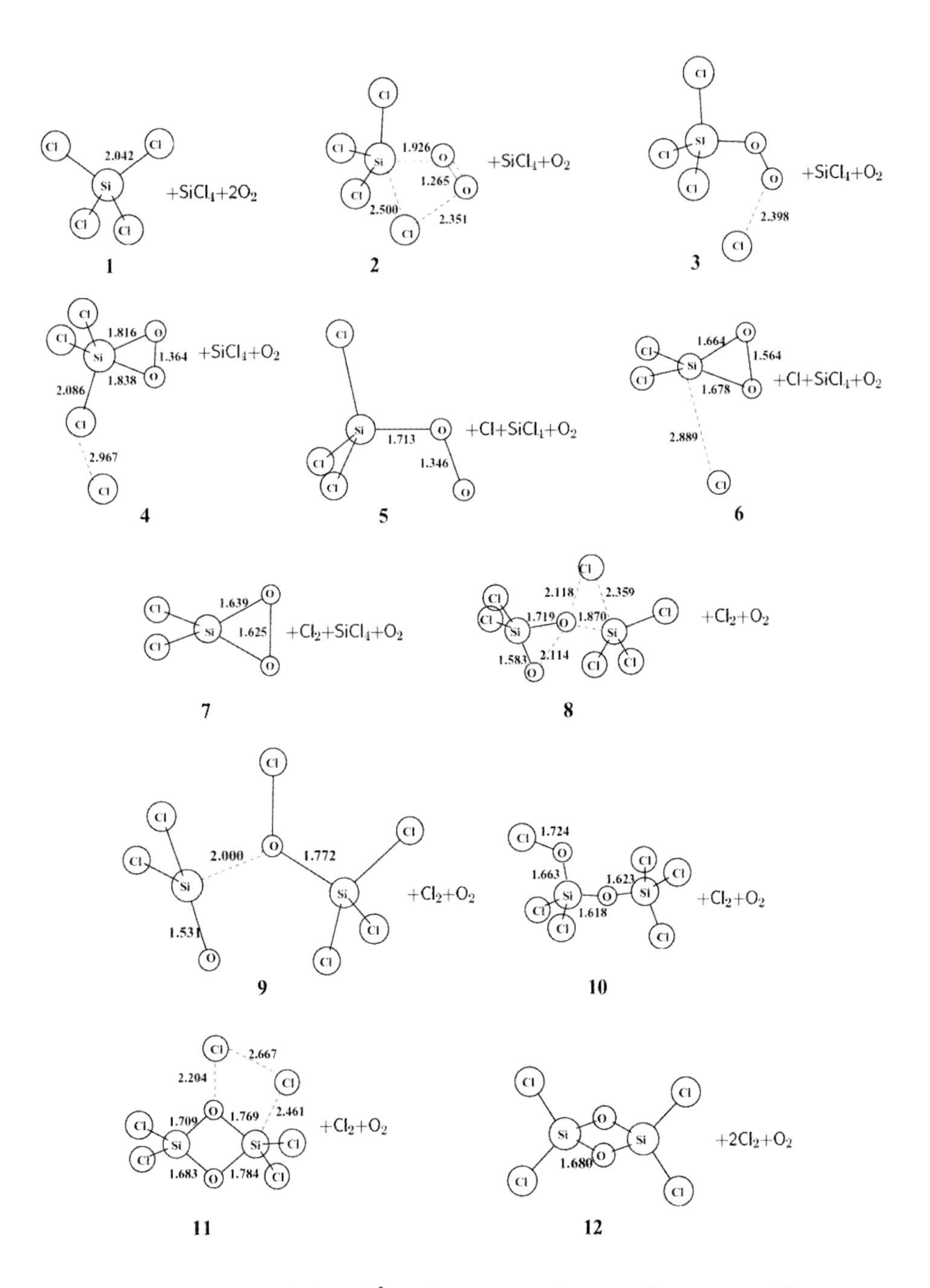

Figure 33.2. Structural data (Å) of reactants, intermediates, transition states, and products for pathway A.

Starting from the relatively stable structure **10**, it is suggested that the chain $Si_3O_2Cl_8$ is formed by insertion of $SiCl_4$ under elimination of Cl_2 and that the larger ring $Si_4O_4Cl_8$ is obtained by dimerization of structure **10** under elimination of 2 Cl_2. The latter ring can be the starting structure for the cage $Si_8O_{12}Cl_8$. The steps and the energy changes calculated with the SINDO1 method [15] have been previously documented.[16]

33.3 Pathways for Silsesquioxane Growth

Among the siloxanes, the silsesquioxanes $Si_nO_{3n/2}R_n$ are of special interest. $Si_8O_{12}R_8$ has a cube-like structure with the eight silicon atoms at the corners and the twelve oxygen atoms bridging the silicon atoms. Many cages of this kind have been synthesized and their properties analyzed.[17–19] Hydridosilsesquioxanes with R = H and n = 8–18 have also been prepared and their structures investigated.[20–24] Most recently, the size of experimentally detected silsesquioxanes was extended to n = 38.[9]

Stimulated by the experimental work and the need for a more specific description of structure and stability, we undertook a comprehensive study of silsesquioxanes. Initially, we calculated selected perchlorosilsesquioxanes with up to 60 atoms [11] with the SINDO1 method.[15] We observed that some of the chosen structures, namely the sodalite cage of $Si_{24}O_{36}Cl_{24}$, and the α-cage $Si_{48}O_{72}Cl_{48}$, did not fit the global trend of stabilization. We therefore repeated the calculations with the newly developed MSINDO method,[25] which is substantially improved in terms of structure and energy compared to the SINDO1 method. In the selection of structures, we initially followed the pattern of fullerene structures, since the silicon atoms of silsesquioxanes form rings which are reminiscent of fullerene rings with a preference for five- and six-membered rings. In order to save computer time, we studied hydridosilsesquioxanes [26] in which the cage structure is the same as for perchlorosilsesquioxanes. It turned out that the isolation rule of five-membered rings that applies to larger fullerenes is not a guiding rule for the stability of silsesquioxanes. Instead, egg-shaped structures are preferred for $Si_{48}O_{72}H_{48}$ and $Si_{60}O_{90}H_{60}$, in which six Si 5-rings are located at both the top and bottom of the egg-shaped cage. (Note that the ring $Si_5O_5Cl_{10}$ is called an Si 5-ring, because it contains five Si atoms).

The observation of such an egg-shaped structure led us to an extension of the idea that such structures could be precursors of nanotubes. We therefore studied even larger silsesquioxanes with 72, 84, 96, 144, 192, and 240 silicon atoms in tube-like shapes.[27] Figure 33.3 shows the three largest structures. We suggested a growth mechanism similar to the C_2 insertion into fullerenes.

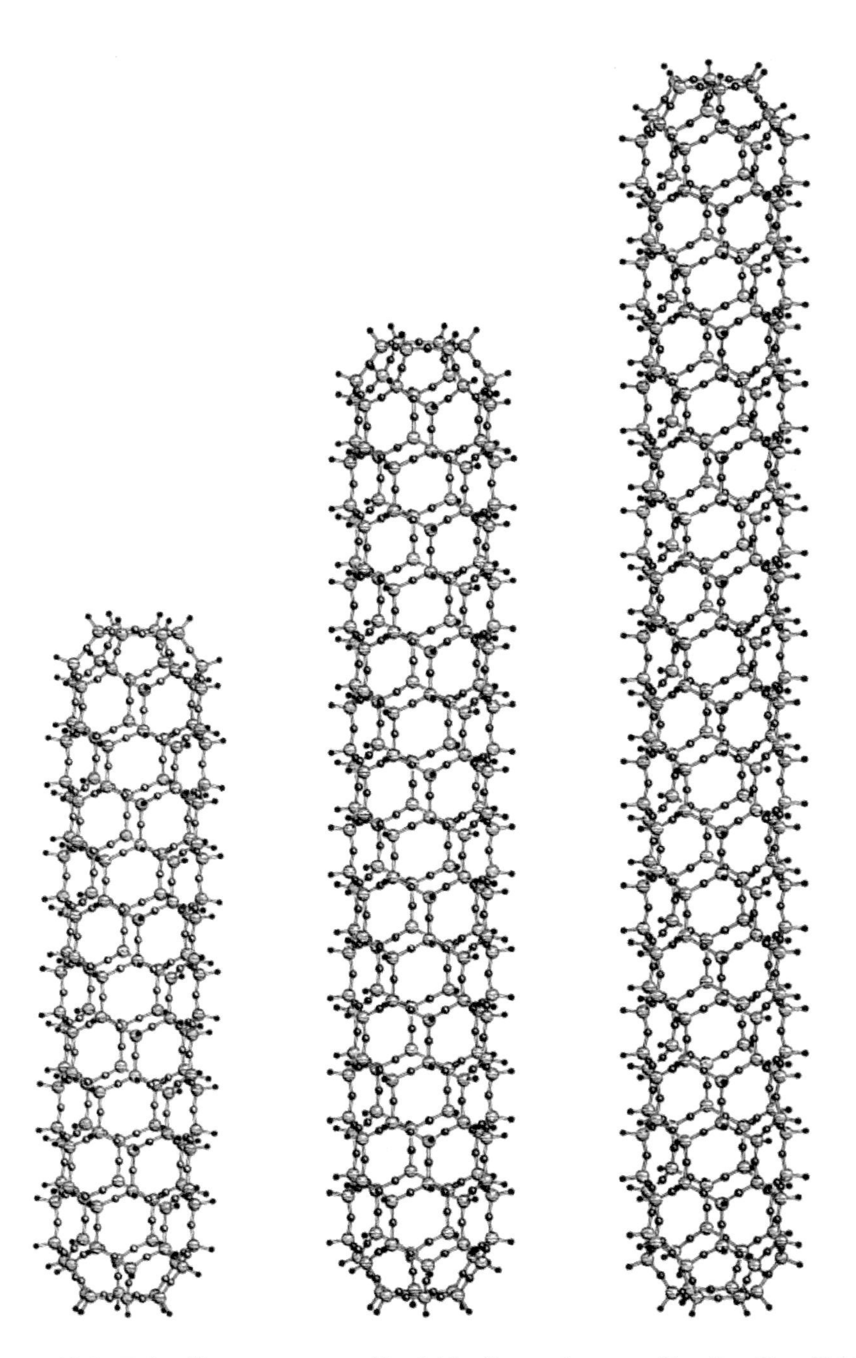

Figure 33.3. Tube-like structures of hydridosilsesquioxanes $Si_{144}O_{216}H_{144}$ (D_{6d}), $Si_{192}O_{288}H_{192}$ (D_{6h}), $Si_{240}O_{360}H_{240}$ (D_{6d}).

In order to test this mechanism, we simulated the growth of $Si_{30}O_{45}H_{30}$ to $Si_{32}O_{48}H_{32}$ by addition of $[Si(OH)(H)Cl]_2O$ and elimination of 2 HCl.[28] These compounds are of the size found [9] in the silsesquioxane mixtures formed by cage rearrangement of octasilsesquioxanes [8] in solution. Figure 33.4 shows the transition states for this process. A topological study showed that such an insertion mechanism with retention of Si 5-rings and 6-rings would cease to be applicable when the necessary local arrangement of 5- and 6-rings is no longer possible. We also found that isomerization involves a much higher barrier and can be excluded.

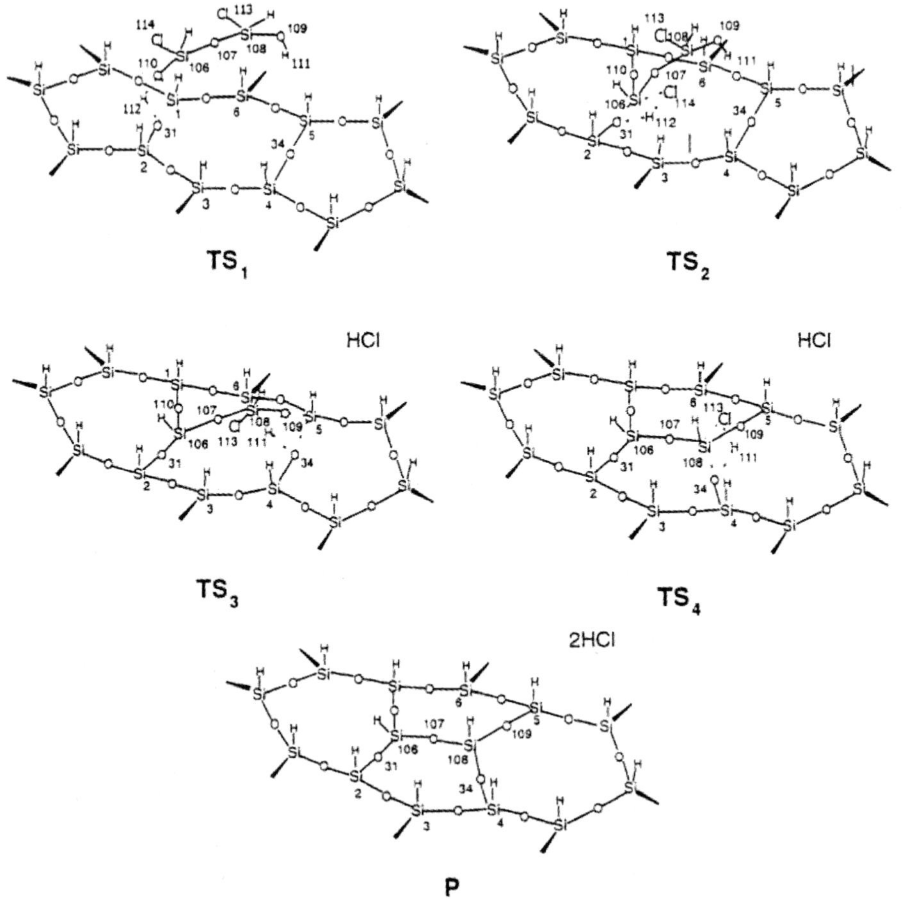

Figure 33.4. Structure of the transition states for the growth of $Si_{30}O_{45}H_{30}$ to $Si_{32}O_{48}H_{32}$ by addition of $[Si(OH)(H)Cl]_2O$.

We therefore decided to abandon the postulate of retention of only Si 5-rings and 6-rings and studied an insertion mechanism which allowed the generation of Si 7-rings. By topological considerations, it could be shown that such 7-rings can be converted to 6-rings in a subsequent step. MSINDO calculations for such non-classical reaction pathways resulted in energy profiles very similar to those for the classical pathways involving Si 5-rings and 6-rings only. The maximum barrier differs only by 0.2 kcal/mol.

33.4 Conclusion

The initiation of the reaction of $SiCl_4$ with O_2 for the formation of siloxanes is quite complicated and involves a large number of steps, including the generation of radicals. This explains the great variety of siloxanes produced in the process. The interesting structures of silsesquioxanes show similarities to the fullerenes. However, their growth pattern differs from that of the fullerenes not only in the violation of the 5-ring isolation rule, but also in the ease with which growth may occur via Si 7-rings.

33.5 References

[1] M. Binnewies, K. Jug, *Eur. J. Inorg. Chem.* **2000**, 1127.
[2] Y. Abe, in *Tailor-made Silicon-Oxygen Compounds*, (Eds.: R. Corriu, P. Jutzi), Vieweg, Braunschweig, Wiesbaden **1996**, p. 305.
[3] W. G. Bruland, M. E. Coltrin, P. Ho, *J. Chem. Phys.* **1986**, *59*, 3267.
[4] J. M. Jasenski, B. S. Meyerson, B. A. Scott, *Ann. Rev. Phys. Chem.* **1987**, *38*, 109.
[5] M. Binnewies, M. Jerzembek, A. Kornick, *Angew. Chem.* **1991**, *103*, 762; *Ange. Chem. Int. Ed. Engl.* **1991**, *30*, 745.
[6] H. Quellhorst, A. Wilkening, M. Binnewies, *Z. Anorg. Allg. Chem.* **1997**, *623*, 1871.
[7] R. West, in *Tailor-made Silicon-Oxygen Compounds*, (Eds.: R. Corriu, P. Jutzi), Vieweg, Braunschweig, Wiesbaden **1996**, p. 3.
[8] E. Rikowski, H. C. Marsmann, *Polyhedron* **1997**, *16*, 3357.
[9] R.-P. Krüger, H. Much, G. Schulz, E. Rikowski, *Monatsh. Chem.* **1999**, *109*, 5534.
[10] K. Jug, D. Wichmann, *J. Mol. Struct. (Theochem)* **1994**, *155*, 313.
[11] K. Jug, D. Wichmann, *J. Mol. Struct. (Theochem)* **1997**, *398–399*, 365.
[12] D. Wichmann, K. Jug, *Chem. Phys.* **1998**, *236*, 87.

[13] M. Junker, A. Wilkening, M. Binnewies, H. Schnöckel, *Eur. J. Inorg. Chem.* **1999**, 1531. See also the chapter 32 by M. Binnewies and N. Stöger in this book.
[14] A. Kumar, T. Homann, K. Jug, *J. Phys. Chem. A* **2002**, *106*, 6802.
[15] D. N. Nanda, K. Jug, *Theor. Chem. Acta* **1980**, *57*, 95; K. Jug, R. Iffert, J. Schulz, *Int. J. Quantum Chem.* **1987**, *32*, 265.
[16] D. Wichmann, Ph. D. Thesis, Universität Hannover 1997.
[17] G. Calzaferri, in *Tailor-made Silicon-Oxygen Compounds*, (Eds.: R. Corriu, P. Jutzi), Vieweg, Braunschweig, Wiesbaden **1996**, p. 149.
[18] A. R. Bassindale, T. E. Gentle, P. G. Taylor, A. Watt, in *Taylor-made Silicon-Oxygen Compounds,* Eds. R. Corriu, P. Jutzi, Vieweg, Braunschweig, Wiesbaden **1996**, p. 171.
[19] P. Jutzi, C. Batz, A. Mutluay, *Z. Naturforsch.* **1994**, *49b*, 1689.
[20] P. A. Agaskar, V. W. Day, W. G. Klemperer, *J. Am. Chem. Soc.* **1987**, *109*, 5554.
[21] H.-B. Bürgi, K. W. Törnroos, G. Calzaferri, H. Bürgy, *Inorg. Chem.* **1993**, *32*, 4914.
[22] K. W. Törnroos, *Acta Cryst. C* **1994**, *50*, 1646.
[23] K. W. Törnroos, H.-B. Bürgi, G. Calzaferri, H. Bürgy, *Acta Cryst. B* **1995**, *51*, 155.
[24] P. A. Agaskar, W. G. Klemperer, *Inorg. Chim. Acta* **1995**, *299*, 355.
[25] B. Ahlswede, K. Jug, *J. Comput. Chem.* **1999**, *20*, 563, 572.
[26] D. Wichmann, K. Jug, *J. Phys. Chem. B* **1999**, *103*, 10087.
[27] K. Jug, D. Wichmann, *J. Comput. Chem.* **2000**, *21*, 1549.
[28] K. Jug, I. P. Gloriozov, *Phys. Chem. Chem. Phys.* **2002**, *4*, 1062.
[29] K. Jug, I. P. Gloriozov, *J. Phys. Chem. A* **2002**, *106*, 4736.

34 Structurally Well-Defined Amphiphilic Polysiloxane Copolymers

Guido Kickelbick*, Josef Bauer, Nicola Hüsing†

34.1 Introduction

In recent years, the study of block copolymers has been an expanding field because of their extraordinary behavior with respect to phase separation or their compatibilizing properties at interfaces.[1] Amphiphilic macromolecules, having an affinity for both 'water' and 'oil', have attracted particular interest and find widespread technological applications.[2] Many of the widely used non-ionic commercially available amphiphilic block polymers typically have either polyether (Pluronic®; PEO-*b*-PPO-*b*-PEO with poly(propylene oxide) (PPO) and poly(ethylene oxide) (PEO) and Brij®; alkyl ether surfactants) or polyhydroxyl(glyceroglycolipid) sugar derivatives as the polar group.[3]

The extraordinary properties, easy functionalization, and hydrophobic nature of poly(dimethylsiloxane) (PDMS) make it attractive for the formation of speciality polymers such as block and graft copolymers. In the case of PDMS, the formation of several types of di- and triblock copolymers (Figure 34.1a) is particularly easily achieved, either using an approach whereby the 'living' chain end is extended in an ionic polymerization or by applying coupling techniques to end-functionalized polysiloxane blocks.[4]

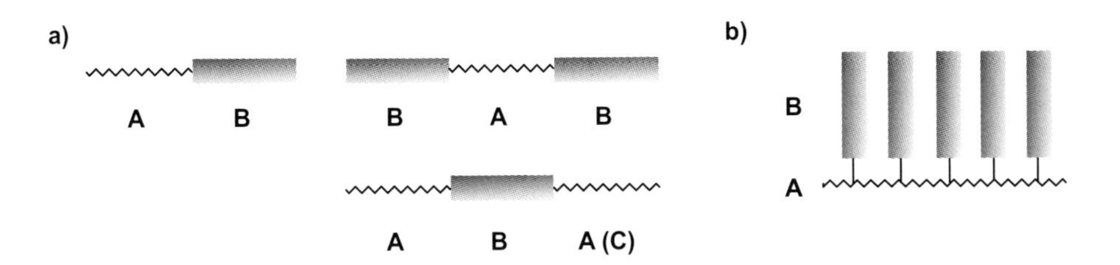

Figure 34.1. Schematic representation of (a) di- or triblock and (b) graft copolymers.

* Corresponding author

† Address of the authors: Institut für Materialchemie, Technische Universität Wien, Getreidemarkt 9/165, A-1060 Wien, Austria.

Another type of siloxane copolymers is graft copolymers, with the polysiloxane constituting either the backbone or the branches (Figure 34.1b). These grafted systems have been prepared, for example, by hydrosilation of Si-H group containing polysiloxanes with ω-allyl-PEO[5] or ω-undecenoyl-PEO.[6]

Block copolymer-type siloxane surfactants are mostly limited to the combination of hydrophobic siloxane backbones and organic hydrophilic substituents grafted onto the siloxane chain. Functionalized polymers with purely inorganic siloxane backbones have to be prepared either from monomers in which the functional group is protected before the commonly used ring-opening polymerization (ROP) or by a post-functionalization of the preformed block polymers.[7] Post-modification of the blocks is an appropriate means of preparing polysiloxane block copolymers with functional side groups that would otherwise deactivate the 'living' end groups in the sequential ROP of two different monomers. Amphiphilic block copolymers with a pure polysiloxane backbone consisting of a hydrophilic sequence formed by a polysiloxane chain with pendant carboxyl groups represent one of the few examples in this field.[8] Eq. 1 shows an example of the typically used post-modification of preformed block and/or graft copolymers via hydrosilation. Further examples of modifications are the introduction of alkyl polyethers as pendant chains into Si-H containing siloxane polymers by hydrosilation with allyl ether glycols[5] or by reaction with allyl glycidyl ether.[9] A subsequent sulfonation was performed to obtain polymer electrolytes.[10]

$-[Si(CH_3)(H)-O]_m-$ + allyl glycidyl ether $\xrightarrow[\text{2. } NaS_2O_5 / H_2O]{\text{1. Pt}}$ $-[Si(CH_3)(CH_2CH_2CH_2O\text{-}CHOHCH_2SO_3^-Na^+)-O]_m-$ (1)

Thio-ene addition is another attractive method for the attachment of organic thioether groups to polysiloxane chains. It is based on the free-radical addition of a thiol group to a vinyl-substituted siloxane or the addition of a terminal alkene to a thiol-substituted siloxane (Eq. 2).[11] Boileau et al.[12] used this reaction to graft different amino acids to the polymer backbone to induce potential interactions with polar surfaces.

$-[Si(CH=CH_2)(H)-O]_m-$ + $HSCH_2CH(NHCOCH_3)CO_2H$ $\xrightarrow{\text{'Pt'}}$ $-[Si(CH_2CH_2SCH_2CH(NHCOCH_3)CO_2H)(H)-O]_m-$ (2)

We were interested in two types of polysiloxane diblock copolymers, both having a hydrophobic PDMS block, the second block consisting either of a hydrophilic PEO block or a hydrophilic polysiloxane block.

34.2 Synthesis of Block Copolymers

34.2.1 Preparation of PDMS-*b*-PEO Copolymers

'Living' anionic ring-opening polymerization is probably the most frequently used method for the controlled synthesis of siloxane polymers with narrow molecular weight distributions ($M_w/M_n < 1.2$) and defined block composition. Strong bases, such as organolithium compounds, and THF as solvent are typically used for the anionic ring-opening of cyclosiloxanes.[13] Applying this technique, anionic polymerization of hexamethylcyclotrisiloxane (D_3) is initiated by butyl lithium, which forms the silanolate $Bu[Si(CH_3)_2O]_3Li$ quantitatively in hydrocarbon solvents. However, polymerization does not occur in the absence of a donor solvent such as THF, glyme, HMPA, or DMSO. When small amounts of these promoters are present, it is possible to obtain PDMS with controlled molecular weight and a narrow molecular weight distribution.

The propagation step is reversible because a reverse back–biting reaction of the active propagation center with its own chain may reproduce the monomer. The back–biting process also leads to the generation of a series of monomer homologues of various ring sizes. The silanolate center may attack another chain as well, leading to chain transfer, which results in chain randomization (Eq. 3). In the absence of acidic contaminants, the reaction proceeds without termination. Thus, the polymerization must be quenched to deactivate the silanolate center.

$$(Me_2SiO)_m \longrightarrow \left[-\overset{CH_3}{\underset{CH_3}{Si}}-O- \right]_x \rightleftharpoons \left[-\overset{CH_3}{\underset{CH_3}{Si}}-O- \right]_y + (Me_2SiO)_n \qquad (3)$$

Linear Polymer (Kinetic Product)

Equilibrium Mixture of Linear Polymer and Cyclics (Thermodynamic Products)

Several methods for the termination of a 'living' anionic polymerization and hence the introduction of a reactive end group onto the polymer chains are available. Si-H-functionalized polysiloxanes are obtained by quenching the 'living' end groups with chlorodimethylsilane (Eq. 4). The homooligomers/

homopolymers serve as the hydrophobic part in the synthesis of the amphiphilic block copolymers discussed below.

$$\mathrm{BuLi} + x\ (\text{hexamethylcyclotrisiloxane}) \xrightarrow[\mathrm{RT/Ar}]{\mathrm{THF}} \mathrm{Bu}{-}\!\left(\mathrm{Si(CH_3)_2}{-}\mathrm{O}\right)_n\!{-}\mathrm{Si(CH_3)_2}{-}\mathrm{O^-Li^+} \xrightarrow[\text{in situ, RT/Ar/THF}]{\mathrm{Cl{-}Si(CH_3)_2{-}H}} \mathrm{Bu}{-}\!\left(\mathrm{Si(CH_3)_2}{-}\mathrm{O}\right)_n\!{-}\mathrm{Si(CH_3)_2}{-}\mathrm{H} \quad (4)$$

A commonly used hydrophilic polymer block in amphiphilic block copolymers is poly(ethylene oxide) (PEO). An end-functionalized PEO block was prepared by an etherification reaction between an OH-terminated polyethylene oxide of a defined block length and allyl bromide under basic conditions.[14,15]

In addition to the PEO blocks, short–chain hydrophiles, such as (allyloxyethoxy)-2-ethyl ether, were prepared by a similar reaction.

Si-H-functionalized polysiloxanes and allyl-functionalized PEO blocks were coupled by hydrosilation reactions using H_2PtCl_6 (Speier catalyst)[16] or $Pt_2[(CH_2CHSiMe_2)_2O]_3$ (Karstedt catalyst) (Eq. 5). This method allows for variation of the block lengths and, therefore, of the relative molar ratios, over a wide range.

$$\mathrm{Bu}{-}\!\left(\mathrm{Si(CH_3)_2}{-}\mathrm{O}\right)_m\!{-}\mathrm{Si(CH_3)_2}{-}\mathrm{H} + \mathrm{CH_2{=}CH{-}CH_2{-}O}\!\left(\mathrm{CH_2CH_2O}\right)_n\!\mathrm{CH_2CH_2{-}O{-}CH_3} \xrightarrow[\text{b) Speier Catalyst / Toluene}]{\text{a) Karstedt Catalyst / THF or}} \mathrm{Bu}{-}\!\left(\mathrm{Si(CH_3)_2}{-}\mathrm{O}\right)_m\!{-}\mathrm{Si(CH_3)_2}{-}\mathrm{CH_2CH_2CH_2{-}O}\!\left(\mathrm{CH_2CH_2O}\right)_n\!\mathrm{CH_2CH_2{-}O{-}CH_3} \quad (5)$$

All the prepared polymers had narrow molecular weight distributions ($M_w/M_n < 1.4$) as measured by size exclusion chromatography (SEC). The completeness of the hydrosilation reactions, as determined by ^{1}H NMR experiments, was typically of the order of 95% or higher. The purity of the synthesized diblock copolymers was > 95%. The molecular weight determined from SEC experiments (Figure 34.2) on the PEO and PDMS homopolymers and the coupled diblock copolymer showed a marked shift to a lower retention volume (thus to a higher molecular weight) after the hydrosilation reaction. This shift, and the absence of signals in the homopolymer range, as well as the disappearance of the allyl and Si-H signals in the ^{1}H and ^{13}C NMR spectra, indicated complete conversion in the coupling reaction.

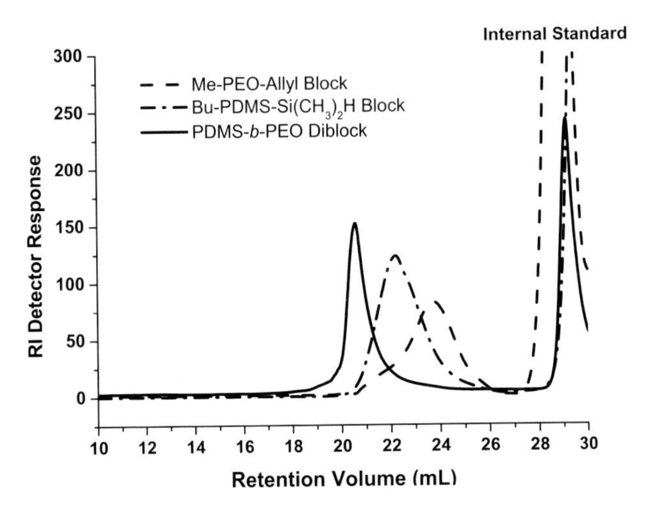

Figure 34.2. Comparison of SEC chromatograms of PEO (M_n = 585 g/mol, PD = 1.15) and PDMS (M_n = 1320 g/mol, PD = 1.24) homopolymers and $PDMS_{18}$-*b*-PEO_{12} diblock copolymer (M_n = 1780 g/mol, PD = 1.34) prepared by hydrosilation (internal standard: diphenyl ether).

34.2.2 Preparation of Poly(methylvinylsiloxane)–Polydimethylsiloxane Diblock Copolymers (PMVS-*b*-PDMS)

Poly(methylvinylsiloxane) (PMVS) was prepared by anionic ring-opening polymerization of either 1,3,5,7-tetramethyl-1,3,5,7-tetravinylcyclotetrasiloxane (D_4^V) or 1,3,5-trimethyl-1,3,5-trivinylcyclotrisiloxane (D_3^V). Cyclotrisiloxanes are generally preferred for this type of polymerization because of their higher ring strain compared to cyclotetrasiloxanes, resulting in a greater ring-opening tendency. For example, the ring strain energy of hexamethylcyclotrisiloxane (D_3) was determined to be between about 8.5 to 16.7 kJ/mol, while octamethylcyclotetrasiloxane (D_4) is practically strain–free (1.05 kJ/mol).[17,18] Furthermore, it was shown that during the polymerization process the equlibrium weight fraction of non-preferred siloxane rings increases when bulkier substituents, e.g. vinyl groups, are present.[19,20]

For the preparation of a vinyl-functionalized polysiloxane block by anionic ring-opening polymerization commercially available D_4^V was used and compared with D_3^V. Because of the diminished ring strain in D_4^V, the ring-opening reaction needs additional activation compared to that of cyclotrisiloxanes.[21,22]

It was shown in the literature that an increased nucleophilicity of the initiating anion leads to a better efficiency of the initiator. For example, Me_3SiCH_2Li proved to be superior to commonly used linear organolithium initiators, especially in combination with cryptands.[23-25] Contrary to the *n*-BuLi/THF system, polymerization of D_4^V was possible when Me_3SiCH_2Li /THF was used as the initiator. However, the resulting polymers revealed broad molecular weight distributions, and an increased formation of larger cycles (D_5^V and D_6^V) throughout the reaction was confirmed by GC/MS analyses. With [12]crown-4 as the promoter, the polymerization showed a much higher conversion (up to 80%) in THF with quite narrow molecular weight distributions of the resulting polymers. After 75% conversion of D_4^V, the formation of larger cycles (D_5^V and D_6^V) by 'backbiting' reactions again became more noticeable by a broadening of both the polymer and monomer signal in SEC measurements, which was also confirmed by GC/MS analysis.[26]

Kinetic studies of the anionic polymerizations of the cyclosiloxanes D_3, D_3^V and D_4^V were performed in THF using *n*-BuLi or Me_3SiCH_2Li as initiators at room temperature. The propagation rate constants increased in the row $D_4^V < D_3 < D_3^V$. Moreover, a small difference of the propagation rates of 1,3,5-trimethyl-1,3,5-trivinylcyclotrisiloxane D_3^V was noticed when different initiator systems were used. The propagation rate k_{px} of D_3^V using *n*-BuLi as the initiator was 1.7 times higher than in the case of Me_3SiCH_2Li.

34.2.3 Amphiphilic Polysiloxane Block Copolymers

Although polysiloxanes have superior properties compared to common organic polymers, the controlled formation of amphiphilic block copolymers consisting of a pure polysiloxane backbone remains essentially unexplored. There have been only a few reports of such polymers in the literature, for example one bearing pendant carboxyl groups on the polysiloxane chain,[8] and another prepared by a one-pot sequential anionic ring-opening polymerization of two cyclotrisiloxanes.[27]

34.2.3.1 Functionalized block copolymers by sequential monomer addition. Sequential addition of monomers to a 'living' anionic polymerizing system is at present the most useful method for the synthesis of well-defined block copolymers.[28,29] An AB diblock copolymer is produced by first completely polymerization of one monomer (A) using an initiator such as butyllithium. The second monomer (B) is then added to the 'living' anions. When the second monomer has reacted completely, a terminating agent is introduced into the reaction mixture and the block copolymer is isolated.[30] This method can be used to synthesize any of the various types of block copolymers (di-, tri-, tetrato multiblocks) by employing the appropriate sequence of additions of the monomers, provided that each propagating anion can initiate the polymerization of the next monomer. This synthetic strategy has already been employed in the synthesis of AB– or ABA–type siloxane block copolymers with controlled structures. Important examples of macromolecules produced by 'living' anionic polymerizations include the copolymers of dimethylsiloxane with diphenylsiloxane, styrene, methyl methacrylate, and several other vinyl monomers.[31,32] Siloxane block copolymers with a pure siloxane backbone in which the substituents are varied in the cyclotrisiloxane, have also been synthesized by sequential 'living' anionic polymerization. Examples are di- and triblock copolymers of dimethylsiloxane and (fluoroalkylmethyl)siloxane,[33] as well as the controlled synthesis of block copolymers of diphenylsiloxane and dimethylsiloxane.[34] However, it would be desirable if larger rings could also be used as starting materials, because cyclic trimers are often not commercially available and their preparation tends to be low-yielding.

The controlled synthesis of polysiloxane block copolymers with pendant functional groups on one of the blocks is not an easy task; the anionic polymerization technique tolerates only certain functional groups because of possible termination reactions with the propagating chain end. In particular, polar groups such as alcohols or carboxylic acids, which are interesting for the introduction of hydrophilic properties, usually give rise to termination reactions due to their acidic character. Two possible methods can be employed to avoid such interactions: the use of protecting groups or the use of organic functionalities that do not interact with the chain end during polymerization and can be transformed into the desired polar functional groups after the block synthesis. An example of the latter methodology is the polymerization of cyclosiloxanes with vinyl groups and their subsequent functionalization.

34.2.3.2 Ring-opening polymerization of D_4^v extended by D_3. Block copolymers from 'living' ring-opening polymerizations are obtained by using a polymer as a macroinitiator and extending it with a second monomer. For the PDMS-*b*-PMVS system, it was found that the extension of a PDMS chain with PMVS by polymerization of D_4^v only leads to the addition of up to five monomeric units (n = 15 to 20) until the equilibrium between polymeric and ring species is reached, while initial polymerization of D_4^v allows for a much better control of the reaction.

Differences in the sequence of polymerization were investigated and it became apparent that a better control is obtained if the slower propagating monomer (D_4^v) is polymerized first. The method permits the synthesis of block copolymers with molecular weight distribution < 1.4 and high block purity.

The kinetically controlled 'living' anionic polymerizations of first D_4^V and then D_3 were carried out in two steps under argon at room temperature. Purified D_4^V was first allowed to react with the initiator in THF for 3 h, in order to quantitatively afford lithium silanolates. Then, the activating agent, [12]crown-4, was added to the stirred mixture. After five days, when up to 80% of the D_4^V ($M_w/M_n \approx 1.40$) had been consumed, a solution of D_3 in THF was introduced (Eq. 8).

$$H_3C{-}Si(CH_3)_2{-}CH_2Li + x\,D_4^V \underset{\text{2. Promoter}}{\overset{\text{1. THF / Ar /RT}}{\rightleftharpoons}} H_3C{-}Si(CH_3)_2{-}CH_2{-}\left(Si(CH_3)(CH{=}CH_2){-}O\right)_m{-}Si(CH_3)(CH{=}CH_2){-}O^- Li^+$$

$$\xrightarrow[\text{2. } Me_3SiCl]{\text{1. } y\,D_3} H_3C{-}Si(CH_3)_2{-}CH_2{-}\left(Si(CH_3)(CH{=}CH_2){-}O\right)_m\left(Si(CH_3)_2{-}O\right)_n{-}Si(CH_3)_2{-}CH_3 \quad (8)$$

The polymerization of D_3 was followed by GC/MS. After 36 h, the reaction was quenched by the introduction of trimethylchlorosilane. About 92% of the D_3 had been converted, while the amount of unconverted D_4^V had not changed significantly. ^{29}Si NMR analyses allowed the determination of the sequence distribution of repeat units, which showed no random copolymerization of D_4^V and D_3 as in the case of diblock copolymers prepared by sequential anionic copolymerization of D_3 extended with D_4^V. Because the polymerization of the first monomer could not be carried out to completion (<100% conversion) without increasing the molecular weight distribution, the second monomer (with faster propagation rates) had to be introduced before the equilibration reaction became established. Therefore, unreacted monomer from the first step was still in solution when the second monomer was added. The risk of random copolymerization can be suppressed if the second monomer has far higher reactivity towards polymerization than the first monomer. The block formed in the second step contained only a few methylvinylsiloxane units, i.e. the 'block purity' was very high.

34.2.4 Functionalization of the PMVS-*b*-PDMS Copolymers

The PMVS-*b*-PDMS diblock copolymers still display purely hydrophobic properties, but the vinyl groups can easily be transformed into a variety of other functional groups. The double bonds of the methylvinylsiloxane blocks were modified either by hydrosilation[35] or by epoxidation.[36] In the latter case, further modification reactions of the polyepoxide block were carried out (Eq. 9).

R-O-O-R

R_3SiH, "Pt"

(9)

The hydrosilation of the double bonds with Karstedt catalyst at 70 °C was quantitative. $HSi(OEt)_3$, $HSi(OMe)_3$ and $HSi(CH_3)_2Cl$ were used for this reaction to obtain a functionalized polysiloxane block. While the trialkoxy-silane groups can be hydrolyzed and used for cross–linking reactions, the chlorosilanes are able to react with a variety of nucleophiles, such as alcohols, carboxylic acids, etc.

Another possibility for the functionalization of the double bond is epoxidation and subsequent ring-opening of the oxirane ring. The epoxide was quantitatively formed by selective oxidation of the double bonds with *m*-chloroperoxybenzoic acid (MCPBA). The obtained epoxide was opened with various mono- or difunctional nucleophiles, such as hydroxide ions, diamines, diols, dicarboxylic acids, hydroxy-functionalized ethers, and carboxylic acid chlorides (Eq. 10).[37] These reactions resulted in the formation of a hydroxide group at one carbon atom and the coupling product with the nucleophile at the other. Nucleophilic attack of the epoxide at the more hindered (inner) carbon atom was not evident from the ^1H NMR spectra. Acid- or base-catalyzed hydrolysis of the epoxidized block copolymers led to the corresponding 1,2-diol side groups. Epoxide ring opening was quantitative. The processing parameters of the ring-opening reactions allowed for the preparation of different materials. High dilution of the block copolymers in the solvent during the ring-opening led to linear polymers, while cross–linked hydrogels were formed at low dilution. To completely suppress cross–linking reactions using multifunctional nucleophiles, a large excess of reactants is necessary and harsh purification conditions should be avoided. For a total prevention of cross–

linking reactions it is advisable to use monofunctional protected ring-opening agents and to remove the protecting groups in a further step.

Hydrolysis

$LiClO_4$, $H_2NCH_2CH_2NH_2$

Pyridine, $(HOOCCH_2)_2$

$LiClO_4$, $HOCH_2CH_2OH$

$LiClO_4$, $HOCH_2CH_2OBu$

Et_4NCl, CH_3COCl

(10)

34.3 Phase Behavior

We investigated the phase behavior of selected samples of short chain PDMS-*b*-PEO (repeating units < 20) diblock copolymers by means of several techniques, such as optical polarization microscopy, surface tension measurements, ^{2}H NMR spectroscopy, small angle X-ray scattering (SAXS), fluorescence spectroscopy, and cryogenic transmission electron microscopy (*cryo* TEM). The samples showed lamellar phases over a wide range of compositions, concentrations, and temperatures (in water), and vesicles at low concentrations (< 0.12 wt%) of the diblock copolymers in water. On systematically increasing the hydrophobic block volume (by lengthening the siloxane block at a constant PEO volume), the lamellar phase region became smaller and destabilized, which correlates with the formation of turbid dispersions at lower concentrations.

34.4 Summary

Polysiloxane–containing amphiphilic block copolymers have been prepared by different approaches. Coupling of an end-functionalized polydimethylsiloxane with a functionalized poly(ethylene oxide) led to the formation of PDMS-*b*-PEO diblock copolymers. The sequential anionic ring-opening polymerization of tetramethyltetravinylcyclotetrasiloxane and hexamethylcyclotetrasiloxane resulted in the formation of a vinyl-substituted diblock copolymer, the vinyl groups of which could be modified by further reactions so as to import amphiphilic character. The phase behavior of short–chain PDMS-*b*-PEO diblock copolymers revealed the preferred formation of lamellar phases by this type of amphiphile

34.5 Acknowledgement

We gratefully acknowledge Wacker-Chemie for their kind donation of chemicals. Furthermore, we like to thank the group of Krister Holmberg and Anders Palmqvist at Chalmers University of Technology for their support in determining the phase behavior of the block copolymers.

34.6 References

[1] D. F. Evans, H. Wennerström, *The Colloidal Domain Where Physics, Chemistry, Biology and Technology Meet.*, VCH, New York, 1994.
[2] S. Förster, M. Antonietti, *Adv. Mater.* **1998**, *3*, 195.
[3] B. Jönsson, B. Lindman, K. Holmberg, B. Kronberg, *Surfactants and Polymers in Aqueous Solution*, J. Wiley, New York, 1998.
[4] T. J. Barton, P. Boudjouk, *Adv. Chem. Ser.* **1990**, *224*, 3.
[5] E. Wu, I. M. Khan, J. Smid, *Polym. Bull.* **1988**, *20*, 455.
[6] Y. Sela, S. Magdassi, N. Garti, *Colloid Polym. Sci.* **1994**, *272*, 684.
[7] G. Belorgey, G. Sauvet in R. G. Jones, W. Ando, J. Chojnowski (Eds.), *Silicon-Containing Polymers*, Kluwer Acad. Publ., Dordrecht, 2000, p 43.
[8] M. Scibiorek, N. K. Gladkova, J. Chojnowski, *Polym. Bull.* **2000**, *44*, 377.
[9] G. B. Zhou, I. M. Khan, J. Smid, *Polym. Prepr. (Am. Chem. Soc., Div. Polym. Chem.)* **1989**, *30(1)*, 416.
[10] G. B. Zhou, I. M. Khan, J. Smid, *Polym. Commun.* **1989**, *30*, 52.
[11] B. Boutevin, F. Guida-Peitrasanta, A. Ratsimihety in R. G. Jones, W. Ando, J. Chojnowski (Eds.), *Silicon-Containing Polymers*, Kluwer Acad. Publ., Dordrecht, 2000, p 79.

[12] S. Abed, S. Boileau, L. Bouteiller, J. R. Caille, N. Lacoudre, D. Teyssiè, J. M. Yu, *Polym. Mater. Sci. Eng.* **1997**, *76*, 45.
[13] L. Wilczek, J. Kennedy, *Polymer* **1987**, *19*, 531.
[14] H. Riemschneider, *Monatsh. Chem.* **1959**, *90*, 787.
[15] T. N. Mitchell, K. Heesche-Wagner, *J. Organomet. Chem.* **1992**, *436*, 42.
[16] J. L. Speier, J. A. Webster, G. H. Barnes, *J. Am. Chem. Soc.* **1957**, *79*, 974.
[17] R. G. Jones, W. Ando, J. Chojnowski (Eds.), *Silicon-Containing Polymers*; Kluwer Acad. Publ., Dordrecht, 2000.
[18] W. A. Piccoli, G. G. Haberland, R. L. Merker, *J. Am. Chem. Soc.* **1960**, *82*, 1883.
[19] J. A. Semlyen in S. Clarson, J. Semlyen (Eds.), *Siloxane Polymers*; PTR Prentice Hall, London, 1993, pp 135.
[20] J. A. Semlyen, *Adv. Polym. Sci.* **1976**, *21*, 41.
[21] S. Hubert, P. Hèmery, S. Boileau, *Makromol. Chem., Macromol. Symp.* **1986**, *6*, 247.
[22] M. Jelinek, Z. Laita, M. Kucera, *J. Polym. Sci. C* **1967**, *16*, 431.
[23] S. Boileau, *Makromol. Chem., Makromol. Symp.* **1993**, *73*, 177.
[24] J. M. Yu, D. Teyssie, R. B. Khalifa, S. Boileau, *Polym. Bull.* **1994**, *32*, 35.
[25] K. Ròzga-Wijas, J. Chojnowski, T. Zundler, S. Boileau, *Macromolecules* **1996**, *29*, 2711.
[26] J. Bauer, N. Hüsing, G. Kickelbick, *J. Polym. Sci. Part A: Polym. Chem.* **2002**, *40*, 1539.
[27] J. Chojnowski, M. Cypryk, W. Fortuniak, K. Kazmierski, M. Scibiorek, K. Rozga-Wijas, *Polym. Prepr.* **2001**, *42(1)*, 227.
[28] M. Morton, L. J. Fetters in C. E. Schildknecht, I. Skeist (Eds.), *Anionic Polymerizations and Block Copolymers*, Wiley, New York, 1977.
[29] M. Morton, *Anionic Polymerization: Principles and Practice*; Academic Press, New York, 1983.
[30] L. J. Fetters, E. M. Firer, M. Defauti, *Macromolecules* **1978**, *10*, 1200.
[31] J. B. Plumb, J. H. Atherton in A. Noshay, J. E. McGrath (Eds.), *Block Copolymers: Overview and Critical Survey*, Academic Press, New York, 1977.
[32] P. C. Juliano, *New Silicon Elastoplastics*, General Electric Co., New York, 1974.
[33] M. J. Owen in D. J. Meier (Ed.), *Block Copolymers, Science and Technology*, Academic Publ., Harwood, 1983; Vol. 3, p 129.
[34] N. V. Gzovidc, J. Ibemesi, C. C. Meier, *Macromol. Symp., Proc.* **1982**, *28*, 168.
[35] Y. Chang, Y. C. Kwon, S. C. Lee, C. Kim, *Polym. Prepr.* **1999**, *40(2)*, 269.
[36] F. Macchia, P. Crotti, M. Chini, *Tetrahedron Lett.* **1990**, *31*, 4661.
[37] J. Bauer, N. Hüsing, G. Kickelbick, *Chem. Commun.* **2001**, 137.

35 Synthesis and Functionalization of Mesostructured Silica-Based Films

Nicola Hüsing*, Beatrice Launay, Guido Kickelbick†

35.1 Introduction to Mesostructured Materials

In the early 1990s, researchers discovered that in addition to single molecules such as tetramethylammonium bromide used for the preparation of zeolites, molecular assemblies, as found in liquid crystals, can be used for templating inorganic matrices.[1,2] With this discovery, research in the field of templating and patterning inorganic materials so as to obtain perfectly periodic, regularly sized and shaped cavities, channels, and layers has expanded dramatically. This supramolecular templating relies on the ability of amphiphilic molecules to self-assemble into micellar structures that, when concentrated in aqueous solutions, undergo a second stage of self-organization resulting in lyotropic liquid crystal-like mesophases. Molecular inorganic species can cooperatively co-assemble with these structure-directing agents (templates) to eventually condense and form the mesoscopically ordered inorganic backbone of the final material (Figure 35.1). The mesostructured nanocomposite is typically either calcined, ozonolyzed, or solvent–extracted to obtain a porous inorganic material in which the pore dimension relates to the chain length of the hydrophobic tail of the template molecule.

There are some excellent review articles on different aspects of mesostructured materials, such as synthesis, properties, and applications.[3-6] Extensive research effort has been devoted to the exploitation of new phases (lamellar, cubic, hexagonal structures), expansion of the pore sizes (about 2–50 nm are accessible), and variable framework compositions (from pure silica, through mixed metal oxides to purely metal oxide-based frameworks, and inorganic–organic hybrid mesostructures). Another research focus is on the formation of mesostructured materials in other morphologies than powders, e.g. monolithic materials and films, which are required for a variety of applications including, but not limited to, sensors (based on piezoelectric mass balances or surface acoustic wave devices), catalyst supports, (size– and shape–selective) filtration membranes or (opto)electronic devices. The current article is focused

* Corresponding author
† Address of the authors: Institut für Materialchemie, Technische Universität Wien, Getreidemarkt 9/ 165, A-1060 Wien, Austria.

on the different methods of film formation and the possibilities for their modification and functionalization.

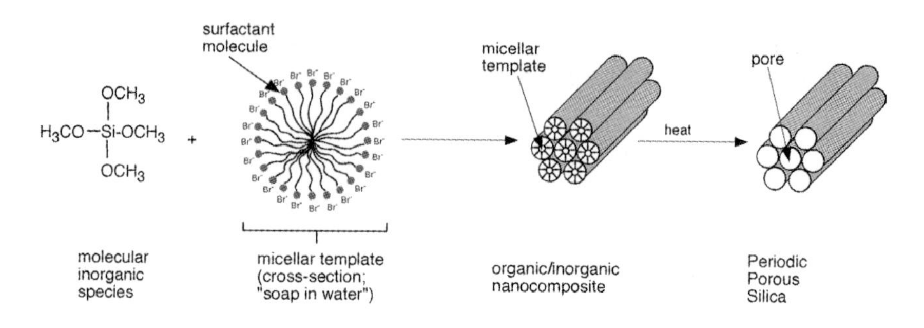

Figure 35.1. Schematic representation of the supramolecular templating (reprinted from ref.[1] with friendly permission of Nature Publishing Group).

35.2 Synthesis of Mesostructured Thin Films

Mesostructured thin silica films can be prepared as freestanding films or supported by a variety of different substrates. In the following, a short description of the different synthesis methods is given; the synthesis and the formation mechanism of mesoporous silica films has been reviewed in detail elsewhere.[4]

35.2.1 Films Grown Spontaneously from Solution (Solid–Liquid, Liquid–Liquid and Liquid–Vapor Interface)

Continuous mesoporous silica films can be grown at interfaces through interfacial silica–surfactant self-assembly processes from surfactant solutions above the critical micelle concentration. These interfaces include substrates such as mica or graphite, which are immersed in an acidic solution of a mixture of surfactant (e.g., hexadecyltrimethylammonium salts) and a silica source such as tetraethoxysilane. Three-dimensional surfactant structures form at the solid–liquid interface in which the final structure is determined by physical (and/or) energy constraints on the first micellar layer (depending on the substrate).[7,8] Following the nucleation of small oriented structures on the surface, these structures grow and eventually coalescence to form a continuous mesostructured film. The films have a thickness of 0.2 to 1 μm, depending on the time for which the substrate is immersed in the solution, which is typically in the range of a couple of hours to weeks. After calcination, the films are truly mesoporous materials, with hexagonally packed one-dimensional channels that run parallel to the film surface. Under similar reaction conditions, freestanding films can be grown at air–water and oil–water interfaces.[9,10]

35.2.2 Films Based on Solvent Evaporation Techniques (Spin–Coating, Dip–Coating, Casting)

Periodic mesoporous continuous thin films can also be prepared by processes that rely on solvent evaporation induced self-assembly (EISA).[11,12] The first detailed study on the process was performed on dip-coated samples and can be described as follows. The starting point is a homogeneous solution of ethanol, water, hydrochloric acid, soluble silica, and surfactant, at a concentration far below that at which micelles or other aggregates are formed. Upon withdrawal of the substrate from the sol, preferential evaporation of the ethanol concentrates the film in water, silica species, and surfactant. Therefore, the surfactant concentration is progressively increasing, resulting in the formation of micelles and, upon further evaporation of ethanol, in the formation of liquid crystal-like mesophases consisting of silica–surfactant co-assemblies. This process allows the formation of a mesostructured nano-composite film within a few tens of seconds (Figure 35.2).

After template removal, a mesoporous material is obtained in which the pores can be arranged in either a hexagonal or a cubic fashion.

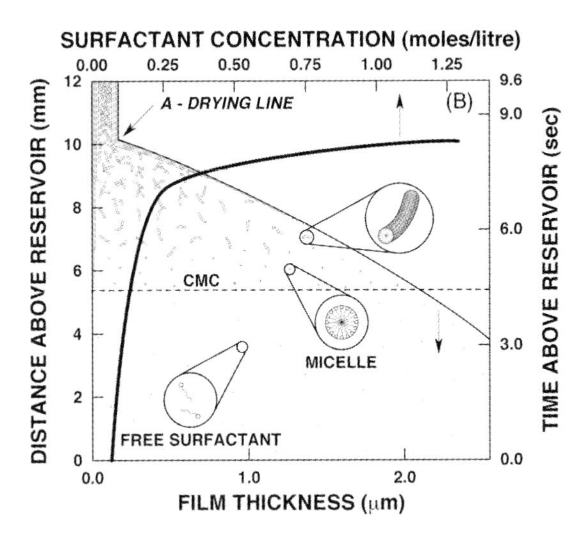

Figure 35.2. Schematic representation of the dip coating process (reprinted from ref.[12] with friendly permission of Wiley-VCH).

This mechanism has been utilized in casting, spin–coating and dip–coating processes, and a variety of films differing in pore size and pore orientation has been synthesized.[13-15] The principle has been investigated extensively for silica-based systems. Recently, it was shown that mesostructured transition metal oxide based systems can also be synthesized by using the EISA process.[16]

Electrodeposition or pulsed-laser deposition techniques are additional possibilities for forming thin mesostructured films. These techniques have mainly been applied to non-silica mesostructured films such as niobia or platinum, but also hexagonally oriented mesoporous silica films were prepared by pulsed-laser deposition.[17]

35.3 Functionalization of Mesostructured Thin Films

For many applications, e.g. in the area of sensing, catalysis or separations, modifications of the inorganic backbone are required to provide a certain specific surface chemistry or active sites on the inner pore surface. Different routes, such as simple inclusion of molecules, or the covalent attachment of functional entities, are applied. Because of the specific synthesis strategy, the structure-directing agent can also be used for functionalization of the mesostructured material. Only *in situ* modifications are discussed; post-synthesis modifications of the final porous material are, in principle, possible, but are beyond the scope of this article.

35.3.1 Modification by Inclusion of Functional Moieties

For the inclusion of functional moieties, control of the location and the orientation of the guest species within the host architecture is of considerable importance. The supramolecular templating approach allows the *in situ* controlled placement of molecules due to the formation of reservoirs of different polarity (the hydrophobic interior of the micelles, the hydrophilic compartment, the inorganic framework) during the synthesis.[18] However, the synthesis conditions must be carefully chosen because the mesostructure formation involves a delicate interplay between micelle formation, condensation of the inorganic framework, and the formation of the film.

It has been shown that mesostructured silica-based films are excellent hosts for photochromic dyes such as spiropyrans and spirooxazines, which are predominantly incorporated into the hydrophobic cores of the formed micelles.[19]

35.3.2 Modification by Co-condensation Reactions

Through covalent attachment of organic groups to the inorganic backbone of mesostructured silica films, the pore surface characteristics, such as polarity, hydrophilic–hydrophobic balance or functionality, can be systematically and deliberately tailored. A direct route for the covalent modification of mesoporous silica–based films is the co-condensation of tetraalkoxysilanes with organotrialkoxysilanes in the presence of a structure-directing agent. Simple organic groups, functional groups (e.g. mercapto, amine, quaternary amine), and more complex systems such as organic dyes (3-(2,4-dinitrophenylamino)propyltriethoxysilane) all covalently attached to an $Si(OMe)_3$ unit, have been used in the modification of thin films.[20-22]

It was shown that the pores can be modified without destruction of the periodic pore structure. However, for template removal calcination cannot be used because the organic groups would be destroyed as well. Thus, a solvent extraction process has to be applied.

The dual approach of co-condensation and controlled placement has been demonstrated by using methacrylate-substituted trialkoxysilanes in the presence of tetraalkoxysilanes and methacrylate-containing organic monomers as precursors. The inorganic network is modified by methacrylate groups, which can be polymerized with the methacrylate monomers that are located within the hydrophobic micelle reservoir during film formation. Polymerization reactions led to a nanocomposite coating, which mimics nacre with alternating organic and inorganic layers.[23]

A different network structure is obtained when the organic groups link two or more $Si(OR)_3$ groups, that is when silanes of the type $(RO)_3Si\text{-}X\text{-}Si(OR)_3$ are employed as precursors. The use of such precursors permits the organic groups to become an integral part of the network rather than just a pendant functionality. This has been demonstrated for different bridged silsequioxanes such as ethane- (a), ethene- (b), benzene- (c), and thiophene- (d) derivatives (Figure 35.3), which are processed either alone or in the presence of a tetraalkoxysilane by using spin or dip coating, or starting from bulk silica-surfactant solutions above the critical micelle concentration.[24,25]

The incorporation of integral organic groups and functionalities can result in synergistic properties promising an unprecedented ability to tune the properties and function of the resulting mesoporous film.

35.3.3 Modification by Functional Surfactant Molecules

In the synthesis of mesostructured thin films, modification or functionalization of the material can also be achieved by varying the template molecule.

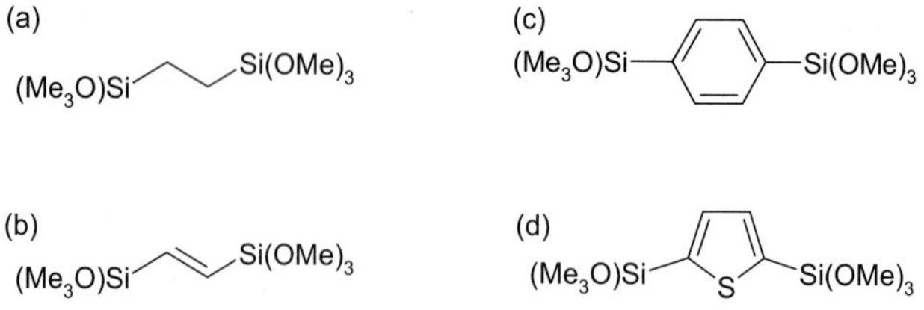

Figure 35.3. Chemical formulas of bridged silsesquioxane monomers employed in the preparation of mesostructured thin films.

This can either already bear a functionality, e.g. a polymerizable group or a metal complex system, resulting in mesostructured nanocomposite materials, or the template can be converted in a second step to functionalize the final (porous) material.

The first approach using a functional template molecule was demonstrated by applying an oligoethylene glycol functionalized diacetylenic surfactant, from which an ordered polydiacetylene/silica nanocomposite film was prepared that exhibited unusual chromatic responses to thermal, mechanical or chemical stimuli.[26]

The application of a 11-ferrocenylundecyl-ammonium bromide/hexadecylammonium bromide surfactant mixture as structure–directing agent resulted in a lamellar mesostructured silica film, which showed electronic conductivity due to electron transport in the ferrocenyl chains.[27] Lyotropic lithium triflate-silicate liquid crystals have been utilized as supramolecular templates in the synthesis of ionically conducting nanocomposite films.[28]

In addition to the synthesis of mesostructured nanocomposite materials, the template molecule can also be used to functionalize the periodic *porous* silica film. This was demonstrated by applying polyethylene glycol containing surfactants that had been pre-reacted with titanium alkoxides.[29] These lyotropic metal-containing surfactants were used in the synthesis of mesostructured silica-titania mixed metal oxide thin films serving two different functions: as structure-directing agents and as moderator of the hydrolysis rate of the titanium alkoxide (Figure 35.4).

The tailored design of the titanium-coordinated surfactant and its application in an evaporation-induced self-assembly process followed by heat treatment enabled the preparation of periodic mesoporous silica-based films with a high loading and good dispersion of tetrahedral titanium atoms within the silica matrix. This approach is also feasible for a variety of other transition metal oxides.

In another approach, inorganic–organic hybrid template molecules such as poly(dimethylsiloxane)-*b*-poly(ethylene oxide) block copolymers were applied (Figure 35.5). These amphiphilic molecules offer the interesting possibility of converting parts of the template molecule into silica, thereby reinforcing and thickening the wall of the inorganic framework.

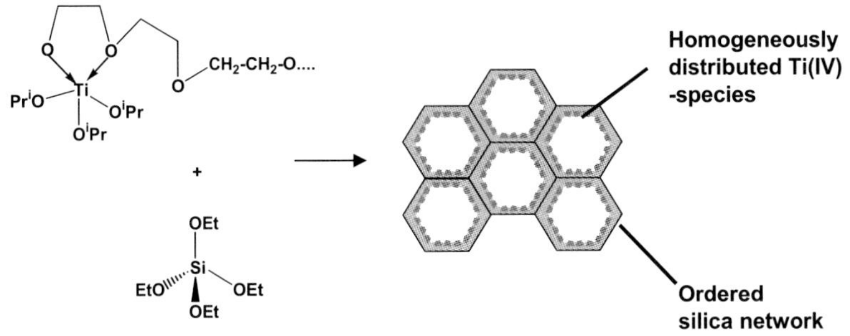

Homogeneously
distributed Ti(IV)
-species
Ordered
silica network

Figure 35.4. Schematic pathway to mesostructured silica / titania mixed oxide films.

52Å
26Å
X 8.4
Intensity
[a.u.]
° 2θ
70.00 nm

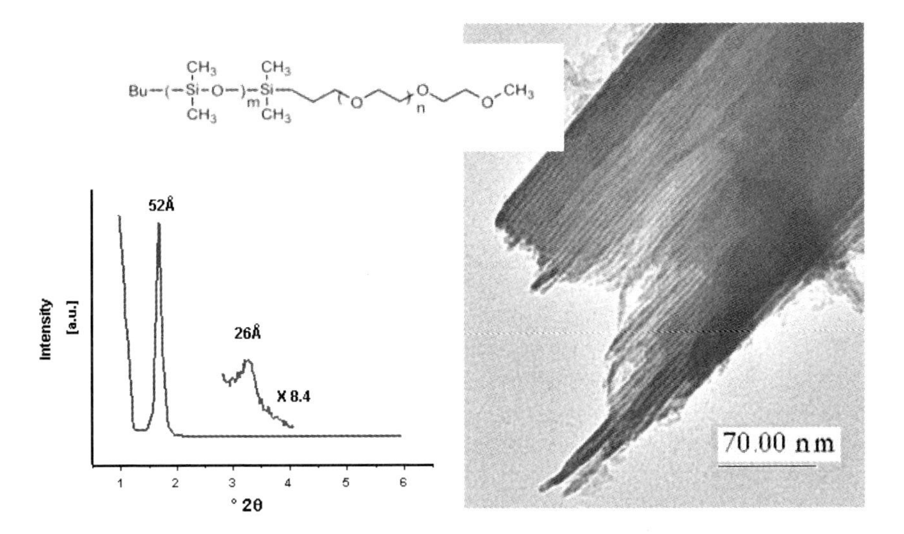

Figure 35.5. X-ray diffraction pattern and TEM image of a $PDMS_{18}$-PEO_{12} templated calcined silica film.

This is the first example of a stable lamellar phase in a material that was heat treated up to 450 °C. At this temperature, the silicone part of the surfactant does not degrade completely, but forms silsesquioxane units condensed to the silica backbone of the film. The silsesquioxane units located on the surface of the pore wall structures render the material hydrophobic and also act as a buffer between the layers, such that no collapse of the lamellar structure can be observed.[30]

35.4 References

[1] J. S. Beck, J. C. Vartuli, W. J. Roth, M. E. Leonowicz, C. T. Kresge, K. D. Schmitt, C. T.-W. Chu, D. H. Olson, E. W. Sheppard, S. B. McCullen, J. B. Higgins, J. L. Schlenker, *J. Am. Chem. Soc.* **1992**, *114*, 10834; C. T. Kresge, M. E. Leonowicz, W. J. Roth, J. C. Vartuli, J. S. Beck, *Nature* **1992**, *359*, 710.

[2] T. Yanagisawa, T. Shimizu, K. Kuroda, C. Kato, *Bull. Chem. Soc. Jpn.* **1990**, *63*, 988.

[3] J. Y. Ying, C. P. Mehnert, M. S. Wong, *Angew. Chem. Int. Ed.* **1999**, *38*, 56.

[4] K. J. Edler, S. J. Roser, *Int. Rev. Phys. Chem.* **2001**, *20*, 387.

[5] K. Moller, T. Bein, *Chem. Mater.* **1998**, *10*, 2950.

[6] A. Corma, *Chem. Rev.* **1997**, *97*, 2373.

[7] H. Yang, A. Kuperman, N. Coombs, S. Mamiche-Afara, G.A. Ozin, *Nature* **1996**, *379*, 703.

[8] I. A. Aksay, M. Trau, S. Manne, I. Honma, N. Yao, L. Zhou, P. Fenter, P. M. Eisenberger, *Science* **1996**, *273*, 892.

[9] H. Yang, N. Coombs, I. Sokolov, G. A. Ozin, *Nature* **1996**, *381*, 589.

[10] S. Schacht, Q. Huo, I. G. Voigt-Martin, G. D. Stucky, F. Schuth, *Science* **1996**, *273*, 768.

[11] M. Ogawa, *J. Am. Chem. Soc.* **1994**, *116*, 7941.

[12] Y. Lu, R. Ganguli, C. A. Drewien, M. T. Anderson, C. J. Brinker, W. Gong, Y. Guo, H. Soyez, B. Dunn, M. H. Huang, J. I. Zink, *Nature* **1997**, *389*, 364; C. J. Brinker, Y. Lu, A. Sellinger, H. Fan, *Adv. Mater.* **1999**, *11*, 579.

[13] D. Zhao, P. Yang, N. Melosh, J. Feng, B. F. Chmelka, G. D. Stucky, *Adv. Mater.* **1998**, *10*, 1380.

[14] H. Miyata, T. Noma, M. Watanabe, K. Kuroda, *Chem. Mater.* **2002**, *14*, 766.

[15] D. Grosso, F. Babonneau, G. J. de A. A. Soler-Illia, P. A. Albouy, H. Amenitsch, *Chem. Commun.* **2002**, 748.

[16] E. L. Crepaldi, G.J. de A.A. Soler-Illia, D. Grosso, C. Sanchez, P.-A. Albouy, *Chem. Commun.* **2001**, 1582; D. Grosso, G.J. de A.A. Soler-Illia, F. Babonneau, C. Sanchez, P.-A. Albouy, A. Brunet-Bruneau, A.R. Balkenende, *Adv. Mater.* **2001**, *13*, 1085; E. L. Crepaldi, D. Grosso, G.J. de A.A. Soler-Illia, P.-A. Albouy, H. Amenitsch, C. Sanchez, *Chem. Mater.* **2002**, *14*, 3316.

[17] K. J. Balkus Jr., A. S. Scott, M. E. Gimon-Kinsel, J. H. Blanco, *Microporous Mesoporous Mater.* **2000**, *38*, 97.

[18] R. Hernandez, A.- C. Franville, P. Minoofar, B. Dunn, J. I. Zink, *J. Am. Chem. Soc.* **2001**, *123*, 1248.

[19] G. Wirnsberger, B. J. Scott, B. F. Chmelka, G. D. Stucky, *Adv. Mater.* **2000**, *12*, 1450.

[20] H. Fan, Y. Lu, A. Stump, S. T. Reed, T. Baer, R. Schunk, V. Perez-Luna, G. P. Lopez, C. J. Brinker, *Nature* **2000**, *405*, 56.

[21] E. M. Wong, M. A. Markowitz, S. B. Qadri, S. L. Golledge, D. G. Castner, B. P. Gaber, *Langmuir* **2002**, *18*, 972.
[22] B. Lebeau, C. E. Fowler, S. R. Hall, S. Mann, *J. Mater. Chem.* **1999**, *9*, 2279.
[23] A. Sellinger, P. M. Weiss, A. Nguyen, Y. Lu, R. A. Assink, W. Gong, C. J. Brinker, *Nature* **1998**, *394*, 256.
[24] Y. Lu, H. Fan, N. Doke, D. A. Loy, R. A. Assink, D. A. LaVan, C. J. Brinker, *J. Am. Chem. Soc.* **2000**, *122*, 5258.
[25] Ö. Dag, C. Yoshina-Ishii, T. Asefa, M. J. MacLachlan, H. Grondey, N. Coombs, G. A. Ozin, *Adv. Funct. Mater.* **2001**, *11*, 213.
[26] Y. Lu, Y. Yang, A. Sellinger, M. Lu, J. Huang, H. Fan, R. Haddad, G. Lopez, A. R. Burns, D. Y. Sasaki, J. Shelnutt, C. J. Brinker, *Nature* **2001**, *410*, 913.
[27] H. S. Zhou, D. Kundu, I. Honma, *J. Eur. Ceram. Soc.* **1999**, *19*, 1361.
[28] Ö. Dag, A. Verma, G. A. Ozin, C. T. Kresge, *J. Mater. Chem.* **1999**, *9*, 1475.
[29] N. Hüsing, B. Launay, D. Doshi, G. Kickelbick, *Chem. Mater.* **2002,** *14*, 2429.
[30] N. Hüsing, B. Launay, J. Bauer, G. Kickelbick, D. Doshi, *J. Sol-Gel Sci. Technol.* **2003,** *26*, 609..

36 Modification of Ordered Mesostructured Materials during Synthesis

Stephan Altmaier, Peter Behrens*

36.1 Introduction

Today's challenge in the synthesis of materials extends further than establishing the correct structure on an atomic scale. Structures on different length scales up to the visible range, built upon each other in a hierarchical fashion, have to be engineered in order to optimize the properties of the materials.[1, 2]

A promising approach to the preparation of designed materials is offered by template-based syntheses, where organic molecules direct the formation of inorganic solids. The organic molecules are designated as structure-directing agents (SDAs). This approach is at least partly inspired by Nature, where biopolymers direct the formation of inorganic matter on a variety of length scales in a highly controlled manner in the process of biomineralization.[3–5] The action of molecular SDAs has been applied extensively in the synthesis of zeolites and related solids.[6–8] On an intermediate scale between the atomic structure of zeolites and the biomineralization process, which extends up to the millimeter scale, neither individual molecules nor polymers are adequate to act as SDAs. For this length scale, the mesoscale (ca. 1 to 20 nm), amphiphilic molecules or surfactants have been found to be suitable SDAs. In aqueous solutions, such molecules can form structures of long-range order, so-called lyotropic phases.[9, 10] Organic–inorganic meso-composites with the typical structures of such lyotropic phases can be formed when precursors of inorganic solids are combined with solutions of such amphiphiles.

The general chemistry of such surfactant–inorganic systems was elaborated using silica as the inorganic component. The materials discovered first by researchers from the Mobil Oil company are the so-called M41S materials.[11–14] In synthesis systems consisting of alkyltrimethylammonium ions, $H_3C(CH_2)_{n-1}N(CH_3)_3^+$ = C_nTMA^+, and a silica precursor such as tetra-ethoxysilane (TEOS), a family of mesostructures with different topologies was discovered (Figure 36.1): hexagonal MCM-41, cubic MCM-48, lamellar phases as well as a slightly disordered cubic phase (LMU-1[15, 16] or KIT-1[17]). These different mesostructures can be discerned by their typical powder X-ray diffraction (PXRD) patterns appearing at low diffraction angles. Other

---

* Address of the authors: Institut für Anorganische Chemie, Universität Hannover, Callinstrasse 9, D-30167 Hannover, Germany

mesostructures were obtained by changing the ionic headgroup of the alkylammonium ions (for example SBA-1[18]) or by switching to other types of amphiphilic molecules.[19–21] The organic component can be removed from such mesostructures either by calcination or by extraction, to obtain mesoporous materials with a narrow pore size distribution (lamellar samples, however, will collapse when the organic SDAs are removed). The inorganic silica structure is amorphous on an atomic scale. It was also shown that the synthesis of mesostructures can, in principle, be extended to a variety of other inorganic compositions when these are able to undergo a kinetically controlled condensation process.[21–25]

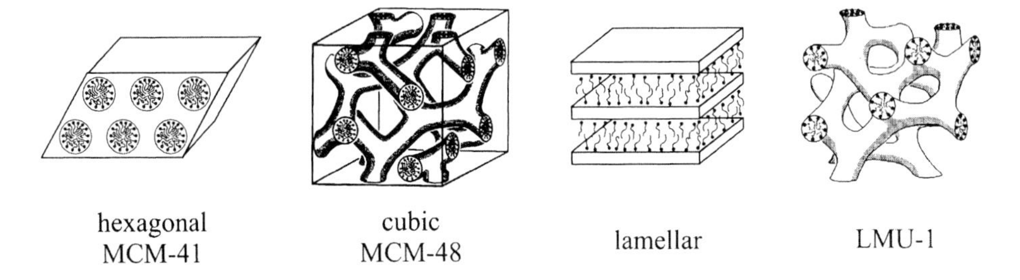

Figure 36.1. Topologies of as-synthesized mesostructures.

The discovery of mesoporous M41S materials has shown that the idea of structure-directed synthesis using organic SDAs is fruitful and can be extended significantly further than the classical syntheses of microporous zeolites and zeotype compounds. With regard to the facts that solutions of surfactants and amphiphiles are possibly the most complex solution systems of *organic* substances in water, and that silica solutions belong to the most intricate *inorganic* solution systems, it is clear that the chemistry of solutions or gels leading to M41S materials has to be very complicated.

The mesoporous materials, which are obtained from the organic–inorganic meso-composites after template removal, have only rarely been applied directly, in spite of their favorable pore characteristics.[26–30] The walls of the mesopores possess silanol groups, as is typical of amorphous silicas. The number of silanol groups is high when the SDAs have been removed by extraction, but is considerably smaller after calcination, due to condensation during this high-temperature process. These silanol groups have only modest acidic character and cannot be used directly in acidic catalysis (in contrast to similar groups in crystalline aluminosilicates).

For most applications, the mesoporous silica materials therefore have to be modified in order to obtain designed reactivities,[29–32] for example by the introduction of metal clusters, by ion-exchange at the silanol groups, or by polymerization of suitable precursors within the channels. Here, we deal with another type of modification, namely the covalent attachment of organic groups to the walls of the mesopores.[31, 32] There are principally two ways to perform such modifications, namely either by grafting functional groups onto the walls of the mesoporous materials after synthesis and template removal or by co-condensation during the formation of the mesostructured solids (Figure 36.2):

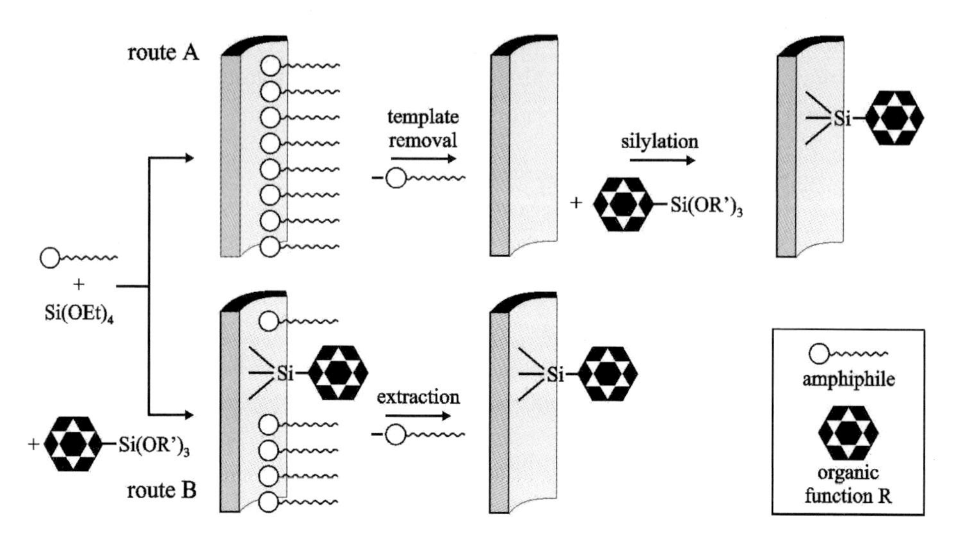

Figure 36.2. Routes for the preparation of organically modified ordered mesostructured materials.

- Post-synthetic modification methods (route A in Figure 36.2)[31–33] rely on the presence of silanol groups on the inside of the pore walls. The simplest type of post-synthetic modification is the hydrophobization of the calcined mesoporous materials by reaction with alkylchloro- or alkylalkoxy-silanes,[34,35] but functional groups and metal complexes can also be attached.[24, 36–40]
- As an alternative route to these post-synthetic modifications, the possibility of introducing pre-designed properties and reactivities into M41S materials during synthesis appears attractive.[31, 32] This route (designated as B in Figure 36.2) can directly yield materials with interesting properties and reactivities. It relies on the co-condensation of silicon alcoholates such as TEOS with trialkoxysilanes $RSi(OR')_3$ (R': CH_3, CH_2CH_3) bearing additional organic residues R. The Si-C bond in the organoalkoxysilanes is resistant towards hydrolysis and it is known that in the normal sol-gel process leading to disordered silica, co-condensation is possible.[41, 42] It was first shown by Burkett et al.[43] and by Macquarrie[44] that this "one-pot

synthesis" approach also functions in the formation of mesostructures. The advantages of this approach lie in the possibility of the direct construction of an organically modified and functionalized meso-composite and in a more uniform distribution of the functional groups over the pore walls, which is possibly not attainable by the post–synthetic modification route.[45] It has been reported that a variety of different functionalities can be introduced into mesostructured materials in this way,[31, 32] and even the simultaneous incorporation of two different functionalities has been claimed.[46]

Addition of a further component to a mixture suitable for the synthesis of M41S materials, which is already very complex, can of course introduce problems. This is especially true when this component may interfere with both the composition of the silicate species in the solution as well as with the ordering within the organic part. $RSi(OR')_3$ molecules, after (partial) hydrolysis of their SiOR' connections, could do so (i) by forming organosilicate anions capable of performing condensation reactions with anions derived from the main silica source, and (ii) by exhibiting hydrophilic–hydrophobic (i.e. quasi-amphiphilic) properties (especially when the residue R is large and nonpolar). The key questions are then:

- To what extent can the main silica source $Si(OR')_4$ be replaced by $RSi(OR')_3$ while still maintaining the geometrical order of the pores and the narrow pore size distributions typical of M41S materials ?
- To what extent is the cooperative ordering process[16, 47, 48] leading to the formation of M41S materials influenced by the addition of $RSi(OR')_3$?
- What degree of functionalization can be attained in the final mesoporous material after removal of the templating SDAs ?

For the removal of the surfactants from organically modified mesostructured silicas, calcination at elevated temperatures is only an option in rare special cases.[49] Instead, extraction of the surfactant is the method of choice.[31, 45, 50] For the extraction, however, the inorganic pore structure has to be stable towards the removal of the surfactant and the chemicals involved in this process, without the stabilization provided by further condensation within the silica walls, which takes place during the high-temperature calcination process, but will not occur during extraction at low temperatures.

In this chapter, we review our work[23, 51–54] on the direct modification of M41S-type materials. A detailed investigation on the influence of the addition of an organosilane was carried out in the synthesis system $C_{14}TMA^+$/TEOS/PTMOS (phenyltrimethoxysilane) in a basic synthesis medium at 110 °C (Section 36.2); here, TEOS was replaced by PTMOS to an extent of 10 %. In the same synthesis system, we also investigated the actual degree of incorporation of the organosilane into the silica walls at different reaction temperatures (Section 36.3).

36.2 Influence of Phenyltrimethoxysilane on the Formation of Mesostructures

We have performed a detailed investigation on the formation of mesostructures in the presence of PTMOS by constructing synthesis field diagrams (SFDs), which show the type of product formed in a certain synthesis system as a function of relevant synthesis parameters, for example temperature or concentrations. In contrast to phase diagrams, which show thermodynamically stable phases as functions of thermodynamic parameters, the shapes of synthesis fields are functions of the reaction conditions leading to certain synthesis products. Phase diagrams will not vary with time, whereas SFDs of kinetically controlled reactions typically change with reaction time. The construction of SFDs has been elaborated with considerable success in the synthesis of zeolites.[55] In our group, we have used this tool for the study of the synthesis of mesostructures.[15, 16, 52, 53] For the coordinate axes of an SFD, we use the concentration of the silica source (TEOS) and the concentration of the surfactant. The diagrams are then constructed by performing a number of individual synthesis experiments, which is large enough to reliably establish the fields of synthesis conditions leading to the different products.

Here, we present SFDs constructed using the $C_{14}TMA^+$ surfactant, basic synthesis conditions (0.33 M KOH), and a reaction time of 2 days. SFDs were obtained with pure TEOS or with a 90:10 mixture of TEOS and PTMOS as a silica source at three different temperatures (90 °C, 110 °C, 130 °C). These six SFDs are shown in Figure 36.3.

The general appearance of SFDs for the preparation of M41S-type materials has been described previously.[15, 16] At low silica concentrations, there is a "solution" region, i.e. there is a minimum silica concentration necessary in order to produce a solid precipitate. At high silica concentrations, disordered mesoporous silica is formed, which does not exhibit any peaks in PXRD, but which is mesoporous, as sorption measurements on the calcined materials show. At intermediate silica concentrations, the ordered (or only slightly disordered) mesostructured phases are formed. At the lower silica concentrations within this regime, we find hexagonal, cubic, and lamellar phases, whereas at somewhat higher concentrations the LMU-1 phase dominates. It is of interest to note that the most important boundary lines run mainly vertical, implying that the type of phase formed depends mainly upon the silica concentration and less upon the surfactant concentration.[15, 16]

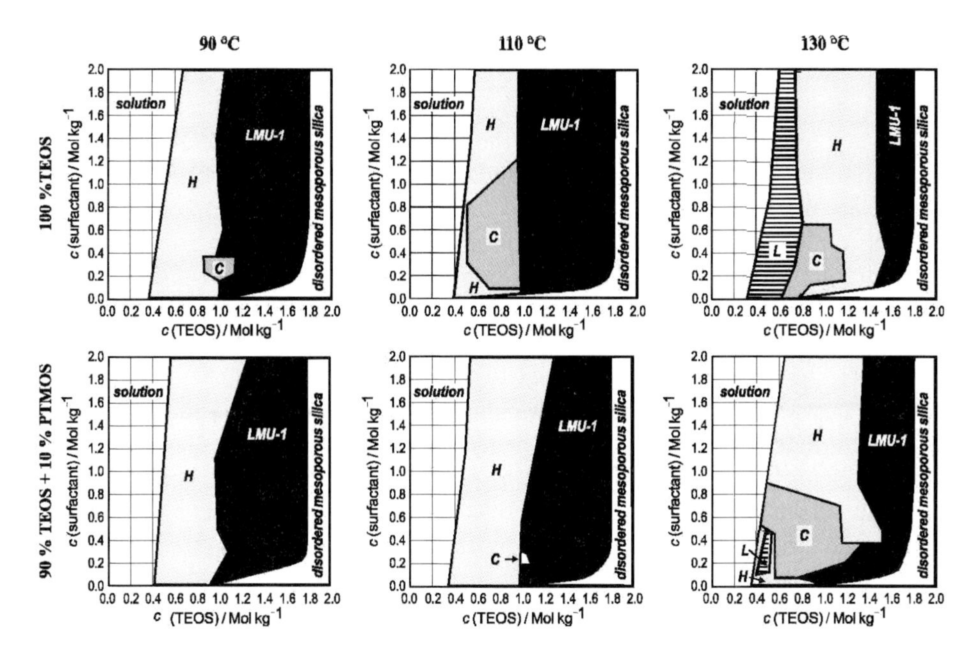

Figure 36.3. Synthesis field diagrams for the synthesis of mesostructures ($C_{14}TMA^+$ surfactant, 0.33 M KOH, 2 days). The reactions were carried out at 90 °C (left column), 110 °C (center column), and 130 °C (right column). Top row: 100 % TEOS as a silica source; bottom row: 90% TEOS and 10 % PTMOS as silica source. *H*: hexagonal MCM-41-type phase, *C*: cubic MCM-48-type phase, *LMU-1*: LMU-1 or KIT-1, *L*: lamellar phase.

When comparing the upper row of three SFDs in Figure 36.3 (obtained with pure TEOS), one sees a clear influence of the synthesis temperature on the formation of the mesostructures.[15, 53] However, here we are more concerned with changes occurring upon the substitution of part of the TEOS by PTMOS. Significant changes can be seen here, too. At 90 °C, the synthesis field of the hexagonal phase extends at the expense of that of the LMU-1 phase and the field of the cubic phase vanishes when PTMOS is present in the synthesis mixture. At 110 °C, the regime of the hexagonal phase again extends further to higher silica concentrations, at the expense of the LMU-1 phase. This is an important result as it shows that the presence of an $RSi(OR')_3$ silane can even enhance the formation of well-ordered structures. The field of the cubic phase is strongly diminished when PTMOS is present. Finally, at 130 °C the field of the cubic phase has grown mainly at the expense of the area of the hexagonal phase. The lamellar phase, prominent near to the solution boundary when pure TEOS is used, occupies only a very small field when PTMOS is present. One can also discern a certain "diagonal relationship" between the SFDs presented in Figure 36.3. The SFD obtained with pure TEOS at 90 °C resembles the one obtained with PTMOS substitution at 110 °C, and there is also some similarity in the distribution of synthesis fields between the SFD constructed using pure TEOS at a temperature of 110 °C and the one obtained with PTMOS at 130 °C.

Although these trends appear to be puzzling at first, they can be summarized based on the general knowledge of the structural chemistry of lyotropic phases [9, 10, 56, 57] and the way in which they can be applied to the corresponding mesostructured phases.[22, 47, 48] This approach is based on the fact that the different mesophases exhibit different curvatures at the organic–inorganic interface. The curvature increases in the sequence from lamellar to MCM-48 to MCM-41. As a disordered MCM-48-type structure, the LMU-1 phase should have a curvature intermediate between that of the lamellar and that of the MCM-48 phase. The type of silica mesostructure formed is then connected to the number of deprotonated silanol groups which balance the charge of the cationic surfactants.[22, 47, 48] With increasing condensation of the silica part of the composite (i.e. with longer reaction times or higher reaction temperatures), the number of silanol groups decreases.

Under these premises, all the changes in the SFDs that occur upon partial substitution of TEOS by PTMOS can be summarized in terms of an increase in curvature, which is caused by the presence of the organosiloxane, or rather by its hydrolysis product. This finding is also in agreement with the decrease in lattice constants often observed upon substitution of TEOS by $RSi(OR')_3$.[58, 59]

A key question now is in what way the PTMOS substituent influences the intricate aggregation processes occurring at the organic–inorganic interface. Under the chemical conditions of its environment, it is clear that the PTMOS molecule enters the organic–inorganic self-assembly process in a hydrolyzed and (at least partially) deprotonated state. Two configurations at the organic–inorganic interface appear probable and the way in which they influence the curvature at the organic-inorganic interface has to be discussed.

Figure 36.4a shows schematically the situation at the interface of a "normal" M41S solid. When the integration of phenylsiloxy moieties into the silica framework by co-condensation occurs as expected, then the phenyl residues attached to the silica walls will extend into the headgroup area of the micellar arrangements and two cases are viable. The phenyl residues could screen the repulsive interactions between the positive charges of the headgroups, so that the distances between these would become smaller, leading to a decrease of the effective headgroup area and concomitantly to a preference for mesostructures with smaller curvatures (Figure 36.4b). However, the phenyl residues will also acquire their own space within the headgroup area, so that the effective headgroup area per cationic surfactant could increase, leading to a greater curvature of the interface (Figure 36.4c).

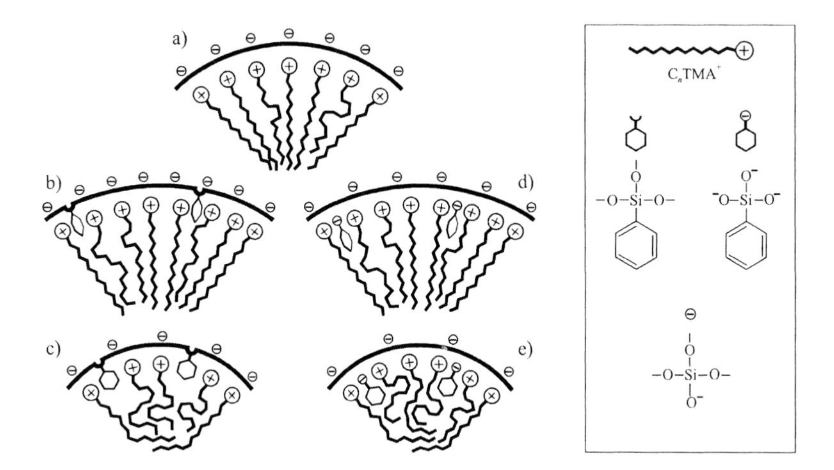

Figure 36.4. Schematic depiction of different configurations at the organic–inorganic interface of mesostructures; a) standard M41S phase; b), c) with phenylsiloxy residues attached to the silica walls; d), e) with hydrolyzed and deprotonated phenylsiloxy anions incorporated into the micellar structures (the phenylsiloxy anions are not necessarily fully deprotonated). Configurations b) and d) lead to a decrease, configurations c) and e) to an increase in curvature at the interface. For further explanation, see text.

When the phenylsiloxy anions do not become incorporated into the silica walls, they might be considered as small quasi-amphiphilic entities which enter the micellar aggregates. The attractive interactions between the positively charged headgroups of the surfactants and the negatively charged siloxy groups could lead to a decrease in the effective head group area and thus to a decrease in curvature (Figure 36.4d). However, this effect could be over-compensated by the decrease in the effective volume of the hydrophobic chains of the surfactant molecules. Due to the fact that the phenyl residues acquire only a very small volume of the interior space of the micelles as compared to the alkyl chains of the surfactant molecules, the chains gain additional space in the interior of the micellar arrangements, so that structures with larger curvatures should become stabilized (Figure 36.4e). On the basis of similar arguments, several synthesis systems have been described where mixtures of surfactants have been used to fine-tune the curvature at the organic–inorganic interface and hence to direct the outcome of syntheses.[60–66] Mixtures of cationic and anionic surfactants have also been applied in this sense.[61, 64] The cases depicted in Figures 36.4b and 36.4d can be ruled out on the basis of the results presented.

Finally, it is also possible that the phenylsiloxy entities do not become incorporated into the mesostructure at all (neither into its organic nor into its inorganic part), but that they only influence the general kinetics of the self-assembly process. In this respect, the observed "diagonal relationship" described above could be explained in term of a kinetic hindrance of the synthesis systems in the presence of PTMOS, which can be overcome by higher temperature. This would explain the observation that the SFDs obtained in the presence of PTMOS resemble those constructed using pure TEOS at a reaction temperature that is 20 °C lower.

Some more information regarding the importance of these possible synthesis mechanisms is presented in the next section.

This part of our work shows that the construction of SFDs can yield valuable information about the phase formation in the synthesis of mesostructured materials and about the underlying mechanisms.[16] However, constructing the SFDs presented here was a tedious and time-consuming task. Therefore, we have developed a parallel synthesis approach, using an autoclave block which allows the simultaneous preparation of 24 samples under hydrothermal conditions and has considerably shortened the time needed to construct an SFD.[52]

36.3 Investigation of the Degree of Incorporation of Phenyltrimethoxysilane into the Silica Mesostructures

In this section, we describe the detailed characterization of certain selected samples, which was carried out in order to gain information with regard to the actual degree of incorporation into the silica walls of the mesostructures. We discuss samples that were obtained by a hydrothermal treatment at 110 °C for 2 days in 0.33 M KOH, using $C_{14}TMA^+$ as surfactant and molar ratios of TEOS:PTMOS of 100:0 (standard MCM-41 sample) and 80:20, and compare them with samples prepared at 25 °C by stirring the ingredients for one day. We deal with the as-synthesized samples as well as with their state after extraction, which is performed using acetic acid in ethanol under reflux. A more detailed account of this work is in preparation.[54]

Figure 36.5 provides some physical data obtained for these mesostructures. The PXRD diagrams of the samples prepared at 110 °C (Figure 36.5a, right) show that the extraction process does not affect the ordered arrangement of the mesopores, not even at a degree of substitution of 20% PTMOS. On the other hand, while the preparation at 25 °C yields samples of similar quality in the absence of PTMOS, the order of the mesostructure is strongly diminished upon substitution of 20 % of the TEOS by PTMOS.

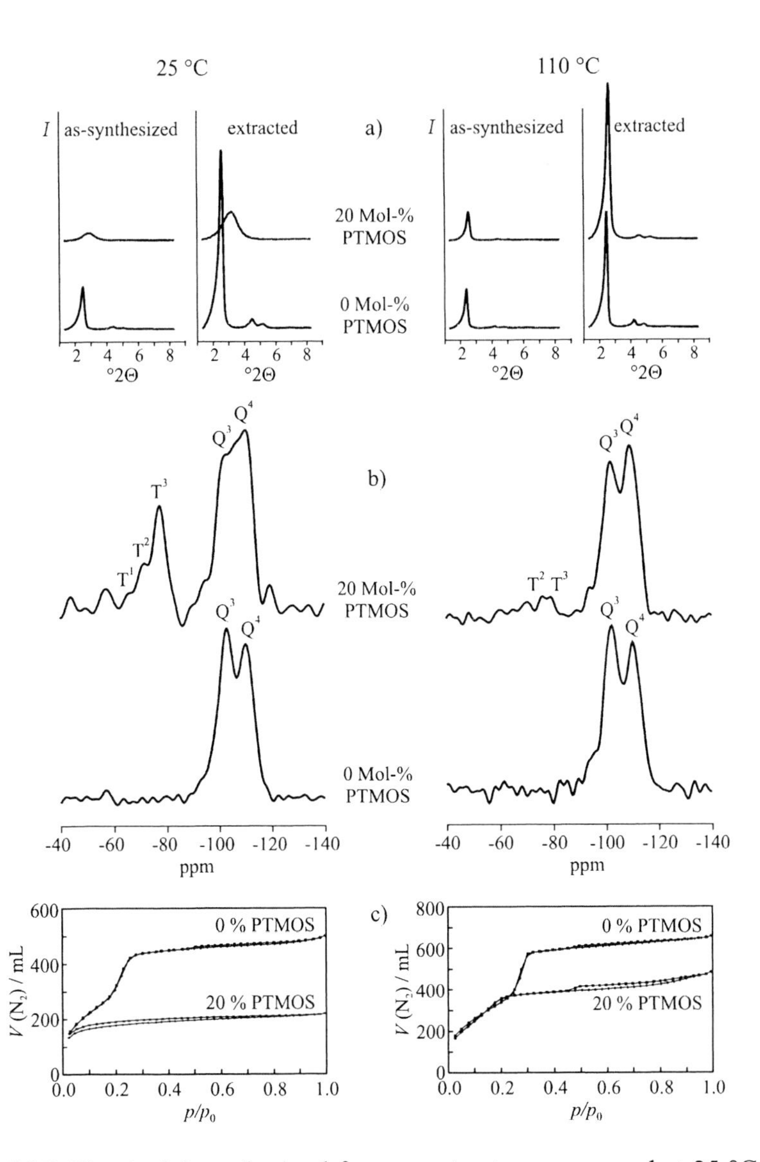

Figure 36.5. Physical data obtained for mesostructures prepared at 25 °C (left) and at 110 °C (right) with molar TEOS : PTMOS ratios of 100 : 0 and 80 : 20; a) powder X-ray diffractograms of as-synthesized and extracted samples; b) ^{29}Si HPDEC NMR spectra of as-synthesized samples; c) N_2 sorption measurements on extracted samples.

Figure 36.5b shows ^{29}Si HPDEC NMR spectra of the samples after extraction. For the samples synthesized without PTMOS, the typical spectra of M41S materials are obtained. The signals at $\delta \approx$ -102 ppm and $\delta \approx$ -110 ppm are due to Q^3 and Q^4 units, respectively ($[SiO_4]$ units attached to three or four other Si atoms, respectively).

The spectra of the samples prepared in the presence of PTMOS look quite different from each other. At a synthesis temperature of 25 °C, the appearance of T peaks indicates that the phenylsiloxy residues have become incorporated very effectively (T^1: $[RSiO_3]$ unit attached to one other silicon atom, appears at ca. -61 ppm; T^2: $[RSiO_3]$ unit attached to two other silicon atoms, ca. -71 ppm; T^3: $[RSiO_3]$ unit attached to three other silicon atoms, ca. -79 ppm). The ratio of the T:Q signals is ca. 1:3 for this sample. On the other hand, for the sample synthesized at 110 °C, T signals are hardly visible, and a quantitative evaluation gives a T:Q ratio of ca. 1:12. Obviously, the phenylsiloxy residues have become incorporated into the silica walls only to a very small degree.

Figure 36.5c shows the results of sorption measurements. Samples prepared in the absence of PTMOS show the highest porosities after extraction; the surface area of 1356 m^2g^{-1} and the pore volume of 1.00 cm^3g^{-1} measured for the sample synthesized at 110 °C correspond to optimal MCM-41 samples. When PTMOS is integrated into the synthesis, the jump in the isotherms, which is characteristic of mesoporosity, vanishes and the characteristics of a type-IV isotherm are lost. Similar observations were made by Bambrough et al. for a phenyl-modified silica.[67] Whereas the 110 °C sample still retains high values of surface area and pore volume, the one synthesized at 25 °C shows a partial collapse of the pore structure, which is in line with the low quality of mesostructural order reflected by the PXRD experiments.

Now, we obviously have a vicious circle. At 25 °C, the phenylsiloxy residues become effectively incorporated into the silica mesostructure; however, this mesostructure is not well-ordered, is not stable against extraction, but partially collapses and exhibits only low porosity. On the other hand, after synthesis at 110 °C, the sample is well-ordered, stable against extraction, and has high porosity, but the amount of phenylsiloxy residues that have become attached to the silica structure is only very small. This result is astonishing, especially in view of the strong influence that the substitution by PTMOS (even at an only 10% level) has on the appearance at the SFD aquired at 110 °C (see Section 36.2).

With regard to the open question from Section 36.2, it can now be stated that the model depicted in Figure 36.4e probably most adequately accounts for the observed increase in curvature of the mesostructures upon substitution of TEOS by PTMOS, because in this model the phenylsiloxy residues are not attached to the silica walls. However, the idea of a general kinetic hindrance of structure formation as a result of the presence of PTMOS, which leads to the observed "diagonal relationship" between the SDFs in Figure 36.3, is also in line with the fact that the phenylsiloxy residues do not become incorporated into the silica structure.

Comparing our results to findings reported in the literature, we note that the problems we have encountered have not been described previously.[31, 32] We will restrict our discussion to one-pot syntheses applying phenyltrialkoxy-silanes. The reaction conditions in the work of Babonneau and co-workers[49, 68] (acidic synthesis system) and of Mercier and Pinnavaia[59] (*n*-octylamine surfactant) are different, and so their results may not be directly comparable to ours. Mann and co-workers[43, 58, 67, 69] as well as Bambrough et al.[67, 70] have described the preparation of phenyl-modified M41S materials in a basic medium at room temperature (followed by filtration and drying at 100 °C). *T* and *Q* signal intensities derived from ^{29}Si NMR measurements[43, 58] of as-synthesized phenyl-modified samples are similar to ours. No PXRD patterns are provided for these products.[43, 58] However, *d* values of four reflections are given for an as-synthesized MCM-41 phase prepared with 20% phenyltriethoxysilane (PTEOS), but only one value is provided for the extracted sample; this could point to problems regarding the stability of the mesostructure towards extraction, as we have observed. Nitrogen sorption curves displayed in ref.[58] appear to be very similar to those presented here, but considerably higher surface areas are tabulated. One-pot syntheses at elevated temperatures have not been reported, except for the synthesis of a cubic MCM-48-type phase synthesized at 100 °C.[69] Although there were indications of the presence of phenyl groups in this material, their amount was not quantified. In summary, the evidence for a highly ordered mesoporous material with a high degree of modification by phenyl groups, synthesized in a basic medium, is scarce, in line with the results presented here.

36.5 Conclusions

This work reports detailed investigations on the "one-pot" approach to the synthesis of organically functionalized mesostructures. As expected, the chemistry of surfactant–silica systems is further complicated by the addition of a trialkoxysilane, and it appears that the formation of mixed silica–organosiloxane mesostructures is difficult at elevated temperatures under hydrothermal conditions in basic media. Room temperature reactions show a higher degree of incorporation of the organosiloxane, but the pore systems of the materials are not stable towards extraction. Therefore, in general, the assembly of organically modified mesostructures from acidic solutions appears preferable. It is possible that mixed silica–organosiloxane mesostructures, which are actually formed at room temperature under basic conditions upon hydrolysis of the precursor molecules, segregate at higher synthesis temperatures due to the hydrolysis-recondensation of siloxane bridges. The latter process is hindered in acidic solutions.

36.6 Acknowledgements

We thank the Fonds der Chemischen Industrie and the Bundesministerium für Forschung und Bildung for general support of our research. Part of this work was carried out at the Ludwig-Maximilians-Universität München. Finally, we would like to thank Ulrich Schubert and Nicola Hüsing for valuable discussions.

36.7 References

[1] I. Soten, G. A. Ozin, *Current Opin. Colloid & Interface Sci.* **1999**, *4*, 325.
[2] G. A. Ozin, *Chem. Commun.* **2000**, 419.
[3] L. Addadi, S. Weiner, *Angew. Chem. Int. Ed. Engl.* **1992**, *31*, 153.
[4] E. Baeuerlein (Ed.), *Biomineralization*, Wiley-VCH, Weinheim, **2000**.
[5] S. Mann, *Biomineralization*, Oxford University Press, **2002**.
[6] R. W. Thompson, in *Molecular Sieves: Science and Technology*, H. G. Karge, J. Weitkamp (Eds.), Vol. 1: *Synthesis*, Springer, Berlin, **1998**, p. 1.
[7] H. Gies, B. Marler, U. Werthmann, in *Molecular Sieves: Science and Technology*, H. G. Karge, J. Weitkamp (Eds.), Vol. 1: *Synthesis*, Springer, Berlin, **1998**, p. 35.
[8] M. E. Davis, *Nature* **2002**, *417*, 813.
[9] A. G. Petrov, *The Lyotropic State of Matter: Molecular Physics and Living Matter Physics*, Gordon and Breach, Amsterdam, **1999**.
[10] H. Stegemeier, *Lyotrope Flüssigkristalle*, Steinkopff, Darmstadt, **1999**.
[11] C. T. Kresge, M. E. Leonowicz, W. J. Roth, J. C. Vartuli, J. S. Beck, *Nature* **1992**, *359*, 710.
[12] J. S. Beck, J.C.Vartuli, W. J. Roth, M. E. Leonowitcz, C. T. Kresge, K. D. Schmitt, C. T. W. Chu, D. H. Olson, E. W. Sheppard, S. B. McCullen, J. B. Higgins, J. L. Schlenker, *J. Am. Chem. Soc.* **1992**, *114*, 10834.
[13] P. Behrens, *Adv. Mater.* **1993**, *5*, 127.
[14] P. Behrens, G. D. Stucky, *Angew. Chem. Int. Ed. Engl.* **1993**, *32*, 696.
[15] A. M. Glaue, Ph.D. Thesis, Ludwig-Maximilians-Universität München, Logos-Verlag, Berlin, **1999**, ISBN: 3-89722-3331-7.
[16] P. Behrens, A. Glaue, Ch. Haggenmüller, G. Schechner, *Solid State Ionics* **1997**, *101-103*, 255.

[17] R. Ryoo, J. M. Kim, C.H. Shin, *J. Phys. Chem.* **1996**, *100*, 17713.
[18] Q. Huo, D. I. Margolese, G. D. Stucky, *Chem. Mater.* **1996**, *8*, 1147.
[19] S. A. Bagshaw, E. Prouzet, T.J. Pinnavaia, *Science* **1995**, *269*, 1242.
[20] D. Zhao, J. Feng, Q. Huo, N. Melosh, G.H. Fredrickson, B.F. Chmelka, G. D. Stucky, *Science* **1998**, *279*, 548.
[21] P. Yang, D. Zhao, D. I. Margolese, B.F. Chmelka, G.D. Stucky, *Nature* **1998**, *396*, 152.
[22] Q. Huo, D. I. Margolese, U. Ciesla, P. Feng, T. E. Gier, P. Sieger, R. Leon, P. M. Petroff, F. Schüth, G. D. Stucky, *Nature* **1994**, *368*, 317.
[23] P. Behrens, *Angew. Chem. Int. Ed. Engl.* **1996**, *35*, 515.
[24] J. Y. Ying, C. P. Mehnert, M. S. Wong, *Angew. Chem. Int. Ed. Engl.* **1999**, *38*, 56.
[25] F. Schüth, *Chem. Mater.* **2001**, *13*, 3184.
[26] A. Sayari, *Chem. Mater.* **1996**, *8*, 1840.
[27] A. Corma, *Chem. Rev.* **1997**, *97*, 2373.
[28] J. C. Vartuli, W. J. Roth, J. S. Beck, S. B. McCullen, C. T. Kresge, in *Molecular Sieves: Science and Technology*, H.G. Karge, J. Weitkamp (Eds.), Vol. 1: *Synthesis*, Springer, Berlin, 1998, p. 97.
[29] F. Schüth, *Stud. Surf. Sci. Catal.* **2001**, *135*, 1.
[30] G. Ove, J. Sjöblom, M. Stöcker, *Adv. Coll. Interf. Sci.* **2001**, *89-90*, 439.
[31] A. Stein, B. J. Melde, R.C. Schroden, *Adv. Mater.* **2000**, *12*, 1403.
[32] A. Sayari, S. Hamoudi, *Chem. Mater.* **2001**, *13*, 3151.
[33] T. Maschmeyer, *Curr. Opin. Solid State Mater. Sci.* **1998**,*3*, 71.
[34] R. Anwander, C. Palm, J. Stelzer, O. Groeger, G. Engelhardt, *Stud. Surf. Sci. Catal.* **1998**, *117*, 135.
[35] R. Anwander, I. Nagl, M. Widenmeyer, G. Engelhardt, O. Groeger, C. Palm, T. Röser, *J. Phys. Chem. B* **2000**, *104*, 3532.
[36] X. Feng, G. E. Fryxell, L.-Q. Wang, A.Y. Kim, J. Liu, K.M. Kemner, *Science* **1997**, *276*, 923.
[37] A. Corma, M. T. Navarro, J. Perez-Pariente, J. Chem Soc. *Chem. Commun.* **1994**, 147.
[38] T. Maschmeyer, F. Rey, G. Sankar, J.M. Thomas, *Nature* **1995**, *378*, 159.
[39] R. Anwander, H.W. Görlitzer, G. Gerstberger, C. Palm, O. Runte, M. Spiegler, *J. Chem. Soc., Dalton Trans.* **1999**, 3611.
[40] D. S. Shephard, W. Zhou, T. Maschmeyer, J. M. Matters, C. L. Roper, S. Parsons, B. F. G. Johnson, M. J. Duer, *Angew. Chem. Intern. Ed. Engl.* **1998**, *37*, 2719.
[41] U. Schubert, N. Hüsing, A. Lorenz, *Chem. Mater.* **1995**, *7*, 2010.
[42] R. J. P. Corriu, *Angew. Chem. Int. Ed.* **2000**, *39*, 1377.
[43] S. L. Burkett, S. D. Sims, S. Mann, *Chem. Commun.* **1996**, 1367.
[44] D. J. Macquarrie, *Chem. Commun.* **1996**, 1961.
[45] M.H. Lim, C. F. Blanford, A. Stein, *J. Am. Chem. Soc.* **1997**, *119*, 4090.

[46] I. Díaz, C. Márquez-Alvarez, F. Mohino, J. Pérez-Pariente, E. Sastre, *J. Catal.* **2000**, *193*, 283.
[47] G. D. Stucky, A. Monnier, F. Schüth, Q. Huo, D. Margolese, D. Kumar, M. Krishnamurty, P. Petroff, A. Firouzi, M. Janicke, B.F. Chmelka, *Mol. Cryst. Liq. Cryst.* **1994**, *240*, 187.
[48] A. Monnier, F. Schüth, Q. Huo, D. Kumar, D. Margolese, G.D. Stucky, M. Krishnamurty, P. Petroff, A. Firouzi, M. Janicke, B.F. Chmelka, *Science* **1993**, *261*, 1299.
[49] F. Babonneau, L. Leite, S. Fontlupt, *J. Mater. Chem.* **1999**, *9*, 175.
[50] S. Inagaki, Y. Sakamoto, Y. Fukushima, O. Terasaki, *Chem. Mater.* **1996**, *8*, 2089.
[51] S. Altmaier, K. Nusser, P. Behrens, in *Organosilicon IV*, N. Auner , J. Weis (Eds.), Wiley-VCH, Weinheim, **2000**, p. 22.
[52] P. Behrens, C. Tintemann, *Stud. Surf. Sci. Catal.* **2001**, *135*, 06-P-25.
[53] S. Altmaier, P. Behrens, submitted to *J. Mater. Chem.*
[54] P. Behrens, S. Altmaier, in preparation for *Chem. Monthly.*
[55] R.M. Barrer, *Hydrothermal Chemistry of Zeolites*, Academic Press, London, **1982**.
[56] J.N. Israelachvili, D. J. Mitchell, B.W. Ninham, *J. Chem. Soc., Faraday Trans.* **1976**, *2*, 1525.
[57] J.N. Israelachvili, *Intermolecular and Surface Forces,* Academic Press, London, **1991**.
[58] S. D. Sims, S. L. Burkett, S. Mann, *Mater. Res. Soc. Symp. Proc.* **1996**, *431*, 77.
[59] L. Mercier, T. J. Pinnavaia, *Chem. Mater.* **2000**, *12*, 188.
[60] Y. Sakamoto, M. Kazeda, O. Terasaki, D. Y. Zhao, J. M. Kim, G. D. Stucky, H. J. Shin, R. Ryoo, *Nature* **2000**, *408*, 449.
[61] F. Chen, L. Huang, Q. Li, *Chem. Mater.* **1997**, *9*, 2685.
[62] R. Ryoo, S. H. Joo, J. M. Kim, *J. Phys. Chem. B* **1999**, *103*, 7435.
[63] R. Ryoo, C. H. Ko, I. Park, *Chem. Commun.* **1999**, 1413.
[64] F. Chen, F. Song, Q. Li, *Microporous Mesoporous Mater.* **1999**, *29*, 305.
[65] J. M. Kim, Y. Sakamoto, Y. K. Hwang, Y. Kwon, O. Terasaki, S. Park, G.D. Stucky, *J. Phys. Chem. B* **2002**, *106*, 2552.
[66] M. Kruk, M. Jaroniec, R. Ryoo, S. H. Joo, *Chem. Mater.* **2002**, *12*, 1414.
[67] C. M. Bambrough, R. C. T. Slade, R. T. Williams, S. L. Burkett, S.D. Sims, S. Mann, *J. Coll. Interface Sci.* **1998**, *201*, 220.
[68] V. Goletto, M. Imperor, F. Babonneau, *Stud. Surf. Sci. Catal.* **2000**, *129*, 287.
[69] S.R. Hall, C. E. Fowler, B. Lebeau, S. Mann, *Chem. Commun.* **1999**, 201.
[70] C.M. Bambrough, R. C. T. Slade, R. T. Williams, *J. Mater. Chem.* **1998**, *8*, 569.

37 Biosilicification – Structure, Regulation of Structure and Model Studies

Carole C. Perry*

37.1 Introduction

Silica, in its various forms (orthosilicic acid ($Si(OH)_4$) through to hydrated amorphous silica ($SiO_n(OH)_{4-2n}$, n = 2–4)), is important in geological and biological processes. Geological minerals may be crystalline or amorphous, porous or non-porous. In contrast, biologically derived silica is routinely amorphous, and yet organisms are able to produce complex micro-architectural features that may contain distinct structural motifs built up from silica particles of characteristic size, degree of hydration, surface chemistry, and particle-particle orientation. How is this done? This chapter describes the structure of siliceous materials found in biological organisms and indicates some of the progress that is being made in the search for the underlying reactions/ interactions that control the process. Model studies to look at specific interactions between biopolymers and simpler molecules containing the "presupposed required functionalities" with silica are also described.

37.2 Structural Chemistry of Biosilicas

The mineral silica, or more specifically hydrated amorphous silica, often referred to as opal, is the second most abundant mineral type formed by organisms, with only the carbonate minerals exceeding it in abundance and distribution. Much of the biogenic silica produced is formed at temperatures of 4° C or lower in the polar oceans.

The morphology of siliceous structures still forms the basis for the taxonomy of a number of groups, including the radiolaria, Figure 37.1. The mineral is used to strengthen cell walls and to provide skeletal features. It is also found as secondary wall associated deposits where adventitious particles including quartz stick to the outside of the organism.[1] Table 37.1 lists the organisms known to process silicon.[2]

Amorphous silica is found in biological organisms as the kinetic barrier to crystallization is of the order of 800 kJ mol^{-1}. Biogenic silica is not a

* Address of the author: Department of Chemistry and Physics, The Nottingham Trent University, Clifton Lane, Nottingham NG11 8NS, U.K.

stoichiometric mineral and the nature (density, hardness, solubility, viscosity) and composition of siliceous structures in biology may vary considerably, being influenced directly and indirectly by a wide range of cellular processes. Although there is flexibility in terms of composition in respect of hydroxyl functionality biological silicas actively exclude other metal ions from their structures.[3]

Figure 37.1. Radiolarians. Reproduced from Haeckel, Ernst, *Art Forms in Nature*, Dover Publications, New York (1974) with the permission of the publisher. Originally published by Ernst Haeckel in *Kunstformen der Natur* (1899-1904). In comparison to organisms such as diatoms and choanoflagellates, much less is known concerning the formation of the skeletons of these organisms as they are much more difficult to culture in the laboratory. Organisms may be up to many hundreds of micrometres in diameter.

If we consider silicas produced by biological organisms we find that there are examples of both gel-like and particulate materials that exhibit variable porosity (Table 37.2). These materials are all formed under conditions of ambient temperature and pressure and at circumneutral pH and yet many of the structures show evidence of long–range structural organization. Even more remarkable is the fact that the biosilicas show no evidence of crystallinity and yet are able to produce cohesive structures over length scales many orders of magnitude greater than the dimensions of the fundamental building blocks.

Although biogenic silica exhibits no long-range crystallographic order (as determined by ultra high resolution transmission electron microscopy and diffraction methods,[4,5] morphological order exists at the microscopic level. The initial distinction that can be made is between gel, continuous, and particulate structures. In true 'gel-like' phases such as costal strips from the marine protozoan, *Stephanoeca diplocostata* Ellis, where the cell is surrounded by a basket–like casing (lorica) built up from 150-180 silica strips (costal rods), the silica phase is a continuous network throughout the material [6] and even on partial dissolution, particles are not visible (Figure 37.2).

In contrast, the silica deposits in many organisms, including higher plants comprise 'particulate' structures where secondary structural motifs are built up from primary particles of different sizes, shapes, and surface chemistries.[7,8]

Table 37.1. Silicon involvement in biological systems.

Type of involvement	Species of silicon required	Known function	Occurrence
As a trace element	Orthosilicic acid	DNA polymerase	Diatoms
		Collagen and glycosaminoglycan synthesis	Vertebrates
		Bone mineralization	Vertebrates
		For growth	Diatoms, some higher plants, vertebrates
Cell wall and skeleton formation	Silicic acid ↓	Protection, support	Diatoms, chryso-phytes, other silicified algae
	Amorphous silica		Choanoflagellates, radiolaria, testaceous, amoebae, other protozoa, sponges, mollusc teeth
Secondary cell wall-associated deposits	Silicic acid ↓ Amorphous silica	Possibly as an alternative to excretion	Some higher plants

Modified from Simpson and Volcani, 1981. [2]

Table 37.2. Adsorption data for selected biogenic silicas.

	Surface area [m^2g^{-1}]	Pore diameter [Å]	Particle size* [Å]
Plant hairs (*Phalaris canariensis*)			
Silica only	240	35	50–100
Intact hairs	0	-	50–100
Diatom (*Navicula pelliculosa*)			
Silica only	100	100	50
Sponge (*Euplectella* Owen)	1	-	30

* Particle sizes were determined by electron microscopy and were for the smallest recognizable particles, not the fundamental particles making up these structures.

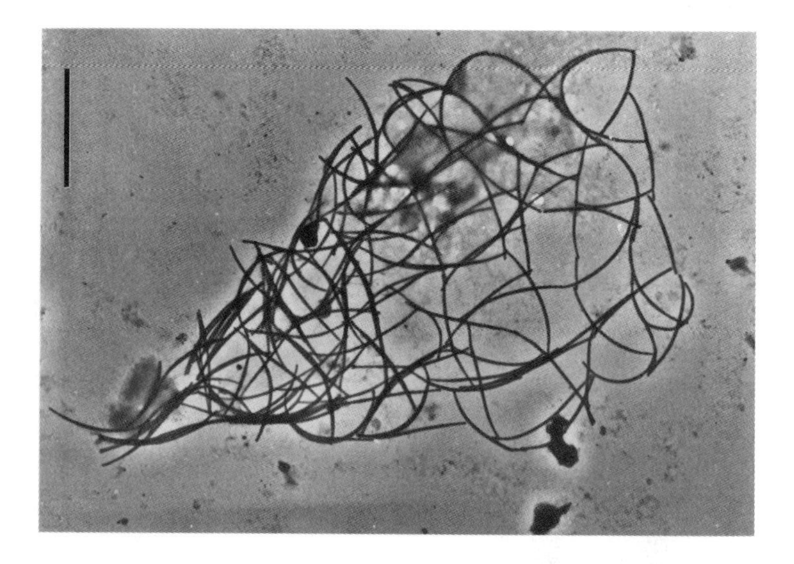

Figure 37.2. Transmission electron micrograph of Stephanoeca diplocostata Ellis, Scale bar = 8 μm (Courtesy of Prof. S. Mann, Bristol University).

The structures have been classified as sheet–like, globular and fibrillar depending on the nature of the aggregation between the particles.[7] The range of particle sizes observed in a particular structural motif is limited. Apart from globular structures where small (5–10 nm) particles coalesce to produce larger particles, all the other motifs are built up from clearly distinguishable particles that remain distinct from one another under conditions of analysis.

The stability of the structural motifs in the electron microscope is considerable, in comparison to industrially prepared precipitated silicas of similar dimensions, perhaps indicating a surface stabilization for the biologically derived silicas.[8] A noticeable feature of particulate silicas formed in the biological environment is the narrow particle size distribution for specific structural motifs and the lack of necking between particles, which is synonymous with controlled nucleation, particle growth, and aggregation.[9] Figure 37.3 shows the sizes and particle size distribution of the smallest particles measurable making up particulate structures in globular silica extracted from the branches of Equisetum species, and examples of nano- and macro-structures for this species are shown in Figure 37.4.

An intermediate structural type is that exhibited by some sponges and diatoms, which we term 'continuous' silica (Figure 37.5). Morphological forms such as fibrillar and hexagonal columnar arrangements have been observed in the development of some diatoms [10,11] that are not discernible in the fully mineralized frustule or coat. It is only when partial dissolution of the diatom skeleton has taken place that the shell can be seen to be built up from 5 nm particles (similar to those observed in particulate arrangements) with further impregnation of siliceous or organic material to give the resultant composite frustule. This final material we would describe as 'continuous' silica.

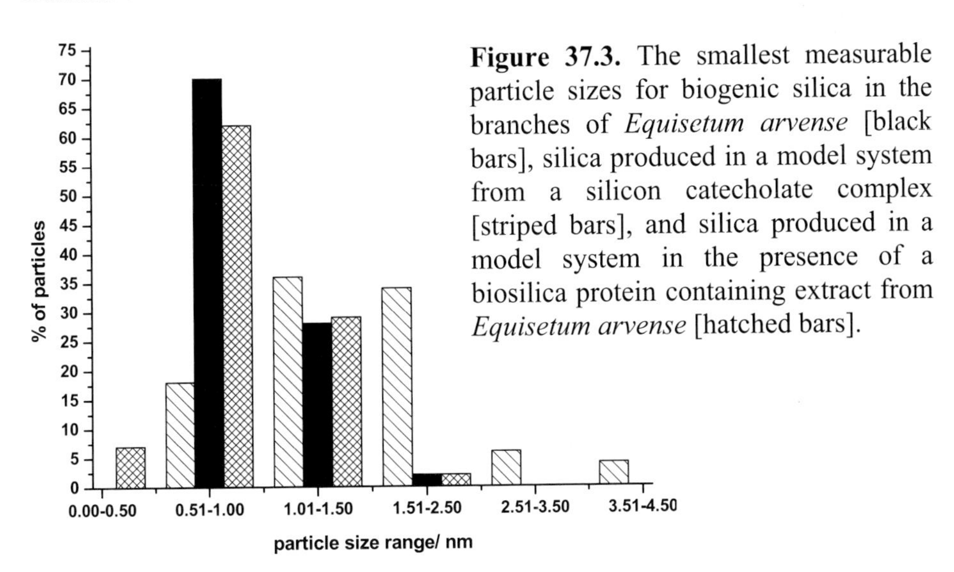

Figure 37.3. The smallest measurable particle sizes for biogenic silica in the branches of *Equisetum arvense* [black bars], silica produced in a model system from a silicon catecholate complex [striped bars], and silica produced in a model system in the presence of a biosilica protein containing extract from *Equisetum arvense* [hatched bars].

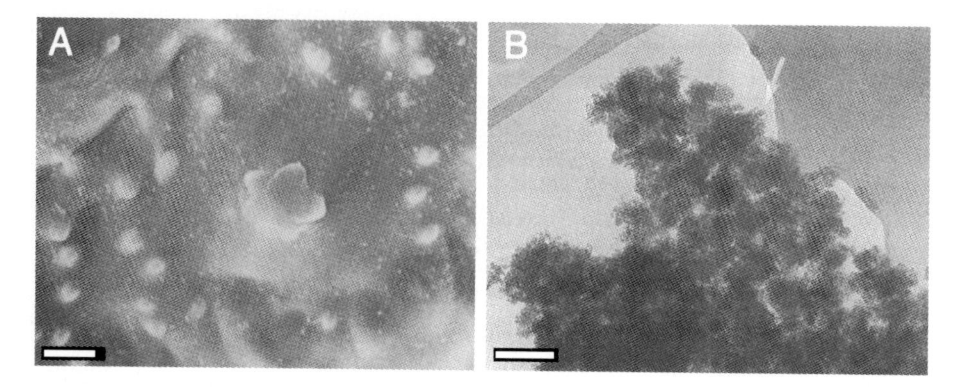

Figure 37.4. Plant silica examples. A. Silicified cell wall structures from the branches of *Equisetum arvense*, Scale bar, 20 μm, B. Silica found inside the stem of *Equisetum arvense* (Scale bar 100 nm).

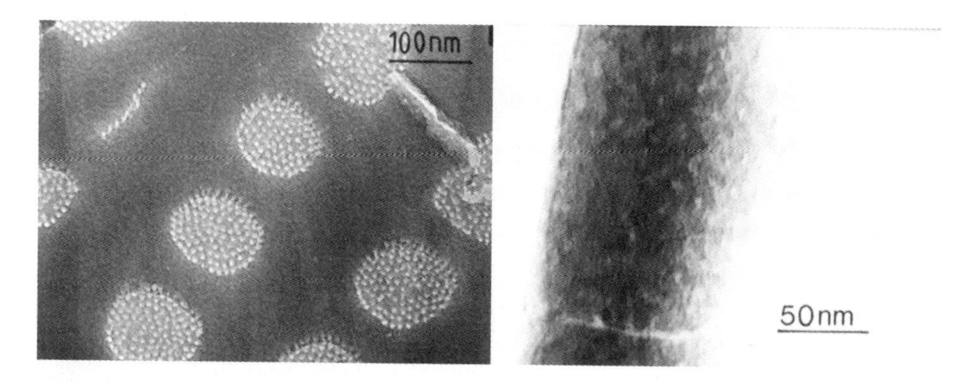

Figure 37.5. Diatom silica from *Navicula pelliculosa*. Left: pore structures. Right: silica network structures.

On a macroscopic scale, the beauty of siliceous structures has fascinated scientists since the advent of light microscopy. The complex, often geometric morphology of these structures still forms the basis for the taxonomy of a number of groups of lower plants and animals, e.g. radiolaria (see Figure 37.1). The architectural design of these intricate structures at the macroscopic level may be mirrored in the ultrafine morphology on the microscopic scale, for example within each individual pore of a diatom frustule (Figure 37.5, left).[12] Clearly, control at this level of sophistication is ultimately the result of genetic processes acting on the regulation and organization of chemical and biochemical reactions at the molecular level. Silica structures are sometimes found to mimic the underlying organic cell wall structure (Figure 37.5, right).

37.3 Identification of Biomolecules in Biological Silica

The finding that proteinaceous matrices can act as templates for the nucleation, growth, and growth regulation of crystalline biominerals has prompted a search for similar regulatory molecules in silicifying organisms such as diatoms, sponges, and higher plants. The chemical nature of the biomolecules extracted from the different organism types show some similarities but there are also many differences in composition and structure.

Studies in the early 1970s on the organic casing of a range of mature diatoms [13] showed that diatom cell wall proteins were consistently enriched in serine and threonine, comprising 16–34 mol%, and glycine, at 11–21 mol%. In contrast, the concentrations of sulfur–containing (cysteine and methionine) and aromatic (phenylalanine and tyrosine) residues were low and the cell wall was depleted in acidic amino acids, namely aspartic and glutamic acid, although at levels ranging between 16 and 23 mol% the material could still be considered relatively rich in acidic residues, consistent with composition data obtained from matrices extracted from crystalline biominerals.

Recent studies performed on one particular diatom, *Cylindrotheca fusiformis* have identified three types of proteinaceous materials. Treatment with EDTA produced a new glycoprotein family, termed frustulins,[14-16] that appear to have a modular structure. Each frustulin contains at least three of five structural elements: (a) a presequence domain, (b) an acidic cysteine rich (ACR) domain with a highly repetitive structure with a common sequence being C-E/Q-G-D-C-D, (c) proline-rich domains, (d) polyglycine domains and (e) a tryptophan-rich domain. The secondary structure of the frustulins is expected to contain both extended helical structural components (proline-rich and polyglycinc domains) and globular structural components (ACR and tryptophan-rich domains) [14] with some but not all frustulins being glycosylated. Frustulins are assembled through non-covalent calcium ion salt bridges.[14] A second class of proteins, referred to as HF–extractable proteins, or HEPs, have been identified as being tightly associated with a substructure of the silica scaffold [17] although all information on sugar content is lost due to the treatments used. The most thoroughly investigated protein, HEP200, displays a bipartite structure with serine being more concentrated in the N-terminal portion where it constitutes ca. 11 mol% of the amino acid composition. A third set of low molecular weight protein containing materials (4-17 kDa molecular weight) have also been isolated from *C. fusiformis* by anhydrous HF treatment. The polycationic peptides have been termed silaffins, [18] and their amino acid composition is particularly rich in basic amino acids, with Lys-Lys and Arg-Arg clusters being spaced in a highly regular manner and the hydroxy-amino acids serine and tyrosine being located principally between the basic clusters. The silaffins contain amino acids which have been subject to post-translational modification with ε-*N,N*-dimethyllysine and oligo-*N*-methyl-propylamine side chain functionalities, both of which increase the positive charge on the molecules in the sub-neutral pH range.[18,19]

Investigations into the nature of an intra-silica matrix have also been performed on sponges and higher plants. The sponge *Tethya aurantia* has a central protein filament which can be dissociated into three similar sub-units, referred to as silicatein α, β, and γ. The amino acid compositions of the purified extracts appear to be remarkably consistent with those for the various diatom species, showing ca. 20 mol% acidic residues, 16–20 mol% serine/threonine and ca. 15 mol% glycine.[20] The more abundant α subunit of the sponge filament proteins has been further characterized and identified as a novel member of the cathepsin L subfamily of papain-like cysteine proteases. A distinctive feature of this subunit is the clustering of the hydroxy amino acids giving rise to serine–rich sequences, which strengthens the idea that they are molecular templates for silicification.

Studies on higher plants, including *Equisetum telmateia* (Great horsetail), and *E. arvense* (Common horsetail), and on hairs found on the lemma of the grass *Phalaris canariensis*, have also identified an organic matrix that is intimately associated with biogenic silica.[21,22] Three extracts have been obtained with distinctive chemical compositions. The most readily released is enriched in respect of serine/threonine (25 mol%), and glycine (20 mol%) and is relatively rich in acidic residues (25 mol%). The second extract has lower levels of glycine and hydroxyl-containing residues, with the glycine being replaced principally by proline, and an increase in the lysine concentration, which doubles from approximately 4 to 8 mol% in *Phalaris* and from 6 to 12mol% in *Equisetum*. A third, insoluble material is also obtained, which contains up to 26 mol% of lysine, high levels of proline, and high levels of aliphatic amino acids. The proportion of acidic amino acids is still around 16%, but the level of hydroxylated amino acids is reduced to around 6%.

37.4 Proposed Mechanisms of Action

The original mechanism proposed for templated silicification was that proposed for diatoms by Hecky et al. [13] and is still referred to today by workers in the field. It was envisaged that the serine/threonine–enriched protein would form the inner surface of the silicalemma, and as such would present a layer of hydroxyls, facing into the silica deposition vesicle, onto which orthosilicic acid molecules could condense. This initial layer of geometrically constrained orthosilicic acid molecules would promote condensation with other orthosilicic acid molecules. The carbohydrate moiety present in the diatom cell wall was assigned a number of possible roles, including action as a physical buffer between the organism and the aquatic environment, providing resistance to chemical and bacterial degradation, and functioning as an ion–exchange medium.

As a consequence of more recent structural and mechanistic studies of silica formation in diatoms, the frustulins are proposed to play a structural role in the organic casing of diatoms rather than a regulatory role in the silicification process.[14] It has been tentatively speculated that the HEP's may regulate morphogenesis in subsets of the girdle bands that are structural components of the developing diatom frustule.[18] A role has been proposed for the low molecular weight silaffins that involves promotion of silicic acid polycondensation and precipitation of biogenic silica under neutral or slightly acidic pH conditions. The presence of the post–translational polyamine modifications would be indispensable for silaffins to exert silica precipitation activity at the sub–neutral pH conditions proposed to exist within the lumen of the silica deposition vesicle.[18,23] A distinctive feature of specific subunits within the silaffins is the clustering of the hydroxy amino acids giving rise to serine-rich sequences, a fact that again strengthens the idea that they are molecular templates for silicification.

For the silicateins extracted from sponges, their structural relationship to the proteases which catalyze hydrolysis reactions (essentially the reverse of the condensation reaction by which orthosilicic acid oligomerizes to produce silica) adds further weight to the hypothesis that they are involved in silica deposition. Additionally, structural variants of the silicatein protein produced using site–directed mutagenesis [24] have shown that specific serine and histidine side chains are required for optimal activity in the catalyzed formation of silica.

Among the materials extracted in studies of plant materials,[22] the highly charged, (very high levels of lysine and acidic groups) and relatively rigid, (high levels of proline) materials are envisaged as playing roles in regulating nucleation. Those materials which contain higher levels of amino acids (such as serine) that could bind to silica by a hydrogen-bonding mechanism would provide a means by which the growth of particles is regulated. It is likely that cell wall proteins and some of the carbohydrate polymers present in the secondary cell wall of these silicifying structures may also be involved in regulating particle growth and in the delineation of spaces with specific shapes in order for structural motifs of silica and macroscopic features to result.

37.5 Studies of the Effect of Biosilica Extracts on the *in vitro* Formation of Silica

Proteinaceous extracts from diatoms, sponges, and higher plants have been used in model studies of silicification in order to identify the role of the extracted biomolecules.

Studies using low molecular weight silaffins from diatoms [17] have shown that these molecules accelerate the silica polycondensation process and aid in flocculation, the biopolymer extract being occluded in the silica formed. These molecules were shown to be able to cause a metastable solution of silicic acid to spontaneously aggregate, with small ca. 50 nm spherical particles being produced in the presence of a mixture of silaffins and larger ca. 500–700 nm particles being produced in the presence of a purified silaffin. The silaffins were found to be effective in promoting silicification at pH values as low as 4, a situation possibly encountered in the lumen of the diatom silica deposition vessel.[23] The provision of highly positively charged side chains was thought to be very important in enabling the silaffins to promote condensation and aggregation under mildly acidic conditions.

Studies on the effect of silicatein α extracted from sponge spicules on polysiloxane synthesis at neutral pH have shown a ca. 10–fold increase in the amount of harvestable silica compared to experiments run in the presence of the denatured silicatein or proteins such as bovine serum albumin and trypsin.[25] A catalytic role for the silicatein in the polycondensation of siloxanes was proposed in a manner analogous to that of serine– and cysteine–based proteases. The suggested mechanism of catalysis was that of a general acid/base catalyst with a specific requirement for both serine and histidine amino acids at the active site.[24] A scaffolding or structure-directing activity was also reported as the polymerized silica was able to form a layer following the contours of the underlying protein fiber, which was not observed on either cellulose or silk fibroin fibers.[25]

Biosilica extracts from *Equisetum arvense* [9,26] have been found to accelerate the rate of formation of small oligomers of silicic acid, and more silica was produced that exhibited a narrower distribution of particle sizes than for silica formed in the absence of the biomolecule extracts (Figure 37.3), with the particle size distribution closely mirroring that found for the biogenic silica from which the biosilica extract was taken. Layered, crystalline materials were also produced that could have arisen from the epitaxial matching of initially formed oligomers on a β-sheet type material, the silica structure continuing to develop from the initial biopolymer-controlled nucleation event.

37.6 Biomimetic Studies of Silica Precipitation

Using information obtained from silica precipitation studies performed using biosilica extracts as a starting point, several groups have been investigating the effect of amino acids, polyamines, and synthesized fragments of the biosilica extracts as solution adjuncts in studies of silica formation from both aqueous (catecholates and diluted solutions of water glass, e.g. [21,27,28]) and non-aqueous precursors (alkoxysilanes and their derivatives. [24,25,29]) The results obtained show that, in accord with results obtained on simple carbohydrate materials, [21] amino acids have little effect on the kinetics of silica formation.[27,28] Homopeptides of arginine (up to Arg_{65}) and lysine (up to Lys_{1005}) [28] affect the silica precipitation process at circumneutral pH, with the effect being related to the length of the polypeptide chain. Electrostatic interactions have been proposed as the principal mode of interaction between negatively charged silicic acid species and positively charged polypeptide chains leading to gelation and/or precipitation according to the system investigated.[28] Other studies using block copolypeptides of cysteine and lysine have shown that silane hydrolysis is enhanced in the presence of such molecules and that certain of the block copolymers investigated are also able to direct the formation of silica morphologies such as spheres and columns.[29] The state of oxidation of the copolymer was also found to influence the morphology of the precipitates formed. A polycationic peptide (derived from the C *fusiformis* silaffin-1 protein chemically synthesized with no post-translational modification of the lysine residues) has recently been used in the generation of a two-dimensional array of silica nanospheres within a hybrid polymer hologram (Figure 37.6).[30]

37.7 Conclusions

The results obtained by various researchers working with both single molecule biosilica extracts and mixtures of molecules clearly demonstrate that some biopolymers and components thereof are able to accelerate the process of silica formation in an *in vitro* situation. The fact that structures not usually formed in aqueous synthetic experiments **can** be formed in the presence of some of these extracts suggests that in biological organisms, additional controls must be exerted during the process of silica precipitation in order to prevent the formation of crystalline phases. It is clearly not a straightforward task to nucleate and control the growth of an amorphous material, generating shape and functionality along the way.

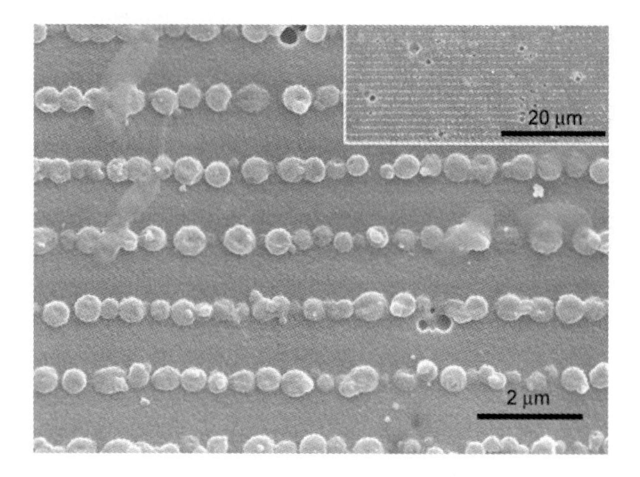

Figure 37.6. Biosilica nanostructure created upon reacting a silane with a peptide-embedded hologram (courtesy of Dr. Morley Stone, Wright Patterson Air Force Base, Ohio, USA).

Much remains to be investigated in our search for an understanding of the mechanisms by which biological organisms regulate mineral formation, particularly a mineral of infinite functionality such as silica. The information obtained in this search will however assist us in our understanding of the essentiality of silicon to life processes and in the generation of new materials with specific form and function for industrial application in the twenty–first century. Already we have seen that the ability to create an ordered organic/inorganic array of silica nanospheres using a polycationic peptide derived from a diatom silaffin to promote localized condensation might be of practical use in the fabrication of photonic devices. [29]

37.8 References

[1] K. Gold, E. A. Morales, *Trans. Am. Microscop. Soc.* **1976**, *95*, 69.
[2] T. L. Simpson, B. E. Volcani (Eds.), *Silicon and Siliceous Structures in Biological Systems*, Springer, New York, 1981.
[3] C. C. Perry, T. Keeling-Tucker, *J. Biol. Inorg. Chem.* **2000**, *5*, 537.
[4] C. C.Perry in S. Mann, J. Webb, R. J. P. Williams (Eds.), *Biomineralisation, Chemical and Biological Perspectives*, VCH, Weinheim, 1989, p. 233.
[5] S. Mann, C. C. Perry in D. Evered and M. O'Connor (Eds.), *Silicon Biochemistry*, CIBA Foundation Symposium 121, J. Wiley & Sons, Chichester, 1986, p. 40.
[6] S. Mann, R. J. P. Williams, *Proc. Roy. Soc. Lond.* **1982**, *B216*, 137.

[7] C. C. Perry, S. Mann, R. J. P. Williams, *Proc. Roy. Soc. Lond.* **1984**, *B222*, 427.
[8] C. C. Perry, E. J. Moss, R. J. P. Williams, *Proc. Roy. Soc. Lond.* **1990**, *B241*, 47.
[9] C. C. Perry, T. Keeling-Tucker, *J. Coll. Polymer Sci.*, in press.
[10] C. W. Li, B. E. Volcani, *Protoplasma* **1985**, *124*, 10.
[11] C. W. Li, B. E. Volcani, *Protoplasma* **1985**, *124*, 147.
[12] C. C. Perry, in B. Jamtveit, P. Meakin (Eds.), *Growth, Dissolution and Pattern Formation in Geosystems*, Kluwer Acad. Publ., Dordrecht 1999, p. 237.
[13] R. E. Hecky, K. Mopper, P. Kilham, E. T. Degens, *Marine Biol.* **1973**, *19*, 323.
[14] N. Kröger, C. Bergsdorf, M. Sumper, *EMBO J.* **1994**, *13*, 4676.
[15] N. Kröger, C. Bergsdorf, M. Sumper, *Eur. J. Biochem.* **1996**, *239*, 259.
[16] W. H. VandePoll, E. G. Vrieling, W. W. C. Gieskes, *J. Phycol.* **1999**, *35*, 1044.
[17] N. Kröger, G. Lehmann, R. Rachel, M. Sumper, *Eur. J. Biochem.* **1997**, *250*, 99.
[18] N. Kröger, R. Deutzmann, M. Sumper, *Science* **1999**, *286*, 1129.
[19] N. Kröger, R. Deutzmann, M. Sumper, *J. Biol. Chem.* **2001**, *12*, 26066.
[20] K. Shimizu, J. Cha, G. D. Stucky, D. Morse, *Proc. Natl. Acad. Sci.* U.S.A. **1998**, *95*, 6234.
[21] C. C. Harrison (now Perry), Y. Lu, *Bull. de l'Institut Océanographique Monaco* **1994**, n° spécial *14*, 151.
[22] C. C. Harrison (now Perry), *Phytochemistry* **1996**, *41*, 37.
[23] E. G. Vrieling, W. W. C. Gieskes, T. P. M. Beelen, *J. Phycol.* **1999**, *35*, 548.
[24] Y. Zhou, K. Shimizu, J. N. Cha, G. D. Stucky, D. E. Morse, *Angew. Chem. Int. Ed..* **1999**, *38*, 780.
[25] J. N. Cha, K. Shimizu, Y. Zhou, S. C. Christiansen, B. F. Chmelka, G. D. Stucky, D. E. Morse, *Proc. Natl. Acad. Sci.* U.S.A. **1999**, *96*, 361.
[26] C. C. Perry, T. Keeling-Tucker,. *Chem. Commun.* **1998**, 2587.
[27] T. Coradin, J. Livage, *Colloids Surf.* **2001**, *21*, 329.
[28] T. Coradin, O. Duruphy, J. Livage, *Langmuir*, **2002**, *18*, 2331.
[29] J. N. Cha, G. D. Stucky, D. E. Morse, T. J. Deming, *Nature* **2000**, *403*, 289.
[30] L. L. Brott, R. R. Naik, D. J. Pikas, S. M. Kirkpatrick, D. W. Tomlin, P. W. Whitlock, S. Clarson, M. O. Stone, *Nature* **2001**, *413*, 291.

Subject Index

MICHIGAN MOLECULAR INSTITUTE
1910 WEST ST. ANDREWS ROAD
MIDLAND, MICHIGAN 48640